Christopher Heath

Practical Anatomy

Salzwasser

Christopher Heath

Practical Anatomy

1. Auflage | ISBN: 978-3-84605-230-3

Erscheinungsort: Frankfurt, Deutschland

Erscheinungsjahr: 2020

Salzwasser Verlag GmbH

Reprint of the original, first published in 1869.

PRACTICAL ANATOMY:

A

MANUAL OF DISSECTIONS.

BY

CHRISTOPHER HEATH, F.R.C.S.

ASSISTANT-SURGEON TO UNIVERSITY COLLEGE HOSPITAL
AND TEACHER OF OPERATIVE SURGERY IN UNIVERSITY COLLEGE, LONDON:
LATE ASSISTANT-SURGEON TO, AND LECTURER ON ANATOMY
AT, THE WESTMINSTER HOSPITAL.

Second Edition.

LONDON:
JOHN CHURCHILL AND SONS,
NEW BURLINGTON STREET.

MDCCCLXIX.

PREFACE TO THE SECOND EDITION.

In preparing a Second Edition of this Manual I have carefully revised every page, and have made such additions—particularly in respect of the anatomy of the brain—as seemed called for. In order to enable the student to test his knowledge after dissection, I have introduced a series of sectional views of the limbs and trunk, and have for this purpose availed myself of the able assistance of Mr. G. E. L. Pearse, Demonstrator of Anatomy at the Westminster Hospital. In illustrating the anatomy of the convolutions of the brain I have employed the drawings illustrating Mr. Marshall's paper on the subject, with that gentleman's permission. In conclusion, I have to thank several teachers of anatomy in various parts of the kingdom for kind suggestions, and to express a hope that the work may prove of service to students of anatomy.

CHRISTOPHER HEATH.

9, Cavendish Place, W.
September, 1869.

PREFACE TO THE FIRST EDITION.

In preparing this work I have constantly borne in mind the requirements of the Student, whose time for anatomical study is limited to two or, at most, three winters, and have endeavoured, as far as is possible, to simplify anatomical details without omitting any. With the view of giving interest and practical importance to anatomical study, I have called attention to those leading facts which bear directly upon the practice of Medicine and Surgery, and especially the operations; and have not scrupled to assist the remembrance of mere anatomical facts by those artificial aids which many years' experience has proved to be useful. The order in which the several regions are dissected is that which I have found most convenient for the Student to pursue.

The Illustrations are from various sources, and I have to acknowledge the liberality of Mr. Erasmus Wilson; MM. Hirschfeld, Leveillé, and Baillière; Mr. Fergusson; Mr. John Wood; and Dr. Savage, for permission to employ drawings from their respective works. The original illustrations are from drawings by Mr. J. T. Gray (formerly Demonstrator of Anatomy at the Westminster Hospital), of recent dissections made under my own superintendence; and my best thanks are due to that gentleman for the care and skill he has displayed. The engravings were entrusted to the experienced hands of Messrs. Butterworth and Heath.

I have to acknowledge much kind and able assistance from Mr. W. F. Teevan, Demonstrator of Anatomy at the Westminster Hospital; Mr. C. W. B. Bell, late Demonstrator, Trinity College, Dublin; and Mr. Gray, in revising the sheets during their passage through the press.

September, 1864.

CONTENTS.

	Page
INTRODUCTION	1

PART I.

DISSECTION OF THE ARM	5
The Axilla	11
The Front of the Upper Arm	22
The Nerves of the Arm	29
The Bend of the Elbow	30
The Parts about the Scapula	32
The Back of the Arm	37
The Front of the Fore-arm	40
The Back of the Fore-arm	52
The Palm of the Hand	65
The Muscles of the Little Finger	71
The Muscles of the Thumb	72
The Ligaments of the Scapula	76
The Shoulder Joint	78
The Elbow Joint	79
The Radio-Ulnar Articulations	82
The Wrist Joint	83

PART II.

DISSECTION OF THE LEG	87
The Front of the Thigh	89

	Page.
Femoral Hernia	92
Scarpa's Triangle	98
Muscles of the Front of the Thigh	103
The Inner side of the Thigh	108
The Buttock	113
The Popliteal Space	123
The Back of the Thigh	126
The Hip Joint	129
The Back of the Leg	132
The Sole of the Foot	143
The Front of the Leg and Foot	154
The Knee Joint	165
The Tibio-Fibular Articulations	171
The Ankle Joint and Articulations of Foot	172

PART III.

Dissection of the Abdomen	179
The Perinæum	181
Posterior Space in both Sexes	183
Anterior Space in the Male	186
Anterior Space in the Female	195
The Abdominal Wall	202
The Anatomy of Inguinal Hernia	213
The Cavity of the Abdomen	224
The Mesenteric Vessels	234
The Cœliac Axis	238
The Diaphragm	245
The Abdominal Aorta	248
The Lumbar Plexus	259
The Viscera of the Abdomen	262
The Pelvis	279
Pelvic Viscera in the Male	300
Pelvic Organs in the Female	309
The Ligaments of the Pelvis	313

PART IV.

	Page.
Dissection of the Head and Neck	317
The Scalp	318
Removal of the Brain	323
The Dura Mater and Sinuses	327
The Posterior Triangle of the Neck	334
The Orbit	344
The Side of the Neck	354
The Anterior Triangle of the Neck	371
The Face	372
Muscles of the Face	377
The Pterygo-maxillary Region	390
The Submaxillary Region	400
The Thorax	407
Great Vessels of the Root of the Neck	430
The Prevertebral Region	452
The Pharynx	455
The Palate	466
The Superior Maxillary Nerve	472
The Nose	473
Meckel's Ganglion	477
The Otic Ganglion	481
The Tympanum	481
The Tongue	484
The Larynx	487
Ligaments of the Vertebræ	497
Ligaments of the Atlas, Axis, and Occiput	500
Articulation of the Ribs with the Vertebræ	504

PART V.

Dissection of the Back and Spinal Cord	507
The Spinal Cord and its Membranes	521

PART VI.

	Page.
DISSECTION OF THE BRAIN	527
The Membranes of the Brain	527
The Vessels of the Brain	529
The Base of the Brain	532
The Exterior of the Brain	541
The Inner Surface of the Hemisphere	544
The Interior of the Brain	546
The Lateral Ventricles	547
The Third Ventricle	552
The Cerebellum	555
The Fourth Ventricle	559

PART VII.

DISSECTION OF THE EYE	562
THE EAR	570

TABLE OF ILLUSTRATIONS.

Fig.			Page
1.	Anterior Muscles of Trunk	Wilson	8
2.	The Axilla	Original	12
3.	Axillary Artery and Branches	Original	17
4.	Nerves of the Axilla	Hirschfeld and Leveillé	18
5.	Veins of the Bend of the Elbow	Wilson	23
6.	Muscles of the Front of Upper Arm	Wilson	25
7.	Incisions for Tying Arteries of Arm	Fergusson	27
8.	Dissection of Front of Upper Arm	Hirschfeld and L.	28
9.	Dissection of Back of Upper Arm	Hirschfeld and L.	34
10.	Section through Upper Arm	*altered from* Béraud	39
11.	Superficial Muscles of Fore-arm	Wilson	42
12.	Superficial Dissection of Front of Fore-arm	H. and L.	44
13.	Deep Muscles of Fore-arm	Wilson	48
14.	Deep Dissection of Front of Fore-arm	Hirschfeld and L.	50
15.	Superficial Muscles of the Back of the Fore-arm	Wilson	55
16.	Deep Muscles of the Back of the Fore-arm	Wilson	59
17.	Dissection of the Musculo-Spiral Nerve	H. and L.	60
18.	Arteries of Back of Fore-arm	Original	62
19.	Section through Fore-arm	*altered from* Béraud	64
20.	Section through Wrist	Original	65
21.	Muscles of the Hand	Wilson	69
22.	Palmar Interossei of Hand	Original	74
23.	Dorsal Interossei of Hand	Original	75
24.	Ligaments of the Scapula	Wilson	77
25.	Ligaments of the Elbow-joint	Wilson	80
26.	Ligaments of the Elbow-joint	Wilson	81
27.	Ligaments of the Carpus	Wilson	84
28.	Diagram of Synovial Membranes of Wrist	Wilson	85
29.	Superficial Dissection of the Groin	Wood	90
30.	Crural Sheath laid open	Wood	93
31.	Section of the Femoral Arch	Wilson	94
32.	Imaginary Section of Crural Canal	Original	95
33.	Irregular Origin of Obturator Artery	Wood	96
34.	Irregular Origin of Obturator Artery	Wood	96
35.	Superficial Dissection of Thigh	Hirschfeld and L.	98
36.	Incisions for Tying the Femoral Artery	Fergusson	99
37.	Muscles of Anterior Femoral Region	Wilson	104
38.	Diagram of the Profunda Artery	Original	109

Fig. Page.
39. Nerves of the Thigh . . . Hirschfeld and Leveillé 110
40. Muscles of the Buttock . . . . Wilson 114
41. Deep Muscles of the Buttock . . . . . . Wilson 117
42. Dissection of the Buttock and Thigh Hirschfeld and L. 121
43. Dissection of the Popliteal Space Hirschfeld and L. 123
44. Section of Thigh at Apex of Scarpa's Triangle Original 129
45. Ligaments of the Hip-joint . . . . . . Wilson 131
46. Superficial Nerves of Back of Leg Hirschfeld and L. 132
47. Superficial Muscles of Back of Leg . . . Wilson 133
48. Deep Muscles of Back of Leg . . . . . Wilson 137
49. Incisions for Tying Posterior Tibial Artery Fergusson 139
50. Deep Dissection of Back of Leg Hirschfeld and L. 140
51. Relations of Parts behind Inner Malleolus H. and L. 142
52. Section of Right Ankle . . . . Original 143
53. Superficial Dissection of the Sole Hirschfeld and L. 144
54. First Layer of Plantar Muscles . . . . . Wilson 145
55. Arteries of the Sole of the Foot . . . . . Wilson 147
56. Second Stage of Dissection of Sole Hirschfeld and L. 148
57. Second and Third Layers of Plantar Muscles Wilson 151
58. Deep Dissection of Sole of Foot Hirschfeld and L. 152
59. Muscles of the Front of the Leg . . . . Wilson 156
60. Incisions for Tying Anterior Tibial Artery Fergusson 158
61. Deep Dissection of the Front of the Leg H. and L. 159
62. Section of Leg in Middle . . . *altered from* Béraud 163
63. Dorsal Interossei of Foot . . . . Original 164
64. Plantar Interossei of Foot . . . . Original 164
65. Anterior View of Knee-joint . . . . . . Wilson 166
66. Posterior View of Knee-joint . . . . . Wilson 167
67. The Inter-articular Ligaments of the Knee . Wilson 169
68. Vertical Section of Knee-joint . . . . . Wilson 170
69. Posterior View of Ankle-joint . . . . . Wilson 171
70. Internal View of Ankle-joint . . . . . Wilson 173
71. External View of Ankle-joint . . . . . Wilson 173
72. Ligaments of the Sole of the Foot . . . . Wilson 174
73. Superficial Dissection of Male Perinæum . . Original 184
74. Deep Dissection of Male Perinæum . . . Original 192
75. Arteries of the Perinæum . . . . Wilson 194
76. The Female Perinæum . . . . Savage 196
77. Muscles of Female Perinæum . Savage and Luschka 199
78. Nerves of Abdominal Wall . Hirschfeld and Leveillé 204
79. Superficial Dissection of the Groin . . . . Wood 207
80. Diagram of the Descent of the Testis . . . Original 210
81. Diagram of the Descent of the Testis . . . Original 211
82. Diagram of the Descent of the Testis . . . Original 211
83. Transversalis Abdominis Muscle . . . . . Wilson 212
84. Dissection of the Inguinal Canal . . . . Wood 214
85. Diagram of the Inguinal Canal . . . . . Original 215
86. Diagram of a Scrotal Hernia . . . . . Original 218
87. Diagram of a Congenital Hernia . . . . Original 218

Fig.			Page.
88.	Diagram of an Infantile Hernia	Original	218
89.	The Abdominal Wall from Within	Wood	223
90.	The Abdominal Viscera *in situ*	Original	225
91.	Diagram of Vertical Section of Peritoneum	Original	229
92.	Transverse Section of Peritoneum	Original	231
93.	Female Pelvic Organs from above	Savage	233
94.	Superior Mesenteric Artery	Wilson	235
95.	Inferior Mesenteric Artery	Wilson	237
96.	The Cœliac Axis	Wilson	240
97.	The Cœliac Axis	*altered from* H. and L.	242
98.	The Portal Vein	Wilson	243
99.	Abdominal Surface of Diaphragm	Wilson	246
100.	Abdominal Aorta and Vena Cava	Henlé	250
101.	Incisions for Tying the Iliac Arteries	Fergusson	255
102.	Lumbar Plexus	Hirschfeld and Leveillé	260
103.	Vertical Section of Stomach	Wilson	263
104.	Mucous Membrane of Stomach	Wilson	264
105.	Perpendicular Section of Stomach	*after* Kölliker	264
106.	A Tubular Gastric Gland	*after* Kölliker	264
107.	Section of a Peyer's Patch	*after* Kölliker	266
108.	Cæcum and Ileo-cæcal Valve	Wilson	268
109.	Mucous Membrane of Colon	Wilson	269
110.	Solitary Gland from the Colon	*after* Kölliker	269
111.	Malpighian Corpuscles of Spleen	Wilson	270
112.	Upper Surface of the Liver	Wilson	271
113.	Under Surface of the Liver	Bonamy	272
114.	Diagram of Circulation of Liver	*after* Kiernan	274
115.	Section of Kidney	Wilson	276
116.	Plan of Structure of Kidney	*after* Kölliker	276
117.	Plan of Renal Circulation	*after* Bowman	278
118.	Pelvic Fascia from the Outside	Original	280
119.	Pelvic Fascia from Behind	Wilson	282
120.	Pelvic Fascia from the Front	Original	283
121.	Side-View of Male Pelvis	Original	284
122.	Side-View of Female Pelvis	Wilson	290
123.	Pelvis with Internal Iliac Artery	Savage	292
124.	Floor of Female Pelvis	Savage	295
125.	Nerves of Pelvis	Hirschfeld and Leveillé	298
126.	Longitudinal Section of Bladder and Penis	Wilson	302
127.	Bladder and Urethra laid open	Wilson	305
128.	Transverse Section of Testicle	Wilson	307
129.	Anatomy of the Testis	Wilson	308
130.	Uterus and Appendages	Wilson	311
131.	Ligaments of the Pelvis and Hip-joint	Wilson	315
132.	Nerves of the Face and Scalp	Hirschfeld and Leveillé	320
133.	Side-View of Sinuses of Skull	Hirschfeld and Leveillé	326
134.	Sinuses and Nerves of Base of Skull	Original	329
135.	Diagram of Carvernous Sinus	Original	332
136.	Diagram of Nerves of Carvernous Sinus	Original	332

Fig.			Page
137.	Superficial Dissection of Triangles of Neck	Original	336
138.	Muscles of Side of Neck	Wilson	339
139.	Incisions for Tying Carotid and Subclavian	Fergusson	341
140.	Diagram of Brachial Plexus	Original	342
141.	The Nerves of the Orbit	Hirschfeld and Leveillé	344
142.	The Vessels of the Orbit	Hirschfeld and Leveillé	346
143.	Side-View of Nerves of Orbit	*after* Arnold	348
144.	The Lenticular Ganglion	Hirschfeld and Leveillé	349
145.	Arteries of Orbit from Outer Side	Hirschfeld and L.	350
146.	Muscles of the Eyeball	Wilson	353
147.	Sterno-clavicular Articulation	Wilson	356
148.	Muscles of the Front of Neck	Wilson	359
149.	The Side of the Neck	Original	360
150.	The Eighth Nerve of Left Side	Hirschfeld and L.	366
151.	Diagram of the Cervical Ganglia	Original	369
152.	Appendages of the Eye	Wilson	373
153.	The Pinna and its Muscles	*after* Arnold	375
154.	The Pinna and its Muscles	*after* Arnold	376
155.	Muscles of the Nose	*after* Arnold	378
156.	Muscles of the Face	Wilson	379
157.	Arteries of the Face and Head	Original	382
158.	Nerves of the Face	Hirschfeld and Leveillé	386
159 A.	Fibro-Cartilages of Nose	Wilson	389
159.	Fibro-Cartilages of Nose	*after* Arnold	389
160.	Pterygoid Muscles	Wilson	392
161.	External Lateral Ligament of Lower Jaw	Wilson	392
162.	Section of Temporo-Maxillary Joint	Wilson	393
163.	Internal Lateral Ligament of Lower Jaw	Wilson	393
164.	Pterygo-Maxillary Region	Hirschfeld and Leveillé	397
165.	Inferior Maxillary Nerve	Hirschfeld and Leveillé	399
166.	Nerves of the Tongue	Hirschfeld and Leveillé	402
167.	Styloid Muscles	Wilson	404
168.	Lingual Artery and Branches	Hirschfeld and Leveillé	405
169.	Diagram of Section of Thorax	Original	410
170.	The Heart and Lungs	Wilson	413
171.	Upper Surface of Diaphragm	Wilson	414
172.	Diagram of Heart *in situ*	Original	416
173.	Right Side of Heart laid open	Wilson	417
174.	Left Side of Heart laid open	Wilson	422
175.	Section of Heart at Level of Valves	Sibson	425
176.	Diagram of Fœtal Circulation	Wilson	426
177.	Diagram of Large Vessels of Thorax	Wilson	428
178.	The Eighth Nerve of Right Side	Hirschfeld and L.	436
179.	The Eighth Nerve of Left Side	Hirschfeld and L.	440
180.	Course and Termination of Thoracic Duct	Wilson	442
181.	Veins of the Trunk and Neck	Wilson	448
182.	Section of Superior Aperture of Thorax	Original	449
183.	Diagram of Structure of Lung	Wilson	451
184.	Prevertebral Muscles	Wilson	453

Fig.			Page
185.	Dissection of Pharynx with Vessels and Nerves	Orig.	456
186.	Diagram of 8th, 9th, and Sympathetic Nerves	H. & L.	458
187.	Side-View of Muscles of Pharynx	Wilson	463
188.	Pharynx laid open	Wilson	465
189.	Muscles of the Palate	Original	467
190.	Superior Maxillary Nerve	Hirschfeld and Leveillé	471
191.	Meatuses of the Nose	Wilson	475
192.	Nerves of the Nose	Wilson	476
193.	Meckel's Ganglion	Hirschfeld and Leveillé	479
194.	Otic Ganglion	Hirschfeld and Leveillé	482
195.	The Tongue	Wilson	485
196.	Posterior View of Larynx	Wilson	489
197.	Side-View of Larynx	Wilson	489
198.	Vertical Section of Larynx	Wilson	492
199.	Anterior Common Vertebral Ligament	Wilson	497
200.	Posterior Common Vertebral Ligament	Wilson	497
201.	Ligamenta Subflava	Wilson	500
202.	Anterior Ligaments of Atlas and Axis	Wilson	500
203.	Posterior Ligaments of Atlas and Axis	Wilson	502
204.	Occipito-Axoid Ligament	Wilson	502
205.	Cruciform Ligament of Atlas and Axis	Wilson	503
206.	Ligaments of Ribs	Wilson	505
207.	Superficial Muscles of the Back	Wilson	510
208.	Deep Muscles of the Back	Wilson	514
209.	Suboccipital Region	Original	516
210.	Section of Neck	*altered from* Béraud	520
211.	Transverse Section of Spinal Cord	Hirschfeld and L.	522
212.	The Spinal Cord	Hirschfeld and Leveillé	525
213.	Arteries of the Brain	Wilson	530
214.	The Base of the Brain	Hirschfeld and Leveillé	534
215.	Lateral View of the Brain	Marshall	541
216.	Upper Surface of the Brain	Marshall	543
217.	Inner Surface of the Brain	Hirschfeld and Leveillé	545
218.	Lateral Ventricles of the Brain	Hirschfeld and L.	548
219.	Descending Cornu of Lateral Ventricle	H. and L.	550
220.	Third Ventricle of Brain	Hirschfeld and Leveillé	554
221.	Fourth Ventricle of Brain	Hirschfeld and Leveillé	557
222.	Under Surface of Cerebellum	Hirschfeld and Leveillé	558
223.	Vertical Section of Brain	Hirschfeld and Leveillé	560
224.	Longitudinal Section of Eyeball	H. Power	563
225.	Choroid Coat and Vasa Vorticosa	*after* Arnold	565
226.	Choroid and Ciliary Processes	Wilson	566

MANUAL OF PRACTICAL ANATOMY.

INTRODUCTION.

By Practical Anatomy is meant the study of Anatomy by dissection of the dead body, in contradistinction to Descriptive Anatomy, taught by lectures, diagrams, and preparations.

In dissecting, there are four principal objects to be constantly borne in mind by the student :—1st. The impression on the memory of those facts of general anatomy taught in the lectures. 2nd. The study of those parts of the body more especially concerned in surgical affections and operations. 3rd. The education of the sense of touch, and of the hand in the use of instruments; and 4th, the education of the eye in the knowledge of the several tissues of the body, in various positions, and under varying circumstances. It is to assist the student in these requirements that the following work is designed, and every effort has been made to present the facts of Anatomy in such a manner, that they may be most easily grasped by the mind and retained by the memory; it will be found also, that as far as is compatible with a work of the kind, attention is drawn to all those points which have especial interest in the practice of medicine or surgery, and directions are given for the performance of many operations which do not seriously interfere with or injure the dissection. The education of the eye is a gradual and tedious process, but one which is pretty certain to be satisfactorily accomplished if the student do but use his hands properly, and therefore a few words on the manual part of dissection may not be out of place.

First, as to the instruments requisite for dissection. A case containing six or eight scalpels, two pairs of scissors, a pair of dissecting forceps, a set of chain-hooks, a blow-pipe, and a probe, will enable the student to make al requisite dissections, supposing that he is allowed th use of a saw and chisel in the dissecting room. Gre variety exists in dissecting cases, both as to form an expense, but so long as the instruments themselves ar strong and good, the simpler the case the better.

Scalpels for dissection are made of two principal shape in one, the edge is bevelled to the point, the back bei straight; in the other, both back and edge are bevelled a point midway between the two. The latter form is pr ferable for most purposes. The blade should not be mo than an inch and a half long, and *never* double-edged, b the material of which the handle is constructed is matter of indifference.

For all ordinary dissection, it will be found most co venient to hold the scalpel like a pen; but for cleaning t fascia off muscles, and following out small nerves, it better to hold it reserved, so that the *back* of the kn may be against the tissue which is to be preserved. making the first incision through the skin of a limb, or any other position where a long incision is required, t knife may, with advantage, be held *under* the hand, which the wrist has more play, and the student has the o portunity of practising a mode of holding the knife, which will find very useful when operating on the living body.

The forceps should be broad at the extremities a coarsely serrated, so that it may retain a firm hold on sm portions of tissue. It is very important that the forcep should not be too strong in the spring, for in that case becomes so fatiguing to the hand that it is impossible continue its use for any length of time.

The forceps should be held lightly between the thumb ıd the first and second fingers of the left hand, which ıay be steadied by resting the little finger on a neighbouring part.

The chain-hooks should be strong, and bent in the direction of the thickness and not of the breadth of the steel, s is sometimes done. These latter are very inferior, eing liable to be unbent under any considerable strain. are should be taken that the chains are firmly linked, nd that the central ring is sufficiently stout to bear any orce that may be applied to it. The scissors should be arge and strong, and it will be found to be advantageous o have one curved pair, which is very useful in preparing the ligaments. Sounds and staffs will be required for the purpose of practising the introduction of the instruments into the bladder; but these are found in most dissecting rooms, together with saws and other large tools requisite for dissection.

The student will do well to bear in mind that he will probably be called upon in after life to operate on the living body, the only true preparation for which is careful dissection; he should therefore, as far as possible, conduct ll his dissections as methodically and with as much care if operating on the living body, and by this means he ill do much to fit himself for his duties as a practical urgeon.

In removing the skin, the first incisions should be nade at once through its whole thickness, and a corner eing held with the forceps, the knife is to be carried ith a sweeping movement beneath it, the edge being owards the skin and the back to the fascia, which should e left smooth and uniform. The under surface of skin eatly reflected is white, and the tissue beneath it more r less yellow.

In cleaning muscles, it is essential to have the fibres stretched and rendered tense, which may be accomplished by moving the limb or applying the hooks. A muscle should invariably be cleaned *along* its fibres; the dissector beginning at one edge and advancing steadily to the other, and thus reflecting a complete layer of fascia; the knife being held with the back to the muscle, to avoid injury to it. The attachments of a muscle (origin and insertion) should be most carefully followed out; but it is important also to clean the fascia from the whole length of the muscle, or it will soon look dirty.

The arteries of a subject are usually injected, and it is impossible to follow out all the minute branches without this assistance; but opportunity should be taken by the advanced student to repeat his dissection upon an uninjected subject, in which the relations and appearance of the parts more closely resemble those of the living body.

The main trunks of the nerves and their principal branches are readily followed out, but their minute ramifications require more time and labour than a student can afford to bestow upon them.

The student should bear in mind that his manual labour is only a part of his duty, and will be thrown away, unless he at the same time study the description of the part upon which he is engaged; he should not therefore carry the dissection further than he can learn the description on the same day, and *at the subject;* and should if possible re-peruse the description in the evening, and always on the next morning, before carrying the dissection any further.

In order to preserve a part, it is essential that the dissector should himself secure the skin around it with a few stitches, and wrap it with damp cloths. These may be dipped in some preservative solution, or common salt may be grated finely over the part. This latter, however, destroys the colour of the part and its smooth appearance. A dissected part should be sponged over daily, when it is uncovered for fresh dissection.

PART I.

DISSECTION OF THE ARM.

[*The Student is requested to read the 'Introduction' before commencing the dissection, unless he has done so previously.*]

BEFORE beginning the dissection, the student should make himself fully acquainted with the external configuration of the part, and the relations of surface-markings to deeper structures; and if he has already dissected this region, he should make the incisions necessary to expose the several arteries in the positions in which they are usually tied, according to the directions which accompany the description of each vessel, taking care not to disturb the tissues unnecessarily, and to stitch up the incisions without delay.

The curves of the clavicle are generally seen with ease; but the finger should be carried along the bone to note any irregularity denoting old fracture, and to trace its articulation with the acromion process of the scapula. The sterno-clavicular joint should also be examined, and if the arm is freely moved, the extensive range of motion in that joint will be better appreciated. The development of the mammary region will vary according to the sex and age of the subject. In the case of a female subject, the advanced student should notice the condition of the nipple and its surrounding areola, as indicating previous pregnancies or the contrary, and may advantageously practise removal of the breast by two elliptical incisions, one above, the other below the nipple, taking great care to remove the whole of the breast, and not to leave any glandular tissue attached to the skin or the deeper structures.

The roundness of the shoulder will be found to depend upon the projection of the head of the humerus beyond the bony arch formed above it by the acromion and clavicle, and in a thin subject the head of the bone and the bicipital groove may be readily felt when the arm is rotated. About a finger's breadth to the inner side of the head of the humerus, and indistinctly felt on account of the muscles attached to it, is the coracoid process, and the relation of these bones in health should be fully appreciated. In a muscular subject, a long, nearly vertical groove or dimple at this point marks the separation between the deltoid and pectoralis major; and another, seen occasionally extending obliquely from near the inner end of the clavicle, marks the division between the sternal and clavicular fibres of the latter muscle. By lifting the arm and drawing it from the body, the anterior and posterior boundaries of the axilla will be made prominent, and the fascia extending from one to the other will be put on the stretch so that the finger cannot be pushed into the armpit; whereas if the arm is brought to the side the fascia is relaxed, and the finger will readily reach the head and neck of the humerus, with the great vessels and nerves to their inner side, and internal to these again, the wall of the thorax, with probably some lymphatic glands.

The brachial artery and nerves can be traced along the inner border of the biceps to the bend of the elbow, and the shaft of the humerus can be readily felt. In front of the elbow is a hollow, corresponding to a space to be afterwards dissected, and occasionally the superficial veins stand out prominently, in which case venæsection should be practised. The prominent internal condyle of the humerus and its relation to the olecranon process of the ulna are readily seen, and the ulnar nerve can be felt b [illegible] wo points of bone. The rounded external condyle, with the ridge leading to it, can be felt in most subjects, as also the head of the radius, which should be rotated, that its relation to the condyle may be better appreciated.

In the fore-arm the radius can be but little felt, except at its lower extremity; but the olecranon process and sharp posterior margin of the ulna are always to be distinguished even in very muscular subjects. It should be observed that the ulna is the more prominent of the two bones at its lower end, whilst the end of the radius is a little lower down than that of the ulna; the styloid process of each can be readily distinguished. The tendons of the flexor carpi radialis and flexor carpi ulnaris are usually prominent, and serve as guides to the radial and ulnar arteries, which may be felt (if injected) to the outer side of each tendon. The metacarpus and phalanges are to be examined, and the advanced student may advantageously amputate one or two fingers (middle or ring). It is to be borne in mind that the transverse markings on either the palmar or dorsal aspect of the fingers, form no guide to the articulation, which is always nearer to the nail. When the finger is bent, the prominence at the articulation is due to the proximal phalanx, and the joint is below that point. The best rule, therefore, in amputating through the phalangeal articulations, is to bend the phalanx which is to be removed to a right angle with the one above it, and to begin the incision at a point midway in the thickness of the upper phalanx, cutting transversely to a corresponding point on the opposite side.

The limb should be carefully bandaged from the fingers to the middle of the upper arm, and only uncovered as may be necessary in the progress of the dissection.

[The arm being drawn away from the side, an incision is to be made half an inch to one side of the median line of the sternum in its whole length, and to be joined at right angles by another running along the whole length of the clavicle on to the acromion process, and thence in a vertical direction half-way down the upper arm. Another incision is to be made transversely from the ensiform cartilage, and must be carried beyond the posterior of the armpit. The large flap of skin thus marked out is to be reflected towards the arm, the dissector on the right side beginning at the lower end of the sternum, and on the left side, at its upper extremity.]

Fig. 1.

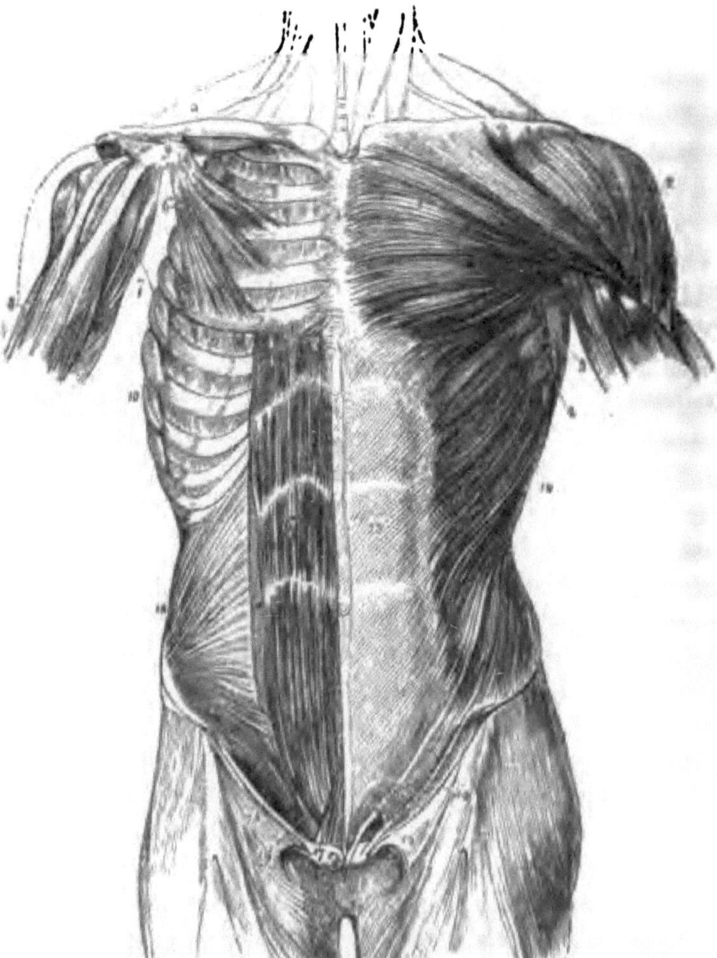

Fig. 1.—Muscles of the anterior aspect of the trunk (from Wilson); on the left side of the body the superficial layer is seen, on the right the deeper layer.

1. Pectoralis major.
2. Deltoid.
3. Anterior border of the latissimus dorsi.
4. Serratus magnus.
5. Subclavius, right side.
6. Pectoralis minor.
7. Coraco-brachialis.
8. Upper part of the biceps, showing its two heads.
9. Coracoid process of the scapula.

Beneath the skin is the subcutaneous fascia, which is often very fatty in the female. The mamma will also be exposed (an incision being carried round the nipple so as to leave it attached to the organ) in the female, or its rudiment in the male subject. In a well-developed body the fibres of the platysma myoides may be seen arising from the fascia below the clavicle.

The **Mamma**, or breast, is a gland of very variable bulk lying upon the great pectoral muscle, from which it is separated by a quantity of loose cellular tissue. Its base is nearly circular, and usually extends from the side of the sternum to the margin of the anterior wall of the axilla, reaching from the third to the seventh rib in the vertical direction. Near its centre and at a point opposite to the fourth rib is the *mamilla* or nipple, around which is the *areola* of discoloured skin, the tint of which becomes darker as pregnancy advances. The nipple may be shown to be pierced by numerous small lactiferous ducts by introducing bristles into their openings, and its skin is thin and delicate.

The gland consists of numerous lobules bound together by tough cellular tissue, and having a quantity of fat interspersed between them. Each lobule is abundantly supplied with bloodvessels and has a separate *galactoferous duct*. These last converge towards the nipple, and beneath the areola become straight and somewhat dilated so as to form lacteal sinuses.

---

10. Serratus magnus, right side.
11. External intercostal muscle of the fifth intercostal space.
12. External oblique.
13. Its aponeurosis: the median line to the right of this number is the linea alba; the curved line to its left, the linea semilunaris; the transverse lines above and below the number, the lineæ transversæ.
14. Poupart's ligament.
15. External abdominal ring.
16. Rectus muscle of the right side brought into view by the removal of the anterior segment of its sheath;* posterior segment of its sheath with the divided edge of the anterior segment.
17. Pyramidalis muscle.
18. Internal oblique.
19. Conjoined tendon of the internal oblique and transversalis.
20. The lower curved border of the internal oblique muscle.

The arteries of the breast are derived from the thoracic branches of the axillary artery, from the intercostal arteries, and from the internal mammary branch of the subclavian, which runs down the back of the sternum.

Cutaneous Nerves and Vessels.—In the superficial fascia on the front of the chest are the *anterior cutaneous* branches of the intercostal nerves and arteries. They appear at regular intervals on each side of the sternum as they pierce the pectoral muscle, and are of small size.

The terminations of the descending branches of the *superficial cervical plexus* may also be found beneath, or piercing, the platysma and crossing the clavicle in the positions their names imply, viz. *sternal*, *clàvicular*, and *acromial*.

[The Pectoralis major is to be put on the stretch by drawing the arm from the side and supporting it at a convenient height, its fibres are then to be cleaned from below on the right, and from above on the left side.]

The **Pectoralis Major** (Fig. 1, 1) consists of a sternal and a clavicular portion separated by a cellular interval. The sternal portion *arises* from the whole length of one side of the sternum and from the cartilages of all the true ribs except the seventh, and is connected below with the aponeurosis of the abdominal muscles. The clavicular portion *arises* from the inner or sternal half of the anterior border of the clavicle, and is separated by another cellular interval from the deltoid muscle. The two portions are united at their *insertion* into the anterior or outer edge of the bicipital groove of the humerus, the muscle being folded near its insertion, so that the sternal fibres pass behind, and reach higher up the bone than, those from the clavicle. The deeper portion of the insertion is more tendinous than the superficial fibres, which are closely connected with the deltoid.

The action of the pectoralis major is to rotate the humerus inwards and then cross it over the chest, and the upper fibres may slightly assist the deltoid in raising the arm. If the arm were fixed above the chest, the muscle would drag the trunk upwards, as in climbing. It may also act as an extraordinary muscle of inspiration when the arm is fixed, as is seen in the case of asthmatic persons. It is *supplied* by the external anterior thoracic nerve and partly by the internal anterior thoracic nerve. In very well developed subjects, muscular slips are not unfrequently found lying at the side of the sternum, and superficial to the pectoralis, with which they may have some connection; but they are more frequently continuous with the sterno-mastoid muscle.

The Axilla.

[If the arm is well drawn from the side the boundaries of the axilla or armpit will be readily seen, and if the skin has not been already removed from the posterior boundary (latissimus dorsi), it should now be turned back. The strong deep fascia will be seen stretching from the pectoralis to the latissimus dorsi, forming the base of the conical space known as the axilla: and the student should again notice that it is impossible to thrust the finger deeply into the space so long as the fascia is stretched by the extension of the arm, but that when the arm is drawn to the side the finger passes into the space, and can feel the axillary artery (if injected), and the head of the humerus partially.]

The *lateral cutaneous* branches of the *intercostal vessels and nerves* are to be found behind the border of the pectoralis major by carefully dividing the fascia parallel to the ribs. The first intercostal nerve has no lateral branch, but six lateral branches from the upper nerves (from 2nd to 7th) will be found appearing at the side of the chest, between the digitations of the serratus magnus. Each lateral cutaneous nerve, except that of the second, divides into an anterior and posterior branch which turn forwards to the pectoral region, and backwards over the latissimus dorsi respectively.

The lateral cutaneous branch of the second nerve passes down the inner side of the arm to join the nerve of

Wrisberg, and is hence called the *intercosto-humeral nerve*. The posterior branch from the third nerve usually accompanies and joins this. These last must be found at once and are to be carefully preserved.

Fig. 2.

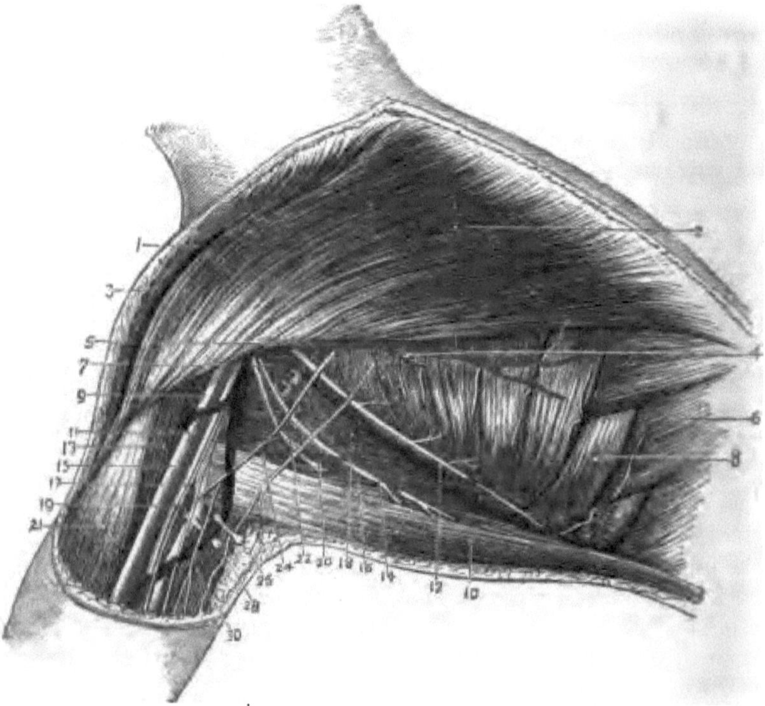

[All the fat is to be removed from the axilla, and the muscles cleaned, so far as they are exposed, care being

Fig. 2.—The Axilla. (Drawn by J. T. Gray.)

1. Cephalic vein.
2. Pectoralis major.
3. Deltoid.
4. Pectoralis minor.
5. Long thoracic artery.
6. Obliquus externus.
7. Alar thoracic artery.
8. Serratus magnus.
9. Median nerve.
10. Latissimus dorsi.
11. Axillary artery.
12. Posterior thoracic nerve.
13. Coraco-brachialis.
14. Long subscapular nerve.
15. Ulnar nerve.
16. Teres major.
17. Internal cutaneous nerve.
18. Subscapularis.
19. Nerve of Wrisberg.
20. Subscapular artery.
21. Biceps.
22. Middle subscapular nerve.
24. Intercosto-humeral nerve.
26. Basilic vein becoming axillary vein.
28. Musculo-spiral nerve.
30. Triceps.

taken not to injure the axillary vessels (particularly the vein) and nerves, which are to the outer side of the space; nor the *posterior thoracic nerve*, which runs vertically upon the serratus magnus, about three inches from the edge of the pectoralis; nor the subscapular vessels and nerves at the edge of the subscapularis.]

The **Axilla** is a conical space between the chest and upper arm, and has the following boundaries:—*In front* is the pectoralis major muscle, with the minor beneath it; *behind* are the subscapularis, latissimus dorsi, and teres major muscles; to the *inner side* is the serratus magnus, with the upper ribs and intercostal muscles; and to the *outer side* the upper part of the humerus and the coraco-brachialis muscle. The *base* of the cone has already been seen to be the axillary fascia stretched between the anterior and posterior boundaries, and the *apex* will now be found with the finger in a triangular interval between the clavicle, the upper border of the scapula, and the first rib. The great vessels and nerves lie along the outer boundary of the space, the artery being between the nervous cords and the vein to their inner side, whilst the branches of the artery take a general direction towards the thorax. On the inner side of the space, and imbedded in fat, are the glands, ten or twelve in number, which receive the lymphatics from the mamma and arm, and which may therefore be extirpated when diseased, or incised when suppurating, without danger to the large vessels.

Opportunity is to be taken, before any further dissection is made, to expose the lower part of the axillary artery thoroughly, as that is the portion of the vessel to which a ligature may be applied.

Surgery.—*To tie the axillary artery:* the arm being drawn away from the trunk, an incision, three inches long, is to be made in the length of the limb, at a distance behind the edge of the pectoral muscle equal to one-third of the whole width of the axilla. On dividing the deep fascia, the large axillary vein will come into view, which must be drawn down and held aside, when the large nervous cords will be seen, and a little dissection will expose the axillary artery placed deeply between them. The aneurism needle is to be passed from below (in this position), *i. e.* between the vein and artery.

[The pectoralis major is to be divided vertically about its middle and reflected, when branches of the acromio-thoracic artery and the external anterior thoracic nerve will be seen entering its under surface. The pectoralis minor will now be exposed, with the long thoracic artery at its lower border.]

The **Pectoralis Minor** (Fig. 1, 6) is beneath the pectoralis major, and therefore assists in forming the anterior boundary of the axilla. It *arises* from the anterior extremities of the third, fourth, and fifth ribs, an inch from their junctions with the cartilages, and is *inserted* into the inner side of the coracoid process of the scapula, thus crossing obliquely the axillary vessels and nerves. The pectoralis minor would draw the coracoid process and scapula downwards, but is principally a muscle of inspiration, taking its fixed point from the coracoid process, and so raising the ribs. It is *supplied* by the internal anterior thoracic nerve.

The **Costo-Coracoid Membrane** is the strong fascia seen immediately above the pectoralis minor, where it covers the axillary vessels and gives passage to the acromio-thoracic artery, the cephalic vein, and the external anterior thoracic nerve. It is attached to the coracoid process, to the lower border of the clavicle, and to the first rib, and when divided will be seen to] cover the axillary vessels and nerves and to enclose the subclavius muscle, being continued for some distance on the vessels, the sheath of which it strengthens.

The **Subclavius** (Fig. 1, 5) is a small muscle which *arises* by a tendon from the cartilage of the first rib, and is *inserted* into the groove on the under surface of the middle third of the clavicle. It is a depressor of the clavicle, and is *supplied* by a special branch of the brachial plexus above the clavicle.

The *sheath* of the axillary vessels is derived from the fascia in the neck, and passes beneath the clavicle. It is strengthened by a process of the costo-coracoid membrane, and is gradually lost upon the vessels.

[The pectoralis minor is to be divided two inches from its insertion, when the axillary vessels and nerves will be fully exposed and are to be cleaned. The small nerve emerging from between the artery and vein, and entering the pectoralis minor, is the internal anterior thoracic.]

The **Axillary Artery** (Fig. 2, 11) is the continuation of the subclavian, and extends from the lower border of the first rib to the lower border of the insertion of the teres major. In its upper part it is placed deeply in the axilla, but its lower portion has been already seen to be comparatively superficial. The artery is divided into three parts, viz. first, or that *above* the pectoralis minor; second, or that *beneath* the pectoralis minor; and third, or that *below* the pectoralis minor.

First part (from the lower border of the first rib to the upper border of pectoralis minor) has *in front* the pectoralis major muscle and the costo-coracoid membrane, and is crossed by the cephalic and acromio-thoracic veins, and external anterior thoracic nerve; *behind*, it has the first intercostal space, with a digitation of the serratus magnus and the posterior thoracic nerve (external respiratory of Bell); to its *inner side* is the axillary vein, and to the *outer* the large cords of the brachial plexus. This part of the artery gives off two branches, the Superior Thoracic and Acromio-thoracic.

Second part (beneath the pectoralis minor) has *in front* the pectoralis major and pectoralis minor; *behind*, at a little distance, is the subscapularis, with the posterior cord of the brachial plexus; to its *inner side* are the vein and the inner cord; whilst on the *outer side* are the coraco-brachialis and outer cord of the plexus. The branches of this part of the artery are the Long Thoracic and Alar Thoracic.

Third part (from the lower border of the pectoralis minor to the lower border of teres major) has the pectoralis major *in front*, except at the lowest portion, and is also embraced by the two heads of the median nerve, which itself lies for a short distance in front of the artery; *behind*, it lies upon the lower part of the subscapularis, the tendon of the latissimus dorsi, and a small portion of the teres major muscle, together with the musculo-spiral and circumflex nerves; to its *inner side* is the axillary vein, with the inner head of the median nerve (for a short distance), the ulnar, the internal cutaneous and lesser internal cutaneous (Wrisberg) nerves; to the *outer side* is the coraco-brachialis muscle, with the outer head of the median, at first, but lower down the entire nerve, and also the external cutaneous nerve, until it pierces the coraco-brachialis. The branches of this portion are the Subscapular, Anterior Circumflex, and Posterior Circumflex.

Branches of the Axillary Artery (Fig. 3).*—

1. The *superior thoracic* (3) is a small branch supplying the side of the chest about the first and second ribs, and anastomosing with the intercostal arteries.

2. The *acromio-thoracic (thoracica-humeraria; thoracic axis)* (1) is a large branch arising from the artery at the upper border of the pectoralis minor, and immediately piercing the costo-coracoid membrane to divide into three sets of branches. One set runs forward to supply the pectoralis major, a second backwards to the acromion process and shoulder, and the third downwards in the groove between the pectoralis major and deltoid by the side of the cephalic vein.

3. The *long thoracic* (5) runs along the lower border of the pectoralis minor to the side of the chest, where it supplies the mamma (and is hence sometimes called the external mammary artery). It anastomoses with the internal mammary, intercostal, and subscapular branches.

4. The *alar thoracic* (3) is a small branch to the axillary

* The branches of the Axillary Artery may be kept in mind by the word SALASAP; thus:—Superior thoracic, Acromio-thoracic, Long thoracic, Alar thoracic, Subscapular, Anterior and Posterior circumflex.

glands. It very seldom arises directly from the axillary, but generally from one of the other branches.

5. The *subscapular* (15) is a large branch which runs along the lower border of the subscapularis muscle to the side of the chest with the long subscapular nerve, and then supplies the adjacent parts, and anastomoses with the long

Fig 3

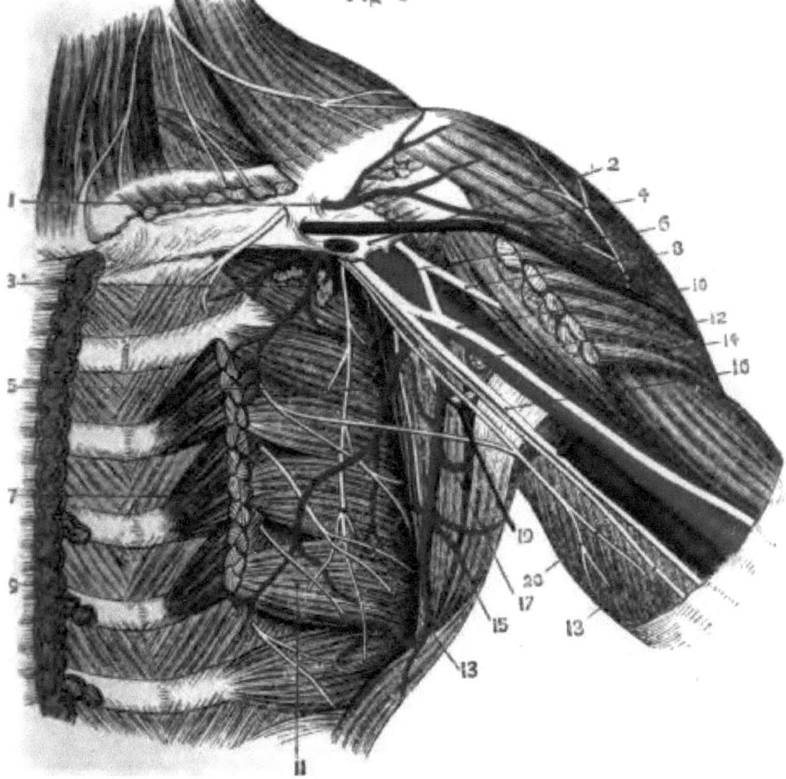

Fig. 3.—The Axillary Artery and its Branches, the Pectoral Muscles being removed (drawn by G. E. L. Pearse).

1. Acromio-thoracic artery.
2. Costo - coracoid membrane with cut axillary vein.
3. Superior thoracic and alar thoracic arteries.
4. Cephalic vein.
5. Long thoracic artery.
6. Axillary artery.
7. Pectoralis minor (cut).
8. Musculo-cutaneous nerve.
9. Pectoralis major (cut).
10. Median nerve.
11. Serratus magnus.
12. Posterior circumflex artery.
13. Posterior thoracic nerve.
14. Pectoralis major.
15. Subscapular artery.
16. Ulnar nerve.
17. Dorsalis scapulæ artery.
18 Internal cutaneous nerve.
19. Circumflex nerve (drawn down).
20. Nerve of Wrisberg joined by intercosto-humeral nerve

thoracic artery. About an inch from its origin this artery gives off a large branch, the *dorsalis scapulæ*, which winds round the lower border of the subscapularis muscle and, after giving an infrascapular branch, disappears through a triangular interval, bounded (as seen from the front) by the long head of the triceps, the teres major, and the subscapularis.

6. The *anterior circumflex* is a small branch which

Fig. 4.

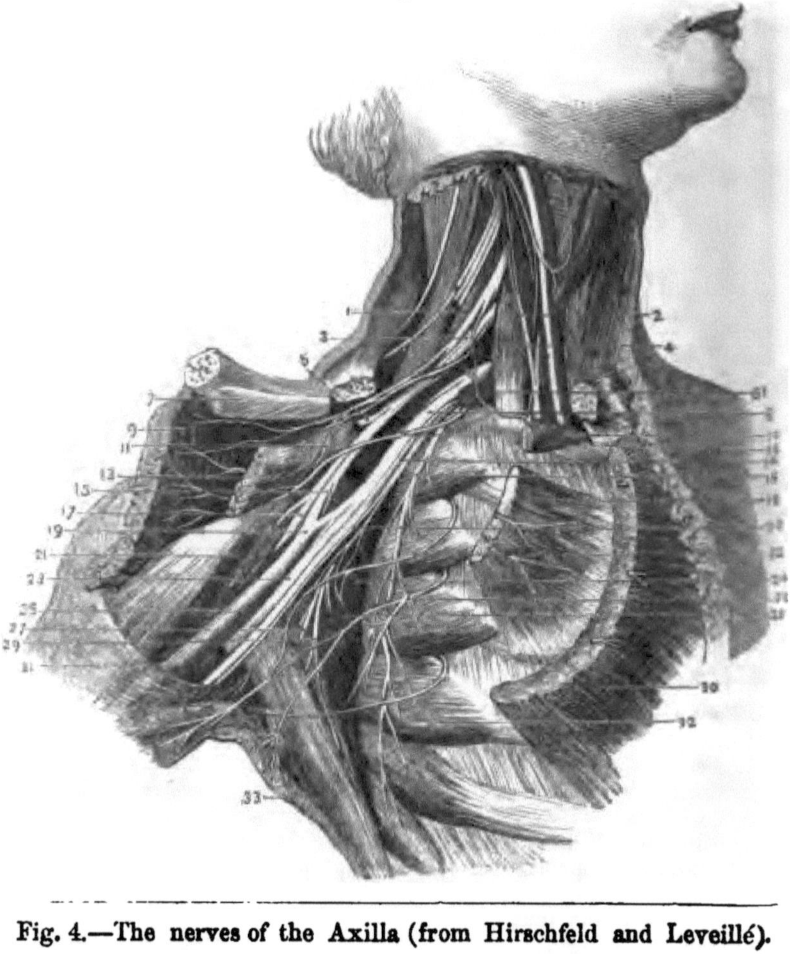

Fig. 4.—The nerves of the Axilla (from Hirschfeld and Leveillé).

1. Scalenus medius.
2. Scalenus anticus.
3. Cord formed by 5th and 6th cervical nerves.
4. 7th cervical nerve.
5. Suprascapular nerve.
6. Subclavian artery (cut).
7. Insertion of subclavius.

winds round the neck of the humerus beneath the coraco-brachialis and biceps, which it supplies, giving also a branch up the bicipital groove to supply the shoulder-joint.

7. The *posterior circumflex* (12) is larger than the anterior, and winds behind the neck of the humerus to supply the deltoid, passing through a quadrilateral space bounded (as seen in front) by the teres major, subscapularis, long head of the triceps, and the neck of the humerus.

The **Axillary Vein** lies to the inner side of the artery for the greater part of its course; it is formed, at a variable point, by the junction of the basilic vein with the *venæ comites* of the brachial artery, and having received branches corresponding to those of the artery, and the cephalic vein immediately below the clavicle, it ends in the subclavian vein.

Brachial Nerves (Fig. 4).—The large cords of the brachial plexus are derived from the 5th, 6th, 7th, and 8th cervical nerves, together with the 1st dorsal, which make their appearance in the posterior triangle of the neck. The nerves coalesce and divide again in an irregular manner (see Posterior Triangle of Neck), but the result is that they enter the axilla as three cords, lying to the outer side of the first part of the axillary artery; lying to the inner and outer sides of and behind the second portion of that vessel; and the several branches derived from these

---

8. Cord formed by 8th cervical and 1st dorsal nerves.
9. Pectoralis major (reflected).
10. Internal anterior thoracic nerve.
11. External anterior thoracic nerve.
12. Origin of subclavius.
13. Pectoralis minor (reflected).
14. Internal cutaneous nerve.
15. Axillary artery (cut).
16. Posterior thoracic nerve (Bell).
17. Musculo-cutaneous nerve.
18. Origin of pectoralis minor.
19. Median nerve.
20. Nerve of Wrisberg.
21. Coraco-brachialis.
22. Intercosto-humeral nerve.
23. Ulnar nerve.
24. Subscapularis.
25. Brachial artery.
26. Lateral cutaneous branch of 3rd intercostal nerve.
27. Middle subscapular nerve.
28. Short subscapular nerve.
29. Long subscapular nerve.
30. Pectoralis major (cut).
31. Basilic vein.
32. Serratus magnus.
33. Latissimus dorsi.

trunks, and now to be examined, more or less surrounding the artery in its third part.*

Brachial plexus formed by 5th, 6th, 7th, 8th cervical, and 1st dorsal nerves . . .	Outer cord . .	External anterior thoracic. External cutaneous. Outer head of median.
	Inner cord. . .	Inner head of median. Ulnar. Internal cutaneous. Lesser internal cutaneous. Internal anterior thoracic.
	Posterior cord .	Musculo-spiral. Circumflex. Three subscapular nerves.

The several branches are to be thoroughly identified, and their relations to the vessels ascertained (*vide* AXILLARY ARTERY); but the description of the majority is more conveniently taken at a later stage of the dissection.

The *External anterior thoracic nerve* (11) is a small branch derived from the outer cord, going to the under surface of the pectoralis major, which it supplies.

The *Internal anterior thoracic nerve* (10) is a small branch from the inner cord, which passes between the axillary artery and vein to supply the pectoralis minor and communicate with the preceding nerve.

The *Circumflex* (Fig. 3, 19) is a large nerve from the posterior cord, and lies behind the axillary artery for a short distance, leaving it then to accompany the posterior circumflex artery through the quadrilateral space already described, and wind round the humerus to the deltoid, to which it will be subsequently traced.

Three *Subscapular nerves* (27, 28, 29) are derived from the posterior cord. The long subscapular nerve is seen at the lower border of the subscapularis muscle accompanying the artery, and ends in the latissimus dorsi, sometimes giving a branch also to the upper part of the teres major. The two shorter subscapular nerves will be better dissected when the arm is removed; they enter the

* The brachial plexus may consist occasionally of only two cords when it enters the axilla; in which case a third cord is formed below the clavicle by the union of branches derived from these two cords, the relations to the artery remaining the same as are given above.

subscapularis and teres major muscles, and supply them.

The *Posterior thoracic nerve* (external respiratory of Bell) (16) lies upon the serratus magnus, which it supplies, and is one of the supra-clavicular branches of the brachial plexus, being derived from the fifth and sixth nerves, with an occasional branch from the seventh; it descends behind the brachial cords and axillary vessels to its distribution upon the surface of the serratus.

The **Serratus Magnus** (Fig. 4, 32) covers the side of the chest, *arising* by eight digitations from the outer surfaces of the eight upper ribs, the first digitation being of large size, and attached to the second as well as to the first rib. The fibres have a general direction backwards to the inner surface of the base of the scapula where they are inserted, but the first two digitations unite to form a thicker portion corresponding to the surface at the upper angle, while the last four form a still larger bundle, corresponding to the surface at the inferior angle of the scapula. These last portions interdigitate with the external oblique muscle of the abdomen. The serratus magnus is a great muscle of inspiration, taking its fixed point from the scapula, and thus raising the ribs. If the muscle has the ribs for its fixed point, the scapula is drawn forward, the bone gliding over the ribs and a quantity of loose cellular tissue which lies upon them. The serratus is *supplied* by the posterior thoracic nerve.

Insertions of the Latissimus Dorsi and Teres Major.— The upper part of the latissimus winds round the lower angle of the scapula, which it envelopes in a sort of fold, and ends in a broad, thin tendon, which passes in front of the teres major to be *inserted* into the bottom of the bicipital groove of the humerus above the level of the teres major. The tendon of the latissimus dorsi is twisted, so that the fibres of the muscle, which were highest at their origin, are lowest at their insertion, and *vice versâ*, and it is the narrowest of the three tendons inserted into the bicipital groove.

The teres major, which is fleshy throughout, passes

behind the tendon of the latissimus being more or less incorporated with it, and is *inserted* into the inner or posterior edge of the bicipital groove, exactly opposite the pectoralis major, a bursa being formed between the insertions of the latissimus and teres.

Opportunity should be taken, when the clavicle is divided by the dissector of the neck, to trace the axillary nerves to their origins, and to study the brachial plexus, the description of which is given with the "Posterior Triangle."

THE FRONT OF THE UPPER ARM.

[An incision is to be made down the front of the arm to three inches below the bend of the elbow, and the skin is to be reflected on each side for a couple of inches, to allow of the dissection of the superficial nerves and veins.]

The **Internal Cutaneous Nerve** (Fig. 5, 15) is to be followed out from the inner cord of the plexus to the fore-arm. It lies at first to the inner side of the axillary artery, and then, piercing the deep fascia at or near the point at which the basilic vein enters, it runs in front of the inner condyle and generally over the median basilic vein, to end in branches to the upper and inner side of the fore-arm. A posterior branch winds above the inner condyle to the back of the fore-arm.

The **Lesser Internal Cutaneous Nerve** (nerve of Wrisberg) (Fig. 4, 20), if it exists, pierces the fascia at a variable point, and after joining with the intercosto-humeral nerve, supplies the inner side of the upper arm.

A *cutaneous branch of the musculo-spiral nerve* appears on the inner side of the upper arm, below the teres major, and is distributed to the skin.

[The veins of the arm are very variable, and the best way of dissecting them therefore, will be to follow both basilic and cephalic veins down to the elbow, and to trace out their radicles.]

The **Median Vein** is near the centre of the fore-arm and divides into two branches, the *median basilic* and *median cephalic*, which pass obliquely inwards and outwards respectively, to assist in forming the basilic and cephalic veins. Piercing the deep fascia near the centre of the bend of the elbow is the *profunda vein*, which joins the median near its point of bifurcation, and forms a communication between the *venæ comites* of the arteries of the fore-arm and the superficial veins, the direction of the current of the blood varying in different subjects.

Fig. 5.

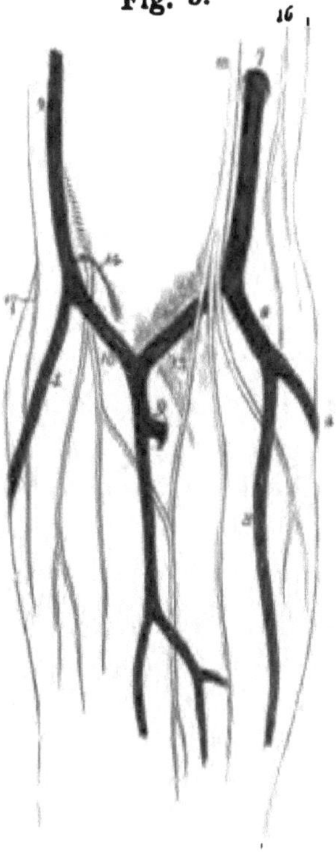

The **Anterior** and **Posterior Ulnar Veins** end on the inner side of the fore-arm in a single trunk, which unites with the median basilic to form the basilic vein.

The **Basilic Vein** (Fig. 5, 6) runs up the inner side of the arm to about its middle, where it pierces the deep fascia, and, at a variable point, unites with the venæ comites of the brachial artery to form the axillary vein. To the inner side of the basilic vein and about an inch above the condyle is a lymphatic gland of small size.

Fig. 5.—Veins of the fore-arm and bend of the elbow (from Wilson).

1. Radial vein.
2. Cephalic vein.
3. Anterior ulnar vein.
4. Posterior ulnar vein.
5. The trunk formed by their union.
6. Basilic vein, piercing the deep fascia at 7.
9. Profunda vein.
10. Median cephalic vein.
11. Median basilic vein.
12. The position of the brachial artery.
13. The semilunar fascia, derived from the tendon of the biceps, which separates the median basilic vein from the brachial artery.
14. External cutaneous nerve, piercing the deep fascia and dividing into two branches which pass behind the median cephalic vein.
15. Internal cutaneous nerve, dividing into branches which pass in front of the median basilic vein.
16. Nerve of Wrisberg.
17. Cutaneous branch of the musculo-spiral nerve.

The **Radial Vein** ascends on the outer side to join the median cephalic and form the *cephalic vein*, which passes up the outer side of the arm, and has already been traced to its termination in the axillary vein.

The **Median Basilic Vein** (Fig. 5, 11) has important relations, since it lies superficial to the brachial artery, and crosses under the internal cutaneous nerve. It is the vessel usually selected for venæsection on account of its size; and protection is afforded to the artery beneath, by the interposition of a process of fascia (bicipital or semilunar) which is connected with the tendon of the biceps, and may now be defined.

The **Median Cephalic Vein** (Fig. 5, 10) is of smaller size and crosses the tendon of the biceps, close to the outer side of which, and beneath the vein, the *external cutaneous nerve* makes its appearance and supplies the adjacent parts. Above and to the outer side of the external cutaneous nerve will be found one or two external cutaneous branches of the musculo-spiral nerve, while close to the deltoid may be found a small twig of the circumflex.

The *deep fascia* of the arm is sufficiently seen in following the superficial vessels and nerves. It is continuous with the fascia of the axilla and fore-arm, and is attached to ridges above the condyles of the humerus, forming the external and internal intermuscular septa.

[The muscles of the front of the arm are now to be cleaned together with the vessels and nerves, but the fascia on the upper part of the muscles of the fore-arm need not be disturbed.]

The **Biceps** (Fig. 6, 7) is the superficial muscle of the upper arm, and arises by two heads which generally unite in the upper third of the arm, but occasionally remain distinct nearly to their insertions. The *long* or *outer* head cannot be fully traced at present, since it *arises* within the capsular ligament of the shoulder joint from the top of the glenoid cavity, and also from the glenoid ligament. Traversing the joint but excluded from the synovial cavity by a reflection of synovial membrane, the tendon enters the bicipital groove of the humerus, and is seen to emerge from the capsular ligament and lie upon the tendon of the latissimus, and between the insertions of the pectoralis major and teres major muscles. The *short* or *inner* head *arises* from the coracoid process of the scapula, in common with, but to the outer side of, the coraco-brachialis; and the two heads having united, the tendon gives off an expansion, called the bicipital or semilunar fascia, to join the fascia on the upper and inner part of the fore-arm (Fig. 8, 16). This must be divided in order to see the tendon of the biceps, which passes deeply to be *inserted* into the back of the bicipital tubercle of the radius, a bursa being placed between the tendon and the upper smooth surface of the tubercle. The tendon becomes flattened near its insertion, and is twisted so that its *edges* become anterior and posterior. The action of the biceps is to supinate the fore-arm, and then flex it upon the upper arm, being a good example of force applied to a lever of the third order. Its action as an extensor of the upper arm is very slight.

Fig. 6.

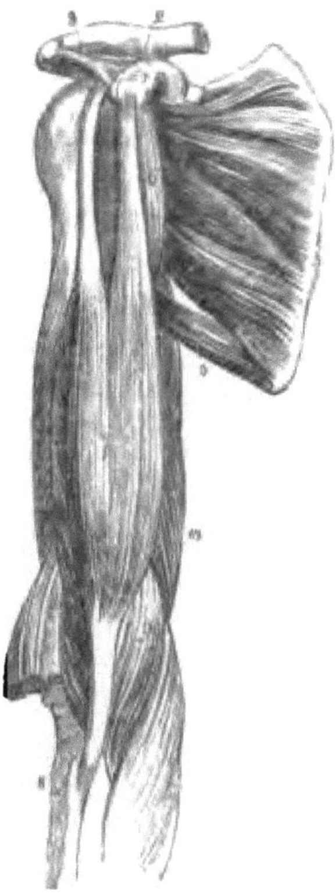

Fig. 6.—Muscles of the anterior aspect of the upper arm (from Wilson).

1. Coracoid process of the scapula.
2. Coraco-clavicular ligament (trapezoid).
3. Coraco-acromial ligament.
4. Subscapularis.
5. Teres major.
6. Coraco-brachialis.
7. Biceps.
8. Upper end of the radius.
9. Brachialis anticus.
10. Internal head of the triceps.

The **Coraco-brachialis** (Fig. 6, 6) *arises* from the tip of the coracoid process, in common with, but to the inner side of, the short head of the biceps. It is generally pierced by the external cutaneous nerve, and is *inserted* on the inner side of the shaft of the humerus about its middle, and exactly opposite the insertion of the deltoid. It draws inward and flexes the humerus.

The **Brachialis anticus** (Fig. 8, 11) *arises* from the front of the shaft of the humerus in its lower half; it has two slips which embrace the insertion of the deltoid, and the fibres have also an extensive attachment to the internal intermuscular septum, and a smaller one to the upper part of the external intermuscular septum above the supinator longus. It is *inserted* into a triangular surface on the front of the coronoid process of the ulna, covering the front of the elbow joint. It is the direct flexor of the fore-arm upon the upper arm.

The biceps, coraco-brachialis, and brachialis anticus are *supplied* by the external or musculo-cutaneous nerve; and the brachialis anticus has an additional supply from the musculo-spiral nerve, which is close to its outer border, lying deeply between it and the supinator longus.

The **Brachial Artery** (Fig. 8, 14) is the direct continuation of the axillary, and extends from the lower border of the teres major to the point of bifurcation into radial and ulnar arteries, which is usually at the bend of the elbow. The artery is superficial, or at least crossed by no muscle, in the whole of its course, and it gradually turns from the inner side of the arm to the middle of the bend of the elbow.

In front it has only the fascia and skin, except that at about the middle of the arm it is generally crossed by the median nerve (but the nerve may pass beneath the vessel); and at the bend of the elbow it is crossed by the bicipital fascia, and has the median basilic vein superficial to it. *Behind* it has first the musculo-spiral nerve, and then rests upon a small portion of the internal head of the triceps; next, on the insertion of the coraco-brachialis; and afterwards upon the brachialis anticus for the rest of its course. To the *outer* side is the median nerve, with the coraco-brachialis muscle for a short distance, and afterwards the biceps muscle, which is the guide to the vessel in applying a ligature. To the *inner* side is the ulnar nerve as far as the internal intermuscular septum, and afterwards, from a varying point, the median nerve, which continues in close relation to the vessel for the rest of its course.

Surgery.—*To tie the brachial artery.* An incision, three inches long, is to be made close to and exposing the inner edge of the biceps in the middle third of the arm. A little dissection parallel to the muscle will expose the median nerve, and the artery will be found in close relation to it, and generally to its outer side in this position, but this will vary in different bodies. The needle is to be passed *from* the nerve, care being taken of the venæ comites.

Fig. 7.

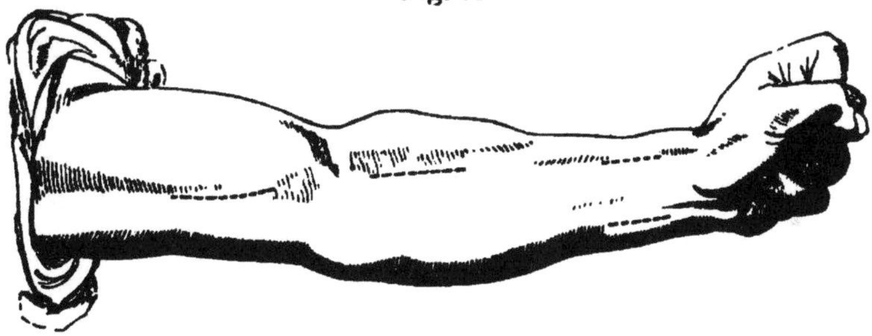

The **Branches** of the brachial artery are—

1. *Muscular*, arising at various points to supply the adjacent muscles.

2. *Nutritious*, which enters the foramen on the inner side of the humerus about the middle of the bone, and has a general direction *towards* the elbow.

3. The *Superior profunda*, coming off from the inner side of the artery at its upper part, and immediately joining the musculo-spiral nerve, which it accompanies between the long and inner heads of the triceps and round the humerus to the outer side, where a branch may

Fig. 7.—Incisions for tying the brachial, radial and ulnar arteries (from Fergusson's 'Surgery').

be now seen with the nerve between the brachialis anticus and supinator longus. It gives off muscular branches (principally to the triceps), and will be afterwards seen to anastomose with the radial recurrent artery.

Fig. 8.

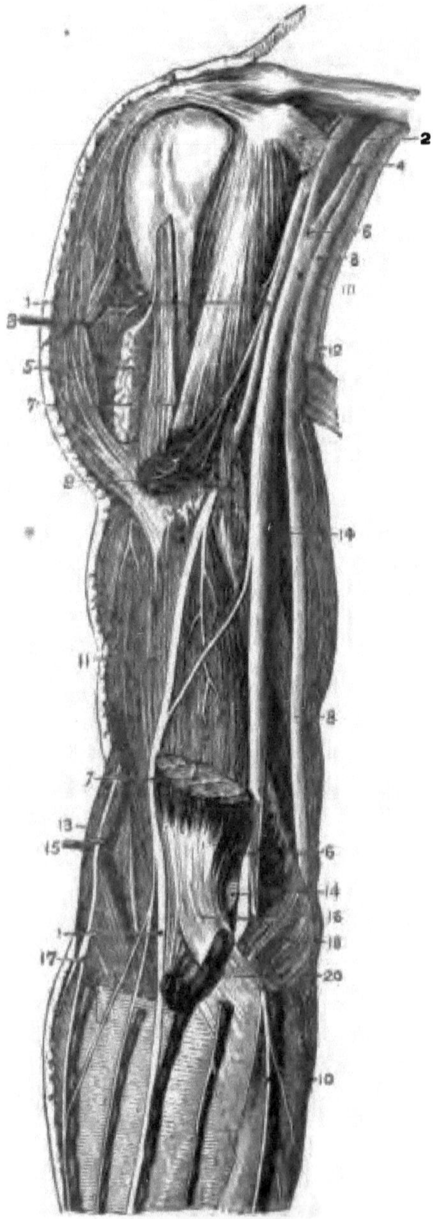

4. The *Inferior profunda*, which arises at a variable distance below the preceding, and passes, with the ulnar nerve, behind the internal intermuscular septum. It runs with the nerve to the groove between the inner condyle and the olecranon, and there joins the posterior ulnar recurrent artery.

5. The *Anastomotica magna*, which arises just above the elbow and runs directly inwards, piercing the internal intermuscular septum and supplying the parts about the elbow. It anastomoses with the inferior profunda and anterior and posterior ulnar recurrent arteries.

Irregularities of the brachial artery.— The

Fig. 8.—Deep dissection of the front of the upper arm (from Hirschfeld and Leveillé).

1, 1. Musculo-cutaneous nerve.
2. Pectoralis minor.
3. Deltoid.
4. Axillary artery.

point of bifurcation is occasionally much higher than the bend of the elbow, so that two large vessels may be found side by side in any part of the arm from this cause, or from a large branch (generally the radial) being given off high up. The two profunda branches not uncommonly arise together, and occasionally there is an axis common to all the branches of the brachial.

Venæ Comites are in close relation with the brachial artery, and receive twigs corresponding to its branches: they unite with the basilic to form the axillary vein.

Nerves of the Arm.

The **Median Nerve** (Fig. 8, 6) is derived from the outer and inner cords of the brachial plexus by two roots, which embrace the axillary artery in the third part of its course. The nerve afterwards lies to the outer side of the axillary artery for a short distance, and continues in the same relation to the brachial artery for about half its length, then, crossing either over or under that vessel, the median nerve lies to its inner side, and continues in the same relation to the bend of the elbow. The median nerve ordinarily gives off no branch in the upper arm, but occasionally communicates with the external cutaneous nerve.

The **External Cutaneous Nerve** (musculo-cutaneous) (Fig. 8, 1) arises from the outer cord of the brachial plexus in common with the outer head of the median, and lies to the outer side of the median nerve and the axillary vessels for a short distance. It then pierces the coraco-brachialis obliquely (hence *n. perforans Casserii*), and having given a branch to that muscle, it lies between the biceps and the brachialis anticus, both of which

---

5. Tendon of pectoralis major.
6, 6. Median nerve.
7, 7. Biceps.
8, 8. Ulnar nerve.
9. Coraco-brachialis.
10. 10. Internal cutaneous nerve.
11. Brachialis anticus.
12. Nerve of Wrisberg.
13. Musculo-spiral nerve.
14, 14. Brachial artery.
15. Supinator longus.
16. Bicipital fascia.
17. Cutaneous branch of musculo-spiral nerve.
18. Pronator radii teres.
20. Median basilic vein

muscles it supplies. It finally becomes cutaneous at the outer side of the tendon of the biceps just above the bend of the elbow, where it has been already seen to lie beneath the median-cephalic vein.

The **Ulnar Nerve** (Fig. 8, 8) arises from the inner cord of the brachial plexus, and lies close to the inner side of the axillary vessels, and afterwards in a similar relation with the upper part of the brachial artery. It leaves that vessel about the middle of the arm, piercing, and then running behind, the internal intermuscular septum, accompanied by the inferior profunda artery, to the interval between the internal condyle and the olecranon process. The ulnar nerve gives no branch in the upper arm.

The **Musculo-spiral Nerve** (Fig. 8, 13) is derived from the posterior cord of the brachial plexus, and lies at first behind the axillary artery and upon the subscapularis muscle with the circumflex nerve. It then rests upon the latissimus dorsi and teres major, and, after giving off a cutaneous branch to the inner side of the arm, winds backwards around the humerus, lying against the inner head of the triceps and disappearing between it and the long head of that muscle. Branches supplying the triceps can now be seen, and in a subsequent dissection the nerve can be followed through the fibres of that muscle. Its lower portion is now visible between the supinator longus and brachialis anticus, to both of which and to the extensor carpi radialis longior, a branch may be traced; the division into *radial* and *posterior interosseous* nerves is also to be dissected out.

The Bend of the Elbow.

[The boundaries of the space in front of the elbow are now to be exposed by removing the fascia from the muscles of the fore-arm to the extent to which the skin has been already reflected, the tendons of the biceps and brachialis are to be thoroughly cleaned, and the fibres of the supinator brevis in the floor of the space carefully dissected.]

The triangle in front of the elbow is bounded *above* by

an imaginary line drawn across the arm about two inches above the condyles, and *below* the apex is formed by the junction of the pronator teres and supinator longus about the same distance below the joint. The *external* boundary is the supinator longus, the *internal* the pronator teres, whilst the *floor* is formed, first, by the lower part of the brachialis anticus, and afterwards by the oblique fibres of the supinator brevis. The relations of the superficial veins and nerves have been already examined (p. 24), and the *contents* of the space are now seen to be—(1) the median nerve, (2) the brachial artery, and (3) the tendon of the biceps, which lie in that order from within outwards (Fig. 12).* The median nerve lies nearly vertically in the space, and disappears between the heads of the pronator teres : the artery usually bifurcates into radial and ulnar upon the insertion of the brachialis anticus, the radial resting against the tendon of the biceps and supinator brevis, and the ulnar lying upon the brachialis till it disappears beneath the pronator teres. The twisting of the biceps tendon has already been referred to.

By slightly displacing the supinator longus outwards, the musculo-spiral nerve will be seen lying between it and the brachialis anticus and dividing into radial and posterior interosseous nerves; but this nerve is not, strictly speaking, in the triangle. The radial nerve may be traced for a short distance upon the supinator brevis, whilst the posterior interosseous can be seen entering its fibres.

It is supposed that by the time the dissection of the bend of the elbow is finished, the subject will be turned, in which case the dissector of the arm should dissect the first two layers of the muscles of the back (*see* DISSECTION OF THE BACK), proceeding afterwards with the following dissections. If necessary, however, the dissection of the front of the fore-arm may be proceeded with first. Since in some dissecting-rooms the dissector of the arm has no part in the back, the following directions are given for the removal of the limb :—

* Artificial Memory, NAT.

[The two superficial layers of the muscles of the back comprise the trapezius, latissimus dorsi, rhomboidei and levator anguli scapulæ. These having been divided, the suprascapular artery and nerve and omo-hyoid muscles are to be traced to the upper border of the scapula, and the posterior scapular artery followed between the rhomboidei and the back of the serratus magnus. If the clavicle has been already divided in the dissection of the neck, the scapula can now be drawn away from the ribs and will be seen to be attached solely by the serratus magnus, between which and the ribs is a quantity of loose areolar tissue which enables the bone to glide smoothly upon the thorax. The serratus magnus having been divided, and the clavicle cut at its middle, if necessary, the axillary vessels and nerves are to be severed after being tied together, and the entire limb removed from the trunk.]

Parts about the Scapula.

[The limb having been placed on the table, it is advisable to identify again the several insertions of muscles on the scapula, cutting each muscle to about an inch in length.]

The trapezius extends along the whole length of the upper border of the spine of the scapula and for the outer third of the posterior border of the clavicle. The omo-hyoid arises from the upper margin of the scapula close to the notch and from the transverse ligament. The levator anguli scapulæ is inserted into that part of the base of the scapula which is above the spine, the rhomboideus minor into the part opposite the spine, and the rhomboideus major into the base below the spine. The serratus magnus is inserted into the inner surface of the base of the scapula for its whole length, but the fibres are much thicker at the upper and lower angles than in the centre.

[A small block is now to be placed beneath the lower border of the scapula so as to put the deltoid on the stretch, and the remainder of its fibres are then to be cleaned, notice being taken of one or two small branches of the circumflex nerve which turn round its posterior border.]

The **Deltoid Muscle** *arises* from the outer half or third of the anterior border of the clavicle and from the lower border of the acromion and spine of the scapula. Its strong coarse fibres converge to a point, and are *inserted* into a rough surface of a triangular shape on the outer side of the humerus in the middle of the shaft, being embraced by the pointed origins of the brachialis anticus, and closely connected with the insertion of the pectoralis major. The action of the deltoid is to raise the arm above the head, and it is an example of power applied to a lever of the third order. It is *supplied* by the circumflex nerve.

[The deltoid is to be divided near its origin and turned down, the circumflex vessels and nerve being preserved. In doing this a large bursa lying between the deltoid and the shoulder joint should be noticed, as its enlargement may be confounded with disease of the joint. The remains of the trapezius muscle are to be cut close to the spine of the scapula, and the thin fascia covering the muscles above and below it removed, the humerus being rotated inwards to put their fibres on the stretch.]

The **Supra-spinatus Muscle** occupies the supraspinal fossa, *arising* from the greater part of the fossa itself, from the upper surface of the spine, and from the fascia covering the muscle. The tendon passes beneath the acromion to be *inserted* into the uppermost facet on the great tuberosity of the humerus, and slightly into the capsular ligament of the shoulder joint. In order to follow the tendon thoroughly, the acromial end of the spine of the scapula is to be divided with the saw, when a bursa will be found between it and the muscle.

The **Infra-spinatus Muscle** (Fig. 9, 1) *arises* from the whole of the infraspinal fossa except the part near the neck of the scapula; from the fascia covering itself, and from the intermuscular septa between it and the teres muscles. It is *inserted* into the middle facet on the great tuberosity of the humerus and into the capsule of the shoulder joint, being blended there with the supra-spinatus and teres minor. The supra- and infra-spinatus muscles are *supplied* by the suprascapular nerve.

Fig. 9.

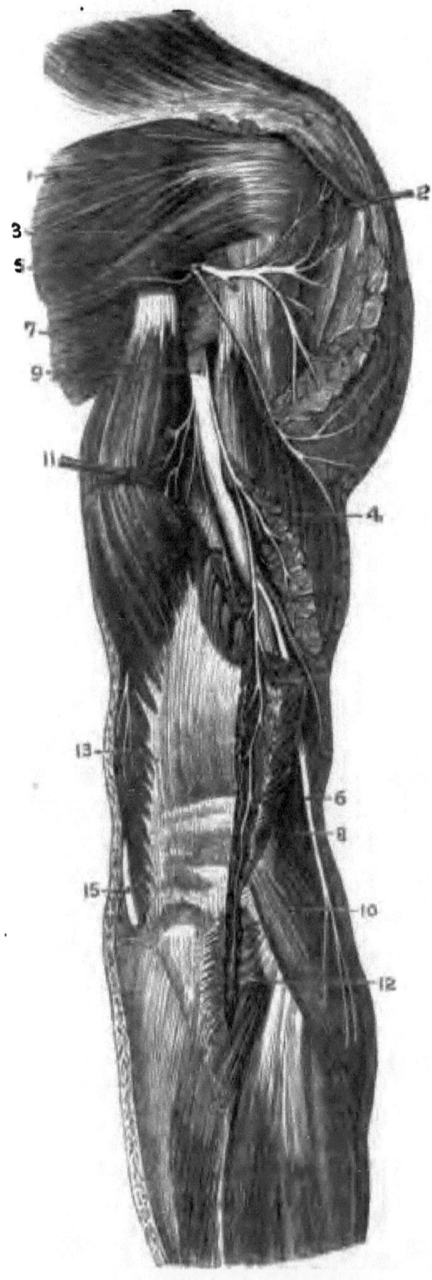

The **Teres Minor Muscle** (Fig. 9, 3) *arises* from the dorsal aspect of the inferior costa or border of the scapula in nearly its whole length, from the fascia covering the muscle, and from the intermuscular septa between it and the infra-spinatus and teres major muscles. It is *inserted* into the lowest facet on the great tuberosity of the humerus and for nearly an inch below it, and also into the capsular ligament of the shoulder. The teres minor is *supplied* by a special branch of the circumflex nerve which should be followed to it.

The supra-spinatus is a feeble *extensor*, and the infra-spinatus and teres minor muscles are *external rotators* of the humerus.

The **Teres Major Muscle** (Fig. 9, 7) *arises* from the triangular surface at the inferior angle of the scapula, and from the intermuscular septa between it

Fig. 9.—Dissection of back of upper arm (from Hirschfeld & Leveillé).

1. Infra-spinatus.
2. Deltoid.
3. Teres minor.
4. Outer head of triceps.

and the teres minor and infra-spinatus muscles. Its close relation to the latissimus dorsi has been already noticed, but its *insertion* into the inner lip of the bicipital groove should be again clearly seen. The teres major is supplied by one of the subscapular nerves, and is an *internal rotator* and *flexor* of the humerus.

The quadrilateral and triangular intermuscular spaces, alluded to in the dissection of the axilla (p. 19), can now be seen *from behind*, when the quadrilateral or upper one will be found to be bounded by the teres major, teres minor, humerus, and long head of triceps; the triangular or lower one, by the teres major, teres minor, and long head of triceps; and the vessels and nerve they transmit are now to be traced out. (A triangular interval between the teres major and the long and external heads of the triceps must not be mistaken for this latter space.)

The **Posterior Circumflex Artery** with its veins, and the **Circumflex Nerve**, appear through the quadrilateral space, and are distributed to the under surface of the deltoid muscle, giving branches to the shoulder-joint; the nerve also supplies the teres minor muscle with a branch (on which there may be found a ganglionic enlargement) and gives a branch or two to the skin of the shoulder.

The **Dorsalis Scapulæ Artery** (dorsal branch of subscapular), does not pass through the triangular space as seen from behind, but winds beneath the teres minor in a groove on the inferior costa of the scapula, to supply the infraspinous fossa and anastomose with the other scapular arteries. It gives off a superficial branch, which runs between the teres major and minor and down to the angle of the scapula, to anastomose with the subscapular and posterior scapular arteries.

---

5. Circumflex nerve.
6. External cutaneous branch of musculo-spiral nerve.
7. Teres major.
8. Supinator longus.
9. Musculo-spiral nerve.
10. Extensor carpi radialis longior.
11. Middle head of triceps.
12. Anconeus.
13. Inner head of triceps.
15. Ulnar nerve.

[The supra-spinatus is to be divided near its insertion, and the fibres cleared out of the supraspinal fossa in order to see the suprascapular artery and nerve and the origin of the omo-hyoid muscle.]

The **Suprascapular Artery** (from the thyroid axis) passes over the transverse ligament of the scapula, and is distributed both to the supraspinous fossa, and also to part of the infraspinous fossa by a branch which winds in front of the spine and anastomoses with the dorsalis scapulæ artery.

The **Suprascapular Nerve** (from the upper part of the brachial plexus) (Fig. 4, 5) passes through the suprascapular notch and beneath the ligament, and is distributed to the supra- and infra-spinatus muscles.

The **Omo-hyoid Muscle** *arises* from about half an inch of the superior border of the scapula immediately behind the suprascapular notch, and from the transverse or posterior ligament over it. Its relations and insertion are seen in the dissection of the neck.

[The limb is to be turned over and the humerus rotated outwards to put the fibres of the subscapularis on the stretch; they should then be cleaned, care being taken of the subscapular nerves entering the muscle. The axillary vessels and nerves should be tied to the coracoid process, so as to bring them as nearly as may be into their proper position.]

The **Subscapularis Muscle** (Fig. 6, 4) is covered in by a thin fascia, and has been already seen to form part of the posterior wall of the axilla. It *arises* from the whole of the venter scapulæ except the part to which the serratus magnus is attached and the portion nearest the neck of the bone, and has several tendinous septa between its fibres which are attached to the ridges on the surface of the bone. It is *inserted* into the lesser tuberosity of the humerus and into the capsular ligament of the shoulder-joint. The subscapularis is an *internal rotator* of the humerus, and is *supplied* by the short subscapular nerves.

The most important action of the supra- and infra-spinatus, teres minor, and subscapularis muscles is their combination to keep the head of the humerus in its proper relation to the glenoid cavity, and thus prevent dislocation. In the present relaxed condition of the parts, the head of the humerus can be drawn away from the glenoid cavity to the full extent of the loose capsular ligament.

The **Subscapular Nerves** (Fig. 4) are branches of the posterior cord of the brachial plexus, and are three in number. The long subscapular nerve has been already seen in the axilla, and can still be traced to the latissimus dorsi; the short subscapular nerves (upper and lower) are now to be followed, the upper to the subscapularis muscle, the lower to give a branch to the subscapularis and supply the teres major.

The **Subscapular Artery** is still to be seen along the lower border of the scapula, and its branches should be thoroughly followed out.

The **Posterior Scapular Artery** is to be found between the serratus magnus and the rhomboidei, and its anastomoses should be defined.

By removing the muscular fibres from both surfaces of the scapula a very abundant network of vessels will be seen, formed by anastomosing branches from the arteries which have been already traced to the scapula; viz., the subscapular with its dorsal branch [axillary], the suprascapular [thyroid axis or subclavian], and the posterior scapular [thyroid axis].

Dissection of the Back of the Arm.

[The fore-arm and hand are to be doubled under the upper arm, which is to be placed on the table with the back upwards, and the scapula is to be drawn down with hooks so as to put the triceps on the stretch.]

While the skin is being removed from the back of the arm a few small cutaneous branches of the *musculo-spiral* nerve may be noticed, and immediately above the elbow one larger branch of the nerve may be seen on the outer side, with a corresponding branch from the *internal cutaneous* nerve on the inner side of the limb.

When the strong deep fascia of the arm has been divided the fibres of the triceps muscle should be cleaned, and a large bursa between the triangular posterior surface of the ulnar and the skin should be noticed.

The **Triceps Muscle** (Fig. 9, 11) has of course three heads—the long or middle, the external, and the internal. The *long* head *arises* from a somewhat triangular rough surface on the inferior border of the scapula immediately below the glenoid cavity. It has been already partly examined both from the front and from behind in relation with certain spaces (p. 35), and its fibres are now seen to be separated by another somewhat triangular interval from the back of the upper third of the humerus, until it joins the fibres of the external head at the junction of the upper with the middle third of the bone. The *outer* head *arises* from immediately below the insertion of the teres minor into the greater tuberosity of the humerus, and from the outer side of the posterior aspect of the bone as far down as the musculo-spiral groove; and slightly from the external intermuscular septum which intervenes between it and the deltoid. The *inner* head *arises* from the whole of the posterior aspect of the humerus below the insertion of the teres major and the musculo-spiral groove, to within half an inch of the elbow-joint on the inner side, and extending to the back of the condyle on the outer side. Its fibres arise on each side from the intermuscular septa which intervene between it and the muscles of the front of the arm. The whole of the fibres converge to be *inserted* into the top of the olecranon process of the ulna, a small bursa intervening between the tendon and the tip of that process, which will be seen when the joint is opened. The triceps is the great *extensor* of the fore-arm upon the upper arm, and its action is that of a force applied to a lever of the first order. It is *supplied* by the musculo-spiral nerve.

[The long head being hooked to the inner side and the musculo-spiral nerve having been put on the stretch, the fibres of the outer head are to be divided sufficiently to follow the nerve, with its artery, round the bone.]

The **Musculo-spiral Nerve** (Fig. 9, 9) has been already shown (p. 20) to be a branch of the posterior cord of the brachial plexus, and has been seen to disappear between the internal and middle heads of the triceps; it can now be followed in the oblique groove on the back of the humerus between the fibres of the muscle, until it reaches the interval between the supinator longus and brachialis anticus. It gives off numerous large branches to the triceps in its course, and afterwards supplies the supinator longus, extensor carpi radialis longior, brachialis anticus (in part) and the anconeus; and the branch to the last muscle should be at once traced through the fibres of the triceps at the back of the external condyle.

Fig. 10.

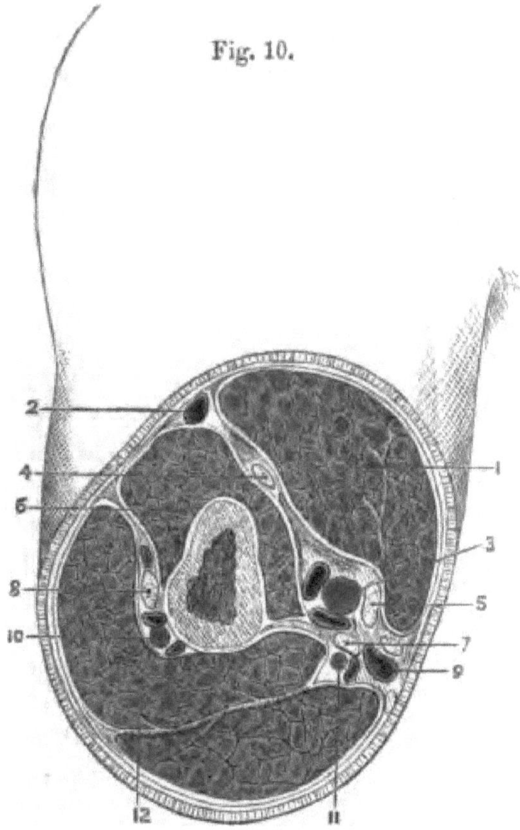

Fig. 10. A section through the middle of the right upper arm (altered from Béraud).

1. Biceps.
2. Cephalic vein.
3. Brachial vessels.
4. Musculo-cutaneous nerve.
5. Median nerve.
6. Brachialis anticus.
7. Ulnar nerve.
8. Musculo-spiral nerve.
9. Basilic vein with internal cutaneous nerves.
10. Superior profunda vessels.
11. Inferior profunda vessels.
12. Triceps with fibrous intersection.

The **Superior Profunda** branch of the brachial artery closely accompanies the musculo-spiral nerve around the bone, and supplies the triceps in its course. At the outer edge of the triceps it gives off a large superficial branch which runs down to the humerus; it then divides into two branches, one going to the back of the elbow beneath the triceps, and the other running between the supinator longus and brachialis anticus to anastomose with the radial recurrent artery.

The Front of the Fore-arm.

[One incision is to be made down the middle of the fore-arm, and another at right angles to it at the root of the thumb, and the flaps of skin are to be turned to each side.]

The **Cutaneous Veins** (Fig. 5) will be found in the superficial fascia, and can be defined without injury to the nerves. The *Anterior Ulnar Vein* is pretty regular in its course, and will be found to commence in one or two small branches about the wrist; then running along the inner side of the fore-arm, it joins the posterior ulnar vein near the elbow and assists in forming the basilic vein.

The *Median vein* is seldom in the centre of the arm at first, but commences near the root of the thumb, coursing obliquely to the centre of the arm near the bend of the elbow, where it is joined by the profunda vein and divides into median basilic and median cephalic branches (p. 23).

The *Radial vein* commences at the back of the hand, and appears on the outer side of the fore-arm about its middle. It has been seen to unite with the median cephalic (p. 24). Besides these named veins there are numerous smaller branches which assist, and occasionally take the place of, the larger ones.

The **Cutaneous Nerves** are now to be followed out.

The *Internal cutaneous nerve* can be traced to the whole of the inner side of the fore-arm, the branches winding round to the back of the limb. It sometimes forms a junction with—

A *cutaneous branch of the Ulnar nerve*, which may, with

care, be found piercing the fascia about a hand's breadth above the wrist, close to the edge of the flexor carpi ulnaris tendon, which can be readily felt.

The *External cutaneous* (musculo-cutaneous) *nerve* is distributed to the outer side of the front of the fore-arm, the branches winding round the edge of the limb to join with the radial nerve.

A *cutaneous branch of the Median nerve* pierces the fascia in the centre of the fore-arm about two inches above the wrist, and passes into the palm of the hand.

The **Deep Fascia** is now to be cleaned, and will be found to be continuous with the deep fascia of the arm above, and with the annular ligament below. It gives numerous intermuscular septa in the fore-arm, which in a thin subject are seen as white lines, running more or less in the length of the limb.

[The muscles from the internal condyle are to be exposed by removing the deep fascia from the front of the arm, and they are then to be cleaned carefully in the direction of their fibres.]

The **Muscles** (Fig. 11) from the inner condyle are five in number, four being *flexors* of the carpus and fingers and one a *pronator* of the fore-arm. Beginning from the outer side, their relative positions are:—1, pronator teres; 2, flexor carpi radialis; 3, palmaris longus (which may be absent); 4, flexor sublimis digitorum; 5, flexor carpi ulnaris. All these muscles have common origins from (1) the internal condyle, (2) the fascia of the fore-arm, and (3) the intermuscular septa derived from it; but three of them, viz. the pronator teres, flexor sublimis digitorum, and flexor carpi ulnaris, have extra bony attachments.

The **Pronator Radii Teres** (Fig. 11, 4) *arises* from the internal condyle above the other muscles; from the fascia of the fore-arm over it; and from the intermuscular septum between it and the flexor carpi radialis. Its second head is from the inner side of the coronoid process of the ulna below the flexor sublimis, and between the two heads is the median nerve. It is

Fig. 11.

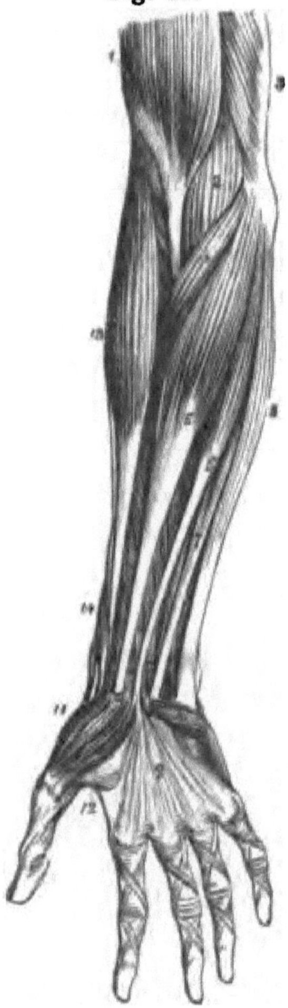

inserted by a broad tendon into the middle of the outer side of the radius, immediately below the supinator brevis. It pronates the hand by rolling the radius on the ulna, and is *supplied* by the median nerve.

The **Flexor Carpi Radialis** (Fig. 11, 5) *arises* from the internal condyle in common with the remaining muscles; from the fascia of the fore-arm; and from the intermuscular septa between it and the pronator teres on one side, and the palmaris longus on the other. It ends about the middle of the arm in a broad tendon, which soon becomes rounded, and disappears at the root of the thumb to pierce the annular ligament and pass through the groove in the trapezium, to be *inserted* into the base of the second, and slightly into the base of the third metacarpal bone. It is *supplied* by the median nerve.

The **Palmaris Longus** (Fig. 11, 6) *arises* from the common attachment to the inner condyle; from the

Fig. 11.—Superficial muscles of the fore-arm (from Wilson).

1. Biceps, with its tendon.
2. Brachialis anticus.
3. Part of triceps.
4. Pronator radii teres.
5. Flexor carpi radialis.
6. Palmaris longus.
7. Part of the flexor sublimis digitorum; the rest of the muscle is seen beneath the tendons of the palmaris longus and flexor carpi radialis.
8. Flexor carpi ulnaris.
9. Palmar fascia.
10. Palmaris brevis.
11. Abductor pollicis.
12. Flexor brevis pollicis.
13. Supinator longus.
14. Extensor ossis metacarpi and extensor primi internodii pollicis, curving around the lower border of the fore-arm.

fascia of the fore-arm; and from the intermuscular septa on each side of it. Its long and narrow tendon passes superficially to be *inserted* into the strong palmar fascia in the centre of the hand, and slightly into that over the muscles of the thumb. It serves to make tense the palmar fascia, and thus protect the deep tissues of the palm, and may slightly flex the hand when largely developed. It is *supplied* by the median nerve.

The palmaris longus is frequently wanting, but it is occasionally largely developed, the whole length of the tendon being covered with strong muscular fibres connected with the flexor carpi radialis.

The **Flexor Carpi Ulnaris** (Fig. 11, 8) *arises* from the common attachment to the internal condyle; from the fascia of the fore-arm; and from the intermuscular septum between it and the adjacent muscle. It has an additional aponeurotic origin from the inner side of the olecranon process and from the posterior ridge of the ulna for two-thirds of its length, this latter origin being common to itself, the flexor profundus digitorum and the extensor carpi ulnaris. The muscle is *inserted* by a flattened tendon (upon which the muscular fibres extend nearly to the wrist) into the pisiform bone, and slightly into the fifth metacarpal bone and the annular ligament. Between the two origins of this muscle the ulnar nerve enters the fore-arm, after winding close behind the internal condyle, and both ulnar artery and nerve will be seen beneath the fascia, close to the outer side of the tendon near the wrist. It is a flexor of the carpus, and is the only one of either the flexors or extensors of the carpus which is inserted directly into one of its bones. It is *supplied* by the ulnar nerve.

The **Radial Artery** (Fig. 12, 17) is the smaller of the divisions of the brachial artery, and the part in the fore-arm extends from the bifurcation at the bend of the elbow to the root of the thumb. It is superficial in the whole of this course, except that at the upper part it is more or less overlaid by the supinator longus muscle, the extent varying with the muscular development of the arm.

It lies at first *between* the supinator longus and the pronator teres, but in the lower half of the arm between the

Fig. 12.

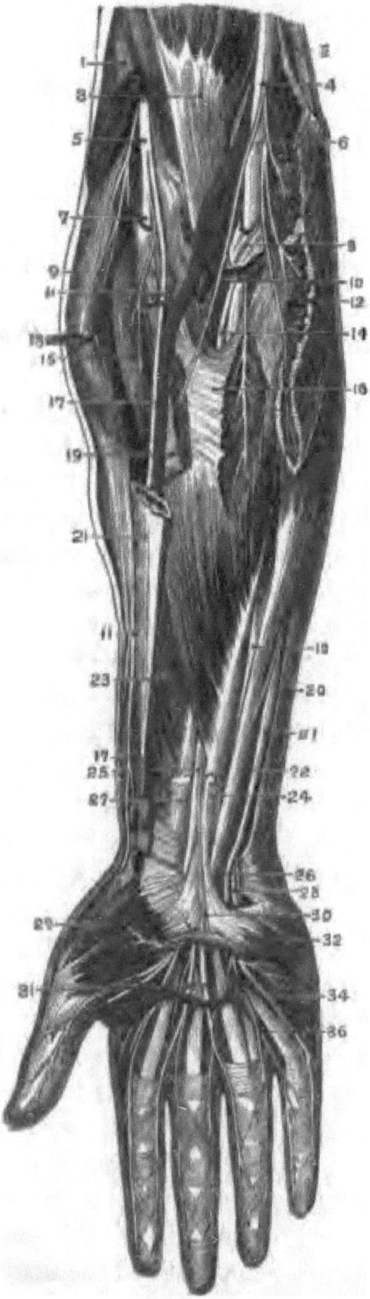

Fig. 12.—Superficial dissection of the front of the fore-arm (from Hirschfeld and Leveillé).

1. Supinator longus (cut).
2. 2. Ulnar nerve.
3. Biceps.
4. Median nerve.
5. Musculo-spiral nerve.
6. Brachial artery.
7. Posterior interosseous nerve.
8. Pronator teres.
9. Supinator brevis.
10. Ulnar artery.
11. 11. Radial nerve.
12. Flexor carpi radialis (cut).
13. Extensor carpi radialis longior.
14. Anterior interosseous nerve.
15. Extensor carpi radialis brevior.
16. Flexor sublimis digitorum.
17. 17. Radial artery.
18. Flexor profundus digitorum.
19. Tendon of pronator teres.
20. Tendon of flexor carpi ulnaris.
21. Tendon of supinator longus.
22. Ulnar artery.
23. Flexor longus pollicis.
24. Tendon of palmaris longus.
25. Median nerve, becoming superficial.
26. Superficial division of ulnar nerve.
27. Tendon of flexor carpi radialis.
28. Deep branch of ulnar nerve.
29. Abductor pollicis.
30. Cutaneous palmar branch of median nerve.
31. Digital branches of median nerve.
32. Palmaris brevis.
34. Superficial palmar arch.
36. Digital branches of ulnar nerve.

tendons of the supinator longus and flexor carpi radialis, which latter tendon is generally taken as the guide to the vessel. To its *outer* side in addition it has the radial nerve, which in the upper third of the arm is at some little distance, in the middle third is in close relation, and in the lower third quits the vessel altogether by passing beneath the supinator longus. The radial artery *lies upon* (1) the tendon of the biceps (though this will depend upon the point at which the bifurcation takes place); (2) the oblique fibres of the supinator brevis; (3) the insertion of the pronator teres; (4) the radial origin of the flexor sublimis; (5) the flexor longus pollicis; (6) the pronator quadratus; and (7) the end of the radius.* It has two venæ comites in close relation with it and gives the following branches:—

Branches.—1. The *radial recurrent* runs transversely beneath the supinator longus and gives ascending and descending branches, the ascending anastomosing with the superior profunda branch of the brachial artery.

2. *Muscular branches* are given off at various points to the adjacent muscles.

3. The *superficialis volæ* arises near the root of the thumb and is of very variable size. It runs forward, and generally beneath some of the fibres of the muscles of the thumb, to complete the superficial palmar arch formed by the ulnar artery.

4. The *anterior carpal* is a small branch which runs across the wrist beneath the deep tendons to join a corresponding branch of the ulnar artery.

Surgery (Fig. 7).—The radial artery is readily tied, about an inch above the wrist, by an incision one inch and a half long, placed midway between the tendons of the flexor carpi radialis and supinator longus, when the vessel will be found lying immediately beneath the deep fascia with the venæ comites in close connection, but with no nerve near. The artery may be exposed in the upper part by an incision two inches long at any point in a line

* The posterior relations of the radial artery are simply the muscles attached to the radius in their order from above downwards.

drawn from the centre of the bend of the elbow to the styloid process of the radius, the knife going at once down to the supinator longus, which must then be carefully turned outwards. The position of the radial nerve to the outer side is to be borne in mind and the ligature passed from it.

[The pronator teres is to be divided about its middle, without injuring the median nerve or the origin of the muscle from the ulna, which can now be thoroughly seen; and the flexor carpi radialis and palmaris longus are to be divided so as to expose thoroughly the flexor sublimis, which is to be cleaned. The skin of the front of one of the fingers is to be carefully removed without interfering with the palm, and the sheath of the tendons dissected out.]

The **Flexor Sublimis Digitorum** (perforatus) (Fig. 12, 16) is the only muscle of the fore-arm which has an origin from the three bones of the arm. It *arises* from the internal condyle of the humerus and slightly from the internal lateral ligament of the elbow; from the intermuscular septa between it and the more superficial muscles; from the inner side of the coronoid process of the ulna above the pronator teres; and from the whole of the oblique line on the front of the radius. Between the origins from the humerus and the radius is the median nerve. The muscle ends in four tendons, of which the two to the second and third fingers lie in front of those to the first and fourth fingers; all pass under the annular ligament and through the palm of the hand to the second phalanges of the four fingers.

The *sheath* of the finger which is exposed should be laid open along the centre, when it will be found to be thick opposite each phalanx, but thin at each joint. The tendon of the flexor sublimis (perforatus) will be seen to be flattened on the first phalanx and split to give passage to the tendon of the flexor profundus (perforans), the two slips of the flexor sublimis being *inserted* into the sides of the second phalanx, and the tendon of the flexor profundus into the base of the third phalanx. The flexor sublimis is a flexor of the phalanges and then of the carpus, and is *supplied* by the median nerve.

The position of the ulnar artery, between the tendon of the flexor carpi ulnaris and the innermost tendon of the flexor sublimis in the lower half of the fore-arm, is now to be noticed, and by a slight separation of the tendons the ulnar nerve can be seen lying close to the ulnar side of the artery. This is the point where the ulnar artery is usually tied.

[The flexor sublimis is now to be divided near its origins and turned down without injuring the median nerve, from which a branch may be traced to the deep surface of the muscle.]

The **Deep Muscles** (Fig. 13) of the fore-arm are the flexor longus pollicis to the radial side, the flexor profundus digitorum to the ulnar side, and the pronator quadratus, a small square muscle with transverse fibres, to be seen above the carpus by drawing aside the tendons. These are now to be cleaned, all vessels and nerves being carefully preserved.

The **Flexor Longus Pollicis** (Fig. 13, 5) *arises* from the whole of the anterior surface of the radius between the oblique line and the attachment of the pronator quadratus, and from the outer half of the interosseous membrane in nearly its whole length. It very generally has a small additional origin, by a slip of very variable size, from the outer side of the coronoid process of the ulna. A single round tendon passes beneath the annular ligament and through the palm of the hand to be *inserted* into the terminal phalanx of the thumb.

The **Flexor Profundus Digitorum** (perforans) (Fig. 13, 4) *arises* from the anterior surface of the ulna between the coronoid process (which it embraces) and the origin of the pronator quadratus, and from the adjacent half of the interosseous membrane; also from the upper two-thirds of the inner surface of the ulna, extending to the olecranon process and the posterior border of the bone, to which an aponeurosis is attached from which some of the fibres arise. The muscle ends in four tendons (of which the outermost alone is quite separate in the fore-arm), which pass beneath the annular ligament, and after giving attachment to the lumbricales muscles in the palm, pierce the tendons of the flexor sublimis, and are *inserted* into the bases of the third phalanges of the four fingers. The two preceding muscles are direct flexors of the thumb and fingers, and afterwards of the carpus. The flexor longus pollicis is *supplied* by the median nerve; the flexor profundus partly by the median and partly by the ulnar nerve.

Fig. 13.

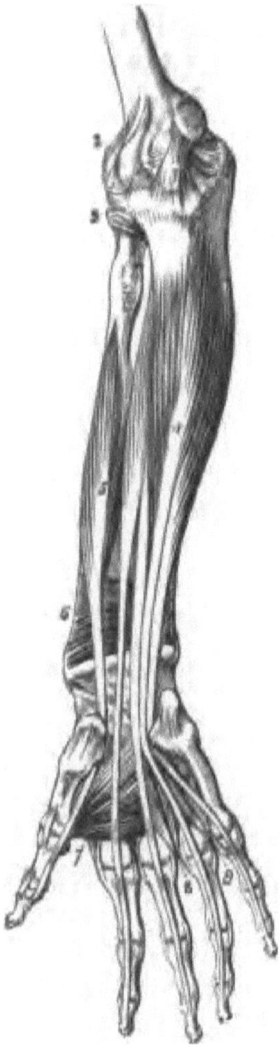

The **Pronator Quadratus** (Fig. 13, 6) is the only muscle in the arm whose fibres are transverse. It *arises* from a somewhat oval surface on the front of the ulna below the origin of the flexor profundus, and is *inserted* into the whole of the front of the radius below the attachment of the flexor longus pollicis. This muscle is a powerful pronator of the fore-arm and is *supplied* by the median nerve.

Fig. 13.—Deep muscles of the fore-arm (from Wilson).

1. Internal lateral ligament of the elbow joint.
2. Anterior ligament.
3. Orbicular ligament of the head of the radius.
4. Flexor profundus digitorum (the lumbricales removed).
5. Flexor longus pollicis.
6. Pronator quadratus.
7. Adductor pollicis.
8. Dorsal interosseous of the middle, and palmar interosseous of the ring finger.
9. Dorsal interosseous muscle of the ring finger, and palmar interosseous of the little finger.

The **Ulnar Artery** (Fig. 14, 16) arises from the bifurcation of the brachial artery at the bend of the elbow, and at once takes an oblique course inwards *beneath* the median nerve and four of the muscles arising from the internal condyle (pronator teres, flexor carpi radialis, palmaris longus, flexor sublimis digitorum). It lies first *upon* the insertion of the brachialis anticus, and then upon the flexor profundus digitorum for the remainder of its course in the fore-arm, being bound down by a process of deep fascia. It is joined in the middle third of its course by the ulnar nerve, which lies to its ulnar side for the rest of its course in the fore-arm. In the lower third of the arm both artery and nerve lie comparatively superficially between the tendons of the flexor carpi ulnaris and the flexor sublimis (but slightly overlapped by the former), until they both pass over the annular ligament into the hand. Two venæ comites are in close relation with the artery.

Surgery (Fig. 7).—The ulnar artery is easily reached in its lower part by an incision two inches long upon the outer edge of the tendon of the flexor carpi ulnaris. The tendon being drawn a little inwards, it would only be necessary to divide the deep fascia to expose the vessel with the nerve to its ulnar side. The operation for tying the vessel in its upper part between the flexor muscles has been abandoned as impracticable on the living body.

Branches.—1. The *Anterior ulnar recurrent* will be found between the brachialis anticus and pronator teres, running up in front of the elbow joint to join the anastomotica magna.

2. The *Posterior ulnar recurrent* is to be followed between the flexor sublimis and flexor carpi ulnaris to the back of the prominent internal condyle, where, after passing between the heads of the flexor carpi ulnaris, it runs in the groove occupied by the ulnar nerve, and anastomoses with the inferior profunda artery.

3. The *interosseous artery* is a short trunk arising about

Fig. 14.

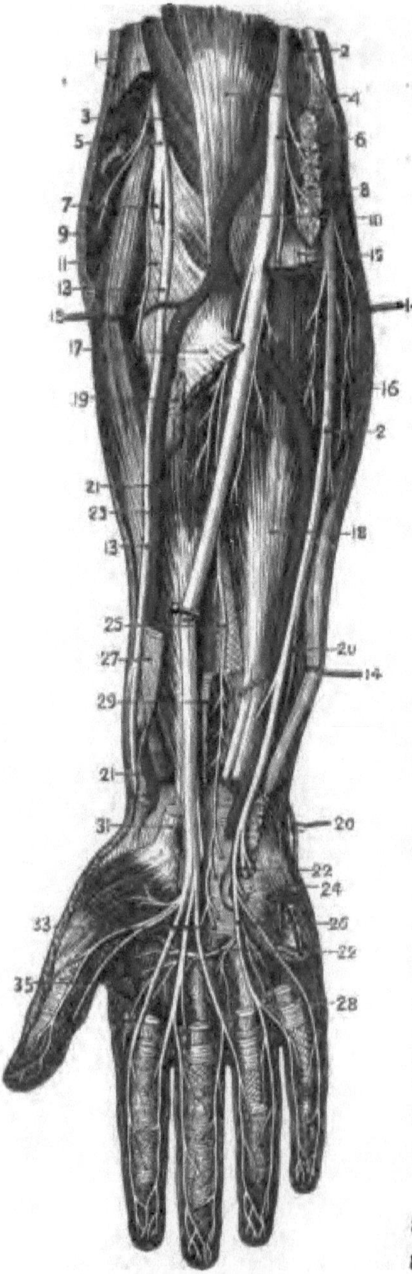

two inches from the commencement of the artery, which is directed backwards to the interosseous space, where it subdivides into anterior and posterior interosseous branches.

The *posterior interosseous* passes between the radius and ulna to the back of the arm, where it will be dissected.

The *anterior interosseous* is to be followed down the front of the interosseous membrane, where it will be found lying with a branch of the median nerve between the flexor longus pollicis and flexor profundus digitorum, until it disappears beneath the pronator quadratus to reach the back of the wrist. The anterior interosseous artery gives off numerous *muscular* branches; a *median* branch to accompany the median nerve (sometimes of large size); and two *nutritious* arteries to the radius and ulna.

4. The *Carpal arteries*, anterior and posterior, supply the front and back

Fig. 14.—Deep dissection of the front of the fore-arm and hand (from Hirschfeld and Leveillé).

1. Supinator longus (cut). 2. Ulnar nerve.

of the carpus, and anastomose with corresponding branches from the radial.*

The **Median Nerve** (Figs. 12 and 14) enters the fore-arm between the heads of the pronator teres, and then passing between the origins of the flexor sublimis digitorum, crosses the ulnar artery to lie between the flexor sublimis and flexor profundus muscles. It is placed superficially between the flexor tendons near the wrist, and passes with them beneath the annular ligament into the hand.

Branches.—The median nerve supplies all the muscles of the front of the fore-arm except the flexor carpi ulnaris and half the flexor profundus, in the following manner:—As soon as it enters the fore-arm it gives branches to the pronator teres, flexor carpi radialis and flexor sublimis digitorum, and, after crossing the ulnar artery, gives off the *anterior interosseous nerve* (Fig. 14, 25). This passes down the front of the interosseous membrane, giving branches

---

3. Brachialis anticus.
4. Biceps.
5. Musculo-spiral nerve.
6. Median nerve.
7. Posterior interosseous nerve.
8. Pronator teres and flexor carpi radialis (cut).
9. Extensor carpi radialis longior (cut).
10. Brachial artery.
11. Supinator brevis.
12. Flexor sublimis digitorum (cut).
13, 13. Radial nerve.
14, 14. Flexor carpi ulnaris.
15. Extensor carpi radialis brevior.
16. Ulnar artery.
17. Radial origin of flexor sublimis digitorum (cut).
18. Flexor profundus digitorum.
19. Tendon of pronator teres.
20, 20. Dorsal branch of ulnar nerve.
21, 21. Radial artery.
22, 22. Deep branch of ulnar nerve.
23. Flexor longus pollicis.
24. Abductor minimi digiti.
25. Anterior interosseous nerve.
26. Digital branches of ulnar nerve.
27. Tendon of supinator longus.
28. One of the lumbricales muscles.
29. Pronator quadratus.
31. Tendon of flexor carpi radialis.
33. Digital branches of median nerve.
35. Adductor pollicis.

---

* Professor Ellis enumerates a metacarpal branch which is usually the continuation of the posterior carpal artery to the back of the 5th metacarpal bone, as will be seen in the dissection of the back of the hand.

to the flexor longus pollicis and the outer half of the profundus digitorum, and then beneath the pronator quadratus (which it supplies), to the front of the wrist-joint, where it gives a branch to the articulation.

A *cutaneous palmar* branch (Fig. 12, 30) arises a short distance above the annular ligament, over which it passes to be distributed to the skin of the palm.

The **Ulnar Nerve** (Fig. 14, 2) enters the fore-arm behind the internal condyle, by passing between the heads of the flexor carpi ulnaris. It lies under cover of that muscle and upon the flexor profundus digitorum for the whole of its course in the fore-arm; and about the middle of the arm comes into close relation with the ulnar artery, and, keeping to its ulnar side, accompanies it over the annular ligament into the palm.

Branches (Fig. 14).—The ulnar nerve gives small articular branches to the back of the elbow, and supplies one and a half of the muscles of the fore-arm, viz., the flexor carpi ulnaris and the inner half of the flexor profundus digitorum.

In the lower third of the fore-arm the nerve gives a *dorsal branch*, (20) which turns backwards beneath the tendon of the flexor carpi ulnaris to be distributed to the back of the little and half the ring finger.

A *cutaneous palmar* branch of small size arises close above the annular ligament, and is distributed to the skin of the palm on the ulnar side.

The **Radial Nerve** (Fig. 14, 13) is seen to arise from the musculo-spiral nerve opposite the elbow. It lies at first upon the supinator brevis, to the outer side of, and at some little distance from, the radial artery, but in the middle third of the arm it is in close relation with the artery on the pronator teres, finally passing beneath the tendon of the supinator longus to the back of the arm and hand.

The Back of the Fore-arm.

[The front of the fore-arm having been dissected, the skin can be reflected from the back of the arm without further incisions, but an incision must be made along the inner border of the hand and across the knuckles, to permit the reflection of the skin from the back of the hand. An incision is also to be made along the thumb and each of the fingers, and the skin reflected, when the cutaneous nerves and veins are to be dissected out of the superficial fascia.]

Cutaneous Nerves.—The back of the fore-arm is supplied at the upper part by the branches of the *musculo-spiral* nerve and *internal cutaneous* nerve, which were seen above the condyles of the humerus in the dissection of the back of the arm (p. 37). In addition, and lower down, there will be found on the outer side of the limb, branches from the *musculo-cutaneous* nerve, and on the inner side, branches of the *internal cutaneous* nerve ; at the wrist, on the outer side, the large *radial* nerve pierces the deep fascia, and, after communicating with the musculo-cutaneous nerve, distributes digital branches to both sides of the thumb, forefinger, and middle finger, and to the radial side of the ring finger, forming in addition a loop across the back of the hand with the following nerve.

The *Dorsal branch of the Ulnar Nerve* appears on the inner side of the wrist, at a point corresponding to the lower end of the ulna, and gives digital branches to both sides of the little finger and the ulnar side of the ring finger, joining the radial nerve across the back of the hand.

The **Superficial Veins** of the hand form an arch across the back of the metacarpus, which joins the radial and posterior ulnar veins at its extremities ; the veins wind round the limb to join the radial and ulnar veins respectively, and one or two branches form communications between them and across the back of the fore-arm.

[The deep fascia is to be removed from the back of the fore-arm and hand, with the exception of a band about an inch wide which is to be left opposite the lower end of the radius, to form the posterior annular ligament. It will be found to be impossible to remove the fascia entirely near the elbow, where it is incorporated with the muscles.]

Muscles of the Back of the Fore-arm (Fig. 15).—The muscles arising from the external condyle of the humerus are either *extensors* or *supinators*, and will be found in the following order beginning from the radial side:—1, supinator longus; 2, extensor carpi radialis longior; 3, extensor carpi radialis brevior; 4, extensor communis digitorum; 5, extensor minimi digiti; 6, extensor carpi ulnaris; there are also two small muscles attached to the bone and not seen at present, viz. 7, anconeus (covered by deep fascia), and 8, supinator brevis (under cover of the long muscles). Arising from the bones of the *fore*-arm alone and appearing from beneath the extensor communis digitorum, will be found four short extensor muscles which hold the following position to one another from the radial side:—1, extensor ossis metacarpi pollicis; 2, extensor primi internodii pollicis; 3, extensor secundi internodii pollicis; 4, extensor indicis.

All the eight muscles attached to the humerus have additional origins from the fascia of the arm or from intermuscular septa derived from it, and two, viz. the extensor carpi ulnaris and supinator brevis, have additional origins from the ulna.

The **Supinator Radii Longus** (Fig. 15, 4) is a long narrow muscle and has been already seen in great part in the dissections of the bend of the elbow and fore-arm. It *arises* from the upper two-thirds of the ridge leading to the outer condyle of the humerus and from the external intermuscular septum of the arm, which separates it from the triceps; and is *inserted* into the base of the styloid process of the radius, its tendon being crossed by two short extensors of the thumb at the annular ligament. It is *supplied* by a special branch of the musculo-spiral nerve, and its action is that of a feeble supinator and flexor of the fore-arm.

The **Extensor Carpi Radialis Longior** (Fig. 15, 5) *arises* from the lower third of the ridge leading to the external condyle and from the external intermuscular septum; its tendon passes beneath the two extensors of the thumb and through the second division of the annular ligament, and having then been crossed by the tendon of the third extensor of the thumb, is *inserted* into the base of the metacarpal bone of the fore-finger (2nd metacarpal bone).

The **Extensor Carpi Radialis Brevior** (Fig. 15, 6) which is beneath the longer extensor, *arises* from the outer condyle of the humerus; from the surface of the external lateral ligament; from the fascia of the fore-arm and from the intermuscular septum between it and the extensor digitorum. Its tendon passes with that of the long extensor beneath the two extensors of the thumb and through the second division of the annular ligament; and lastly beneath the third extensor of the thumb, to be *inserted* into the base of the metacarpal bone of the middle finger (3rd metacarpal bone).

The **Extensor Communis Digitorum** (Fig. 15, 8) *arises* from the external condyle with the preceding muscle; from the

Fig. 15.

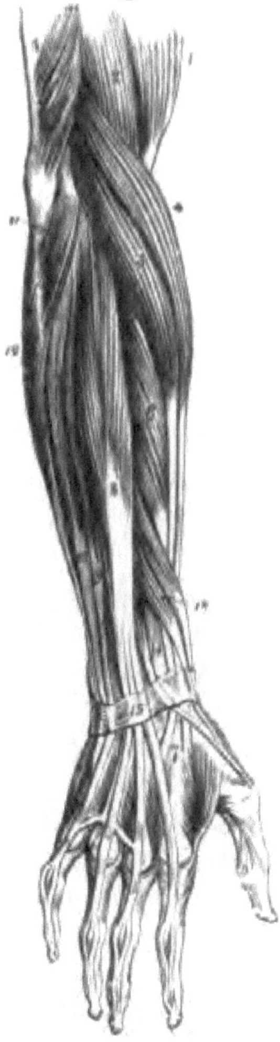

Fig. 15.—Superficial muscles of the back of fore-arm (from Wilson).

1. Biceps.
2. Brachialis anticus.
3. Lower part of the triceps, inserted into the olecranon.
4. Supinator longus.
5. Extensor carpi radialis longior.
6. Extensor carpi radialis brevior.
7. Tendons of insertion of these two muscles.
8. Extensor communis digitorum.
9. Extensor minimi digiti.
10. Extensor carpi ulnaris.
11. Anconeus.
12. Flexor carpi ulnaris.
13. Extensor ossis metacarpi and extensor primi internodii pollicis lying together.
14. Extensor secundi internodii pollicis.
15. Posterior annular ligament. The tendons of the common extensor are seen on the back of the hand, and their mode of insertio[n on] the dorsum of the fi[ngers]

fascia of the fore-arm, and from the intermuscular septa on each side of it. It ends in three tendons, which pass through the fourth division of the annular ligament with the extensor indicis, and the innermost having subdivided, the four tendons thus formed are *inserted* into the fingers in the following way :—On the back of the first phalanx of each finger the tendon expands, and is joined by slips from the lumbricales and interossei muscles ; this expansion is stronger at the margins than in the middle, and its sides are prolonged to the third phalanx, whilst the middle triangular portion is attached to the base of the second phalanx. This division is artificial and must be made with the scalpel, since the three parts are originally united. The tendon of the *extensor indicis* joins the ulnar side of the first or outermost tendon at the first phalanx. The three innermost tendons of the extensor are joined together by tendinous slips on the back of the hand, and that to the little finger is joined by the two divisions of the extensor minimi digiti above and below the point of union with the other tendons.

The **Extensor Minimi Digiti** (Fig. 15, 9) might be taken as a part of the common extensor, but is separated from it by an intermuscular septum. It *arises* from the external condyle; from the fascia and from the intermuscular septa on each side; and its tendon after passing through the fifth division of the annular ligament, is generally divided, both slips being *inserted* into the common expansion on the first phalanx of the little finger, but the innermost reaching further forward than its fellow (Fig. 18, 19).

The **Extensor Carpi Ulnaris** (Fig. 15, 10) *arises* from the external condyle ; from the fascia and from the intermuscular septa on each side; and has an additional origin from an expansion attached to the posterior border of the ulna (common to it, the flexor carpi ulnaris and the flexor profundus digitorum), the fibres covering but not being attached to all the inner part of the posterior surface of the bone. The tendon passes through the sixth division of the annular ligament and behind the styloid process of the ulna, to be *inserted* into the base of the metacarpal bone of the little finger (5th metacarpal bone).

The action of the extensors is implied by their names, but it should be noted that all the extensors of the *carpus* are inserted into the *metacarpus.* The extensor carpi radialis longior is *supplied* by the musculo-spiral nerve; all the others by the posterior interosseus nerve.

The **Anconeus** (Fig. 15, 11) is a small triangular muscle on the outer side of the elbow, which is apparently continuous with the triceps, but is separated by a narrow cellular interval. It is covered by the deep fascia, which must be removed to expose it, when a branch of nerve from the musculo-spiral to the muscle is to be looked for and preserved. It *arises* from the back of the external condyle of the humerus, and from the fascia of the arm, and spreads out to its fleshy *insertion* on the triangular surface upon the outer side of the olecranon and upper third of the ulna. The anconeus is an extensor of the fore-arm, and is *supplied* by a branch of the musculo-spiral nerve, which pierces its upper margin.

[The long extensors must be divided about the middle, and turned aside, in order to see the deep muscles thoroughly, but the supinator longus should be left uncut. The supinator brevis will require careful dissection beneath the origins of the other muscles, and the arm must be fully pronated in order to stretch the fibres whilst being cleaned. The posterior interosseus nerve piercing the muscle is to be carefully preserved and its branches traced to the extensors.]

The **Supinator Brevis** (Fig. 17, 15) has very oblique fibres, which are covered at the upper part by a tendinous expansion or intermuscular septum, from which some of them take their origin. It *arises* from the external condyle; from the external lateral ligament of the elbow; from the orbicular ligament of the radius; from the triangular space below the lesser sigmoid cavity of the ulna, and from the margin of bone below it as fa the attachment of the anconeus. Its fibres sweep round the radius and are *inserted* into all the anterior and outer part of that bone above the oblique lines, with the exception of the bicipital tubercle and the posterior surface of bone above and below it. The supinator is pierced by the posterior interosseus nerve (which *supplies* it) and is separated from the extensor ossis metacarpi pollicis by the posterior interosseous artery. It is a powerful supinator of the radius on the ulna.

The **Extensor Ossis Metacarpi Pollicis** (Fig. 16, 6) is usually the only one of the special extensors of the thumb which has an origin from both bones of the fore-arm, though occasionally the next muscle has a double origin. It *arises* from the middle of the posterior surface of the radius for about two inches, the origin reaching across the bone from the interosseous line to the attachments of the supinator brevis and pronator teres; from a corresponding portion of the posterior surface of the interosseous membrane; and from two inches of the middle of the outer side of the posterior surface of the ulna, between the attachments of the supinator brevis and anconeus above, and the extensor secundi internodii below. The muscle "crops out" between the extensors of the carpus and fingers, and having crossed the extensor carpi radialis longior and the brevior obliquely, it passes on the outer side of the styloid process of the radius, through the first division of annular ligament and over the radial artery, to be *inserted* into the base of the metacarpal bone of the thumb (1st metacarpal bone).

The **Extensor Primi Internodii Pollicis** (Fig. 16, 7) *arises* from the back of the radius immediately below the extensor ossis metacarpi pollicis and close to the interosseous line, for a space two inches long and half an inch wide; also from the radial half of the interosseous membrane for the corresponding distance. Its tendon lies to the ulnar side of the tendon of the extensor of the metacarpal bone in its whole course, passing through the same division of the annular ligament, and is prolonged to be *inserted* into the base of the first phalanx of the thumb.

The **Extensor Secundi Internodii Pollicis** (Fig. 16, 8) *arises* from the middle of the outer half of the posterior surface of the ulna (between the origins of the extensor ossis metacarpi pollicis above, and the indicator below), and from the ulnar half of the corresponding portion of the interosseous membrane. Its tendon becomes superficial just above the annular ligament, through which it passes very obliquely in a separate division (the third); then crossing the tendons of the extensor carpi radialis longior and brevior and the radial artery, it runs along the ulnar side of the extensor primi internodii to be *inserted* into the base of the terminal phalanx of the thumb. This muscle varies very much in size.

Fig. 16.

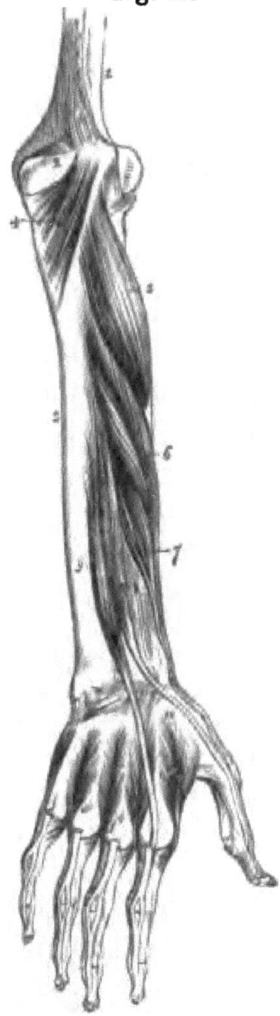

The **Extensor Indicis** (indicator) (Fig. 16, 9) arises from the posterior surface of the ulna below the preceding muscle (by the size of which its attachment is influenced), and slightly from the interosseous membrane. Its tendon is covered by those of the common extensor, and run-

Fig. 16.—Deep muscles of the back of fore-arm (from Wilson).

1. Humerus.
2. Olecranon.
3. Ulna.
4. Anconeus.
5. Supinator brevis.
6. Extensor ossis metacarpi pollicis.
7. Extensor primi internodii pollicis.
8. Extensor secundi internodii pollicis.
9. Extensor indicis.
10. First dorsal interosseous muscle. The other three dorsal interossei are seen between the metacarpal bones of their respective fingers.

Fig. 17.

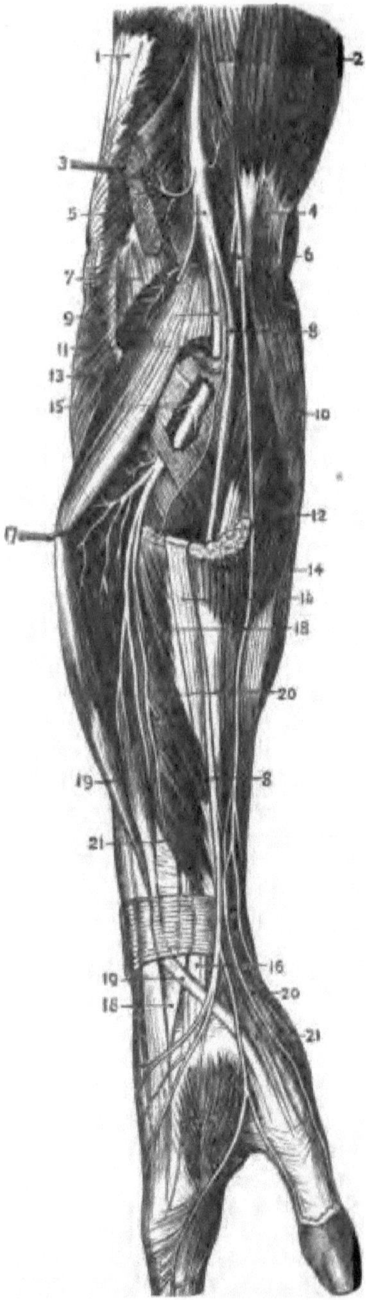

ning through the fourth division of the annular ligament with them, is *inserted* into the common expansion on the back of the first phalanx of the fore-finger. The actions of the special extensors are implied by their names. They are all *supplied* by the posterior interosseous nerve.

The **Posterior Interosseous Artery** (Fig. 18, 8) arising from the interosseous, a branch of the ulnar artery, reaches the back of the arm by passing between the radius and ulna above the interosseous membrane. It appears between the supinator brevis and extensor ossis metacarpi pollicis (usually), giving muscular branches to the extensor muscles and a recurrent branch to the elbow joint; and terminates by joining the posterior carpal arteries and also the terminal branch of the *anterior interosseous artery*, which will be found to reach the back of the arm by piercing the interosseous membrane immediately above the lower end of the radius. The *recurrent* branch runs upon the supinator brevis to the outer

Fig. 17.—Dissection of the musculo-spiral nerve and its branches (from Hirschfeld and Leveillé).

1. Triceps. 2. Brachialis anticus.

and back part of the elbow, where it anastomoses with the superior profunda and other articular branches.

The **Posterior Interosseous Nerve** (Fig. 17, 9) is a division of the musculo-spiral, which has been already seen to divide on the outer side of the space in front of the elbow and beneath the supinator longus. The posterior interosseous nerve can now be traced running obliquely in the fibres of the supinator brevis, and at its lower border breaking up into muscular branches and an articular branch to the wrist. The *muscular branches* supply the extensor carpi radialis brevior, extensor communis digitorum, extensor minimi digiti, the three special extensors of the thumb and the indicator; and the supinator brevis is supplied by a branch or two as the nerve passes through it, and must be divided to see them. The *articular branch* to the wrist reaches the interosseous membrane between the extensors of the first and second phalanges of the thumb and runs to the back of the carpus, where a ganglionic enlargement may sometimes be found.

The **Radial Artery** at the wrist (Fig. 18, 18).—After leaving the front of the lower end of the radius (p. 45) the radial artery lies against the external lateral ligament of the wrist-joint, and beneath the extensors of the metacarpal bone and first phalanx of the thumb. It then winds to the back of the carpus, to the interval between the metacarpal bones of the thumb and fore-finger, where it is crossed by the extensor of the second phalanx of

---

3. Supinator longus.
4. Biceps.
5. Musculo-spiral nerve.
6. Musculo-cutaneous nerve.
7. Origin of extensor carpi radialis longior.
8. Radial nerve.
9. Posterior interosseous nerve.
10. Pronator radii teres.
11. Origin of extensor carpi radialis brevior.
12. Flexor carpi radialis.
13. Anconeus.
14. Tendon of supinator longus.
15. Supinator brevis.
16, 16. Tendon of extensor carpi radialis longior.
17. Extensor communis digitorum.
18, 18. Tendon of extensor carpi radialis brevior.
19, 19. Extensor secundi internodii pollicis.
20, 20. Extensor ossis metacarpi pollicis.
21, 21. Extensor primi internodii pollicis.

Fig. 18.

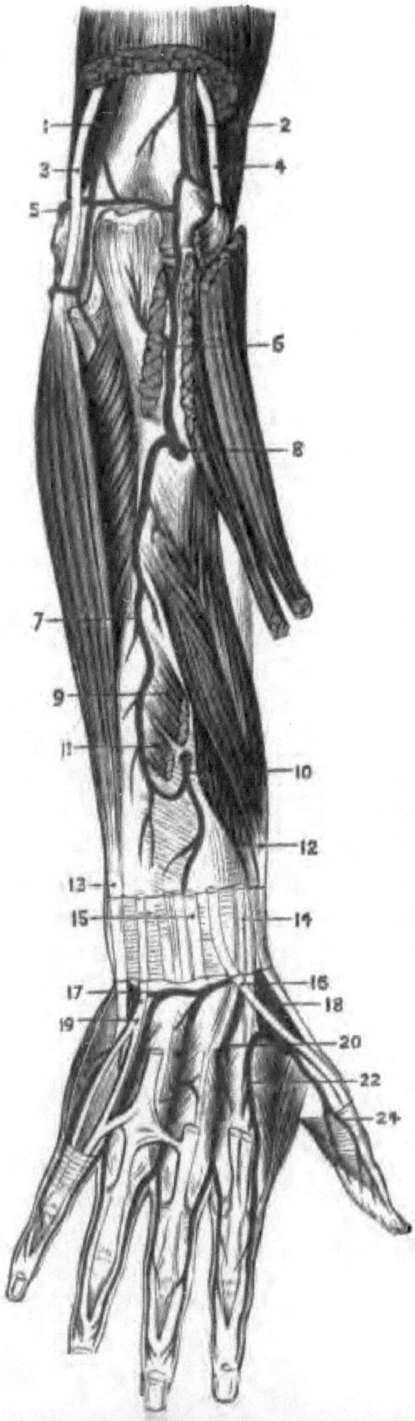

the thumb, and lastly passes into the palm of the hand between the two heads of the abductor indicis (first dorsal interosseous muscle).

The **Branches** (Fig. 18) of the radial artery at the wrist very commonly vary, but should be five in number.

Fig. 18.—Arteries of the back of the fore-arm (drawn by G. E. L. Pearse).

1. Inferior profunda.
2. Superior profunda.
3. Ulnar nerve.
4. Musculo-spiral nerve.
5. Anastomotica magna.
6. Recurrent interosseous artery.
7. Descending interosseous artery.
8. Posterior interosseous artery.
9. Extensor secundi internodii pollicis.
10. Anterior interrosseous artery.
11. Extensor indicis.
12. Extensor ossis metacarpi and primi internodii pollicis.
13. Extensor carpi ulnaris.
4. Extensor carpi radialis longior and brevior.
15. Extensor communis digitorum and indicis.
16. Extensor secundi internodii pollicis.
17. Posterior carpal artery.
18. Radial artery.
19. Extensor minimi digiti.
20. Metacarpal artery.
22. Art. dorsalis indicis.
24. Art. dorsalis pollicis.

1. The *Posterior carpal* (17) is a small branch which runs transversely close upon the bone immediately below the annular ligament, to join the posterior carpal branch of the ulnar and form an arch. From this, two dorsal interosseous arteries are given, which run on the third and fourth interosseous muscles, and receive the perforating arteries which appear between their heads of origin.

2. The *Metacarpal* (20) is the first dorsal interosseous artery, and closely resembles the others. It runs on the second interosseous muscle, and at the root of the fingers often joins the digital branch from the front of the hand.

3 and 4. The *Dorsales pollicis* (24) are two small arteries which are distributed on the metacarpal bone of the thumb; one is often wanting.

5. The *Dorsalis indicis* (22) is a similar branch, which runs along the metacarpal bone of the index finger. This is frequently replaced by a large artery which runs along the metacarpal bone and divides into branches to the forefinger and thumb, being in fact the radialis indicis, a branch of the radial in the palm, arising earlier than usual.

The *Posterior carpal branch* (Fig. 18, 17) of the ulnar artery will be found appearing beneath the extensor carpi ulnaris, and having completed the arch formed by the posterior carpal of the radial, it ends in a branch to the ulnar side of the fifth metacarpal bone.

The **Posterior Annular Ligament** (Fig. 15, 15) has six sub-divisions, which should be carefully examined and compared with the grooves on the bones. The first division corresponds to the groove on the outer side of the styloid process of the radius, and transmits the tendons of the extensor ossis metacarpi pollicis and extensor primi internodii pollicis: the second division corresponds to the groove on the inner side of the styloid process and transmits the tendons of the extensores carpi radialis, longior and brevior: the third division is placed obliquely, and corresponds to the oblique groove on the be radius; it transmits the tendon of the extensor secundi internodii pollicis: the fourth division, corresponding to the broad groove on the radius, transmits the tendons of

Fig. 19.

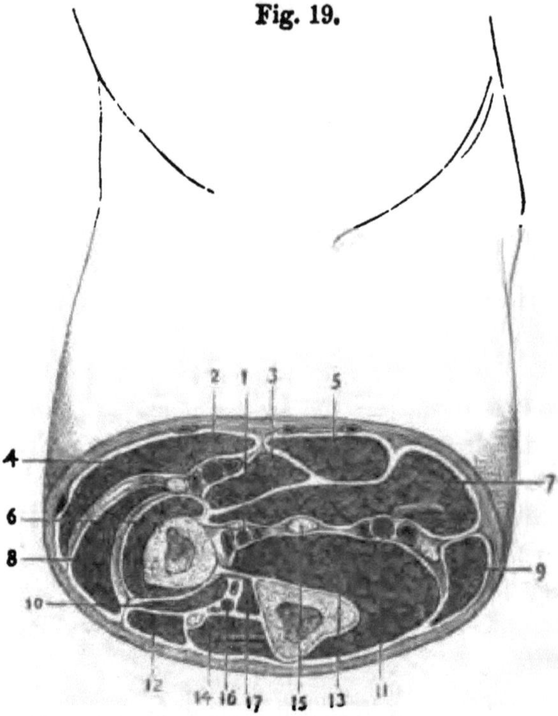

the extensor communis digitorum and of the extensor indicis: the fifth division corresponds to a very slight groove on the edge of the radius (or is placed between the bones), and transmits the tendon of the extensor minimi digiti: the sixth division corresponds to the

Fig. 18.—A section through the middle of the right fore-arm (altered from Béraud).

1. Anterior interosseous vessels and nerve.
2. Radial vessels and nerve.
3. Pronator teres.
4. Supinator longus.
5. Flexor carpi radialis.
6. Supinator brevis.
7. Flexor sublimis digitorum.
8. Extensores carpi radialis longior and brevior.
9. Flexor carpi ulnaris.
10. Extensor ossis metacarpi pollicis.
11. Ulnar vessels and nerve.
12. Extensor communis digitorum.
13. Flexor profundus digitorum.
14. Extensor carpi ulnaris.
15. Median nerve.
16. Posterior interosseous vessels and nerve.
17. Extensor secundi internodii pollicis.

groove at the back of the styloid process of the ulna, and transmits the tendon of the extensor carpi ulnaris. Thus five divisions correspond to grooves on the radius and one to that on the ulna, and they may be remembered

R U

by the formula 22121 : 1. (Fig. 20).

The dorsal interosseous muscles, four in number, can be most conveniently dissected with those of the palm of the hand.

Fig. 20.

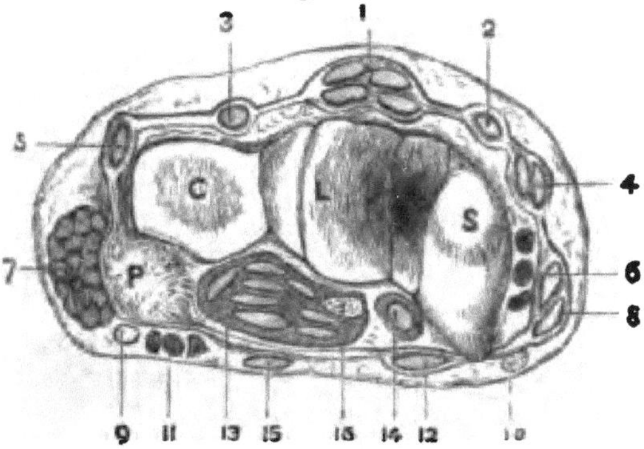

PALM OF THE HAND.

Surface-Marking.—The delicacy of the skin of the palm will depend very much upon the previous occupation of the individual, but in every case it will be found to present no hairs and to be ribbed by the rows of papillæ, upon which the orifices of the sweat-ducts can be

---

Fig. 20.—A section of the right wrist between the rows of carpal bones; the skin being removed (drawn by G. E. L. Pearse).

1. Tendons of extensor communis digitorum.
2. Tendon of extensor secundi internodii pollicis.
3. Tendon of extensor minimi digiti.
4. Tendons of extensor carpi radialis longior and brevior.
5. Extensor carpi ulnaris.
6. Radial vessels.
7. Portion of abductor minimi digiti.
8. Tendons of extensor ossis metacarpi and primi internodii pollicis.
9. Ulnar nerve.
10. Radial nerve.
11. Ulnar vessels.
12. Tendon of flexor carpi radialis.
13. Tendons of flexor sublimis and profundus.
14. Tendon of flexor longus pollicis.
15. Tendon of palmaris longus.
16. Median nerve.

s. Scaphoid bone.
L. Lunar bone.
C. Cuneiform bone.
P. Pisiform bone.

seen with a magnifying glass. If decomposition has advanced rapidly, the cuticle will probably be detached in part, when upon examining its deep surface, depressions corresponding to the papillæ will be found. A fulness on each side of the palm corresponds to the special muscles of the thumb and little finger respectively, and the terms "thenar" and "hypo-thenar" are sometimes applied to them. The palm presents three curved lines, the proximal and distal ones being curved in opposite directions, whilst the middle runs obliquely across the palm to join the proximal line at the outer side of the hand. It will be found on dissection that the point to which the superficial palmar arch reaches corresponds pretty accurately with the centre of this middle line, whilst the point of bifurcation of the digital arteries is midway between the distal or anterior line and the web of the fingers.

[An incision is to be made down the middle of the palm to the roots of the fingers, where another is to be drawn across the width of the whole hand, and the two flaps of skin are to be dissected up and turned in opposite directions. It will be found best to pay little attention to the unimportant cutaneous palmar branches of nerves, and to carry the knife at once through the fine granular fat of the palm down to the glistening palmar fascia and annular ligament. The fascia covering the muscles of the thumb and little finger is much thinner than that in the centre of the hand, and will require care for its preservation. The skin may be at once reflected from over the outer side of the hand, but upon the inner side a small transverse muscle (palmaris brevis) is to be found just below the pisiform bone and is to be preserved with the piece of skin into which it is inserted.]

The **Palmar Fascia** (Fig. 11, 9) consists of three portions. The central is triangular in shape and is attached to the annular ligament by its apex, (into which the palmaris longus is inserted,) whilst it expands at the roots of the fingers to become connected with the sheaths of the flexor tendons. Opposite the heads of the metacarpal bones it splits into four parts, and numerous transverse fibres will be found strengthening the membrane at these points, where also the digital vessels and nerves become subcutaneous. Each of the four slips which have been mentioned finally subdivides to be attached to the sides of the four sheaths of the flexor tendons. The lateral portions of the palmar fascia are much thinner, and simply form a covering for the muscles of the thumb and little finger.

The **Palmaris Brevis** (Fig. 12, 32) is a small muscle placed transversely below the pisiform bone and immediately beneath the skin. It *arises* from the inner edge of the central fascia and is *inserted* into the skin on the inner side of the hand.

[The central portion of the palmar fascia is to be detached from the annular ligament and turned down, care being taken not to injure the vessels and nerves which are close beneath, and these are to be cleaned.]

The **Superficial Palmar Arch** (Fig. 12, 34) is the direct continuation of the ulnar artery, which reaches the hand by passing over the annular ligament. The vessel is subject to very great variations, but if regular, forms an arch with the convexity forwards, reaching to about midway between the annular ligament and the root of the middle finger. The arch is completed on the radial side by the *superficialis volæ* branch of the radial artery, which generally pierces a few of the fibres of the muscles of the thumb. From the convexity of the arch, and principally from its ulnar side, are given off four *digital arteries*, the first of which runs to the ulnar side of the little finger, whilst the three others bifurcate about half an inch from the web of the fingers to supply the radial side of the little finger, both sides of the ring, both sides of the middle, and the ulnar half of the index finger. These branches should be traced along the sides of the fingers, and will be found to inosculate in the pulp of the terminal phalanx, being accompanied by the digital nerves; but the relation between them differs in the palm and in the fingers,—in the former the arteries are superficial to, in the latter they are deeper than, the nerves.

Each of the three outer digital arteries at its point of bifurcation receives a communicating branch from the deep arch (radial artery), and the branch to the fore-finger inosculates at its tip with a branch on the other side of the finger also derived from the radial (radialis indicis).

The **Profunda Artery** is a small communicating branch arising from the ulnar artery as soon as it has crossed the annular ligament. It passes deeply with an accompanying branch of the ulnar nerve between the abductor and flexor brevis minimi digiti, and will be found to complete the deep palmar arch of the radial artery.*

The superficial palmar arch is subject to considerable variation, one of the most common irregularities being its non-completion by the superficialis volæ, in which case it generally joins the *radialis indicis* at the root of the index finger.

The **Ulnar Nerve** (Fig. 12, 26) accompanies the ulnar artery over the annular ligament, lying to its ulnar side, and immediately divides into a superficial and a deep portion. The *deep branch* (28) accompanies the profunda branch of the ulnar artery between the muscles of the little finger, all of which it supplies, and will be seen again in the deep dissection of the palm. The *superficial division* subdivides into two digital branches which pass to the fingers, the inner one running on the ulnar side of the little finger, and the other bifurcating at the roots of the fingers into branches for the radial side of the little and ulnar side of the ring finger, the latter uniting with a branch of the median nerve. The digital nerves can be traced along the sides of the tendinous sheaths, lying superficially to the arteries and joining one another in the pulp of the finger, where the corpuscles of touch (Paccini) are developed like little buds upon a twig. The ulnar nerve gives a small branch to the palmaris brevis muscle.

---

* It will be observed that the number of fingers supplied by the ulnar artery is the complement of that supplied by the nerve, and that the same rule holds with respect to the radial artery and the median or radial nerves, thus,—

Ulnar artery	3½ fingers.	}
Ulnar nerve	1½ fingers.	}
Radial artery	1½ fingers.	}
Median or Radial nerve . .	3½ fingers.	}

The **Anterior Annular Ligament** (Fig. 21, 1) is the broad band of fascia binding down the flexor tendons at the wrist, and is only a thickened portion of the common fascia of the limb. It is attached to the scaphoid bone and to the ridge of the trapezium on the outer side, and to the unciform process of the unciform bone and slightly to the pisiform bone on the inner side. The ulnar artery and nerve and the cutaneous palmar nerves cross it, and also the tendon of the palmaris longus in part. (The tendon of the flexor carpi ulnaris sends an expansion over the ulnar artery and nerve, which must not be mistaken for the annular ligament itself.) When divided in the middle, it will be seen to be perforated by the tendon of the flexor carpi radialis, and to have beneath it the median nerve and the tendons

Fig. 21.

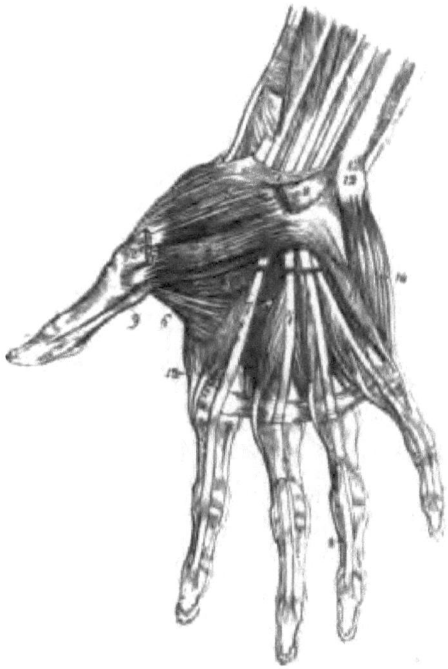

Fig. 21.—Muscles of the hand (from Wilson).

1. Annular ligament.
2, 2. Origin and insertion of the abductor pollicis muscle.
3. Opponens pollicis.
4. Superficial portion of the flexor brevis pollicis.
5. Deep portion of the flexor brevis pollicis.
6. Adductor pollicis.
7, 7. The lumbricales muscles, arising from the deep flexor tendons, upon which the figures are placed. The tendons of the flexor sublimis have been removed.
8 Insertion of one of the tendons of the deep flexor.
9. The tendon of the flexor longus pollicis, passing between the two portions of the flexor brevis to the last phalanx.
10. Abductor minimi digiti.
11. Flexor brevis minimi digiti.
12. Pisiform bone.
13. First dorsal interosseous muscle, the abductor indicis.

of the flexors of the thumb and fingers. A quantity of loose bursal tissue will be found beneath the annular ligament and closely connected with the tendons and median nerve; its use is to facilitate the movements of the tendons, and it occasionally becomes diseased, when fluid is developed in it in considerable quantity, and forms a fluctuating tumour above and below the annular ligament, often containing numerous rice-like bodies. This tissue must be carefully dissected away.

The **Median Nerve** (Fig. 14, 33) passes beneath the annular ligament superficially to the tendons, and divides into two trunks which subdivide into four *digital nerves*. The first or outermost, after giving a small branch to supply some of the short muscles of the thumb, bifurcates into branches to supply the two sides of the palmar aspect of the thumb; the second digital nerve supplies the radial side of the index finger after giving a small twig to the first lumbricalis muscle; the third, after supplying the second lumbricalis, bifurcates near the root of the finger to supply the ulnar side of the index and the radial side of the middle fingers; the fourth bifurcates to supply the ulnar side of the middle finger and the radial side of the ring finger, this last branch joining the branch from the ulnar nerve previously seen.

[The ulnar artery is to be divided beyond the origin of the profunda and the arch turned down as far as possible, but the nerve may be left uncut. The median nerve is to be divided at the wrist and turned down, and the flexor tendons with the lumbricales muscles cleaned and traced to their insertions. The sheaths of the flexor tendons have been already described (p. 46), but at least one more should be laid open in order that the tendons may be well seen.]

The **Tendons of the Flexor Sublimis**, four in number, pass beneath the annular ligament and lie immediately beneath the median nerve in the palm. Each tendon becomes somewhat flattened, and on the first phalanx splits into two portions which are *inserted* into the sides of the second phalanx, giving passage to the tendons of the flexor profundus between them.

The **Tendons of the Flexor Profundus** (Fig. 21, 8) also four in number, lie beneath the corresponding superficial tendons, which they perforate over the first phalanges; they then pass on to be *inserted* into the bases of the third phalanges.

The **Lumbricales** (Fig. 21, 7) are four little muscles connected with the deep flexor tendons, and bear some resemblance to a worm (lumbricus). The first lumbricalis, counting from the radial side (and sometimes the second), *arises* from the radial side of a single tendon, but the others *arise* by distinct slips from the adjacent sides of two deep tendons. The four muscles end in small tendons which wind to the radial side of the four fingers, and, passing beneath the transverse ligament of the metacarpus, are *inserted* into the expansion of the extensor tendons on the back of the first phalanges. The lumbricales act as flexors of the second phalanges and extensors of the third phalanges (Huxley). The first and second have been seen to be *supplied* by branches of the median nerve; the third and fourth receive branches of the deep branch of the ulnar nerve, which will be seen in the deeper dissection.

The **Tendon of the Flexor Longus Pollicis** (Fig. 21, 9) is at the same level as the tendons of the flexor profundus, and can be traced between the halves of the flexor brevis pollicis to the base of the terminal phalanx of the thumb. It lies in a fibrous sheath of its own, but the synovial sheath communicates with the one beneath the annular ligament.

[The tendons of the flexor profundus are to be divided (but not the flexor pollicis) and turned down with the lumbricales; the muscles of the little finger and thumb are to be carefully dissected.]

MUSCLES OF THE LITTLE FINGER.

The **Abductor Minimi Digiti** (Fig. 21, 10) is the most

superficial and the innermost of these. It *arises* from the pisiform bone and slightly from the tendon of the flexor carpi ulnaris, and is *inserted* into the inner side of the base of the first phalanx.

The **Flexor Brevis Minimi Digiti** (Fig. 21, 11) is often inseparably united with the abductor; but the deep branches of the ulnar artery and nerve which pass between the two will serve as guides to the point of separation. It *arises* from the unciform process of the unciform bone and from the annular ligament, and is *inserted* with the abductor into the base of the first phalanx.

The **Adductor Minimi Digiti** (opponens) is the deepest of the set, and *arises* from the unciform process close to the preceding muscle and from the annular ligament. It spreads into a triangular shape and is *inserted* into the inner margin of the shaft of the fifth metacarpal bone, thus resembling the opponens pollicis as to insertion, although there is no power of "opposition" in the little finger.

MUSCLES OF THE THUMB.

The **Abductor Pollicis** (Fig. 21, 2) is a slender muscle, and is the most superficial of the series. It *arises* from the ridge on the trapezium, and from the annular ligament. It is *inserted* into the outer side of the base of the first phalanx of the thumb with the outer half of the flexor brevis. It must be divided to see the following;—

The **Opponens Pollicis** (Fig. 21, 3) *arises* from the front of the trapezium and from the annular ligament, and is *inserted* into the outer border of the shaft of the metacarpal bone of the thumb. Its action is to "oppose" the thumb to the fingers, an action peculiar to man and monkeys, the muscle being sometimes called the *flexor ossis metacarpi pollicis.*

The **Flexor Brevis Pollicis** (Fig. 21, 4) consists of two portions, between which the tendon of the long flexor is placed, one head being more or less conjoined with the adductor, and the other with the abductor pollicis. The *outer* head *arises* from a considerable part of the annular ligament and from the lower part of the trapezium. The *inner* head *arises* from the front of the os magnum, the base of the middle metacarpal bone, and the sheath of the flexor carpi radialis, which is attached to that bone. The two heads are *inserted* into the sides of the base of the first phalanx of the thumb, and into the sesamoid bones which are developed at this point.

The **Adductor Pollicis** (Fig. 21, 6) *arises* from the anterior surface of the whole length of the shaft of the middle metacarpal bone, and is *inserted* into the inner side of the base of the first phalanx of the thumb with one part of the flexor brevis, the muscle forming a triangle with its base to the ulnar side. The muscle necessarily covers the muscles in the first and second interosseous spaces (which may be seen in part at its anterior border), and will have to be divided subsequently to expose them thoroughly.

The **Deep Branch of the Ulnar Nerve** (Fig. 14, 22) is to be traced out, and will be found to arise from the ulnar trunk near the pisiform bone, and to pass between the flexor brevis and the abductor minimi digiti, giving branches to them and to the opponens. It then forms an arch across the bases of the metacarpal bones, accompanying the deep palmar arch of the radial artery and supplying the two innermost lumbricales muscles and the seven interossei muscles (palmar and dorsal); and ends by supplying the adductor and the inner half of the flexor brevis pollicis.*

The **Deep Palmar Arch**, the termination of the radial artery, will be seen appearing between the flexor brevis and the adductor pollicis muscles; but by dividing the adductor near its origin and turning it aside, the entire arch can be traced from the point at which it enters the palm, between the heads of the abductor indicis.

* It may assist the student in remembering the distribution of the ulnar nerve if he notices how it is governed by the number 1½; thus, the nerve supplies one and a half muscles in the fore-arm; gives branches to one and a half fingers on both palmar and dorsal aspects; and lastly, supplies one and a half muscles of the thumb.

The deep arch has a slight convexity towards the fingers, and is placed upon the bases of the metacarpal bones, being completed by the deep branch of the ulnar artery.

Before terminating in the deep palmar arch the radial artery gives off two branches, viz:—

1. *Princeps pollicis*, a large branch of uncertain origin, and frequently arising from the radial at the back of the abductor indicis, which passes along the metacarpal bone of the thumb, and then bifurcates to supply both sides of the thumb superficially.

2. *Radialis indicis*, which runs on the palmar aspect of the abductor indicis to the forefinger, of which it supplies the radial side, giving a branch of communication to the superficial palmar arch and anastomosing at the tip of the finger with the digital artery derived from the superficial arch.

Fig. 22.

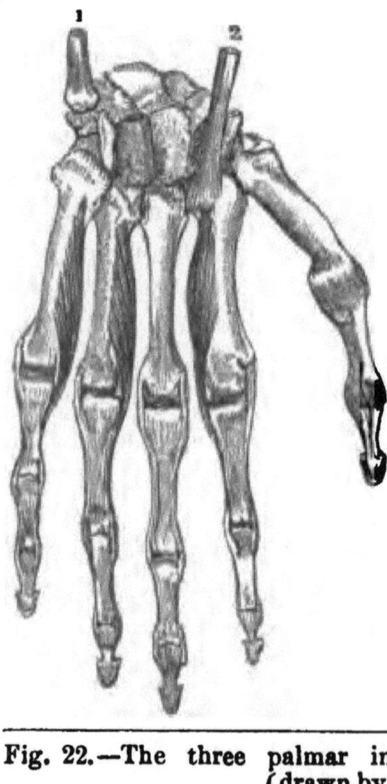

The deep arch gives some small *recurrent* branches to the carpus; three *perforating* arteries to the dorsum, which pass between the heads of the three inner interossei muscles; and three *interosseous arteries*, which run on the three palmar interosseous muscles to the roots of the fingers, and then anastomose with the digital branches of the superficial arch at their points of bifurcation.

The tendon of the *Flexor carpi radialis* can now be easily traced through the groove in the trapezium to its insertion into the base of the second metacarpal bone.

Fig. 22.—The three palmar interosseous muscles of the hand (drawn by J. T. Gray).

1 Tendon of flexor carpi ulnaris. 2. Tendon of flexor carpi radialis.

The **Transverse Metacarpal Ligament** (Fig. 27, 14) passes in front of the heads of the metacarpal bones, and is closely connected with the sheaths of the flexor tendons. It must be divided to follow out the palmar interossei, which pass beneath it.

The **Interosseous Muscles** are seven in number, three palmar and four dorsal. Two of the palmar interossei are readily seen, but the adductor pollicis must be entirely detached from the middle metacarpal bone in order to expose the remaining one.

The *Three palmar interossei* (Fig. 22) are penniform muscles placed in the three metacarpal spaces between the fingers, and are numbered 1st, 2nd, and 3rd from the radial side. They are *ad*-ductors of the fingers to an imaginary line drawn through the long or middle finger, and are arranged as follows :—The 1st palmar interosseous muscle *arises* from the ulnar side of the 2nd metacarpal bone and is *inserted* into the ulnar side of the corresponding first phalanx ; the 2nd *arises* from the radial side of the 4th metacarpal bone and is *inserted* into the radial side of the corresponding first phalanx ; the 3rd *arises* from the radial side of the 5th metacarpal bone and is *inserted* into the radial side of the corresponding first phalanx.

The *Four dorsal interossei*

Fig. 23.

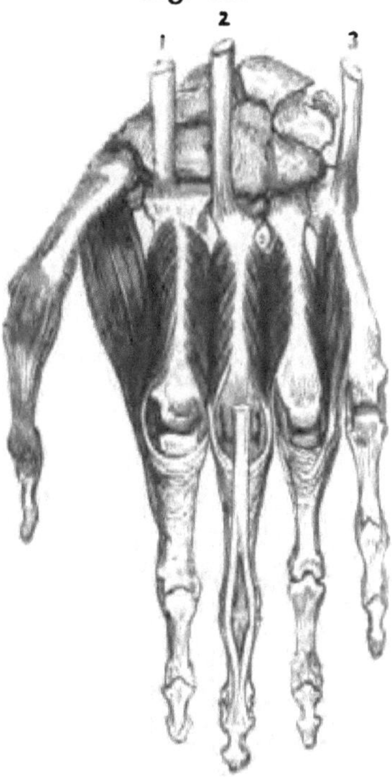

Fig. 23.—The four dorsal interosseous muscles of the hand (drawn by J. T. Gray).

1. Tendon of extensor carpi radialis longior.
2. Tendon of extensor carpi radialis brevior.
3. Tendon of extensor carpi ulnaris.

(Fig. 23) are bi-penniform muscles *arising* from the adjacent sides of the metacarpal bones, and therefore showing by the sides of the palmar muscles to some extent, but are best dissected from behind. They are *ab*-ductors from an imaginary line drawn through the middle or long finger, and are therefore *inserted* as follows:—The 1st (abductor indicis) into the radial side of the base of the first phalanx of the forefinger; the second into the radial side of the first phalanx of the middle finger; the 3rd into the ulnar side of the same phalanx; the 4th into the ulnar side of the first phalanx of the ring finger.

[The interosseous muscles of the hand should be compared with those of the foot, when the same arrangement will be found to exist in both cases, with the exception that, in the foot, the imaginary line is drawn through the long or *second* toe, and that the muscles are arranged in corresponding relation.]

Ligaments of the Scapula.

[The muscles about the scapula are to be followed to their exact insertions and cut short; the deltoid and the remains of the trapezius must be carefully removed in order that the ligaments of the scapula and clavicle may be seen.]

The **Acromio-Clavicular Articulation** (Fig. 24) is a simple arthrodial joint, though occasionally there is a fibro-cartilage developed in it, dividing the synovial cavity into two parts.

The *Superior acromio-clavicular ligament* (Fig. 24, 1), is a quadrilateral band of short strong fibres more or less connected with the fibres of the trapezius and deltoid. It is in reality continuous with the *inferior acromio-clavicular ligament*, which is similar in shape but placed below the joint.

The **Coraco-Clavicular Ligament** (Fig. 24, 2) is really but one ligament twisted upon itself so as to present two surfaces, to which the names Conoid and Trapezoid have been given; if the spine of the scapula has been divided, this twisted arrangement can be readily undone.

The *Conoid Ligament* is the more posterior portion, and resembles a cone with the base upwards. It is attached to the posterior and inner part of the base of the coracoid process *below*, and to the tubercle on the under surface of the clavicle and part of the adjacent bone *above*.

Fig. 24.

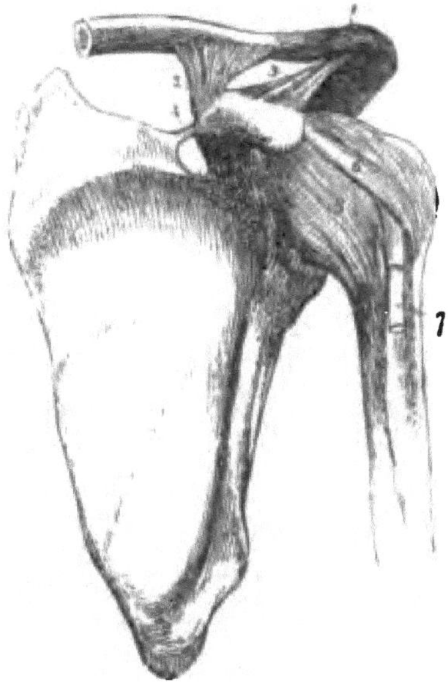

The *Trapezoid ligament* might be more suitably called rhomboid from its shape. It is a quadrilateral band of fibres attached to the line on the posterior part of the upper surface of the coracoid process *below*, and to the line leading to the tubercle on the under surface of the clavicle *above*. It is anterior and external to the conoid ligament.*

The **Coraco-Acromial Ligament** (Fig. 24, 3) is a triangular band attached to the outer side of the coracoid process, and by its smaller end to the tip of the acromion process, and is often divided into two portions. It arches above the shoulder-joint and prevents dislocation upwards.

The **Transverse Ligament** (Fig. 24, 4) is a short band converting the suprascapular notch into a hole, and giving origin to part of the omo-hyoid muscle. The supra-

* The relation of the ligaments may be remembered by the letters *a e* in tr*a*p*e*zoid, which is *a*nterior and *e*xternal to the conoid.

Fig. 24.—Ligaments of the scapula and shoulder-joint (from Wilson).

1. Superior acromio-clavicular ligament.
2. Coraco-clavicular ligament (trapezoid).
3. Coraco-acromial ligament.
4. Transverse ligament.
5. Capsular ligament.
6. Coraco-humeral ligament.
7. The long tendon of the biceps issuing from the capsular ligament and enterin— — bicipital groove.

scapular artery goes over the ligament, but the nerve beneath it.

The movements between the scapula and clavicle are very slight, but the two bones move freely together upon the thorax, the sterno-clavicular joint (*vide* side of neck) admitting of movement in every direction. The scapula is raised by the trapezius (upper part), levator anguli scapulæ and rhomboidei muscles; being again depressed by the weight of the arm, assisted probably by the lower fibres of the trapezius and (indirectly) by the latissimus dorsi. It is drawn forward by the serratus magnus and backward by the rhomboidei. The rotation of the scapula upon the ribs, by which the arm is raised beyond a right angle with the trunk, is due to the trapezius and to the strong lower fibres of the serratus magnus; the upper fibres of the latter muscle and the pectoralis minor acting in the opposite direction.

The Shoulder Joint.

[The muscles surrounding the shoulder are to be carefully removed, those which are inserted into the tubercles of the humerus being dissected as close to the capsular ligament as possible without injuring it, and then cut short.]

The shoulder is an *enarthrodial* or ball-and-socket joint, and has therefore a capsular ligament. It is surrounded and strengthened by the following muscles:—*above* by the supra-spinatus, *below* by the long head of the triceps, *in front* by the subscapularis, and *behind* by the infra-spinatus and teres minor.

The shoulder-joint admits of the following movements —flexion, extension, abduction, adduction, rotation, and circumduction. The humerus is *flexed* upon the scapula y the weight of the limb, by the action of the pectoralis jor, latissimus dorsi, and two teres muscles, and slightly the triceps. It is *extended* by the deltoid and supranatus, *abducted* by the deltoid, and *adducted* by the pectois major and coraco-brachialis muscles. Rotation of the umerus upon the glenoid cavity is produced *outwards* by the infra-spinatus and teres minor, *inwards* by the subscapularis, teres major, latissimus dorsi, and pectoralis major muscles. The last two muscles are antagonistic in the fact that the pectoralis major draws the arm across the chest after rotating it, whilst the latissimus dorsi draws it behind the back.

The **Capsular Ligament** (Fig. 24, 5) is seen to be loose, allowing partial dislocation of the humerus now that all the muscles are divided, and rough, owing to the insertion of some of the muscles of the scapula into it. There is frequently an opening on its inner side by which the bursa of the subscapularis communicates with the articular cavity. It is attached *above* to the outer margin of the glenoid cavity of the scapula, and *below* to the anatomical neck of the humerus, and is pierced at its lower margin by the long tendon of the biceps. It is thickened in front by a band of fibres attached to the root of the coracoid process and called the *Coraco-humeral ligament.*

The tendon of the biceps is to be followed into the articulation by laying open the capsular ligament, when it will be found to be surrounded by a tube of synovial membrane, and having become flattened, to be attached to the glenoid ligament and upper part of the glenoid cavity.

The **Glenoid Ligament** is a fibrous ring continuous with the tendon of the biceps and surrounding the glenoid cavity, which it therefore deepens.

There is a single **Synovial Membrane** in the shoulder-joint, which is reflected over the articular surfaces and capsular ligament; it forms a tube around the tendon of the biceps which is prolonged into the bicipital groove.

THE ELBOW JOINT.

[The muscles about the elbow are to be removed, and great care must be taken in detaching the supinator brevis not to remove or damage the external lateral and the orbicular ligaments. Beneath the triceps close to the joint may be found a little slip of muscular fibre attached to the lower end of the humerus and the back of the joint; this is the *subanconeus*, which is however often wanting or but slightly developed. All the muscles of the fore-arm and hand should now be removed to facilitate the examination of the wrist, etc. In removing the deep muscles of the fore-arm care must be taken not to injure the interosseous membrane.]

Fig. 25.

The articulation of the humerus and ulna is a good example of *ginglymus*. or hinge-joint, and therefore has lateral ligaments; the articulation of the upper end of the radius is an example of *diarthrosis rotatorius*, the head being surrounded by a ring partly of bone and partly of ligament; and the articulation between the head of the radius and the outer condyle of the humerus is an example of simple *arthrodia*. The ligaments of the elbow are united together, and form one general capsule, but may be divided into anterior, posterior, external lateral, and internal lateral.

The elbow-joint has *in front* the brachialis anticus with the brachial artery and median nerve; *behind* are the triceps and anconeus muscles. To the *inner* side are the muscles arising from the internal condyle, and the ulnar nerve with the inferior profunda artery lies upon the internal lateral ligament. *Externally* the muscles arising from the external condyle with the musculo-spiral nerve and superior profunda artery are in relation, and the supinator brevis is closely connected with the external lateral ligament.

Fig. 25.—Ligaments of the elbow-joint; inner side (from Wilson).

1. Anterior ligament.
2. Internal lateral ligament.
3. Orbicular ligament.
4. Oblique ligament.
5. Interosseous ligament.
6. Internal condyle of the humerus, which conceals the posterior ligament.

The fore-arm is *flexed* by the biceps and brachialis anticus, and indirectly by the flexors of the wrist and fingers. It is *extended* by the triceps and anconeus, and indirectly by the extensors of the wrist and fingers. Rotation of the radius upon the ulna and humerus, and consequently pronation and supination of the hand are produced as follows,— *Pronation* by the pronator quadratus and pronator teres, *supination* by the biceps, supinator brevis, and supinator longus.

Fig. 26

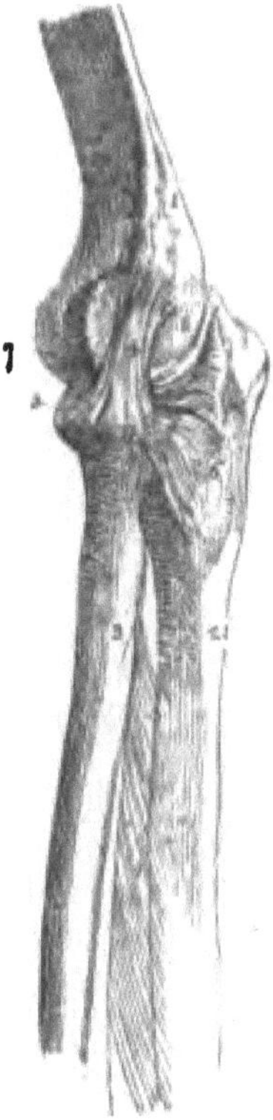

The **Anterior ligament** (Fig. 25, 1) is a broad membrane attached to the humerus immediately above the coronoid fossa, and to the edge of the coronoid process of the ulna, and partly to the orbicular ligament.

The **Posterior ligament** (Fig. 26, 8) is thin and loose, and is attached to the margin of the olecranon fossa of the humerus, and to the edge of the olecranon process of the ulna.

The **Internal lateral ligament** (Fig. 25, 2) is triangular, the apex being attached to the prominent internal condyle, and the base being attached all along the inner margins of the coronoid and olecranon processes, and bridging across the notch between them.

The **External lateral ligament** (Fig. 26, 4) is a short

Fig. 26.—External view of the elbow joint (from Wilson).

1. Humerus.
2. Ulna.
3. Radius.
4. External lateral ligament inserted inferiorly into (5) the orbicular ligament.
6. Posterior extremity of the orbicular ligament, spreading out at its insertion into the ulna.
7. Anterior ligament, scarcely apparent in this view of the articulation.
8. Posterior ligament, thrown into folds by the extension of the joint.

thick band attached to the outer condyle, and radiating slightly at its attachment to the upper border of the orbicular ligament of the radius.

The **Synovial Membrane** will be seen by removing the anterior ligament, and will be found to be reflected between the humerus, radius, and ulna, and also into the small joint between the head of the radius and the lesser sigmoid cavity. The articular surfaces of all the bones are encrusted with cartilage, but occasionally those of the coronoid and olecranon processes are separated by a groove across the bottom of the sigmoid cavity.

Radio-Ulnar Articulations.

Superior (Fig. 26, 5).—This is a part of the elbow joint, and consists of the *Orbicular ligament*, a strong flat band of fibres which is attached to the extremities of the lesser sigmoid cavity, and gives insertion to the external lateral ligament of the elbow, which must be removed to see the orbicular ligament satisfactorily.

Middle (Fig. 25, 5).—The *Interosseous ligament* or membrane is the great bond between the shafts of the bones of the fore-arm, its fibres running obliquely downwards from the radius to the ulna. It is attached to the sharp interosseous borders of both bones in all their lower part, a space being left between the bones above, through which the posterior interosseous vessels go. The membrane has a large opening in it near the lower part for the anterior interosseous artery, and one or two smaller ones for its branches.

The *Round* or *oblique ligament* (Fig. 25, 4) is very rarely seen, but is a band attached to the outer side of the coronoid process of the ulna and to the radius below the bicipital tubercle. Its direction is therefore the reverse of that of the interosseous membrane.

Inferior (Fig. 27, 2) is formed by an inter-articular fibro-cartilage (which will be seen when the joint is opened), and by *anterior* and *posterior* ligaments, which are short fibrous bands passing between the extremities of the radius and ulna.

The Wrist Joint.

This is a hinge-joint, capable of considerable lateral movement, and has anterior, posterior, and two lateral ligaments.

The wrist joint has in *front* the radial artery, the tendons of the flexor longus pollicis, flexor carpi radialis, palmaris longus, flexor sublimis and profundus with the median nerve, ulnar artery and nerve, and flexor carpi ulnaris. To the *outer* side, the tendons of the extensor ossis metacarpi and primi internodii pollicis with the radial nerve. *Behind* the extensor carpi radialis longior and the brevior, extensor secundi internodii pollicis, extensor communis digitorum and indicis and minimi digiti, and the extensor carpi ulnaris (Fig. 20).

The **Anterior** and **Posterior ligaments** (Fig. 27, 3) are broad bands attached to the front and back of the lower end of the radius, and to the front and back of the first row of carpal bones, being united with the ligaments of the carpus.

The **External lateral ligament** (Fig. 27, 4) is a thick band attached to the styloid process of the radius and to the outer side of the scaphoid and trapezium.

The **Internal lateral ligament** (Fig. 27, 5) is longer than the external, and is attached to the styloid process of the ulna and to the upper surface of the cuneiform bone.

The **Bones of the Carpus** (Fig. 27, 6) are bound together by vertical and transverse *dorsal* and *palmar*, and at each side by slender *lateral* ligaments, which will not repay a special dissection; also by *interosseous* ligaments which bind together the several bones of each row (except the pisiform which has a separate capsular ligament), which will be seen when the joints are opened.

The **Metacarpal Bones** (Fig. 27, 8) (with the exception of the first) are connected with the second row of the carpus by *dorsal* and *palmar* ligaments, and the four metacarpal bones are joined together by *t*[illegible]rsal and palmar ligaments.

The first metacarpal bone has a separate capsular ligament connecting it with the trapezium, this joint being enarthrodial (Fig. 27, 9). The bones of the carpus and metacarpus are capable only of a slight gliding or *arthrodial* movement with the exception of the metacarpal bone of the thumb, which is capable of flexion, extension, adduction, abduction and circumduction. The movements of the wrist are due to the actions of the flexors and extensors of the carpus and phalanges.

Fig. 27.

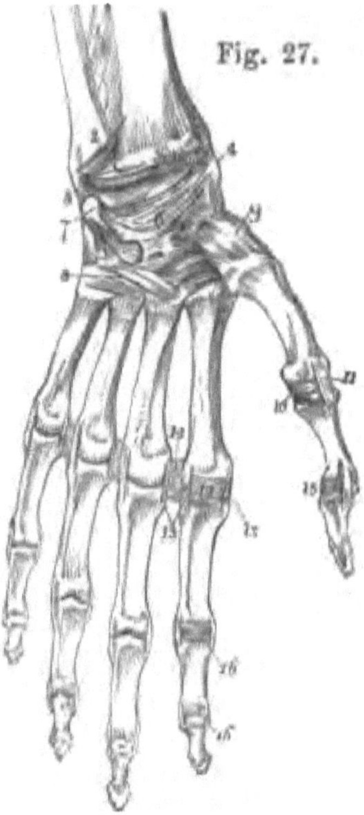

[The synovial membranes of the wrist are to be opened by the following incisions on the dorsal aspect of the hand; one between the lower ends of the

---

Fig. 27.—Ligaments of the anterior aspect of the wrist and hand (from Wilson).

1. Lower part of the interosseous membrane.
2. Anterior inferior radio-ulnar ligament.
3. Anterior ligament of the wrist-joint.
4. External lateral ligament.
5. Internal lateral ligament.
6. Palmar ligaments of the carpus.
7. Pisiform bone, with its ligament.
8. Ligaments connecting the second row of carpal bones with the metacarpals, and the metacarpals with each other.
9. Capsular ligament of the carpo-metacarpal articulation of the thumb.
10. Anterior ligament of the metacarpo-phalangeal articulation of the thumb.
11. One of the lateral ligaments of that articulation.
12. Anterior ligament of the metacarpo-phalangeal articulation of the index finger; this ligament has been removed in the other fingers.

radius and ulna, (care being taken not to divide the fibro-cartilage), and three transverse incisions carried between the lower end of the radius and the carpus, the two rows of carpal bones, and the carpus and metacarpus respectively. The articulations of the pisiform bone and of the first metacarpal bone are also to be opened.]

Fig. 28.

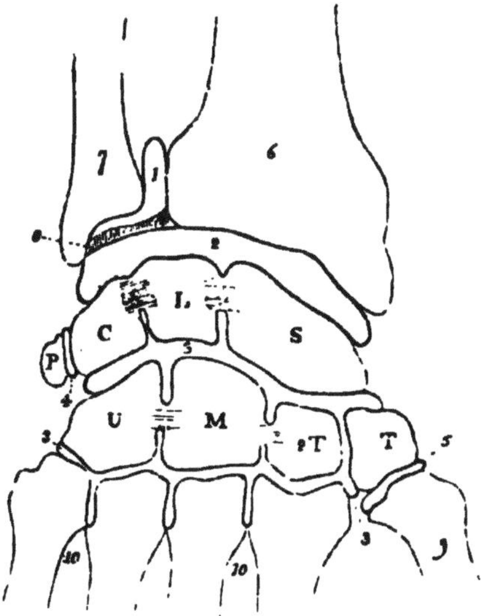

The Synovial Membranes of the Wrist (Fig. 28) are five in number. The 1st or *mem-*

---

13. Lateral ligaments of the same articulation: the corresponding ligaments are seen in the other articulations.
14. Transverse ligament connecting the heads of the metacarpal bones.
15. Anterior and one lateral ligament of the phalangeal articulation of the thumb.
16. Anterior and lateral ligaments of the phalangeal articulations of the index finger; the anterior ligaments are removed in the other fingers, but the lateral ligaments remain.

Fig. 28.—Diagram showing the disposition of the five synovial membranes of the wrist joint (from Wilson).

1. Membrana sacciformis.
2. Second synovial membrane.
3, 3. Third, or large synovial membrane.
4. Synovial membrane between the pisiform and cuneiform bones.
5. Synovial membrane of the metacarpal articulation of the thumb.
6. Lower extremity of the radius.
7. Lower extremity of the ulna.
8. Interarticular fibro-cartilage.
S. Scaphoid bone.
L. Semilunar.
C. Cuneiform; interosseous ligaments are seen passing between these three bones.
P. Pisiform.
T. Trapezium.
T 2. Trapezoid.
M. Os magnum.
U. Unciform; interosseous ligaments are seen connecting the os magnum with the trapezoid and unciform.
9. Base of the metacarpal bone of the thumb.
10, 10. Bases of the o[illegible] carpal bones.

brana sacciformis is between the lower ends of the radius and ulna, and extends between the ulna and the triangular fibro-cartilage; the 2nd is between the radius and the under surface of the triangular fibro-cartilage above, and the three outer bones of the first row of the carpus below; the 3rd is between the first and second rows of carpal bones, passing between the several bones as well; it is prolonged between the bones of the second row to the metacarpal bones, and is reflected over the bases of the four inner bones as well as between them; the 4th is between the cuneiform and pisiform bones; the 5th between the trapezium and the metacarpal bone of the thumb.

The **Triangular fibro-cartilage** (Fig. 28, 8) is best seen by removing the carpus altogether. It is attached by its base to the margin of the radius, between the surfaces for articulation with the ulna and the carpus, and passes transversely between the lower end of the ulna and the carpus, to be attached by its apex to the root of the styloid process of the ulna.

The **Metacarpus and Phalanges** (Fig. 27) are connected by anterior, posterior, and lateral ligaments.

The *Anterior ligaments* are strong bands passing from the head of the metacarpal bone to the base of the phalanx, and closely connected with the transverse metacarpal ligament and the sheaths of the tendons.

The *Posterior ligament* is formed by an expansion of the extensor tendons.

The *Lateral ligaments* are strong bands which are attached to the sides of the heads of the metacarpal bones, and pass obliquely to the anterior margins of the phalanges.

Each joint has a separate synovial membrane.

The **Phalangeal Articulations** are similar to those between the metacarpus and phalanges.

PART II.

DISSECTION OF THE LEG.

[*The Student is requested to read the 'Introduction' before commencing the dissection, unless he has done so previously.*]

BEFORE beginning the actual dissection, the student should make himself fully acquainted with the external configuration of the part and the relations which superficial appearances bear to deeper structures. If he has already dissected this region he should also make the incisions necessary to expose the several arteries in the positions in which they are usually tied, according to the directions which accompany the description of each vessel, taking care not to disturb the tissues unnecessarily, and to stitch up the incisions without delay.

External appearances.—The fold of the groin separates the abdomen from the thigh, and the finger, if carried along it, will recognise Poupart's ligament stretched tightly across when the limb is fully extended. In the same position also the front of the thigh will be convex owing to the large extensor muscles, but if the knee be slightly bent and the thigh abducted, a shallow depression will be seen immediately below the groin corresponding to Scarpa's triangle, and Poupart's ligament will be felt to become relaxed. Lymphatic glands of variable size can be felt along, or a little above, the line of Poupart's ligament; whilst below it, and more or less in a vertical direction, will be found others which belong strictly to the thigh. In the centre of the hollow on the front of the thigh can be felt the femoral artery, which may be followed for about a third of the length of the thigh in a thin subject. A subcutaneous vein (saphena) is occasionally to be seen, when enlarged, on the inner side of the limb, and the course of others joining it is occasionally visible. If the body should be the subject of femoral hernia, probably the hernial tumor will be found occupying the upper part of the hollow in the front of the thigh, and reaching more or less along the lower border of Poupart's ligament. In such a case, if the hernial protrusion be returned, the finger can be readily passed into the enlarged saphenous opening and up into the crural ring behind Poupart's ligament.

The crest and anterior superior spine of the ilium will be readily seen and felt, the anterior inferior spine less so, on account of the muscles covering it; the spine and crest of the pubes can also be distinguished. The great trochanter is easily recognised from three to four inches below the crest of the ilium, and should be thoroughly manipulated that its relation to the upper border of the pelvis and surrounding parts may be clearly appreciated during the movements of the limb; and the two sides of the body should be compared if any morbid change about the hip is suspected.

An important diagnostic sign of a healthy condition of the neck of the femur is that when the limb is rotated the trochanter describes part of the arc of a circle, which is not the case when fracture has occurred. When an impacted fracture or absorption of the neck of the femur has taken place, the arc is much smaller and the movement less complete. The head of the femur may be felt behind the trochanter when the limb is rotated, and in a very thin subject it may be detected in front, when the finger is thrust deeply into Scarpa's triangle.

The condyles of the femur and the patella are to be examined, and it should be noticed how large a part of the articular end of the femur is uncovered by the patella when the knee is flexed. The ligament of the patella and its attachment to the tibia are to be noticed, and the finger carried along the subcutaneous surface of the tibia, the shin, down to the ankle. The head of the fibula is prominent, and the hand carried to the back of the knee will recognise the outer and inner hamstrings attached to it and to the tibia, and when the knee is flexed may feel the popliteal artery behind the joint. The lower third of the fibula is subcutaneous and terminates in the prominent external malleolus, behind which the peroneal tendons can be felt. On the inner side, the inner malleolus is to be examined, with the tendons behind it and the Achilles tendon attached to the heel.

The foot is to be moved freely, when it will be found that the amount of lateral motion in the ankle joint is greatest when the toe is thoroughly pointed. In the same position the broad upper articular surface of the astragalus becomes subcutaneous and is readily seen beyond the margin of the tibia.

In the foot, the following points of practical utility should be thoroughly recognised; first, on the inner side, the tuberosity of the scaphoid bone, to which the tendon of the tibialis posticus may be traced, and which is the guide in Chopart's amputation of the foot; secondly, the slighter prominence of the internal cuneiform bone and base of the first metatarsal bone in front of it, to which the tendon of the tibialis anticus may be traced, and which serve as a guide in Hey's amputation; and thirdly, on the outer side, the prominent base of the fifth metatarsal bone, which serves the same purpose. The toes are commonly much distorted owing to the long pressure of ill-fitting shoes, and the metatarso-phalangeal articulation of the great toe is not unfrequently diseased, forming a bunion.

The Front of the Thigh.

[An incision is to be made along Poupart's ligament and carried down the inner border of the thigh, for half its length, and this is to be joined by another across the limb at that point; the flap of skin thus marked out is to be carefully reflected to the outer side of the

Fig. 29.

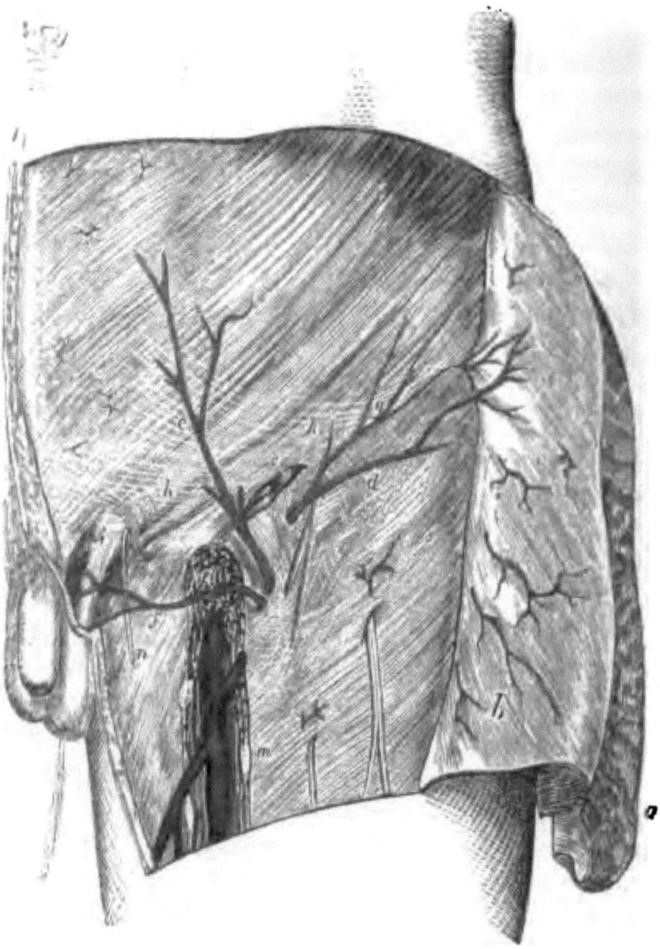

Fig. 29.—Superficial dissection of the inguinal and femoral regions (from Wood 'On Rupture').

a. Superficial layer of fascia (reflected).
b. Deeper layer of fascia (reflected) (the superficial vessels being left attached to the external oblique).
c. Inguinal lymphatic glands.
d. Superficial circumflex iliac artery.
e. Superficial epigastric artery.
f. Superior external pudic artery.
g. Poupart's ligament.
h. Intercolumnar fascia.
i. External abdominal ring.
k. Arciform fibres of external oblique.
l. Internal saphena vein.
m. Femoral lymphatic glands.
n. Ilio-inguinal nerve.
o. Saphenous opening.

The *Superficial fascia* of the thigh is continuous with the superficial fascia of the abdomen over Poupart's ligament, and generally contains a good deal of fat, particularly in the female subject. It is divisible into two layers in the upper part of the thigh, and the deep layer will be seen by reflecting the superficial layer in the same way as the skin but not so far down, the separation between the two layers of fascia being marked by some small vessels which may be at once examined.

The **Superficial Vessels** (Fig. 29) are branches of the femoral artery with their accompanying veins, and are three in number, viz. :—

The *Superficial epigastric artery* (*e*), which arises from the femoral immediately below Poupart's ligament, and passes upwards and inwards over the front of the abdomen.

The *Superficial circumflex iliac artery* (*d*), a small branch running outwards a little below Poupart's ligament, and—

The *Superior external pudic artery* (*f*), running inwards, over the spermatic cord to the scrotum, or to the labium in the female.

The *Veins* correspond in course and will be seen to open into the saphena vein.

The *Lymphatic glands* will also be found between the layers of superficial fascia, and are arranged in two rows, one along the groin, to which the lymphatics of the penis and scrotum pass, and another below the groin in the direction of the thigh, into which the lymphatics of the limb empty themselves.*

The *Deep layer of superficial fascia* is more membranous than the superficial layer, and will be best seen by raising it from the deep fascia beneath, beginning about four inches below the groin and reflecting it on to Poupart's ligament. Between the two is the saphena vein,

* The relation of the lymphatics to these glands is to be borne in mind, since the position of a bubo will vary according to the part primarily affected,—genitals or foot. The student is reminded that the lymphatics of the testicle open into the lumbar glands.

and the deeper layer of superficial fascia will be found bound down to the point at which the vein disappears, the saphenous opening. The deep layer of superficial fascia does not pass over Poupart's ligament like the superficial layer, but is bound down to it, thus helping to direct a large femoral hernia along the groin.

The *Deep fascia of the thigh* (fascia lata) is now exposed, and is to be dissected as far as is necessary for the study of the parts concerned in femoral hernia, the several nerves piercing it being left for subsequent examination.

Femoral Hernia.

[The saphena vein is seen to disappear through an opening in the fascia lata about an inch and a half below Poupart's ligament. This is the saphenous opening, which is obscured by the attachment of a portion of the superficial fascia, called from the numerous perforations in it *cribriform*. If the cribriform fascia be carefully removed together with the small veins, the artificially produced saphenous opening will then be clearly seen.]

The **Saphenous Opening** (Fig. 29, *o*) varies considerably in different subjects, and is generally best seen in the female. It is an oval opening half an inch wide, produced by a splitting and folding of the fascia lata at this point; the outer edge of the opening being formed by the *iliac* portion of the fascia which is attached to the whole length of Poupart's ligament, whilst the inner and deeper part is formed by the *pubic* portion of the fascia lata which, passing beneath the femoral vessels, binds down the pectineus muscle and is attached to the pectineal line. The outer border, which can be made sharp and semicircular with the scalpel, is known as the *falciform* (sickle-shape) margin (Burns), and over the lower part of this the saphena vein passes to open into the femoral vein. The superior extremity of the margin, where it joins Poupart's ligament and overlies the femoral vessels, is sometimes called Hey's ligament. The saphenous opening is the external aperture through which femoral hernia when of large size passes, and might therefore well be called the external femoral ring, although strictly speaking there is in the undissected condition no definite ring, but merely a weak point in the fascia through which the hernia protrudes, pushing before it the cribriform fascia. The position of the thigh has considerable influence upon the condition of the saphenous opening, and its edges will be found to be relaxed when the thigh is flexed towards the median line of the abdomen, the position in which the patient is placed when the "taxis" is applied for the reduction of a hernia.

[An incision is to be made from the lower end of the saphenous opening transversely outwards for two inches, and a corresponding one immediately below Poupart's ligament, and the piece of fascia thus marked out and including the falciform process is to be reflected to the outer side.]

Fig. 30.

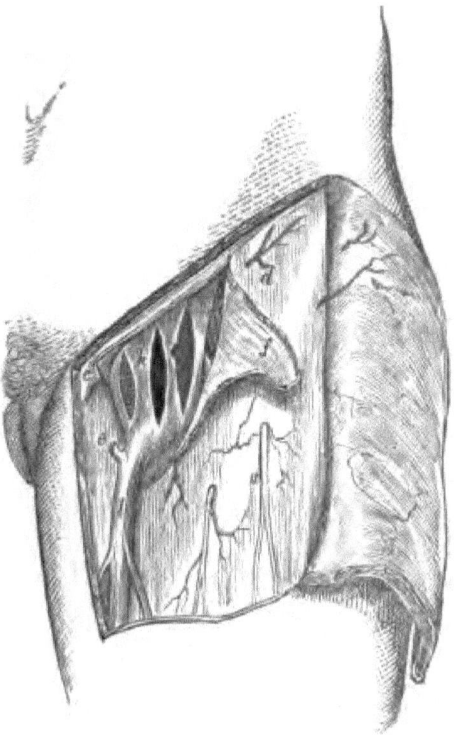

Sheath of the Vessels (Fig. 30, *k*).—Beneath the fascia lata in the upper part of the thigh will now be seen a delicate fascia, containing a little fat and covering the femoral vessels, called the *sheath of the femoral vessels*.

Fig. 30.—Crural sheath laid open (from Wood 'On Rupture').

a. Middle cutaneous nerve.
c. Placed to inner side of Gimbernat's ligament.
d. Iliac portion of fascia lata.
e. Pubic portion of fascia lata.
f. Margin of saphenous opening (turned back).
k. Femoral sheath opened by three incisions.
l. Saphena vein.

This sheath is a tube, broader above than below, and becoming gradually lost upon the vessels, which is continuous with the fascia transversalis and fascia iliaca of the abdomen beneath Poupart's ligament. Three vertical incisions, one in the centre and one at each side of the tubular sheath, will enable the dissector to see that it is divided by two slender septa into three compartments,

Fig. 31.

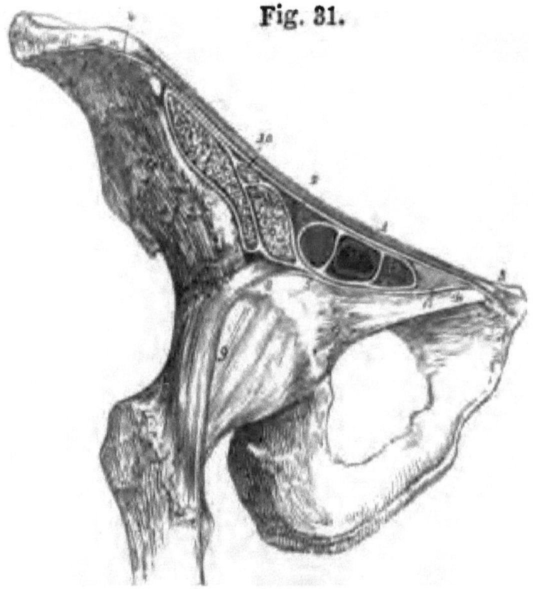

the femoral artery occupying the outermost, the femoral vein the middle, and the innermost or smallest one being occupied only by a lymphatic gland. This last division of the femoral sheath is the *crural canal*.

The **Crural Ring** (Fig. 31, 12).—If the finger be passed upwards along the crural canal it will enter the *Crural* or *Femoral Ring* beneath Poupart's ligament,

Fig. 31.—Section of the structures which pass beneath the femoral arch (from Wilson).

1. Poupart's ligament.
2, 2. Iliac portion of the fascia lata, attached along the margin of the crest of the ilium, and along Poupart's ligament as far as the spine of the pubes (3).
4. Pubic portion of the fascia lata, continuous at 3 with the iliac portion, and passing outwards behind the sheath of the femoral vessels to its outer border at 5, where it divides into two layers; one is continuous with the sheath of the psoas (6) and iliacus (7); the other (8) is lost upon the capsule of the hip-joint (9).
10. The anterior crural nerve.
11. Gimbernat's ligament.
12. The femoral ring, within the femoral sheath.
13. Femoral vein.
14. Femoral artery; the two vessels and the ring are surrounded by the femoral sheath.

displacing a little piece of subperitoneal fat which occupies it and is termed the *septum crurale.* The boundaries of the crural ring can be better felt than seen, and are, *in front,* Poupart's ligament or crural arch (with occasionally a distinct band of fascia transversalis beneath it, called the *deep crural arch*); *behind,* the ilio-pectineal line and body of the pubes; *externally,* the femoral vein separated by the septum; and *internally,* the sharp margin of Gimbernat's ligament. The crural ring is the aperture through which femoral hernia leaves the abdomen, and the point at which strangulation most commonly occurs; the finger should therefore thoroughly explore it.

Fig. 32.

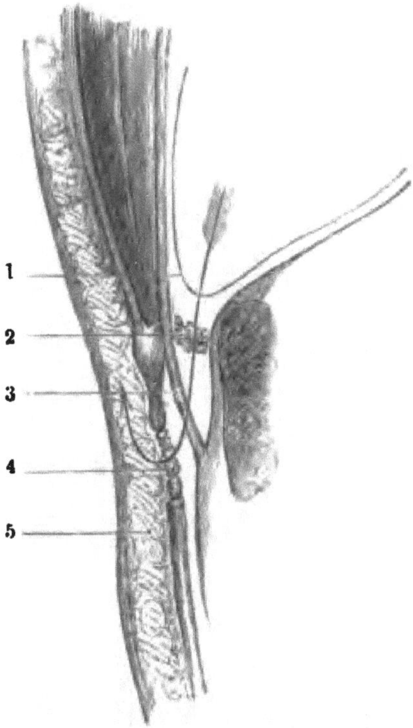

The **Crural Canal** (Fig. 30) is the short canal along which femoral hernia *descends* from the crural ring to the saphenous opening; but as soon as the hernia has forced its way through that spot, it *ascends,* owing to the close attachment of the superficial fascia below that point, and lies along Poupart's ligament, or, if of large size, may stretch the fascia sufficiently to mount over the ligament and simulate an inguinal hernia.

The **Coverings of a Femoral Hernia** (Fig. 32) will be

Fig. 32.—Imaginary section of crural canal to show the course and coverings of a femoral hernia (drawn by J. T. Gray).

1. Peritoneum.
2. Septum crurale.
3. Femoral sheath.
4. Cribriform fascia.
5. Integuments.

readily learned if the course it takes has been thoroughly understood. In its descent the intestine pushes before it (1) peritoneal sac, (2) septum crurale, (3) femoral sheath, (4) cribriform fascia, (5) superficial fascia and skin. It must not be supposed that the surgeon will meet with these various coverings in operating upon a strangulated femoral hernia, since he usually pays no attention to any of them until he reaches the sac or peritoneum, which is to be recognised by its thin bluish appearance and by the fluid generally to be seen through its slightly transparent wall.*

Fig. 33.

Fig. 34.

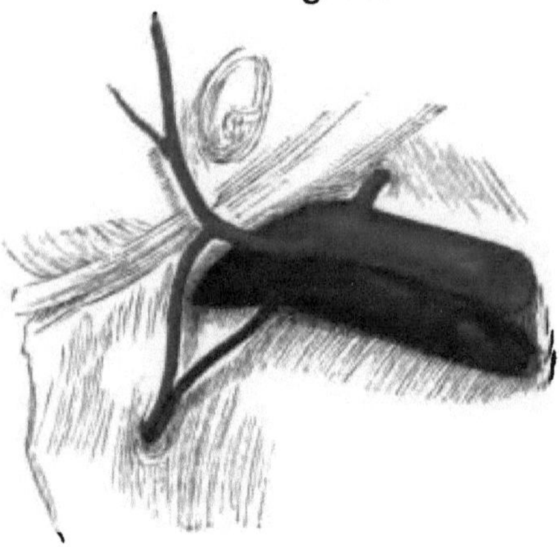

Surgery. — *Operation for strangulated femoral hernia.* The point of stricture is very rarely at the saphenous opening, since it becomes so much enlarged in an old hernia as to offer no resistance; and the skin and superficial structures having been divided, (usually by a vertical incision) the forefinger can therefore be readily passed along the crural canal on the inner side

* The student is particularly warned against the common error of supposing the sac of a hernia to be smooth and glistening on its *exterior.*

Fig. 33.—Irregular origin of obturator artery from epigastric. 1st variety external to crural ring (from Wood 'On Rupture').

Fig. 34.—Irregular origin of obturator artery from epigastric. 2nd variety internal to crural ring (from Wood 'On Rupture').

of the hernia to the crural ring, where the stricture will be found. This may be most satisfactorily and safely relieved by cutting cautiously inwards with a hernia knife, so as to notch or divide Gimbernat's ligament; but if this should not prove sufficient, the sac must be cautiously opened and the stricture divided from within it.

The only possible danger which can be met with in the deep incision, is an abnormal distribution of the obturator artery, which, if it arise from the epigastric artery and wind close to the *inner* side of the neck of the sac, might be divided and give rise to troublesome hæmorrhage. Fortunately this vessel, if it exist, is seldom damaged by a cautious use of the knife, and, as it is impossible to ascertain its presence beforehand, its existence may be ignored in practice.

[The cutaneous nerves of the upper part of the front of the thigh are to be examined with the fascia lata, before it is removed to expose Scarpa's triangle.]

The *Fascia lata* of the thigh is a dense white membrane attached to Poupart's ligament and the crest of the ilium above, and to the pubes internally, and will be afterwards seen to enclose and support all the muscles of the thigh.

Cutaneous Nerves (Fig. 35).—The *Ilio-inguinal nerve*, which emerges from the external abdominal ring, gives one or two small branches to the thigh.

The *crural branch of the Genito-crural nerve* pierces the fascia at the level of, but external to, the saphenous opening, and joins the middle cutaneous nerve. It is very seldom satisfactorily seen unless traced from the lumbar plexus.

The *External cutaneous nerve* (3) is to be found in a fold of fascia lata close to the anterior superior spine of the ilium, and pierces the fascia at a variable point, to be distributed by an anterior and a posterior branch to the outer side of the thigh.

The *Middle cutaneous nerve* (5) (from the anterior crural), pierces the fascia about the middle of the thigh, and will be afterwards traced as far as the knee.

Fig. 35.

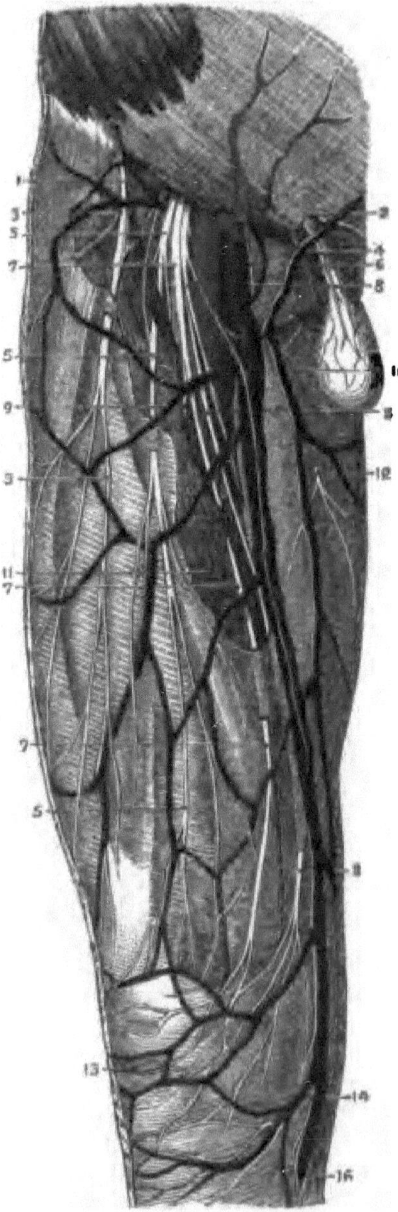

The *Internal cutaneous* and *Long Saphenous nerves* are not seen at present.

SCARPA'S TRIANGLE.

[The fascia lata is to be removed from the hollow below Poupart's ligament as far as the skin has been reflected, and the muscles, vessels, and nerves thus exposed, are to be as thoroughly cleaned as is possible without disturbing their relations.]

Scarpa's triangle is the space in the upper third of the front of the thigh, containing the first part of the femoral vessels. The *base* of the triangle is upwards and is formed by Poupart's ligament; the *outer border* is formed by the sartorius muscle, which crosses the thigh obliquely; the *inner border* corresponds to the margin of the abductor longus; and the *apex* of the triangle is formed by the meeting of the two above-named muscles. The space has been seen to be covered in by the superficial fasciæ and fascia lata, and its *floor* will now be seen to be

Fig. 85.—Superficial dissection of the front of the thigh (from Hirschfeld and Leveillé).

1. Poupart's ligament.
2. Superficial branches of femoral artery.
3. External cutaneous nerve.

formed by the psoas and iliacus on the outer side, the pectineus and adductor longus on the inner side, with (in some subjects) a small portion of the adductor brevis between the two.

Scarpa's triangle *contains* the upper part of the femoral artery and vein; the origin of the profunda artery with its vein; and the anterior crural nerve with some of its branches. The saphena vein is not one of the contents of the space although generally seen to the inner side of the femoral vessels, since it originally was superficial to the fascia lata.

Fig. 36.

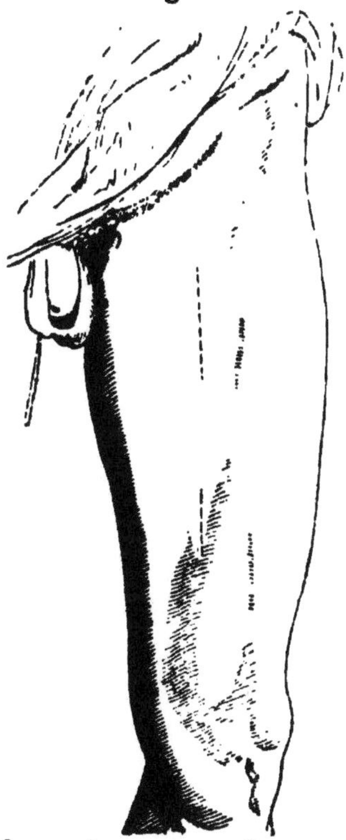

The **Femoral Artery in Scarpa's Triangle** (Fig. 35, 4).—The femoral artery is the continuation of the external iliac artery, and extends from Poupart's ligament to the popliteal space. The part now exposed is that above the sartorius, and is usually nearly half of the whole length of the vessel. When the thigh is slightly flexed and abducted, the artery runs from a point midway between the symphysis pubis and the anterior superior iliac spine (or a little to the inner side of the centre of Poupart's ligament) through the middle of

4. Femoral artery.
5, 5, 5. Middle cutaneous nerve.
6. Femoral vein.
7, 7, 7. Outer division of internal cutaneous nerve.
8, 8, 8. Inner division of internal cutaneous nerve.
9. Branch to sartorius muscle from internal cutaneous.
10. Saphena vein.
11. Sartorious muscle.
12. Cutaneous branch of obturator nerve.
13. Plexus patellæ.
14. Patellar branch of long saphenous nerve.
16. Long or internal saphenous nerve.

Fig. 36.—Incisions for tying the femoral artery (from Fergusson's 'Surgery').

the triangle in a line with the prominent tubercle on the inner condyle of the femur; but when the thigh is fully extended it will be seen to wind slightly to the inner side of the limb. This portion of the artery is *covered* only by the superficial fasciæ and fascia lata, except near Poupart's ligament, where it has been seen to be enclosed in the femoral sheath. One or two branches of the internal cutaneous nerve cross the vessel at variable points. *Behind* the artery from above downwards are, the psoas (with a portion of the femoral sheath); the pectineus, but separated from the femoral artery by the profunda artery and vein and the femoral vein; and the adductor longus. To the *outer* side throughout is the anterior crural nerve, but separated at the upper part by a small piece of the psoas; and the long saphenous branch is in close relation at the lower part of the space. The femoral vein is to the *inner* side near Poupart's ligament, but lower down is *behind* the artery, forming one of its posterior relations.

Surgery (Fig. 36).—From its superficial position, the femoral artery can be readily felt during life, and compression may be effected in any part of the triangle, but most satisfactorily against the margin of the pubes. The artery has been tied above the origin of the profunda, but with unsatisfactory results,—deligation at the lower part of the triangle having been very successful. The latter operation is readily performed through an incision, from three to four inches long, in the middle of the lower part of the space, and in the direction of the limb, which is to be abducted. The incision should be carried at once through the fascia so as to expose the fibres of the sartorius at the lower part of the wound, which are readily recognizable by their oblique direction. The muscle being turned a little outwards, the cellular sheath of the vessels will be seen, and the artery can be easily isolated and secured. The needle should be passed from within outwards, great care being taken not to include the vein which is behind, or the long saphenous nerve which is to the outer side of the vessel.

Branches (Fig. 29).—The small (1) *superficial epigastric;* (2) *superficial circumflex iliac;* and (3) *superior external pudic* have been already examined. A small branch (4) the *inferior external pudic* arises from the upper part of the artery and passes beneath the fascia lata to the pubes.

The largest branch, (5) the *profunda* artery, arises from one or two inches below Poupart's ligament, and is now seen lying first to the outer side of, and then behind, the femoral artery and vein, and resting upon the pectineus. It is the great artery to the muscles of the thigh, and will be afterwards dissected.

The *Femoral vein* at the lower part of Scarpa's triangle lies a little to the outer side of the artery. It then ascends behind it, and after being joined by the profunda vein, lies to the inner side of the femoral artery at the upper part of the space, where it receives the saphena vein and the veins corresponding to the four small branches of the artery.

The **Anterior Crural Nerve** (Fig. 39, 5) is derived from the 2nd, 3rd and 4th nerves of the lumbar plexus, and enters Scarpa's triangle beneath Poupart's ligament, between the psoas and iliacus muscles. It divides into a superficial and deep part, the superficial giving off the middle and internal cutaneous nerves and supplying the sartorius; the deep supplying muscular branches, and the long or internal saphenous nerve which becomes cutaneous at the inner side of the knee.*

[The limb being extended, the line of incision on the inner side of the thigh is to be prolonged to three inches below the knee and then carried across the limb, when the skin thus marked out is to be reflected to the outer side, and the superficial fascia to the same extent, all the cutaneous nerves being carefully left at their points of emergence from the fascia lata.]

The *Bursa patellæ* is the subcutaneous synovial sac in

* This last is often described as coming from the superficial division, but the above is the more common arrangement.

front of the knee-cap, which will be seen by making a vertical incision into it. When inflamed, this bursa forms what is commonly known as "house-maid's knee."

The *Fascia lata* or deep fascia of the thigh is now thoroughly exposed. It has been seen to be attached to Poupart's ligament and the crest of the pubes, and can now be followed to the crest of the ilium. The fascia is strongest on the front and outer side of the thigh, and on each side of the prominent extensor muscles may be seen a white line marking the positions of its external and internal intermuscular septa. The fascia is attached to the condyles of the femur below, and is continued over the patella to the heads of the tibia and fibula, being incorporated with the several tendons attached to those parts.

Piercing the fascia lata at several points are the cutaneous nerves of the thigh, some of which have been already seen, but the following are now to be traced to their distributions and then to their origins, the fascia being removed as may be necessary for the latter purpose.

The *Internal cutaneous nerve* (Fig. 35, 7, 8) is a branch of the anterior crural, and divides at a variable point into two branches, *outer* and *inner*. The *outer* branch frequently perforates the sartorius muscle, which it supplies, and becomes cutaneous in the lower third of the thigh, the branches reaching as far as the patella. The *inner* branch having crossed the femoral artery appears close to the saphena vein a little above the knee, and its branches unite with those of the other branch and with the internal saphenous nerve. Occasionally a junction between this nerve and the obturator may be traced on the inner side of the middle of the thigh.

The *Long* or *Internal Saphenous nerve* (the course of which will be afterwards seen) becomes cutaneous at the inner side of the knee joint, where it lies at first in front of the saphena vein, and immediately sends a large branch forward to the skin of the patella (*ramus cutaneus patellæ*), which unites with other cutaneous nerves to form what has been called the *plexus patellæ*.

The *Superficial veins* of the thigh are very variable, but they either open into the saphena vein separately, or form one or more separate branches which join the main trunk at or near the saphenous opening.

MUSCLES OF THE FRONT OF THE THIGH.

[The remains of the fascia lata are to be removed and the cutaneous nerves turned aside. The muscles of the front of the thigh will be put on the stretch by flexing the knee, except the sartorius, which must be stretched with hooks in order to clean it. All the loose fascia is to be removed and each muscle cleaned in the direction of its fibres, special care being taken to expose clearly the origin and insertion of each.]

Connected with the fascia lata on the outer side of the thigh is the **Tensor Fasciæ or Vaginæ Femoris** (Fig. 37, 4) which is to be exposed by removing the fascia below the crest of the ilium. The muscle *arises* from the outside of the anterior superior spine of the ilium and from a small portion of the crest; and the fibres pass downwards and backwards to be *inserted* into the fascia lata below the great trochanter, the fascia being split at this point to receive the muscular fibres. The tensor vaginæ femoris is, as its name implies, a tensor of the fascia of the thigh, and thereby supports and assists the other muscles of the limb; it is also an internal rotator of the femur. It is *supplied* by a branch of the superior gluteal nerve, which may be seen entering the deep surface of the muscle when it is reflected.

The **Sartorius** (Fig. 37, 5) is the longest muscle of the body, and when dissected lies loosely upon the subjacent muscles.

It *arises* from the anterior superior spine of the ilium and from the notch below it, and, crossing obliquely over the thigh, winds to the inner side of the knee and is *inserted* into the broad and subcutaneous surface below the inner tuberosity of the tibia. The insertion of the sartorius is by a broad tendinous expansion, which is incorporated with the fascia of the limb and overlaps the insertions of the gracilis and semi-ten-

muscles (of which the latter is the lower one), a bursa intervening between them. The sartorius (tailor-muscle) is a flexor of the leg upon the thigh and a flexor and rotator outwards of the thigh upon the pelvis, thus crossing the legs, and hence the name. It is *supplied* by the anterior crural nerve.

Fig. 37.

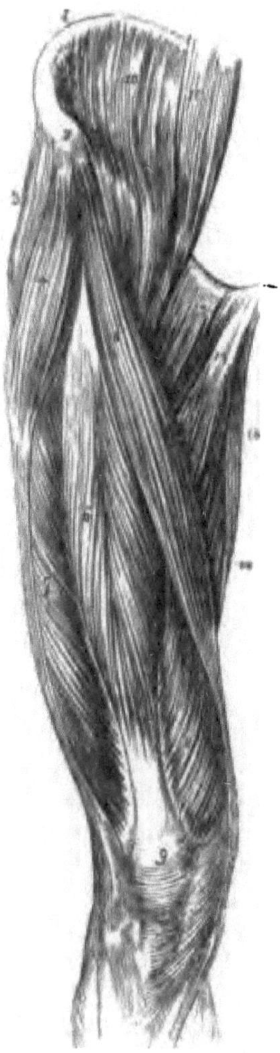

The **Quadriceps Extensor** is the fleshy mass on the front of the thigh consisting of the rectus femoris superficially, with the crureus beneath it and the vastus externus and internus on either side.

The **Rectus Femoris** (Fig. 37, 6) has no attachment to the femur, but stretches over it from the pelvis to the patella. It *arises* by a double tendinous origin, from the anterior inferior spine of the ilium and from the rough depression above the acetabulum. The two heads unite to form a fusiform bipenniform muscle, which is tendinous on its under surface at the lower part, and passes

Fig. 37.—Muscles of the anterior femoral region (from Wilson).

1. Crest of the ilium.
2. Its anterior superior spinous process.
3. Gluteus medius.
4. Tensor vaginæ femoris; its insertion into the fascia lata is shown inferiorly.
5. Sartorius.
6. Rectus femoris.
7. Vastus externus.
8. Vastus internus.
9. Patella.
10. Iliacus internus.
11. Psoas magnus.
12. Pectineus.
13. Adductor longus.
14. Part of the adductor magnus.
15. Gracilis.

to be inserted into the upper part of the patella in conjunction with the other extensor muscles.

Beneath the rectus will be seen the *External circumflex* branch of the profunda artery. This arises from the outer side of the vessel near its origin, and runs transversely between the branches of the anterior crural nerve, upon the vastus externus, but beneath the sartorius and rectus femoris, to the outer side of the thigh, where it divides into ascending, transverse, and descending sets of branches. The *ascending* branches pass beneath the tensor fasciæ to anastomose with the gluteal artery; the *transverse* set supply the vastus externus, and, turning round the limb, anastomose with the sciatic artery; whilst the *descending* branch passes between the vastus externus and the cureus to the knee joint, where it anastomoses with the superior external articular branch of the popliteal artery.

The **Vastus Externus** (Fig. 37, 7) is the largest of three muscles in front of the femur. It arises from the root of the great trochanter and the line leading to the linea aspera; from the outer side of the linea aspera and adjacent part of the shaft of the femur; from two-thirds of the outer division of the same line; and from the external intermuscular septum. Its fibres end in a broad aponeurosis which is inserted partly into the outer side of the patella, blending with the rectus, and partly into the head of the tibia, forming a portion of the anterior ligament of the knee joint and being blended with the fascia.

The **Vastus Internus** (Fig. 37, 8) is more or less united with the crureus. It *arises* from the inner side of the linea aspera and from the adjacent bone as high as the insertion of the iliacus below the lesser trochanter; from the upper half of the inner division of the linea aspera; and from the internal intermuscular septum. Its tendon resembles that of the vastus externus, and is *inserted* into the inner side of the patella and the head of the tibia, forming part of the anterior ligament of the knee, and being blended with the adjacent tendons and fascia. The origin of the lower part of the muscle cannot be seen quite distinctly at present, since its inner border is united with the tendon of the adductor longus to form a tendinous covering to the femoral vessels (Hunter's canal).

The **Crureus** covers the front of the femur between the two vasti muscles, with which it is partially united.

It *arises* from the anterior surface of the femur, from the anterior intertrochanteric line to about four inches from the lower end of the bone, and is *inserted* into the patella with the other extensor muscles. Its surface is tendinous, to allow free movement of the rectus upon it.

The small subcrureus muscle will be best dissected with the knee joint.

The quadriceps extensor is the great extensor of the leg upon the thigh, and is therefore in constant requisition in standing and walking. The rectus in addition flexes the thigh upon the pelvis when the leg is fully extended; also when the leg is fixed it steadies the pelvis upon the femur or draws it forward. The quadriceps extensor is *supplied* by the anterior crural nerve.

Surgery.—These powerful muscles occasionally fracture the patella transversely by their violent and sudden action, the bone being snapped across the condyles of the femur and the upper fragment afterwards drawn up by the contraction of the muscles. In order to relax the muscles and bring the fragment down, it is necessary to 'extend the leg upon the thigh and flex the thigh upon the pelvis.'

The **Femoral Artery in Hunter's Canal** (Fig. 39).—Hunter's canal is formed by a tendinous expansion derived from the adductors longus and magnus and the vastus externus, which encloses the femoral vessels and long saphenous nerve in the middle third of the thigh. The canal begins at the crossing of the sartorius muscle and ends at the opening in the adductor magnus, where the artery enters the popliteal space, being covered by the sartorius in its whole length. On laying open the canal, the femoral artery will be found to have its vein behind and to the outer side until it reaches the opening in the adductor magnus, when the nerve crosses the artery to pierce the tendon and become cutaneous on the inner side of the knee.

The *Opening in the adductor magnus*, as it is commonly called, is in reality an opening between the round tendon of the adductor magnus on the inner side, and the fibres of the vastus internus and the femur on the outer side, the tendinous fibres of the adductor forming the remainder of its circumference. The only structures passing through it are the femoral artery and vein.

The *Anastomotica magna* is the only vessel arising from the femoral artery in Hunter's canal, but its branches not unfrequently come off at two or more separate places. The anastomotica has a superficial and a deep branch: the superficial accompanying the saphenous nerve to the skin, the deep branch entering the vastus internus, through which it descends to the knee to anastomose with the articular branches of the popliteal artery.

Surgery (Fig. 36).—Deligation of the femoral artery in Hunter's canal is not frequently performed in the present day, but the vessel is readily reached by an incision to the inner side of the vastus internus in the middle third of the thigh, by which the outer edge of the sartorius muscle is met with. The sartorius is turned inwards, when the tendinous canal is brought into view, and by laying it open the artery will be exposed. It is more convenient to pass the aneurism needle from without inwards than the reverse, the vein lying to the outer side.

The **Anterior Crural Nerve** (Fig. 39, 5) supplies branches to all the muscles of the front of the thigh except the tensor fasciæ femoris, and also gives a branch to the pectineus on the inner side of the limb. The branch to the sartorius has been already seen to arise from the internal cutaneous nerve, and the remaining branches can now be traced from the deep division. The nerve to the vastus internus, which arises close to the long saphenous nerve, is sometimes (but badly) called the "short saphenous nerve." The nerve to the pectineus crosses transversely beneath the femoral vessels to the surface of the pectineus muscle, and gives a branch to the front of the hip joint.

INNER SIDE OF THE THIGH.

[The superficial muscle of the inner side of the thigh (adductor longus) is already partially seen; to expose the rest of the region, an incision is to be made along the rami of the pubes and ischium, and the skin is to be reflected, when the gracilis will be brought into view.]

A few cutaneous branches to the skin of the inner side of the thigh will be found appearing below the adductor longus. They are derived from the obturator nerve.

The **Adductor Longus** (Fig. 37, 13) *arises* by a round tendon from the front of the os pubis immediately below the crest, and expands to be *inserted* into the middle third of the inner border of the linea aspera. It is *supplied* by the obturator nerve.

The **Gracilis** (Fig. 39, 26) is the most internal of the muscles of the thigh, and *arises* by a broad thin tendon from the rami of the pubes and ischium close to the symphysis. It is a long riband-shaped muscle, and is *inserted* into the inner surface of the tibia beneath the sartorius, but higher than the semi-tendinosus. It is *supplied* by the obturator nerve.

[The adductor longus is to be divided near its origin and turned down, in order to expose the adductor brevis and the insertions of the pectineus and the psoas and iliacus. The branch of the obturator nerve to the muscle is to be preserved, and the profunda vessels are to be cleaned as far as exposed, with the deep muscles.]

The **Pectineus** (Fig. 37, 12) has been already seen in relation with the femoral artery and forming part of the floor of Scarpa's triangle. It *arises* from the upper part of the triangular surface immediately in front of the ilio-pectineal line, and slightly from the line itself, and winds to the back of the femur to be *inserted* into the line leading from the trochanter minor to the linea aspera. It is *supplied* by the anterior crural nerve.

[The pectineus should be divided and turned aside, in order to bring the deeper parts into view.]

Insertion of the Psoas and Iliacus.—The muscles form a single tendon, along the inner border of which, however, muscular fibres are continued for its whole length. The *insertion* of the tendon is into the back of the lesser trochanter, and the muscular fibres are prolonged for a short distance below that point.

The Adductor Brevis (Fig. 39, 18) has the superficial division of the obturator nerve lying upon it, and the deep division of the same nerve appearing between it and the adductor magnus. It *arises* from the front of the pubes below the adductor longus and between the gracilis and obturator externus, and is *inserted* into the upper part of the linea aspera and the line from it to the lesser trochanter, behind the pectineus and adductor longus and overlapping both muscles. It is *supplied* by the obturator nerve.

The action of the adductors is implied by their name, but they act at the same time as external rotators of the thigh. The power of adduction can be carried beyond the middle line so as to cross the thighs over one another.

Fig. 38.

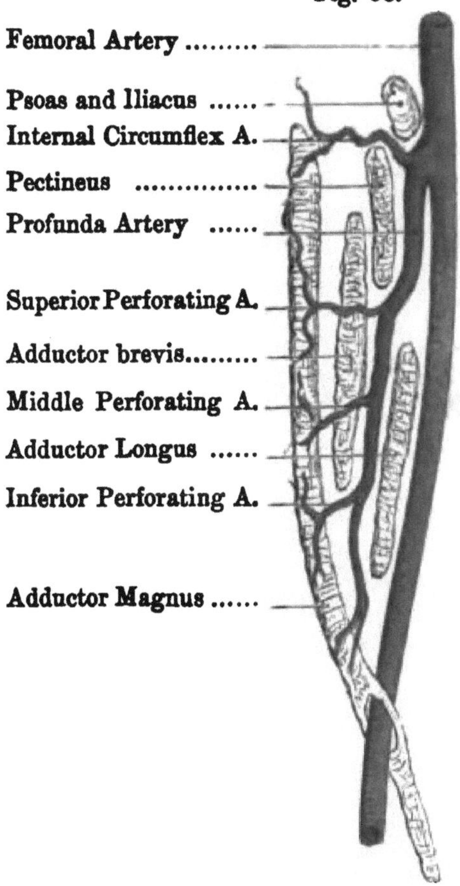

Fig. 38.—Diagram of the profunda artery of the thigh (drawn by J. T. Gray).

Fig. 39.

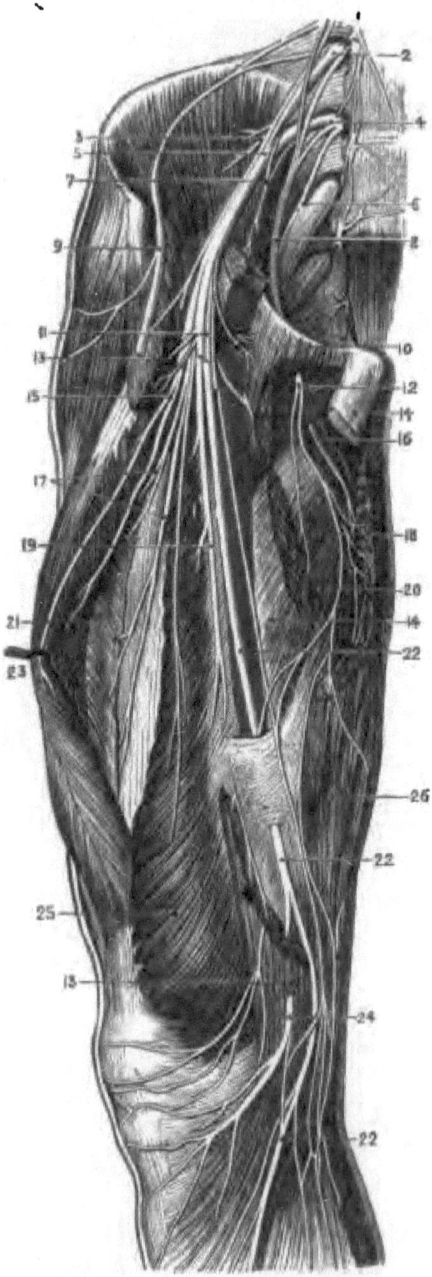

The gracilis is in addition a flexor of the leg. All the muscles of the inside of the thigh are supplied by the obturator nerve except the pectineus, which receives its nerve from the anterior crural.

The **Profunda Artery** (Fig. 38) has been seen to arise from the femoral artery in Scarpa's triangle about two inches below Poupart's ligament. It lies at first to the outer side of the femoral artery and against the psoas, but then winds on to the pectineus behind the femoral artery, which is separated from it by both the profunda and femoral veins. The profunda next winds between the borders of the pectineus and adductor longus, and lies against the adductor brevis; then passing behind the adductor longus, it runs on to the adductor magnus, where it ends. The

Fig. 39.—Nerves of the thigh (from Hirschfeld and Leveillé).
1. Gangliated cord of sympathetic. 2. Third lumbar nerve.

profunda vein is superficial to its artery in the whole of its course.

The **Branches of the Profunda** are the external and internal circumflex, three perforating arteries, and a terminal branch, besides branches to the muscles on the inside of the thigh.

1. The *External circumflex artery* arises from the outer side of the profunda near its origin, and has been already seen in the dissection of the thigh (p. 105).

2. The *Internal circumflex artery* (Fig. 38) arises from the inner side of the profunda near its origin, and passing beneath the pectineus, reaches the obturator externus muscle above the adductor brevis, when it divides into two branches. One passes superficially to the adductor brevis, to supply the muscles on the inner side of the thigh; the other passes beneath the adductor brevis, and after giving a branch to the hip-joint, ends in a branch which will be found between the quadratus femoris and adductor magnus in the dissection of the buttock.

3. The *Perforating arteries* (Fig. 38) which are three in number, pierce the adductor muscles to reach the hamstrings, which they supply on their deep surfaces. The *upper* and *middle* arteries perforate the adductors brevis and magnus; the *lowest* arises below the level of the adductor brevis, and therefore perforates only the adductor magnus. The nutritious artery to the femur generally comes from the middle perforating artery, and

---

3. Branches to iliacus internus.
4. Fourth lumbar nerve.
5. Anterior crural nerve.
6. Lumbo-sacral nerve.
7. Branch to psoas.
8. Obturator nerve.
9. External cutaneous nerve (cut).
10. Nerve to pectineus.
11. Superficial division of anterior crural nerve (cut).
12. Superficial division of obturator nerve.
13, 13. Sartorius muscle.
14, 14. Adductor longus.
15. Branch to rectus.
16. Deep division of obturator nerve.
17. Branches to vastus externus and crureus.
18. Adductor brevis.
19. Branch to vastus internus.
20. Adductor magnus.
21. Vastus externus.
22, 22. Internal saphenous nerve.
23. Rectus femoris.
24. Patellar branch of saphenous nerve.
25. Vastus internus.
26. Gracilis.

when entering the foramen in the linea aspera has a direction from the knee joint.

4. The *Terminal branch* (Fig. 38) ends in the fibres of the adductor magnus.

[The adductor brevis is to be divided in order to expose the deep portion of the obturator nerve, the adductor magnus, and the obturator externus, which are to be cleaned.]

The **Obturator Nerve** (Fig. 39, 8) is the last branch of the lumbar plexus, being derived from the third and fourth lumbar nerves. It leaves the pelvis at the upper part of the obturator foramen with the obturator artery, and immediately divides into superficial and deep portions. The superficial division lies upon the adductor brevis and supplies the adductor longus, the adductor brevis, and the gracilis, giving off a few cutaneous branches and an articular branch to the hip-joint. The deep division lies on the adductor magnus, which it supplies together with the obturator externus, and gives an articular branch to the back of the knee.

An *Accessory obturator nerve* is occasionally found passing from the lumbar plexus over the pubes beneath the pectineus. When this exists it more or less replaces the superficial division of the obturator nerve.

The **Adductor Magnus** (Fig. 39, 20) *arises* from the lower part of the descending ramus of the pubes; from the ascending ramus of the ischium; and from the outer side of the tuberosity. Its upper fibres pass with different degrees of obliquity to be *inserted* into the back of the great trochanter of the femur, immediately below the quadratus femoris, into the whole length of the linea aspera and a small portion of its inner division. The lower fibres form an almost distinct muscle, and end in a round tendon which is *inserted* into the tubercle above the inner condyle of the femur, and, by means of an expansion, into the lower part of the line leading to it. It is this tendon which bounds the "opening in the adductor magnus." It is *supplied* by the obturator nerve.

Above the upper border of the adductor magnus a portion of the **Obturator Externus** can be seen. It *arises* from the margin of the anterior half of the obturator foramen, and from the corresponding portion of the outer surface of the obturator membrane. The fibres are directed backwards, and end in a tendon, which passes in a groove below the acetabulum to be *inserted* into the digital fossa of the femur, as will be seen in the dissection of the buttock.

The obturator externus is an external rotator of the thigh and is *supplied* by the obturator nerve.

The obturator artery is best seen after the removal of the thigh from the pelvis, in the dissection of which it is included.

THE BUTTOCK.

[The body being turned on its face, the buttocks are to be raised by blocks beneath the pelvis until the thigh can be well flexed with the knee on the table, or, if preferred, the legs may be allowed to hang over the end of the table.]

The dissector should recognise the crest of the ilium, which will bound his dissection superiorly, also the sacrum and coccyx, and the tuberosity of the ischium, which can be felt through the gluteus maximus. The relation of the trochanter to the several parts of the pelvis in the different positions of the limb is to be noted also, as being of great practical importance in the diagnosis of fractures and dislocations. The "fold of the buttock" formed by the lower border of the gluteus maximus will be seen when the muscle is stretched by rotating the thigh inwards, but will vary according to the muscular development of the subject. The prominences of the inner and outer hamstrings and the hollow of the ham will be readily seen, and it should be noted how much deeper the space appears when the knee is flexed, and how readily under these circumstances, the popliteal vessels can be felt, owing to the relaxation of the fascia.

[An incision is to be made along the crest of the ilium

and down the middle of the sacrum to the tip of the coccyx, when, if the perinæum and thigh have been dissected, it will only be necessary to make another cut across the upper part of the back of the thigh to allow the large flap of skin to be reflected outwards.]

The amount of fat beneath the skin of the buttock varies considerably, and if there is a large quantity it will be a waste of time to look for cutaneous nerves.

Fig. 40.

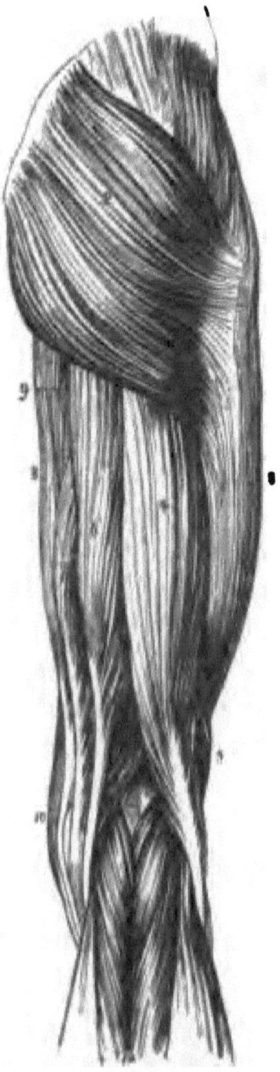

The **Cutaneous Nerves** are descending and ascending.

The *descending* branches will be found passing over the middle of the crest of the ilium, and are the iliac branches of the last dorsal and first lumbar (ilio-hypogastric) nerves, the former being in front of the latter. There are a few branches also derived from the posterior divisions of the lower lumbar and upper sacral nerves, which pass obliquely from near the middle line.

The *ascending* branches turn round the lower border of the gluteus maximus and are branches of the lesser sciatic nerve.

[The portion of deep fascia covering the gluteus medius and seen in front of the gluteus maximus, is to be cleaned, and then the gluteus maximus itself, the fibres being put on the stretch by flexing and inverting the thigh, and the dissector beginning at the upper border of the muscle on the right, and its lower border on the left limb.]

Fig. 40.—Muscles of the posterior femoral and gluteal region (from Wilson).

1. Gluteus medius. 2. Gluteus maximus.

The **Gluteus Maximus** (Fig. 40, 2) *arises* from the rough triangle between the posterior third of the crest of the ilium and the superior curved line on the dorsum ilii;* from the side of the lower part of the sacrum, and the side of the coccyx; from the tendinous expansion over the back of the sacrum; and from the surface of the great sacro-sciatic ligament, which will be seen when the muscle is divided. The fibres run downwards and outwards, and the upper two-thirds of the muscle are *inserted* into the fascia lata covering the great trochanter, the lower third being attached to the rough ridge leading from the back of the trochanter major to the linea aspera of the femur.

The gluteus maximus is one of the most important muscles for maintaining the erect position of the body by fixing the pelvis upon the thigh. When the pelvis is the fixed point, the muscle is a powerful extensor of the thigh, and at the same time rotates it outwards. It is *supplied* by the inferior gluteal branches of the small sciatic nerve.

[The gluteus is to be divided near its origin and turned down, when the extent of its attachment to the fascia will be better seen. Some branches of the superficial gluteal artery at the upper, and of the sciatic at the lower part of the muscle must necessarily be divided, but the nerves going to its under surface from the small sciatic should be preserved if possible.]

Three *bursæ* will be found in connection with the gluteus maximus; one between its fascia and the great trochanter; a second between the fascia and the vastus externus; and the third over the tuberosity of the ischium, which is remarkable for its change of relation to the muscle; since when the body is in the erect posture

* In describing the glutei muscles the dorsum ilii is considered to have *three* curved lines, and not *two* as given by some authors.

---

3. Vastus externus covered in by fascia lata.
4. Long head of biceps.
5. Short head.
6. Semi-tendinosus.
7. Semi-membranosus.
8. Gracilis.
9. Part of the inner border of the adductor magnus.
10. Edge of sartorius.
11. Popliteal space.
12. Gastrocnemius; its two heads.

the muscle covers the tuber and its bursa, but when in the sitting posture, the muscle slips back and the bursa becomes subcutaneous.

The **Parts beneath the Gluteus Maximus** (Fig. 42) can now be seen although covered by loose cellular tissue which is to be afterwards removed. They are from above downwards:—1, gluteus medius; 2, a small piece of the lower border of the gluteus minimus; 3, gluteal vessels and nerve (not to be seen distinctly at present); 4, pyriformis muscle; 5, sciatic vessels and nerves (great and small); 6, pudic vessels and nerve, and nerve to obturator internus; 7, tendon of the obturator internus with the gemelli muscles above and below it; 8, tendon of the obturator externus; 9, quadratus femoris; 10, great sacrosciatic ligament; 11, tuber ischii and muscles attached to it.

[These structures are to be cleaned as far as possible at once, and will be examined in succession. In order to clean the muscles, the thigh must be well rotated inwards, and in the case of the gluteus medius, it must be crossed beneath the opposite limb.]

The **Gluteus Medius** (Fig. 42, 2) *arises* from the dorsum ilii between the superior and middle curved lines, and from the fascia covering its anterior part. The fibres converge to be *inserted* into the outer surface of the great trochanter, extending obliquely from its posterior superior angle downwards and forwards. The anterior border of the gluteus medius is united with the minimus, which lies beneath it. The insertion of the gluteus medius is best seen when the muscle has been divided.

[The gluteus medius is to be divided close to its origin without injuring the gluteus minimus, the separation from which is marked by an artery; and the anterior borders of the muscles having been separated with the scalpel, the gluteus medius is to be turned down without injuring the vessels and nerves beneath. A bursa will be found on the great trochanter beneath the tendon.]

The **Gluteus Minimus** (Fig. 41, 8) *arises* from the dorsum ilii between the middle and inferior curved lines, being united by its anterior border with the gluteus medius. It is *inserted* into the anterior surface of the great trochanter.

Fig. 41.

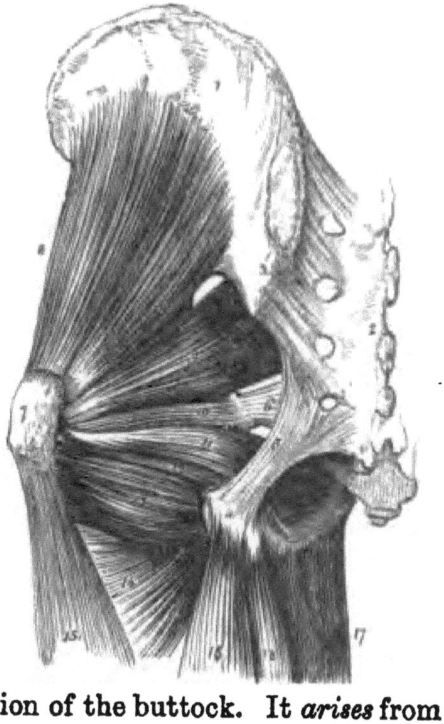

The two smaller glutei muscles assist in maintaining the erect posture by fixing the pelvis on the femur, or if the pelvis be fixed they act as abductors and rotators of the thigh, the gluteus minimus and the anterior portion of the medius rotating inwards, and the posterior portion of the medius rotating outwards. They are both *supplied* by the superior gluteal nerve.

The **Pyriformis** (Fig. 41, 9) is only partly seen in the dissection of the buttock. It *arises* from the front of the lateral mass of the sacrum between the 1st, 2nd, 3rd, and 4th sacral foramina, and passing out of the great sacro-sciatic foramen ends in a round tendon, which is *inserted* into a distinct depression at the posterior part of

---

Fig. 40. Deep muscles of the gluteal region (from Wilson).

1. Ilium.
2. Sacrum.
3. Posterior sacro-iliac ligaments.
4. Tuberosity of the ischium.
5. Great sacro-sciatic ligament.
6. Lesser sacro-sciatic ligament.
7. Trochanter major.
8. Gluteus minimus.
9. Pyriformis.
10. Gemellus superior.
11. Obturator internus, passing out of the lesser sacro-sciatic foramen.
12. Gemellus inferior.
13. Quadratus femoris.
14. Adductor magnus.
16. Biceps.
15. Vastus externus.
17. Gracilis.
18. Semi-tendinosus.

The tendon of the obturator externus should appear between the gemellus inferior and the quadratus femoris.

the upper margin of the great trochanter. The muscle is occasionally divided into two parts by the passage of a portion of the great sciatic nerve through its fibres. The pyriformis is an external rotator of the thigh and is *supplied* by branches of the sacral nerves within the pelvis.

In order to see the remaining muscles satisfactorily the sciatic nerves should be drawn aside over the tuberosity of the ischium.

The **Obturator Internus and Gemelli** (Fig. 41, 11).—Only the tendon of the obturator internus is now seen, the muscle *arising* from the inner surface of the pelvis in front of and behind the obturator foramen, and also from the inner surface of the obturator membrane. The tendon winds out of the lesser sacro-sciatic foramen, and being joined by the gemelli, passes forwards beneath the tendon of the pyriformis (with which it is more or less united) to be *inserted* into a depression at the anterior part of the upper margin of the great trochanter. The gemelli (twin muscles) are placed above and below the tendon of the obturator internus and are *inserted* into it, but one or both may be absent; the superior *arises* from the spine of the ischium and adjacent portion of the lesser sciatic notch; the inferior *arises* from the upper and back part of the tuber ischii.

If the tendon of the obturator internus be divided near its insertion without disturbing the gemelli, and drawn up from the margin of the sciatic foramen, a large bursa will be found, which, when opened, will be seen to lubricate the under surface of the tendon and the margin of the lesser sacro-sciatic foramen. The under surface of the obturator tendon is subdivided into from three to five small tendons, and the margin of the bone is encrusted with cartilage and grooved to correspond with these.

The obturator internus and gemelli are rotators outwards of the thigh. The obturator is *supplied* by a special branch from the upper part of the sacral plexus which also gives a branch to the superior gemellus; the inferior gemellus is *supplied* by the nerve to the quadratus from the sacral plexus.

The **Tendon of the Obturator Externus** (Fig. 42, 12) will be found lying deeply between the inferior gemellus and the quadratus muscle. The origin of the muscle has been already seen (p. 113) and the tendon, after passing through the groove below the acetabulum, is *inserted* into the digital fossa at the root of the great trochanter. The obturator externus is a rotator outwards of the thigh, and is *supplied* by the obturator nerve.

The **Quadratus Femoris** (Fig. 41, 13), so named from its quadrilateral shape, *arises* from the outer side of the tuberosity of the ischium external to the hamstring muscles, and passes horizontally to be *inserted* on the back of the great trochanter immediately above the adductor magnus, with which it is sometimes continuous. The point of insertion has been termed the 'linea quadrati' (though a distinct line seldom if ever exists at the spot), which must not be confounded with the well-marked posterior inter-trochanteric line.

The quadratus femoris is an external rotator of the thigh and is *supplied* by a special branch from the sacral plexus, which also gives a twig to the gemellus inferior.

The **Gluteal Artery** (Fig. 42 3) is a branch of the posterior division of the internal iliac artery, and emerges from the pelvis through the great sacro-sciatic foramen above the pyriformis. It appears at the lower border of the gluteus minimus, and divides into a superficial and a deep portion. The *superficial* division is distributed to the under surface of the gluteus maximus; the *deep* subdivides into superior and inferior branches.

The *superior deep gluteal artery* runs along the middle curved line of the ilium between the gluteus medius and minimus, both of which it supplies, and ends at the anterior border of the ilium by anastomosing with the superior branch of the external circumflex artery of the thigh. The *inferior deep gluteal artery* crosses the gluteus minimus to the great trochanter to supply the parts in its neighbourhood.

Venæ comites accompany the branches of the artery and open into the internal iliac vein.

The **Sciatic Artery** is a branch of the anterior division of the internal iliac artery, and emerges from the pelvis through the great sacro-sciatic foramen below the pyriformis. It gives large *inferior gluteal* branches to the gluteus maximus, and *muscular* branches to the other smaller muscles of the neighbourhood; and anastomoses with both the external and internal circumflex arteries of the profunda femoris. Its named branches are (1) the *coccygeal* which pierces the great sacro-sciatic ligament to supply the parts about the coccyx; (2) the *comes nervi ischiadici* which runs upon the great sciatic nerve for a variable distance; and (3) the *branch to the quadratus* (Ellis) which passes beneath the obturator internus and gemelli to the quadratus and supplies the hip-joint.

The veins corresponding to the sciatic artery open into the internal iliac vein.

The **Pudic Artery** is only seen in its course over the spine of the ischium, as it winds out of the great sacro-sciatic foramen below the pyriformis to re-enter the pelvis by the lesser sacro-sciatic foramen above the tendon of the obturator internus. The artery is accompanied by its vein and by two nerves, the pudic nerve lying to its inner side and the nerve to the obturator internus to its outer side.

A branch of artery may be found both at the upper and

Fig. 42.—Dissection of the buttock and back of the thigh (from Hirschfeld and Leveillé).

1. Gluteus maximus.
2. Gluteus medius.
3. Gluteal artery and nerve.
4. Gluteus minimus.
5. Nerve to obturator internus.
6. Pyriformis.
7. Pudic nerve.
8. Small sciatic nerve.
9. Great sacro-sciatic ligament.
10. Obturator internus and gemelli.
11. Inferior gluteal nerve from small sciatic.
12. Tendon of obturator externus.
13. Inferior pudendal nerve (Soemmering).
14. Quadratus femoris.
15. Gracilis.
16. Great sciatic nerve.
17. Adductor magnus.
18. Insertion of gluteus maximus.
19. United origins of semi-tendinosus and biceps.
20. Short head of biceps.
21. Semi-membranosus.
22. Tendon of biceps.
23. Tendon of semi-tendinosus.
24. External popliteal nerve.
25. Internal popliteal nerve.
26. Communicans fibularis nerve.
27. Popliteal artery.
29. Gastrocnemius.
31. Communicans tibialis nerve.

lower borders of the quadratus. The upper is a branch of either the internal circumflex or obturator arteries accompanying the tendon of the obturator externus, and the lower, which is more constant, is one of the terminal branches of the internal circumflex artery.

Fig. 42.

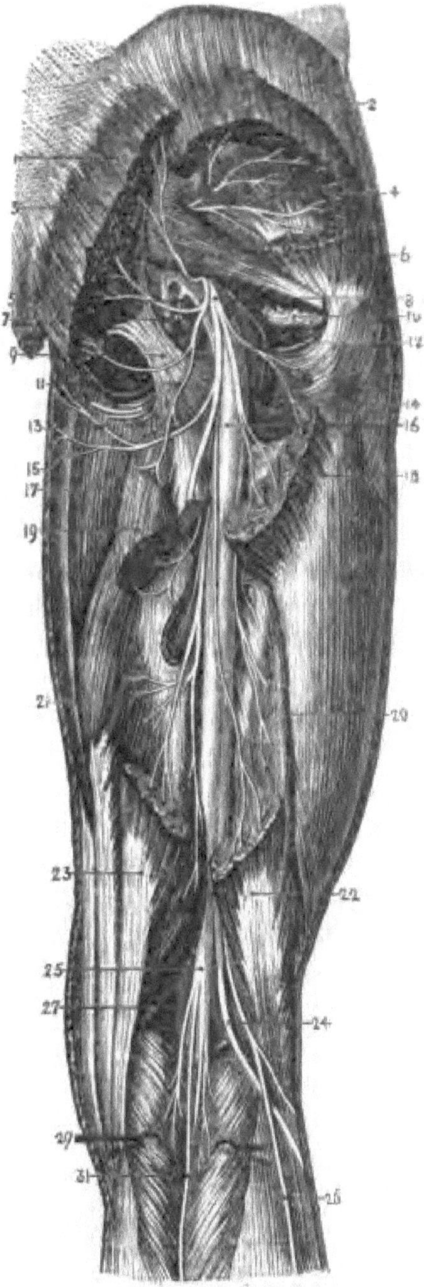

The **Gluteal Nerve** (Fig. 42, 3) (superior) is a branch of the lumbo-sacral cord, and emerges from the pelvis with the gluteal artery above the pyriformis. It divides into two branches, which lie between the gluteus medius and minimus and supply them, giving a branch forwards to the tensor fasciæ femoris.

The **Small Sciatic Nerve** (Fig. 42, 8) is one of the main branches of the sacral-plexus, and appears at the lower border of the pyriformis to become the cutaneous nerve of the back of the thigh. As soon as it emerges from the pelvis it gives cutaneous branches to the skin over the gluteus maximus, and *inferior gluteal* branches to the under surface of the muscle itself. It then gives cutaneous branches to the thigh, one of the largest of which winds below the tuberosity of the ischium to the perinæum and is called the *inferior pudendal* or nerve of Soemmering; and the nerve itself may be traced as far as the popliteal space.

The **Great Sciatic Nerve** (Fig. 42, 16) is the largest trunk of the sacral plexus, and appears at the lower border of the pyriformis (or occasionally through its fibres). It runs midway between the tuberosity of the ischium and the great trochanter, and disappears beneath the hamstring muscles. Properly speaking this nerve gives no branches in this region, but frequently the nerves to the quadratus and hip-joint are derived from it instead of from the sacral plexus itself.

The **Nerve to the Obturator Internus** (Fig. 42, 5) arises from the upper part of the sacral plexus, and leaves the pelvis below the pyriformis. taking the same course as the pudic vessels and nerve over the spine of the ischium, but lying to their outer side. As it winds into the lesser foramen it gives a branch to the superior gemellus, and then supplies the fibres of the obturator within the pelvis.

The **Nerve to the Quadratus** (Fig. 42) arises from the lower part of the sacral plexus, and passes beneath the obturator internus tendon to the quadratus and inferior gemellus. It lies against the back of the hip-joint, to which it gives a branch.

The **Pudic Nerve** (Fig. 42, 7) is seen lying on the spine of the ischium internal to the pudic artery. If the inferior hæmorrhoidal nerve should happen to arise separately from the sacral plexus it will be found accompanying the pudic in this position.

The dissection of the popliteal space should be made before the skin on the back of the upper part of the thigh is removed, so that the relations of the parts in the ham may be undisturbed.

THE POPLITEAL SPACE.

[A vertical incision is to be made in the middle line of the limb at the back of the knee, extending for six inches above and the same distance below the joint. A transverse incision at each end of this will allow the skin to be fully reflected. A few branches of the small sciatic nerve may be found in the superficial fascia, which is to be removed in order to expose the deep fascia.]

Deep or Popliteal Fascia.—This is continuous with the fascia lata, and stretches across the popliteal space to protect the vessels and nerves beneath. When the limb is extended it is tightly stretched, but is at once relaxed on bending the knee.

The fascia being removed, the hollow of the ham will be brought into view filled with loose fat, which is to be removed to expose the pop-

Fig. 43.

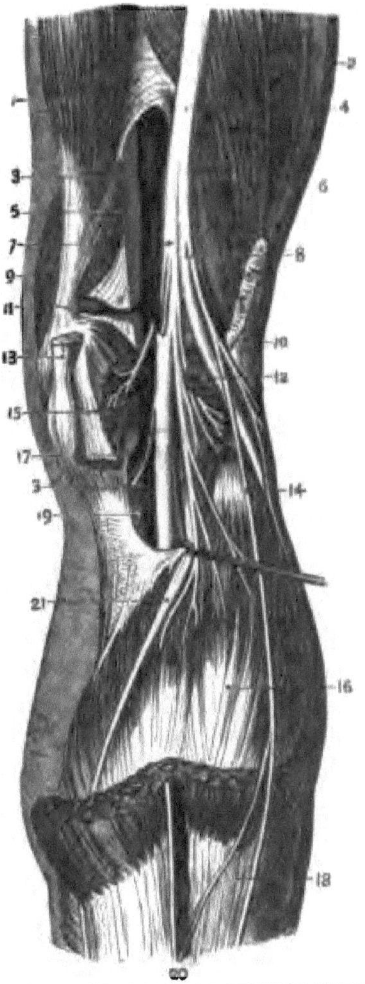

Fig. 43.—Deep dissection of the popliteal space (from Hirschfeld and Leveillé).

1. Adductor magnus.
2. Vastus externus.
3. Popliteal vein.
4. Great sciatic nerve.
5. Popliteal artery.
6. Short head of biceps.
7. Internal popliteal nerve.
8. External popliteal nerve.
9. Vastus internus.
10. Long head of biceps (cut).
11. Superior internal articular artery.
12. Outer head of gastrocnemius.
13. Tendon of semi-membranosus.
14. Communicans peronei nerve.
15. Inner head of gastrocnemius.
16. Soleus.
17. Inferior internal articular artery.
18. Gastrocnemius.
19. Popliteus.
20. External saphenous vein and nerve.
21. Tendon of plantaris.

liteal vessels and nerves. Imbedded in the fat of the popliteal space are some lymphatic glands, and it may be noted that there are no glands in the limb below this point. The boundaries of the space are to be carefully cleaned, care being taken, at the lower part, of two small nerves and a vein which run on the back of the leg.

The **Popliteal Space** (Fig. 40, 11) is a lozenge-shaped cavity at the back of the knee, which in the dissected state extends for some distance above and below the joint. Its upper part is bounded *externally* by the tendon of the biceps; *internally* by the tendons of the semi-tendinosus and semi-membranosus muscles. Its lower part has one of the fleshy heads of the gastrocnemius muscle on *each* side, and on the *outer* side there is, in addition, the small plantaris muscle.

The *floor* of the space is formed at the upper part by that portion of the back of the femur which is between the divisions of the linea aspera; next by the posterior ligament of the knee-joint; and at the lower part by the popliteus muscle and the fascia covering it. These last can be only felt between the heads of the gastrocnemius, but if the inner head of that muscle be divided near its origin and turned down, the popliteus will be sufficiently displayed.

The **Internal Popliteal Nerve** (Fig. 43, 7) is the most superficial of the important structures in the ham. It is one of the divisions of the great sciatic nerve, and passes with a very slight degree of obliquity from the outer to the inner side of the ham, extending through the whole length of the space and disappearing beneath the gastrocnemius muscle. It thus crosses from the outer to the inner side of the popliteal artery, which it accompanies to the lower border of the popliteus muscle. Its branches are (1) *articular* which both pierce the posterior surface of the knee-joint and accompany the inferior internal articular artery; (2) *muscular* to the gastrocnemius, soleus, plantaris, and popliteus; and (3) the *external saphenous nerve* or *ramus communicans tibialis*, which lies in the groove between the heads of the gastrocnemius (sometimes beneath a few of its fibres) and will be traced in the dissection of the back of the leg.

The **External Popliteal Nerve** (Fig. 43, 8), the other division of the great sciatic, is altogether under cover of the biceps at the upper part of the space, but becomes visible close to its tendon near the head of the fibula. It gives a communicating branch (*ramus communicans peronei vel fibularis*) down the back of the leg to join the external saphenous nerve.

The **Popliteal Artery** (Fig. 43, 5) is the continuation of the femoral artery, and extends from the opening in the adductor magnus to the lower border of the popliteus muscle, where it divides into anterior and posterior tibials. It lies at the bottom of the popliteal space, resting upon the back of the femur, the posterior ligament of the knee, and the popliteus muscle, and has a direction from the inner side of the femur to the middle of the leg. Its vein lies superficially in its whole course, but first to its outer side and afterwards to its inner side, whilst the internal popliteal nerve is still more superficial and also crosses the artery from without inwards. The commencement of the popliteal artery is under cover of the semi-membranosus, the middle of the vessel is in the hollow of the ham, uncovered by muscles, and its lower part is beneath the gastrocnemius and plantaris muscles. A branch of the obturator nerve may occasionally be found on the upper part of the artery.

The **Branches** (Fig. 43) of the popliteal artery are muscular (both to the hamstrings and muscles of the calf) and articular.

1. The *Superior muscular* branches arise from the upper part of the artery, and are distributed to the adjacent hamstring muscles.

2. The *Articular* branches are five in number, viz., two superior, two inferior, and one azygos.

The *superior* arteries, internal and external, arise just above the condyles of the femur, around which bone they wind to the front of the knee-joint, to supply it and to anastomose with the anastomotica magna and the descending branch of the profunda respectively. The *inferior* articular arteries arise on the popliteus, and do not correspond on the two sides of the limb. The internal (the lower of the two) winds below the inner tuberosity of the tibia and beneath the internal lateral ligament; the external passes altogether above the head of the fibula but beneath the external lateral ligament, to the front of the joint. The *azygos* arises about the middle of the artery and pierces the posterior ligament to join the other branches.

3. The *Inferior muscular* branches (sural) are distributed to the muscles of the calf.

Surgery.—The popliteal artery may be readily compressed by the hand or a tourniquet. The operation of tying the popliteal artery for aneurism is never performed, but it might be necessary to reach the vessel if wounded; in which case an incision to the inner side of the space would enable the operator to reach the middle of the artery without injury to the nerves.

The **Popliteal Vein** (Fig. 43, 3) is formed by the junction of the anterior and posterior tibial veins at the lower border of the popliteus. The vein lies superficially to the artery and to its inner side, at the lower part of the space, but crosses to the outer side above. It receives branches corresponding to those of the artery, and at a variable point, the *external saphena vein*, which lies on the back of the leg between the heads of the gastrocnemius, opens into it.

Some lymphatic glands are placed close to the artery, into which lymphatics, from the back of the leg, discharge themselves.

THE BACK OF THE THIGH.

[The skin left on the back of the thigh is to be removed, and the small sciatic nerve traced as far as convenient, after which the fascia should be removed and the hamstring muscles, with the sciatic nerve, cleaned.]

The **Hamstring Muscles** (Figs. 40 and 42) are three in number, viz. the biceps on the outer side, and the semi-tendinosus and semi-membranosus on the inner side. The biceps and semi-tendinosus, which arise in common, are superficial to the semi-membranosus near the tuber ischii, and the semi-tendinosus maintains this relation to the semi-membranosus throughout. The short head of the biceps will be found on the outer side of the lower part of the femur.

The **Biceps** (Fig. 42, 19) or outer hamstring *arises* by its long head from the anterior or lower of the two surfaces on the tuber ischii, in common with the semi-tendinosus. It crosses the great sciatic nerve, and by diverging from the semi-tendinosus in the middle of the thigh, forms the outer boundary of the popliteal space. The short head (20), which is fleshy, *arises* from the outer side of the linea aspera, as high as the insertion of the gluteus maximus, and from its outer division to within two inches of the condyle, and the two heads having united opposite the lower end of the femur, the muscle is *inserted* into the outer side of the head of the fibula. The tendon splits to enclose the long external lateral ligament, and one portion of it therefore intervenes between the long and the short external lateral ligaments of the knee-joint.

The **Semi-tendinosus** (Fig. 42, 19) *arises* by fleshy fibres in common with the tendinous origin of the biceps from the anterior of the two surfaces on the back of the tuber ischii. It forms one of the inner boundaries of the popliteal space, and ends in a long slender tendon, which lies on the semi-membranosus and passes beyond it to be *inserted* into the anterior or inner side of the tibia below the inner tuberosity, lying beneath the tendon of the sartorius and below that of the gracilis. Its fibres are crossed by a tendinous intersection.

The **Semi-membranosus** (Fig. 42, 21), though one of the internal hamstrings, is external to the biceps at its origin. It *arises* by a strong tendon from the posterior or upper surface on the tuber ischii and, passing beneath the origins of the biceps and semi-tendinosus, spreads into a broad tendon from which the muscular fibres arise to form a thick belly. It is *inserted* by a strong tend-

which passes horizontally beneath the internal lateral ligament of the knee to the groove in the side of the inner tuberosity of the tibia, and gives off two expansions, one to form the posterior ligament of the knee-joint and the other to cover the popliteus muscle.

The hamstring muscles flex the leg upon the thigh and then extend the thigh upon the pelvis, as in walking or running; their most important action, however, is when the leg is their fixed point, and they then prevent the pelvis and trunk from falling forward, and thus maintain the erect posture of the body. When, taking the pelvis as their fixed point, they act upon the leg or (the knee being fixed) upon the thigh, their power is applied to a lever of the third order; when taking their fixed point below they act upon the pelvis, their power is applied to a lever of the first order. The hamstrings are *supplied* by the great sciatic nerve.

The **Great Sciatic Nerve** (Fig. 42, 16) is continued from the buttock and, after leaving the quadratus femoris, lies on the posterior surface of the adductor magnus and passes beneath the biceps. At a variable point, but generally about the middle of the thigh, the nerve divides into the internal and external popliteal trunks which have been already traced.

Muscular branches to the hamstring muscles are given off by the nerve in this part of its course.

Beneath the sciatic nerve will be seen the posterior surface of the adductor magnus, and piercing it at various points are the terminations of the *perforating arteries* from the profunda, which are now seen to supply the hamstring muscles (*v.* p. 111). The termination of the internal circumflex artery, between the quadratus and adductor magnus can now also be more fully traced out.

[Before the subject is re-turned, the gluteus minimus, pyriformis, and obturator internus should be turned aside and the back of the hip-joint dissected, but the joint itself should not be opened. When the subject is replaced on its back, the front of the joint is also to be dissected removing the psoas and iliacus and the remains of the ctineus, after which it may be opened and the limb moved.]

Fig. 44.

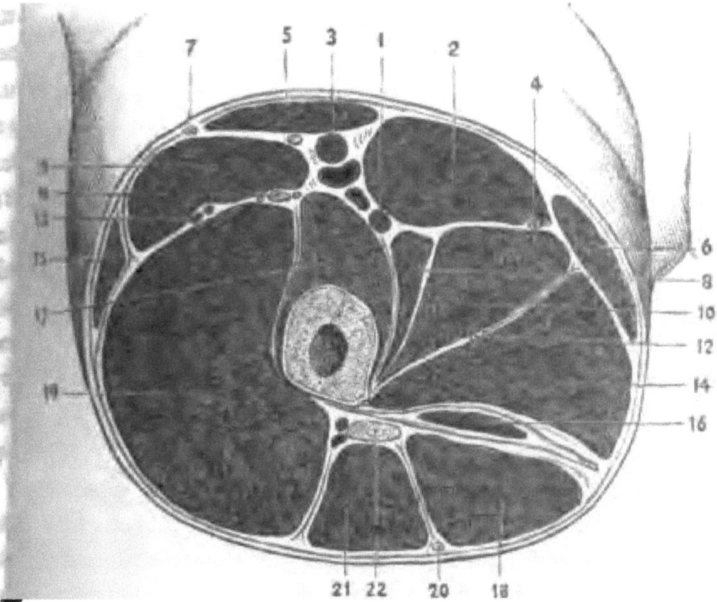

The Hip Joint.

The **Hip Joint** is the best example in the body of enarthrosis or ball-and-socket joint, and of necessity has a capsular ligament besides other internal ligaments.

The hip joint has *in front* the united psoas and iliacus muscles and the femoral vessels; *behind*, the pyriformis, obturator internus and gemelli muscles, with the sciatic nerves and vessels lying upon them; *above* are the gluteus

Fig. 44.—Section of the right thigh at the apex of Scarpa's triangle (drawn by G. E. L. Pearse).

1. Profunda vessels.
2. Adductor longus.
3. Femoral vessels.
4. Superficial obturator nerve.
5. Sartorius.
6. Gracilis.
7. External cutaneous nerve.
8. Pectineus.
9. Rectus femoris.
10. Adductor brevis.
11. Anterior crural nerve.
12. Deep obturator nerve.
13. External circumflex vessels.
14. Adductor magnus.
15. Tensor vaginæ femoris.
16. Semi-membranosus.
17. Vastus internus and crureus.
18. Semi-tendinosus.
19. Vastus externus.
20. Small sciatic nerve.
21. Biceps femoris.
22. Great sciatic nerve.

medius and gluteus minimus, and *below* the tendon of the obturator externus. The hip-joint admits of the movements of flexion, extension, abduction and adduction, rotation, and circumduction. *Flexion* is performed by the psoas and iliacus, and indirectly by the rectus femoris: *extension* by the gluteus maximus, and indirectly by the hamstrings. The femur is *adducted* by the adductor muscles and pectineus, which at the same rotate the bone outwards; it is *abducted* by the gluteus medius, minimus, and tensor vaginæ femoris. Rotation *outwards* is due to the pyriformis and obturator muscles, rotation *inwards* to the gluteus minimus and tensor vaginæ femoris.

The **Capsular Ligament** (Fig. 45, 8) when seen from behind, will be found to be attached to the margin of the acetabulum, but to reach only about halfway down the posterior surface of the neck of the femur, the point to which it reaches varying in different subjects. In front, the capsule extends from the margin of the acetabulum to the anterior inter-trochanteric line, and is strengthened by an oblique band of fibres extending from the anterior inferior spinous process to the lesser trochanter, to which the name *ilio-femoral* ligament has been given (Fig. 31). A bursa intervenes between the tendon of the psoas and the front of the hip joint, which occasionally communicates with the synovial cavity through the capsular ligament.

It should be noticed that the posterior part of the capsular ligament is put on the stretch in adduction of the limb, and being thin, is always torn in dislocation of the head of the femur backwards. The anterior part is put on the stretch in abduction of the thigh, and the ilio-femoral band serves both to check extreme abduction and to prevent the head from starting from the cotyloid cavity, hence the rarity of dislocation forwards.

The capsular ligament having been divided, the thigh will be seen to be held by the so-called *ligamentum teres*, which will be brought into view by abducting the limb. In doing this a squeaking noise will probably be heard. which is due to the sudden separation of the moist articular surfaces of the femur and pelvis, which are covered with cartilage. The inter-articular ligament (teres) being divided, the limb can be removed.

Fig. 45.

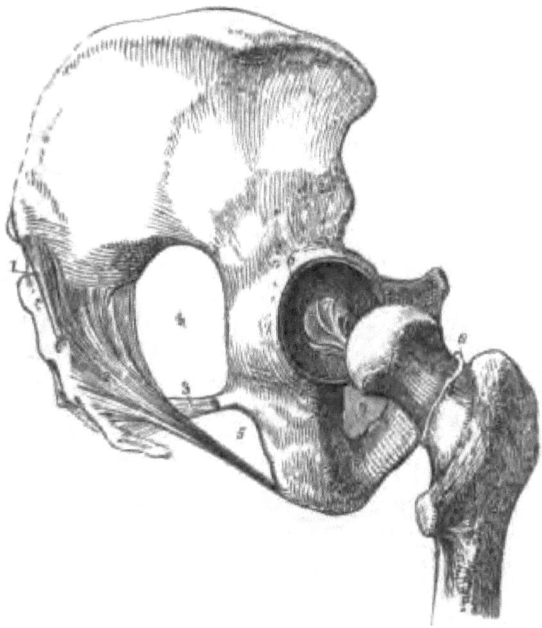

The *single synovial membrane* of the joint will be seen to be reflected over the inner surface of the capsular ligament on to the neck of the femur, and to be prolonged over the ligamentum teres to the bottom of the acetabulum, which it lines. In the bottom of the cavity it is loosely folded over some fat, and is of a darker colour than elsewhere, and has been termed incorrectly the Haversian gland.

The **Ligamentum Teres** (Fig. 45, 7) is not round, but prismatic or flattened. It is attached to the lower part of the non-articular portion of the acetabulum, and to the pit in the head of the femur. This ligament, though it would assist in preventing absolute dislocation, particularly forwards, has but little effect upon the movements of the joint, since it has been shown conclusively that it is only put upon the stretch during rotation outwards and flexion of the thigh.

The **Cotyloid Ligament** (Fig. 45, 6) is a fibro-carti-

Fig. 45.—Ligaments of the pelvis and hip-joint (from Wilson).

1. Oblique sacro-iliac ligament.
2. Great sacro-sciatic ligament.
3. Lesser sacro-sciatic ligament.
4. Great sacro-sciatic foramen.
5. Lesser sacro-sciatic foramen.
6. Cotyloid ligament.
7. Ligamentum teres.
8. Cut edge of capsular ligament.
9. Obturator membrane.

laginous ring, thinner at its free-edge than at the point of attachment, which is fixed to the brim of the cotyloid cavity and serves to deepen it, fitting closely also against the head of the femur. That part of it which is carried across the notch on the inner side of the margin of the acetabulum is called the *transverse ligament*, and beneath it the articular vessels pass to the interior of the joint.

Fig. 46.

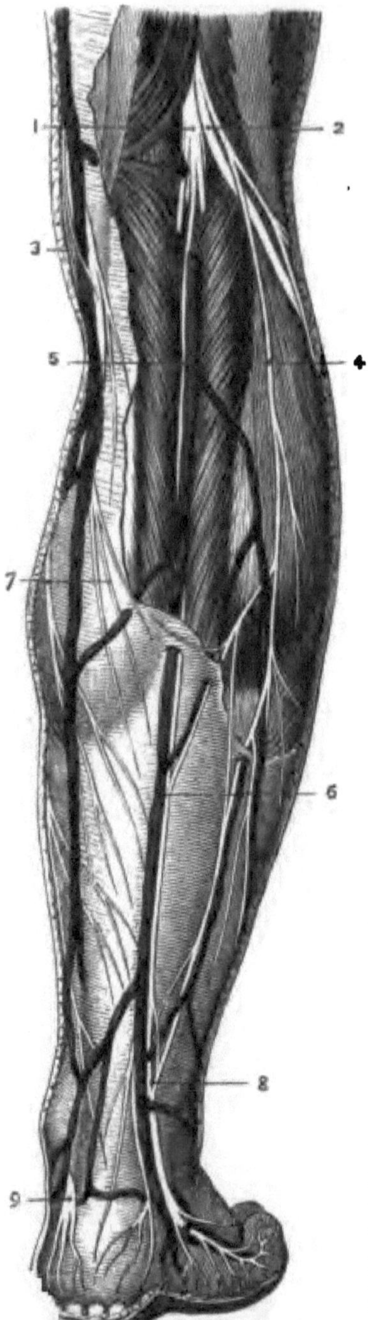

Fig. 46.—Superficial nerves of the back of the leg (from Hirschfeld and Leveillé).

1. Internal popliteal nerve.
2. External popliteal nerve.
3. Internal saphenous vein.
4. Nervus communicans fibularis.
5. Nervus communicans tibialis.
6. External saphena vein.
7. Internal saphenous nerve.
8. External saphenous nerve.
9. Posterior tibial nerve.

THE BACK OF THE LEG.

[A vertical incision through the centre of the remaining portion of skin on the back of the leg, joined by a short transverse cut at the heel, will permit the reflection of the skin to each side.]

The **External Saphenous Nerve** (Fig. 46, 8) (communicans tibialis) will be found piercing the deep fascia of the leg about halfway down the limb, to join the *communicans fibularis* (4), which becomes cutaneous about the same level. The nerve thus formed runs along the outer side of the tendo Achillis, to pass behind the external malleolus to the outer side of the foot.

A branch or two of the *internal saphenous nerve* (7) may be found on the inner side of the back of the leg.

The **External Saphena Vein** (Fig. 46, 6) appears behind the external malleolus close to the saphenous nerve, and running up the centre of the leg, pierces the deep fascia at the point where the communicans tibialis nerve emerges; it has been already traced to the popliteal vein. The saphena vein not unfrequently runs superficially to the upper part of the popliteal space before it joins the popliteal vein.

[The deep fascia of the leg, which is continuous with that of the thigh, is to be removed, except near the internal malleolus, where the internal annular ligament is to be preserved. The muscles of the calf are then to be dissected, being put on the stretch by supporting the knee and drawing the toes towards the front of the leg.]

The **Gastrocnemius** (Fig. 47, 4), the superficial muscle of the calf, has a double-headed *origin* from the back of the femur immediately above the condyles. A bursa sometimes communicating with the knee-joint lies between the inner head and the tendon of the semimembranosus. The two heads

Fig. 47.

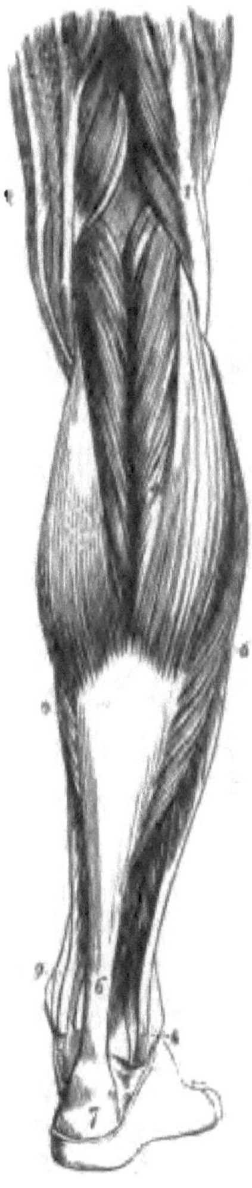

Fig. 47.—Superficial muscles of the back of the leg (from Wilson).

1. Tendon of biceps.
2. Tendons of inner hamstrings.
3. Popliteal space.
4. Gastrocnemius.
5. 5. Soleus.
6. Tendo Achillis.
7. Tuberosity of os calcis.
8. Tendons of the peroneus longus and brevis.
9. Tendons of the tibialis posticus and flexor longus digitorum.

are the lower boundaries of the popliteal space, and unite to form one large muscle, the fibres of which end about the middle of the leg in a broad tendon, which contracts near the heel into the thick rounded *tendo Achillis* (6). This, which is the common insertion of the gastrocnemius and the subjacent soleus, expands again slightly, to be *inserted* into the lower part of the posterior surface of the calcaneum, a bursa intervening between it and the upper part of the same bone.

[The remaining head of the gastrocnemius being carefully divided near its origin, and the muscle turned down as far as possible, the plantaris and soleus with the popliteus covered by its fascia will be brought into view.]

The **Plantaris** (Fig. 43, 21) *arises* from above the external condyle of the femur and from the line leading to it, and the muscle lies internal to the outer head of the gastrocnemius. Its fleshy fibres are about three inches long, but its tendon is the longest in the body and is very slender. It is placed between the gastrocnemius and soleus, and crosses obliquely so as to get to the inner side of the tendo Achillis, by the side of or in common with which it is *inserted* into the os calcis. It is analogous to the palmaris longus of the upper extremity.

The **Soleus** (Fig. 47, 5) is tendinous on its surface, to allow the corresponding tendinous expansion on the under surface of the gastrocnemius to move freely upon it, and between the two is a quantity of loose areolar tissue. It *arises* from the oblique line of the tibia and from the inner border of the tibia in its middle third; also from the posterior surface of the head and upper third of the fibula, and from a deep tendon which arches over the tibial vessels and nerve. The fibres end in a broad tendon, which joins that of the gastrocnemius in the lower third of the leg to form the tendo Achillis, the *insertion* of which has been seen. The muscular fibres are inserted into the deep surface of the tendon to within two inches of the heel.

By dividing the fibres of the soleus vertically and with care, a tendinous expansion on the deep surface of the muscle will be brought into view, which stretches across the posterior tibial vessels and nerve, and protects them from pressure during the movements of the limb. This deep tendon is an important guide in the operation of tying the posterior tibial artery, and must be divided to reach it.

The muscles of the calf extend the foot, *i.e.*, point the toes, the gastrocnemius in addition flexing the leg upon the thigh; or, when taking their fixed point below, they fix the leg upon the foot, the gastrocnemius also steadying the thigh. They are *supplied* by the internal popliteal nerve.

The form of club-foot known as 'talipes equinus' is due in part to contraction of the muscles of the calf, and division of the tendo Achillis is practised for its relief.

[The plantaris is to be divided and the soleus cut near its attachments to the tibia and fibula, when it and the gastrocnemius are to be turned down and the bursa between the tendon and the os calcis noticed. The posterior tibial vessels and nerves, partially covered by the deep layer of fascia, will now be exposed, and beneath them the deep muscles of the leg, with the popliteus covered by its fascia above.]

The **Popliteus** (Fig. 48, 6) is covered by a strong fascia which is derived from the tendon of the semi-membranosus, upon which the lower part of the popliteal vessels and nerve rests. This must be removed to expose the oblique fibres of the muscle, and an incision must be made through the coverings of the knee immediately in front of the external lateral ligament to see its tendinous origin. The popliteus *arises* from within the general capsule of the knee (though outside the synovial cavity) by a round tendon, which is fixed to a distinct pit on the outer side of the external condyle, below and a little in front of the tubercle for the attachment of the external lateral ligament. The tendon passes through a groove in the bone and beneath the external lateral ligament, and after quitting the joint the fleshy fibres are developed, which are *inserted* into nearly the whole of the triangular surface on the back of the upper part of the tibia above the oblique line.

The popliteus is a flexor of the leg upon the thigh, and has a tendency to rotate the tibia inwards, w checked by the arrangement of the crucial ligaments within the knee. When these ligaments are destroyed or relaxed in disease very considerable rotation of the tibia takes place. The popliteus is *supplied* by a branch of the internal popliteal nerve.

[The incomplete intermuscular layer of fascia of the leg being removed, the posterior tibial vessels and nerve are to be cleaned, when the subjacent muscles will be found in the following positions:—flexor longus digitorum most internally, tibialis posticus in the middle, and flexor longus pollicis to the outer side.]

The **Flexor Longus Digitorum** (Fig. 48, 7) *arises* from the posterior surface of the tibia below the oblique line and internal to the attachment of the tibialis posticus (from which it is separated by an indistinct vertical ridge, and an intermuscular septum attached to it), the lower fibres of the muscle reaching to within three inches of the lower end of the tibia. It ends in a single tendon, which lies to the outer side of that of the tibialis posticus at the ankle, and passes through a separate division of the internal annular ligament to the sole of the foot.

The **Tibialis Posticus** (Fig. 48, 8) is the only one of the deep muscles which is attached to both bones of the leg, between which it therefore lies, its tendinous surface forming septa between it and the adjacent muscles. It *arises* from the posterior surface of the tibia, below the oblique line and external to the flexor longus digitorum, to within two inches of the ankle; from nearly the whole length of the interosseous membrane; and from the whole of the internal surface of the fibula immediately behind the interosseous line,* as well as from the intermuscular septa on each side, which are continuous with

* The internal surface here spoken of is not the surface called internal by Cloquet, Cruveilhier, Quain, and Holden, which includes the surface in front of the interosseous line,—nor that considered internal by Ward and Gray, which extends from the interosseous line to the posterior border of the bone; but is the surface existing only in the middle of the bone, between the interosseous line and an oblique line which cuts it off from the posterior surface. The fibula may be best considered as having a quadrilateral shaft with anterior, external, posterior, and internal (partial) surfaces.

the aponeurosis over it. The muscle is overlapped by the flexor digitorum, and between the tibial and fibular origins the anterior tibial vessels pass. The fibres end in a single broad tendon, which passes beneath and then to the inner side of that of the flexor digitorum at the ankle, running through a groove at the back of the internal malleolus and in the most internal division of the annular ligament to the sole of the foot, where it is *attached* to the tuberosity of the scaphoid bone and gives slips to all the other bones of the tarsus, except the astragalus.

Fig. 48.

The **Flexor Longus Pollicis** (Fig. 48, 9) is a well-marked bi-penniform muscle *arising* from the posterior surface of the fibula below the origin of the soleus to within an inch of the ankle; and also from the intermuscular septa between it and the peronei muscles on the outer side of the bone and the tibialis posticus internally. The muscle ends in a tendon which lies to the outer side of the posterior tibial vessels and nerve, and, passing through a separate division of the

Fig. 48.—Deep layer of muscles of the back of the leg (from Wilson).

1. Lower extremity of the femur.
2. Ligamentum posticum Winslowii.
3. Tendon of the semi-membranosus muscle.
4. Internal lateral ligament of the knee joint.
5. External lateral ligament.
6. Popliteus muscle.
7. Flexor longus digitorum.
8. Tibialis posticus.
9. Flexor longus pollicis.
10. Peroneus longus.
11. Peroneus brevis.
12. Tendo Achillis divided near its insertion into the os calcis.
13. Tendons of the tibialis posticus and flexor longus digitorum, just as they are about to pass beneath the internal annular ligament of the ankle; the interval between the latter tendon and the tendon of the flexor longus pollicis is for the posterior tibial vessels and nerve.

annular ligament, winds to the sole of the foot through an oblique groove at the back of the astragalus and another below the lesser process of the os calcis. The flexor longus pollicis and flexor digitorum are direct flexors of the toes and extensors of the foot, *i.e.*, they point the toes. The tibialis posticus is also an extensor of the foot, which it tends to draw inwards. The three muscles are *supplied* by the posterior tibial nerve.

Surgery.—The form of club-foot called 'talipes varus' which causes the patient to walk on the outside of the foot, is due in part to the contraction of the tibialis posticus and sometimes of the flexor digitorum. The operation of dividing these tendons is performed immediately above the internal malleolus, a blunt-pointed tenotomy knife being used after the first incision to avoid all danger of wounding the posterior tibial artery.

The **Posterior Tibial Artery** (Fig. 50, 21) is the direct continuation of the popliteal artery, and begins at the point of bifurcation of that vessel, which is usually the lower border of the popliteus muscle. In the upper part of its course the posterior tibial artery lies between the superficial and deep muscles of the back of the leg, but below the origin of the soleus it is only bound down by the intermuscular layer of fascia, and lies by the side of the tendo Achillis. It rests upon (in the position for dissection) the tibialis posticus for two-thirds of its length, but then upon the flexor digitorum; and at the ankle lies upon the tibia between the tendons of the flexor digitorum and flexor longus pollicis. The vessel afterwards passes through a division of the annular ligament to the interval between the inner malleolus and the heel, where it divides into internal and external plantar arteries. The posterior tibial nerve has a close relation to the artery throughout, lying first to its inner side and then crossing to its outer side, in which relation it continues to the foot. The posterior tibial artery has two *venæ comites*, which join those of the anterior tibial artery to form the popliteal vein.

Surgery (Fig. 48).—The operation of tying the posterior tibial artery is seldom performed except on the dead body, but the vessel may be reached in three parts of its course. Behind the malleolus the artery is readily exposed by a semi-lunar incision two inches long, made three-quarters of an inch behind the margin of the bone, the edge of the knife being directed towards the tibia so as to divide the internal annular ligament. At the lower part of the leg the vessel may be reached by a vertical incision a little to the inner side of the tendo Achillis, and will be found lying upon the flexor digitorum. In the middle of the leg the operation is one of some difficulty and somewhat damages the dissection if performed. The best mode of proceeding is to make a vertical incision four inches long, parallel to, and half an inch behind, the edge of the tibia; the gastrocnemius slips aside and the soleus is then to be divided, especial care being taken to cut through the tendon on its under surface but to go no deeper. A little dissection close beneath this tendon and towards the fibula will expose the posterior tibial artery lying upon the tibialis posticus.

Fig. 49.

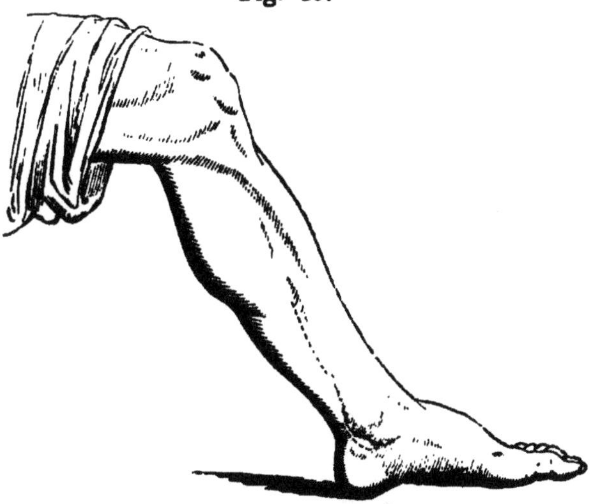

The **Branches** (Fig. 50) of the posterior tibial artery are (1) peroneal; (2) muscular branches; (3) nutritious to the tibia; (4) a communicating branch; and (5) internal calcanean.

Fig. 49.—Incisions for tying the posterior tibial artery (from Fergusson's 'Surgery').

Fig. 50.

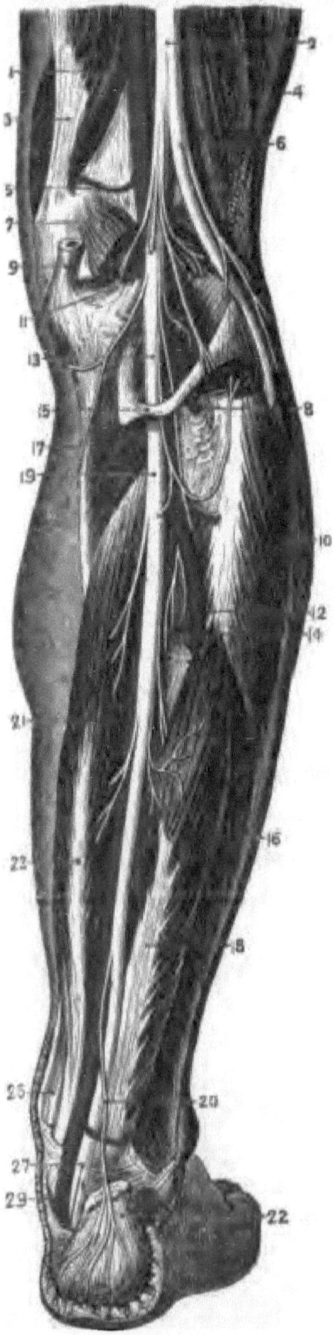

1. The *Peroneal* (12), always of considerable size and sometimes larger than the posterior tibial, comes off about two inches below the popliteus. It lies first between the tibialis posticus and flexor longus pollicis, but afterwards pierces the fibres of the flexor pollicis and runs close to the fibula to near its lower end. In this part of its course it gives *muscular* branches, and the *nutritious* artery to the fibula which runs *from* the knee-joint. It then gives a good-sized *anterior peroneal* branch to the front of the leg between the tibia and fibula below the interosseous membrane, which will be afterwards dissected; and a *communicating* branch to join that from the posterior tibial. The peroneal artery ends below the external malleolus by anastomosing with the malleolar and plantar arteries.

2. *Muscular* branches are given to the deep muscles and also to the soleus.

3. The *nutritious* artery to the tibia is of large size and enters the foramen on the posterior surface of that bone, having a direction *from* the knee-joint.

4. A *Communicating* branch runs transversely imme-

diately above the ankle to join a corresponding branch of the peroneal artery.

5. The *Internal Calcanean* is a branch of variable size and may be replaced by two or more separate arteries. It arises from the posterior tibial whilst beneath the internal annular ligament, which it pierces to supply the inner side of the os calcis with the muscles attached to it, and to anastomose with the other arteries in the neighbourhood.

The **Posterior Tibial Nerve** (Fig. 50, 19) is the continuation of the internal popliteal trunk, and passes with the artery beneath the soleus to lie on the deep muscles of the back of the leg. It is placed at first to the inner side of the posterior tibial artery, but crosses that vessel in the upper part and lies to its outer side at the ankle in the same division of the annular ligament. The nerve divides into internal and external plantar nerves for the sole of the foot, and the division not unfrequently takes place above the inner malleolus, in which case the plantar nerves are usually found on either side of the artery.

The posterior tibial nerve gives *muscular* branches to the tibialis posticus, flexor longus digitorum, and flexor longus pollicis, and a *cutaneous* nerve to the heel (20) which pierces the fascia to the inner side of the tendo Achillis, and may be traced to the skin of the sole of the foot.

Fig. 50.—Deep dissection of the back of the leg (from Hirschfeld and Leveillé).

1. Popliteal artery.
2. Great sciatic nerve.
3. Adductor magnus.
4. Biceps.
5. Superior internal articular artery.
6. External popliteal nerve.
7. Gastrocnemius (cut).
8. Anterior tibial artery.
9. Tendon of semi-membranosus.
10. Peroneus longus.
11. Sural arteries and nerves.
12. Peroneal artery.
13. Internal popliteal nerve.
14. Tibialis posticus.
15. Portion of soleus.
16. Peroneus brevis.
17. Popliteus.
18. Flexor longus pollicis.
19. Posterior tibial nerve.
20. Calcanean branch of posterior tibial nerve.
21. Posterior tibial artery.
22. Tendo Achillis.
23. Flexor longus digitorum.
25. Tendon of tibialis posticus.
27. Plantar nerves.
29. Plantar arteries.

Parts behind the Inner Malleolus (Fig. 51). Although the several parts behind the internal malleolus are only continuations of those which have been already dissected, especial notice should be taken of them in this part of their course, as their relations are important.

Fig. 51.

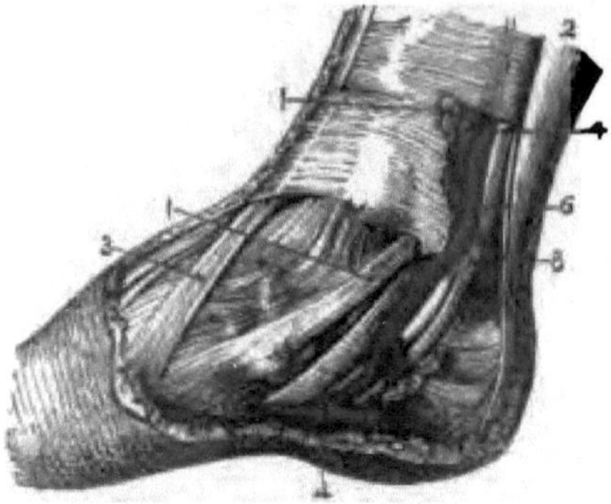

A portion of the skin on the inner side of the heel and sole of the foot is to be reflected if necessary, so as to expose the whole breadth of the internal annular ligament, which stretches from the inner malleolus to the os calcis and gives origin by its lower border to one head of the abductor pollicis.

The divisions in the annular ligament are four in number, and are occupied as follows, beginning from the tibia. In the first division is the tibialis posticus; in the second the tendon of the flexor longus digitorum; in the third the posterior tibial vessels and nerve, though this nerve is generally divided; and in the fourth the tendon of the flexor longus pollicis,* but this last lies so deeply in order to pass in the grooves in the astragalus and os calcis that it is not readily seen from the side. Each of the tendons is lubricated by a distinct synovial membrane.

* *Artificial Memory;* (for which the author is indebted to a St. George's Hospital student), Timothy Does Vex All Very Nervous Pupils (tibialis, digitorum, vein, artery, vein, nerve, pollicis).

Fig. 51.—Relations of parts behind the inner malleolus (from Hirschfeld and Leveillé).

1, 1. Tibialis posticus.
2. Tendo Achillis.
3. Tibialis anticus.
4, 4. Flexor longus digitorum.
6. Posterior tibial artery.
8. Posterior tibial nerve.

The tendon of the flexor longus pollicis is too deeply placed to be shown in this view.

The Sole of the Foot.

Fig. 52.

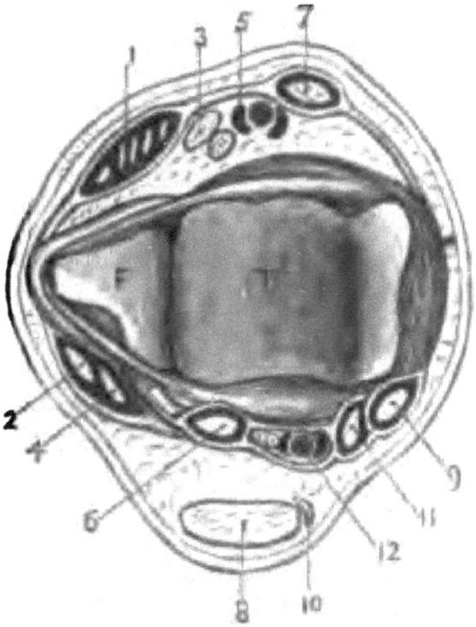

[The foot being raised on a block so that the sole may be fully exposed, an incision is to be made down its centre from the heel to the bases of the toes, the knife being carried at once down to the glistening plantar fascia through the thick lobulated fat which is always found beneath the skin in this region. A transverse incision at the roots of the toes will allow the skin and fat to be reflected to each side, which may be readily accomplished if the knife is kept close to the fascia and parallel to its fibres. Near the toes care must be taken of the divisions of the plantar fascia and of the digital nerves which appear between them; but no attempt need be made to trace out the minute cutaneous branch from the posterior tibial nerve near the heel.]

In the description the terms superficial, deep, &c., necessarily refer to the position of the parts as seen in the dissection, and not to that they would hold if the body were in the upright position.

The **Plantar Fascia** (Fig. 53, 2) closely resembles the fascia of the palm, and is divisible into three portions. The central, which is the strongest, is attached to the under surface of the great tuberosity of the os calcis, but expands as it passes forwards to nearly the entire width of the foot. It splits into five portions opposite the heads of the metatarsal bones, and at this point distinct trans-

Fig. 52.—Section of the right ankle (drawn by G. E. L. Pearse).

1. Extensor longus digitorum and peroneus tertius.
2. Peroneus longus.
3. Extensor proprius pollicis.
4. Peroneus brevis.
5. Anterior tibial vessels and nerve.
6. Flexor longus pollicis.
7. Tibialis anticus.
8. Tendo Achillis.
9. Tibialis posticus.
10. Plantaris.
11. Flexor longus digitorum.
12. Posterior tibial vessels and nerve.

verse fibres will be seen passing across the foot. The five processes are attached to the sheaths of the flexor tendons of each toe, and between them the several digital vessels and nerves pass.

A *transverse ligament* has been described in the skin of the web of the toes passing over the digital vessels and nerves. This is hardly worthy of a special description, as it is merely a slight extra development of the cutis, and has nothing to do with the true plantar fascia.

The central portion of the fascia gives origin to muscular fibres by its deep surface, and also by two intermuscular septa which will subsequently be seen on each side of the flexor brevis digitorum. The lateral portions of the plantar fascia are much thinner than that in the centre, and merely cover the muscles of the great

Fig. 53.

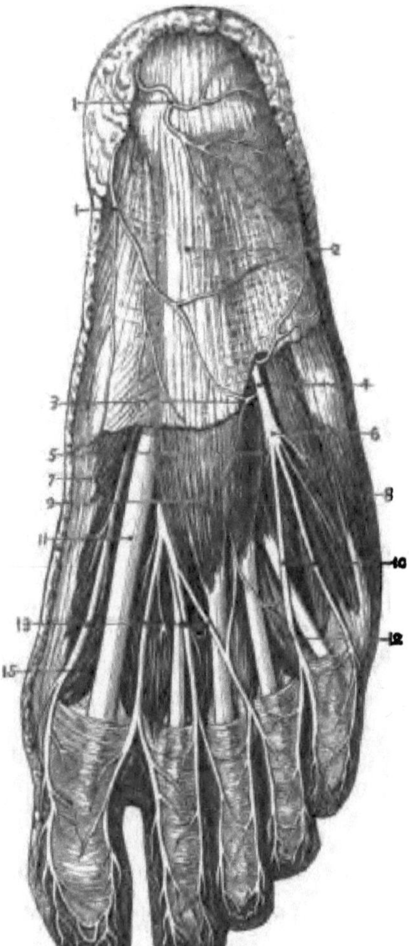

Fig. 53.—Superficial dissection of the sole of the foot (from Hirschfeld and Leveillé).

1, 1. Cutaneous plantar branches of posterior tibial nerve.
2. Plantar fascia.
3. Cutaneous branch of external plantar nerve.
4. External plantar nerve.
5. Its deep division.
6. Its superficial division.
7. Abductor pollicis.
8. Abductor minimi digiti.
9. Flexor brevis digitorum.
10. Digital branches of external plantar nerve.
11. Tendon of flexor longus pollicis.
12. One of the lumbricales muscles.
13. Digital branches of internal plantar nerve.
15. Flexor brevis pollicis.

and little toes on the inner and outer sides of the foot.

[The central portion of the fascia is to be cut across about two inches from the heel, and may then be turned down over the toes without injuring the digital nerves, when the flexor brevis digitorum will be exposed in part, and the intermuscular septa on each side may be seen. The lateral portions of the fascia are to be removed, and the abductor pollicis and abductor minimi digiti cleaned.]

The **First Layer of Muscles** consists of the flexor brevis digitorum in the middle, the abductor pollicis to the inner side, and the abductor minimi digiti to the outer side.

The **Flexor Brevis Digitorum** (Fig. 54, 5) *arises* from the great tuberosity of the os calcis; from the plantar fascia covering it (which has therefore been left undisturbed); and from the intermuscular septa on each side. It divides into four tendons, which pass through tendinous sheaths on the under surfaces of the four smaller toes, and are *inserted*, each by two slips, into the bases of their second phalanges. Passing between the subdivisions of each tendon is the corresponding

Fig. 54.

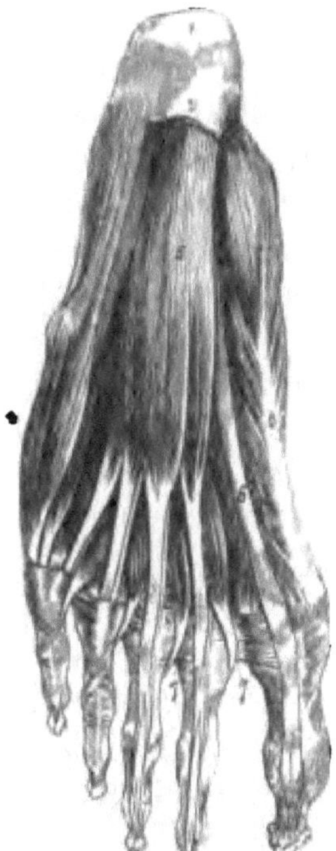

Fig. 54.—First layer of muscles of the sole of the foot (from Wilson).

1. Os calcis.
2. Posterior part of the plantar fascia divided transversely.
3. Abductor pollicis.
4. Abductor minimi digiti.
5. Flexor brevis digitorum.
6. Tendon of the flexor longus pollicis.

7, 7. Lumbricales. On the second and third toes the tendons of the flexor longus digitorum are seen passing through the bifurcation of the tendons of the flexor brevis digitorum.

tendon of the long flexor, which goes to the terminal phalanx, and thus the arrangement is precisely similar to that found in the hand, and the flexor brevis might be spoken of as *perforatus*, and the flexor longus as *perforans*.

The sheaths of the tendons also resemble those of the hand, but are smaller. Beginning over the heads of the metatarsal bones, each sheath is joined by a process of plantar fascia (which splits to be attached to each side of it) and then extends to the base of the terminal phalanx. The sheath is strong opposite each phalanx, but thinner over the joints, so as not to impede their motion; and when divided, will be found to be lined by a synovial membrane which is also reflected over the tendons.

The **Abductor Pollicis** (Fig. 54, 3) *arises* by two heads: one from the internal annular ligament, the other from the greater tuberosity of the os calcis (internal to the origin of the flexor brevis digitorum) and slightly from the plantar fascia. The muscle ends in a broad tendon, which is joined by the inner head of the flexor brevis pollicis, and is *inserted* into the inner side of the base of the first phalanx of the great toe.

The **Abductor Minimi Digiti** (Fig. 54, 4) *arises* from the lesser tuberosity of the os calcis; from the greater tuberosity in front of the flexor brevis digitorum; from the plantar fascia and septum; and is attached to the subjacent ligamentous structures of the foot as far forward as the fifth metatarsal bone. Its tendon is *inserted* into the outer side of the base of the first phalanx of the little toe.*

The actions of the superficial muscles of the foot are sufficiently indicated by their names, but it should be remarked that abduction and adduction are from and to the median plane of the foot and not of the body.

The flexor brevis digitorum and abductor pollicis are *supplied* by the internal plantar nerve; the abductor minimi digiti by the external plantar nerve.

* Mr. John Wood has described an *abductor ossis metatarsi quinti* which is occasionally found beneath the abductor and attached to the base of the fifth metatarsal bone.

[To bring the deeper parts into view, the best way will be to remove with the saw a thin slice of the os calcis, including the attachments of the superficial muscles, when, if the attachment of the abductor pollicis to the internal annular ligament be divided, they can be turned forwards, and the plantar nerves and vessels will be seen.]

Fig. 55.

The **Plantar Arteries** (Fig. 55) internal and external, are the terminal branches of the posterior tibial, and enter the foot beneath the abductor pollicis and in close relation with the plantar nerves. The internal plantar artery, which is the smaller of the two, is accompanied by the large internal plantar nerve, the large external plantar artery going with the smaller external nerve.

The *Internal plantar artery* (6) is of small size, and runs forward to the great toe, where it sometimes gives digital arteries to the great and one side of the second toe. Two *venæ comites* accompany it.

The *External plantar artery* (7) runs obliquely outwards across the foot between the flexor brevis digitorum and the accessorius muscle, and turning round the outer border of the latter muscle, takes a deep course inwards towards the great toe, where it will be subsequently traced.

Fig. 55.—Arteries of the sole of the foot; the first and part of the second layer of muscles having been removed (from Wilson).

1. Under and posterior part of the os calcis; to which the origins of the first layer of muscles remain attached.
2. Musculus accessorius.
3. Long flexor tendons.
4. Tendon of the peroneus longus.
5. Termination of the posterior tibial artery.
6. Internal plantar artery.
7. External plantar artery.
8. The plantar arch, giving off four digital branches.

It is accompanied by the external plantar nerve, which also gives a deep branch across the foot, and by *venæ comites;* and supplies the adjacent muscles, giving a branch to the outer side of the foot to anastomose with the peroneal artery.

Fig. 56.

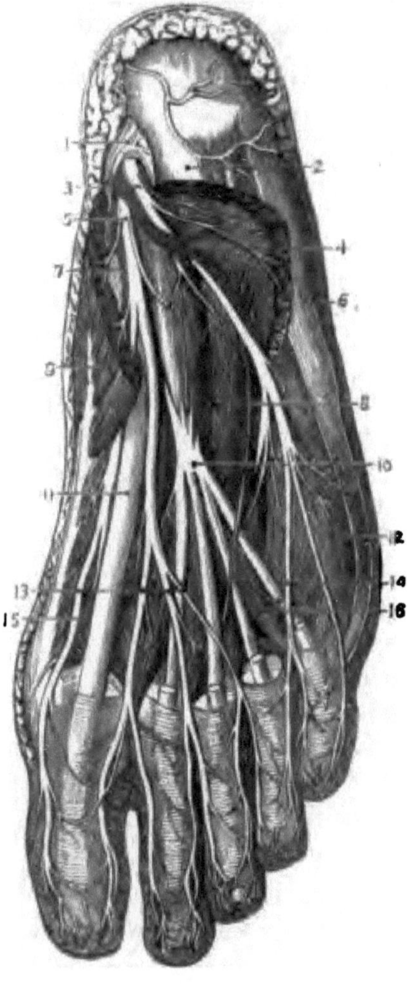

The **Plantar Nerves** (Fig. 56) internal and external, are the branches of the posterior tibial nerve; their relation to the arteries has been mentioned.

The *Internal plantar nerve* (5) is the larger of the two, and after supplying the abductor pollicis and flexor brevis digitorum, gives off a digital branch to the inner side of tho great toe, and breaks up into three other digital branches; each of these bifurcates at the roots of the toes, and they thus *supply* the outer side

Fig. 56.—Second stage of dissection of sole of foot (from Hirschfeld and Leveillé).

1. Internal annular ligament.
2. Flexor brevis digitorum (cut).
3. External plantar nerve.
4. External plantar artery.
5. Internal plantar nerve.
6. Abductor minimi digiti.
7. Internal plantar artery.
8. Accessorius muscle.
9. Abductor pollicis.
10. Flexor longus digitorum.
11. Flexor longus pollicis.
12. Flexor brevis minimi digiti.
13. Digital branches of internal plantar nerve.
14. Digital branches of external plantar nerve.
15. Flexor brevis pollicis.
16. One of the lumbricales.

of the great toe, both sides of the second and third toes, and half the fourth toe. The nerve also gives muscular branches to the flexor brevis pollicis and the two inner lumbricales muscles.

The *External plantar nerve* (3) crosses the foot with the external plantar artery, giving branches to the abductor minimi digiti and accessorius in its course. At the outer border of the accessorius it divides into superficial and deep portions, the deep accompanying the artery. The superficial portion *supplies* the flexor brevis minimi digiti and the interosseous muscles in the fourth or outermost space. It then divides into two digital branches,—one to the outer side of the little toe, and the other bifurcating to supply the inner side of the little and half the next toe.

A slender branch of communication between the internal and external plantar nerves may sometimes be found.

It may be noticed that the nerves of the sole of the foot closely resemble those of the palm of the hand in their distribution, the internal plantar corresponding to the median, and the external plantar to the ulnar nerve; the deep branches of the latter nerves also correspond.

The Second Layer of Muscles includes the tendons of the flexor longus digitorum and flexor longus pollicis, with the small accessorius and lumbricales muscles, all of which should now be cleaned.

The tendon of the **Flexor Longus Digitorum** (Fig. 57, 3) is single at the annular ligament and lies to the inner side of the vessels and nerves, but it will now be seen to cross beneath them and over the tendon of the flexor longus pollicis to the centre of the foot, where it has the fibres of the accessorius attached to its deep surface. It then divides into four tendons which give origin to the lumbricales muscles and pass beneath the digital nerves to the four small toes, where they enter the digital sheaths, lying beneath the tendons of the short flexor. Each deep tendon then passes through the slit in the superficial tendon and is *inserted* into the base of the third or terminal phalanx.

The **Accessorius** (Fig. 56, 8) *arises* from the under su

face of the os calcis by two heads, between which the long plantar ligament is seen. The inner head, which is fleshy and the larger of the two, is attached to the hollow in front of the greater tuberosity; the outer head, which is tendinous, is attached to the bone in front of the lesser tuberosity and to the long plantar ligament. The muscle is *inserted* into the deep surface of the tendon of the flexor longus digitorum.

The **Lumbricales** (Fig. 54, 7) are four small muscles which are so named from their fancied resemblance to earth worms. They *arise* from the tendons of the flexor longus digitorum,—the most internal from one, and the rest from two tendons each,—and wind to the inner sides of the four small toes to reach their dorsal aspects, where they join the expansions of the extensor tendons.

The action of the accessorius is to bring the line of traction of the flexor tendons into the centre of the foot. The use of the lumbricales is doubtful. The accessorius is *supplied* by the external plantar nerve; the two inner lumbricales by the internal plantar and the two outer lumbricales by the external plantar nerve.

The tendon of the **Flexor Longus Pollicis** (Fig. 57, 4) after lying in the groove below the lesser process of the os calcis, runs forward to the great toe, lying beneath the plantar vessels and nerves and being crossed by the tendon of the flexor digitorum, to which it is connected by a few fibres. The tendon lies between the two heads of the flexor brevis pollicis and then enters a sheath, in which it is enclosed until its *insertion* into the base of the terminal phalanx of the great toe.

[The long tendons and the accessorius muscle are to be cut about the middle of the foot and turned down, when the muscles of the third layer are to be cleaned.]

The **Third Layer of Muscles** consists of the flexor brevis pollicis, adductor pollicis, flexor brevis minimi digiti, and transversus pedis.

The **Flexor Brevis Pollicis** (Fig. 57, 5) *arises* by a narrow tendinous origin from the under surface of the external cuneiform bone and adjacent surface of the cuboid, and is connected with the insertion of the tibialis posticus. It consists of two heads, the fibres of which diverge from each other and separate near the head of the metatarsal bone, to be *inserted* into opposite sides of the base of the first phalanx of the great toe with the abductor and adductor respectively. Each head has a sesamoid bone developed in its tendon, entering into the formation of the metatarso-phalangeal joint, which must be opened to see them distinctly. The tendon of the long flexor lies between the two heads of this muscle.

Fig. 57.

The **Adductor Pollicis** (Fig. 57, 6) *arises* from the bases of the second, third, and fourth metatarsal bones and from the sheath of the peroneus longus tendon; and is *inserted* into the outer side of the base of the first phalanx of the great toe, with the outer head of the flexor brevis.

The **Flexor Brevis Minimi Digiti** (Fig. 57, 7) *arises* from the under surface of the projecting base of the fifth metatarsal bone, and is *inserted* into the outer side of the base of the first phalanx of the little toe with the abductor.*

* An *opponens minimi digiti* attached to the metatarsal bone may occasionally be found.

Fig. 57.—Third and part of the second layer of muscles of the sole of the foot (from Wilson).

1. Divided edge of the plantar fascia.
2. Musculus accessorius.
3. Tendon of the flexor longus digitorum.
4. Tendon of the flexor longus pollicis.
5. Flexor brevis pollicis.
6. Adductor pollicis.
7. Flexor brevis minimi digiti.
8. Transversus pedis.
9. Interossei muscles, plantar and dorsal.
10. Sheath of the tendon of the peroneus longus.

The **Transversus Pedis** (Fig. 57, 8) is of variable size and consists of three or four small bundles, which *arise* from the transverse metatarsal ligament over the heads of the three or four outer metatarsal bones. It is *inserted* into the outer side of the first phalanx of the great toe with the adductor and part of the flexor brevis pollicis.

The action of the flexor brevis pollicis and flexor minimi digiti is obvious; the adductor pollicis acts towards the median plane of the foot, and the transversus draws the toes together and narrows the foot. The flexor brevis pollicis is *supplied* by the internal plantar nerve, the other muscles by the external plantar nerve.

[The adductor and flexor brevis pollicis are to be cut near their origins and

Fig. 58.

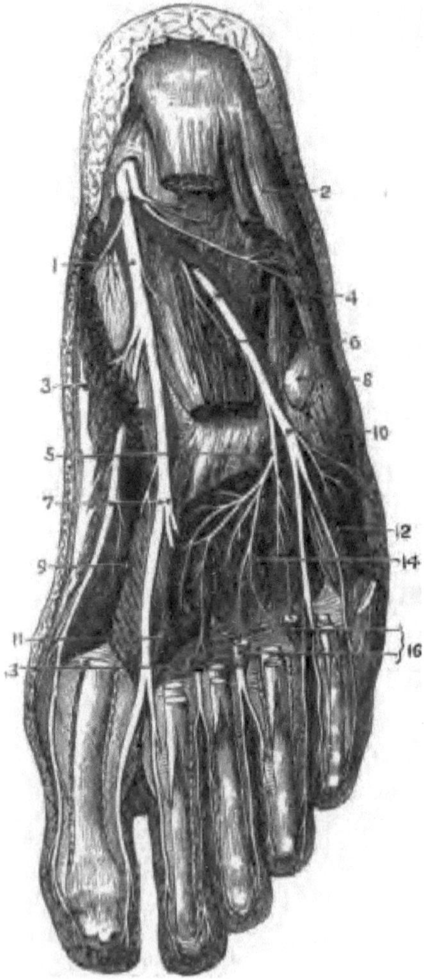

Fig. 58.—Deep dissection of the sole of the foot (from Hirschfeld and Leveillé).

1. Internal plantar nerve.
2. Abductor minimi digiti.
3. Abductor pollicis.
4. External plantar nerve.
5. Its deep division.
6. Accessorius muscle.
7. Digital branches of internal plantar nerve.
8. Tendon of peroneus longus.
9. Flexor brevis pollicis.
10. Superficial division of external plantar nerve.
11. Adductor pollicis.
12. Flexor brevis minimi digiti.
13. Transversus pedis.
14. Interossei muscles.
16. Two outer lumbricales.

turned down, when the plantar arch of the external plantar artery with the accompanying nerve is to be fully traced out.]

The **External Plantar Artery** (Fig. 55, 8) after turning round the accessorius muscle takes a deep course beneath the adductor and flexor brevis pollicis, but upon the interosseous muscles, to the base of the metatarsal bone of the great toe. This deep portion of the artery has been called the *plantar arch*, which is completed by a branch of the anterior tibial artery entering the sole between the first and second toes.

Branches. Three small *posterior perforating arteries* are given off by the plantar arch, which pierce the interosseous muscles to anastomose with the interosseous arteries on the dorsum.

Four *digital artesies* arise from the anterior part of the arch; the outer one goes to the outer side of the little toe, the others bifurcate to supply the adjacent sides of two toes each, and at the point of bifurcation send *anterior perforating* arteries to the back of the foot. The digital arteries run on the sides of the toes with the digital nerves, and anastomose in the pulp of the terminal phalanx.

The *communicating branch of the dorsalis pedis* artery enters the sole between the heads of the first dorsal interosseous muscle, and besides completing the plantar arch, gives off the *arteria magna pollicis* which bifurcates to supply the adjacent sides of the great and next toe; and also a small branch to the inner side of the great toe.

The **External Plantar Nerve** (Fig. 58, 5).—The deep portion of the nerve accompanies the artery beneath the adductor, in which it ends. It gives small branches in its course to the two outer lumbricales, to the transversus pedis, and to all the interosseous muscles both plantar and dorsal, except those of the fourth or outermost interosseous space, which are supplied by the superficial portion.

The **Fourth Layer of Muscles** consists of the plantar

interossei, which will be better seen when the dorsal interossei are dissected.

Front of the Leg and Foot.

[The skin remaining on the front and outer side of the leg and on the back of the foot is to be removed, and the superficial veins and nerves dissected out of the superficial fascia.]

The **Superficial Veins** form an arch across the back of the foot ending in the internal and external saphena veins on opposite sides.

The *internal saphena vein* is to be traced in front of the internal malleolus to the inner side of the knee.

Cutaneous Nerves. On the outer side of the leg are a few cutaneous branches from the *external popliteal nerve.*

The **Musculo-Cutaneous Nerve** (Fig. 61, 13) pierces the deep fascia about the lower third of the outer side of the leg, and at once divides into two branches, which however often appear at different points. The internal division (15) gives branches to the inner side of the great toe, the outer side of the second, and the inner side of the third toe; frequently also giving a small branch to join the anterior tibial nerve between the first and second toes. The outer division (17) supplies the outer side of the third toe, both sides of the fourth, and either one or both sides of the fifth toe.

The **External Saphenous Nerve** (Fig. 61, 21) winds round the external malleolus from the back of the leg, and is distributed to the outer side of the little toe; or occasionally to both sides of the little toe and half the next, joining the musculo-cutaneous nerve.

The **Anterior Tibial Nerve** (Fig. 61, 16) appears in the interval between the great and the second toes; the adjacent sides of which it supplies, often joining with a branch of the musculo-cutaneous nerve.

The **Internal Saphenous Nerve** lies in front of the internal malleolus with the saphena vein, which it accompanies. It may be traced along the inner side of the foot, but is of small size.

The **Fascia of the Leg** is dense and white, and is attached to the tibia on the inner side and to the fibula on the outer side, forming an intermuscular septum between the extensor and the peroneal muscles. It gives origin to muscular fibres at the upper part of the leg, where it should therefore be allowed to remain undisturbed; the rest of the fascia should be removed, except a thickened band over the ankle joint—the *anterior annular ligament*,—and a similar one below the external malleolus, which is called the *external annular ligament*. The muscles, vessels, and nerves are to be cleaned.

The **Extensor Muscles** (Fig. 59) are the tibialis anticus to the inner side; the extensor longus digitorum with an extra slip (the peroneus tertius) to the outer side; and between the two, the extensor proprius pollicis appearing about the middle of the leg.

The **Tibialis Anticus** (Fig. 59, 3) *arises* from the upper two-thirds of the outer surface of the shaft of the tibia and from its outer tuberosity; from the inner half of the interosseus ligament for the same distance; and from the fascia covering the muscle. The fibres end in a broad tendon, which becomes narrower near the ankle and passes through the most internal division of the annular ligament, to be *inserted* into the under surface of the internal cuneiform bone and the base of the metatarsal bone of the great toe. The tibialis anticus is a flexor of the foot and draws up its inner side. It is one of the tendons which helps to produce the form of club-foot called 'talipes varus.' It is *supplied* by the anterior tibial nerve.

The **Extensor Longus Digitorum** (Fig. 59, 4) *arises* from the upper three-fourths of the anterior surface of the fibula (see note, p. 136) and from the adjacent interosseous membrane; from the outer tuberosity of the tibia close to the tibialis anticus; from the fascia upon the upper part of the muscle; and from the intermuscular septum between the extensors and the peronei. The muscular fibres are continued upon the tendon to the annular ligament, through the outer division of which it passes and at once divides into four tendons for the four smaller toes. The *insertion* of the extensors into the toes is similar to that of the extensors of the fingers, but on a smaller scale. The extensor tendons form an expansion on the back of the first phalanx of each toe, which is joined by the tendons of the corresponding lumbricalis and interossei muscles; this is continued forwards, the central triangular portion going to the second phalanges and the lateral stronger pieces to the third phalanges.

Fig. 59.

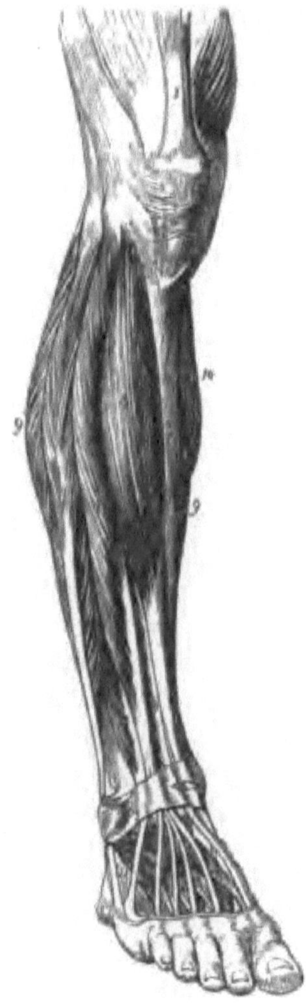

The **Peroneus Tertius** (Fig. 59, 6) is really a part of the extensor longus digitorum. It *arises* below the extensor from the lower fourth of the anterior surface of the fibula; slightly from the interosseous membrane; and from the septum between it and the peroneus brevis. Its tendon passes through the same division of the annular

Fig. 59.—Muscles of the front of the leg (from Wilson).

1. Quadriceps extensor inserted into the patella.
2. Subcutaneous surface of the tibia.
3. Tibialis anticus.
4. Extensor longus digitorum.
5. Extensor proprius pollicis.
6. Peroneus tertius.
7. Peroneus longus.
8. Peroneus brevis.
9, 9. Borders of the soleus muscle.
10. Part of the inner belly of the gastrocnemius.
11. Extensor brevis digitorum; the tendon in front of the figure is that of the peroneus tertius; that behind it, the peroneus brevis.

ligament as the extensor digitorum, and is *inserted* into the upper surface of the base of the fifth metatarsal bone.

The **Extensor Proprius Pollicis** (Fig. 59, 5) appears between the tibialis anticus and the extensor digitorum in the middle of the leg. It *arises* from the middle two-fourths of the anterior surface of the fibula, behind the extensor longus digitorum; and from the adjacent part of the interosseous membrane. Its tendon passes beneath the annular ligament, and, crossing the anterior tibial artery, is *inserted* into the terminal phalanx of the great toe.

The extensors of the toes also act as flexors of the foot, *i.e.* bend it upwards. The peroneus tertius assists in this latter action. The rare form of club-foot 'talipes calcaneus' is produced by the action of these muscles. They are all *supplied* by the anterior tibial nerve.

The **Extensor Brevis Digitorum** (Fig. 59, 11) is the only muscle of the back of the foot. It *arises* from the upper surface of the greater process of the os calcis; from the calcaneo-astragaloid ligament; and from the lower border of the anterior annular ligament. It ends in four tendons which pass obliquely across the foot to be *inserted* into the four inner toes, joining the general expansion of the extensor tendons, except on the great toe, where the tendon is inserted separately into the base of the first phalanx after crossing the dorsal artery of the foot.

The extensor brevis is *supplied* by the anterior tibial nerve.

The **Anterior Tibial Artery** (Fig. 61, 2) is a branch of the popliteal, and enters the front of the leg between the tibia and fibula above the interosseous membrane. The artery lies at first upon the interosseous membrane between the tibialis anticus and extensor longus digitorum, and then between the tibialis anticus and extensor proprius pollicis; but in the lower part of the leg it winds on to the tibia, and becomes superficial above the ankle joint between the tendons of the tibialis anticus and extensor proprius pollicis. It then passes beneath the annular ligament and is crossed by the extensor proprius pollicis, after which it changes its name to *dorsal artery of the foot.*

Two *venæ comites* accompany the artery; the anterior tibial nerve lies to its outer side as far as about the middle of the leg, then crosses it, and in most cases again gets to its outer side at the ankle.

Fig. 60.

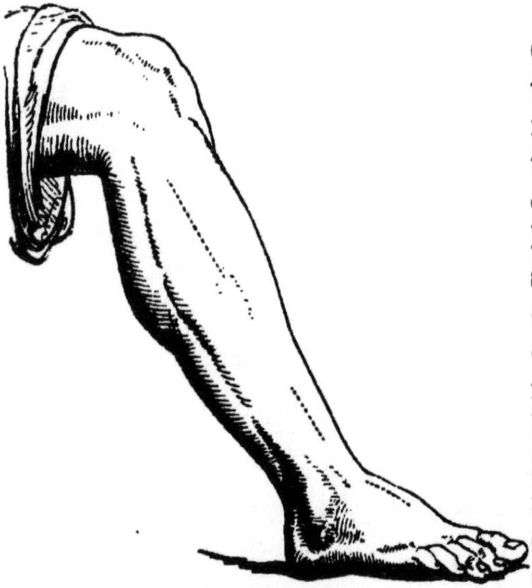

Surgery (Fig. 60). The anterior tibial artery may be readily tied above the ankle by an incision three inches long upon the outer side of the tendon of the tibialis anticus and parallel to it. The artery will be found upon the front of the tibia between the tendons of the tibialis anticus and the extensor proprius pollicis, with the nerve to its outer side.

Fig. 60.—Incisions for tying the anterior tibial artery and the dorsal artery of the foot (from Fergusson's 'Surgery').

Fig. 61.—Deep dissection of the front of the leg (from Hirschfeld and Leveillé).

1. External popliteal nerve.
2. Anterior tibial artery.
3. Musculo-cutaneous nerve.
4. Anterior tibial nerve.
5. Peroneus longus.
6. Tibialis anticus.
7. Extensor longus digitorum.
8. Anterior annular ligament.
9. Peroneus brevis.
10. Tendon of extensor proprius pollicis.
11. Extensor proprius pollicis.
12. Dorsal artery of foot.
13. Point at which musculo-cutaneous nerve pierces the fascia and divides.
14. Tendon of tibialis anticus.
15. Internal branch of musculo-cutaneous nerve.
16. Cutaneous branch of anterior tibial nerve.
17. External branch of musculo-cutaneous nerve.
19. Deep branch of anterior tibial nerve.
21. External saphenous nerve.
23. Extensor brevis digitorum.

The artery may also be reached on the dead body in the upper part of its course, but it lies so deeply between muscles that the operation is seldom, if ever, undertaken on the living subject. As there is no definite guide to the outer border of the tibialis it is well to make the incision a little obliquely and not less than four inches long, beginning about two inches below the head of the fibula, at a point midway between the bone and the outer tuberosity of the tibia. The fascia should be divided in the same direction as the skin, when, if the fore-finger is introduced into the wound, it will probably pass into the cellular interval between the two muscles, there being no intermuscular septum or white line to guide the operator, though these are commonly mentioned. The tibialis anticus and extensor digitorum must be drawn apart, when the artery will be found upon the interosseous membrane with the nerve to its outer side.

Fig. 61.

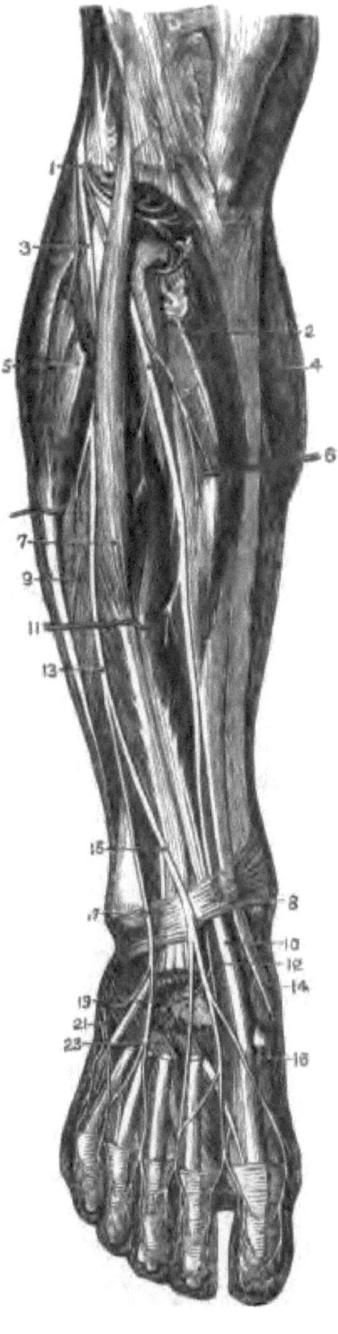

Branches of the anterior tibial artery :—

The *Recurrent branch* arises as soon as the artery reaches the front of the leg, and winds through the fibres of the tibialis anticus to the front of the knee to anastomose with the articular arteries.

Muscular branches to the adjacent muscles arise at various points.

The *Malleolar arteries,* internal and external, pass beneath the tendons to the malleoli, in the neighbourhood of which they are distributed. The external anastomoses with the anterior peroneal artery.

The *Anterior Peroneal* branch of the peroneal artery (page 140) appears between the tibia and fibula through an opening in the lower part of the interosseous membrane, running beneath the peroneus tertius to the outer malleolus to supply it and to anastomose with the external malleolar artery.

The **Dorsalis Pedis Artery** (Fig. 61, 12) is the continuation of the anterior tibial from the annular ligament to the base of the metatarsal bone of the great toe, opposite which it divides into the communicating branch to the sole of the foot and the dorsal artery of the great toe. It lies at first superficially between the tendons of the extensor proprius pollicis and the extensor longus digitorum, but is crossed near the point of bifurcation by the innermost tendon of the extensor brevis digitorum. The artery lies upon the bones of the tarsus and their dorsal ligaments, and has the anterior tibial nerve to its outer side. Two *venæ comites* accompany the vessel.

Surgery (Fig. 60.) The dorsal artery may be tied in the upper part of its course by an incision on the outer side of, but parallel to, the tendon of the extensor proprius pollicis. It occasionally happens however that the artery is displaced to the middle of the foot beneath the tendons of the extensor digitorum.

Branches of the dorsalis pedis artery.

The *Tarsal artery* arises immediately below the annular ligament, and crosses the foot beneath the extensor brevis digitorum to anastomose with the arteries about the external malleolus.

The *Metatarsal artery* arises lower than the preceding and near the bases of the metatarsal bones. It also runs outwards beneath the extensor brevis digitorum, and gives off three *interosseous* branches to the outer spaces. These run forward upon the interossei muscles and bifurcate at the roots of the toes to supply the adjacent sides of two toes each, the outer one giving off a branch also to the outer side of the little toe. The anterior and posterior perforating arteries of the plantar arch join these interosseous arteries at their origins and points of bifurcation.

The *Communicating branch* passes between the heads of the first dorsal interosseous muscle to the sole of the foot, where it joins the plantar arch (p. 153). The *Dorsalis Pollicis (vel hallucis) artery* runs forward upon the great toe and, after giving a branch to its inner side, bifurcates to supply the adjacent sides of the first and second toes.

The **Anterior Tibial Nerve** (Fig. 61, 4) is a branch of the external popliteal nerve, and reaches the anterior tibial artery by piercing the fibres of the extensor longus digitorum. It lies to the outer side of the artery in the upper part of the leg, but crosses it about the middle, reaching its outer side again at the ankle. Its muscular relations are the same as those of the artery, with which it passes under the annular ligament. In the leg it supplies the four extensor muscles, and as soon as it enters the foot it divides into two branches. The outer branch (19) is distributed to the extensor brevis digitorum and the ankle joint; the inner branch (16) lies to the outer side of the dorsal artery, and becomes cutaneous in the interval between the great toe and next, the adjacent sides of which it supplies.

[The peroneal muscles are to be exposed on the outer side of the leg by removing the fascia covering them, when a strong intermuscular septum will be found attached to the fibula on each side, separating them from the muscles of the front and back of the leg. A portion of the fascia is to be left below the malleolus to form an internal annular ligament.]

The **Peroneus Longus** (Fig. 59, 7) *arises* from the upper third of the outer surface of the fibula; from the intermuscular septa on each side, and from the fascia. It ends in a strong tendon which lies superficially to that of the peroneus brevis behind the external malleolus and under the external annular ligament, and the tendon then passes in a sheath of fascia along the outer side of the os calcis to reach the cuboid bone, around which it turns to the sole of the foot. In the sole the tendon lies in the groove of the cuboid bone and in a fibrous sheath ('sheath of peroneus longus') which is now to be opened, when the tendon can be traced to its *insertion* on the outer side of the base of the first metatarsal bone. The sheath of the tendon is formed by fibres of the long plantar ligament and is lined by a synovial membrane. The tendon has a sesamoid cartilage or bone developed in its fibres where it turns round the border of the cuboid bone.

The **Peroneus Brevis** (Fig. 59, 8) lies beneath the tendon of the peroneus longus, and *arises* from the middle third of the outer surface of the fibula and from the intermuscular septa on each side. It lies upon the lower part of the fibula, and its tendon winds behind the external malleolus where it is lodged in a groove beneath the tendon of the peroneus longus. It then runs along the outer side of the os calcis in a sheath of fascia in front of the long tendon, and is *inserted* into the posterior extremity of the fifth metatarsal bone.

The peronei act as extensors of the foot, *i. e.*, they point the toes. Both muscles, but especially the peroneus longus, draw up the outer side of the foot as seen in the form of club-foot called 'talipes valgus.' Both the peronei are *supplied* by the musculo-cutaneous nerve.

The **External Popliteal Nerve** (Fig. 61, 1) is to be followed through the fibres of the peroneus longus, to which it was traced in the dissection of the popliteal space. The nerve lies close to the fibula and gives off a *recurrent articular branch* and a *muscular branch*, which pierce the extensor longus digitorum to reach the upper part of the tibialis anticus and the front of the knee. It then divides into *anterior tibial*, (4) which pierces the extensor longus digitorum, and *musculo-cutaneous*, (3) which, after giving branches to the peroneus longus and brevis, appears between the peroneus longus and extensor longus digitorum, and finally becomes cutaneous by piercing the fascia.

Fig. 62.

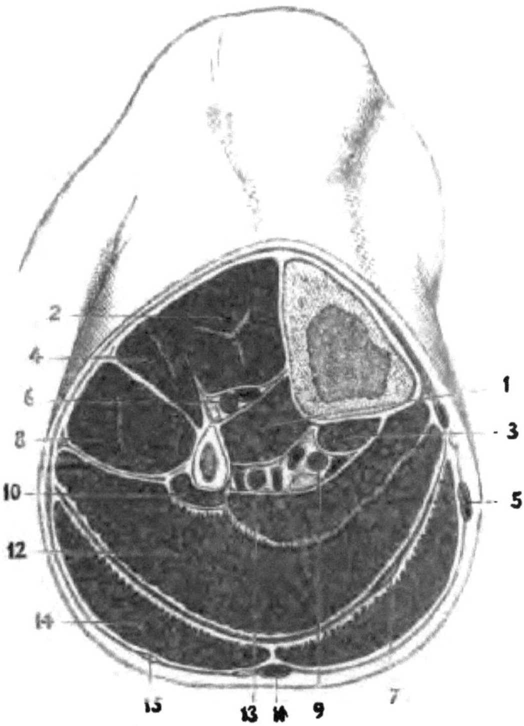

[The tendons on the back of the foot are to be divided or held aside, and the transversus pedis in the sole is to be removed to permit a clear view of the interossei.]

Beneath the transversus will be found the *transverse ligament of the metatarsus* (Fig. 72, 12) from which its fibres arise, and over which the digital vessels and nerves

---

Fig. 62.—A section of the right leg in the upper third (altered from Béraud).

1. Tibialis posticus.
2. Tibialis anticus.
3. Flexor longus digitorum.
4. Extensor longus digitorum.
5. Internal saphenous vein.
6. Anterior tibial vessels and nerve.
7. Tendon of plantaris.
8. Peroneus longus.
9. Posterior tibial vessels and nerve.
10. Flexor longus pollicis.
11. External saphenous vein and nerve.
12. Soleus with fibrous intersection.
13. Peroneal vessels.
14. Gastrocnemius.
15. Communicans peronei nerve.

pass. It connects the heads of the metatarsal bones together and crosses the tendons of the interossei muscles.

The **Interosseous Muscles** are seven in number, viz. four dorsal and three plantar, the former being strictly between the bones, but the latter lying rather on their under surfaces. The dorsal are bi-penniform muscles and are readily seen; the plantar are simply penniform and are a little obscure, owing to the fibres of the dorsal interossei appearing by their sides.

The *four dorsal interossei* (Fig. 63) *arise* from the adjacent sides of the metatarsal bones in nearly the whole length of their shafts, and are *inserted* into the bases of

Fig. 63.

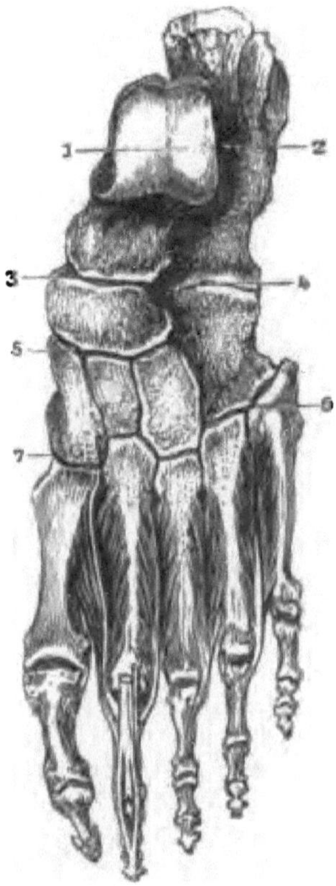

Fig. 64.

Fig. 63.—The Dorsal Interosseous muscles of the Foot. (J. T. Gray.) The figures refer to the seven synovial membranes.

Fig. 64.—The Plantar Interosseous muscles of the Foot. (J. T. Gray.)

he first phalanges in such a manner as to abduct from an imaginary line through the second (or long) toe. Thus the 1st is inserted into the *inner* side of the first phalanx of the second toe; the 2nd into the *outer* side of the first phalanx of the second toe; the 3rd and 4th into the *outer* sides of the first phalanges of the third and fourth toes.

The *three plantar interossei* (Fig. 64) *arise* from the inner sides of the third, fourth, and fifth metatarsal bones, and are *inserted* into the corresponding sides of the first phalanges of the same toes. They thus act as adductors towards the second toe. The interossei are *supplied* by the deep branch of the external plantar nerve.

The interossei of the foot should be compared with those of the hand, in which they are a little more complicated, owing to their being arranged on each side of an imaginary line through the middle or long finger.

The Knee Joint.

The knee is a good example of ginglymus or hinge-joint, the movement taking place between the femur and the tibia, and the patella gliding upon the former bone. It is remarkable for having interarticular fibro cartilages, and its synovial membrane is the most extensive in the body.

The knee joint has *in front* the expansion of the extensor muscles which help to form its capsular ligament, and in front of the patella is a bursa. *Behind* the joint are the popliteal artery, vein, and nerves, with the two heads of the gastrocnemius and the popliteus. To the *inner* side of the joint, but posterior to it are the tendons of the sartorius, gracilis, semi-tendinosus, and semi-membranosus; and the biceps occupies a similar position on the *outer* side.

The ligaments of the knee may conveniently be divided into extra- and intra-articular.

The **Extra-articular Ligaments** are anterior, posterior, two lateral, and capsular.

The **Anterior Ligament** (Fig. 65, 3) is the *ligamentum patellæ*, which is a broad band reaching from the lower margin of the patella to the tubercle of the tibia, and forming part of the insertion of the quadriceps extensor. The expansions of the vasti muscles on each side form an additional covering to the front of the joint and become incorporated with its capsule. Between the ligament and the head of the tibia is a bursa, which must not be confounded with the proper *bursa patellæ* in front of the bone.

Fig. 65.

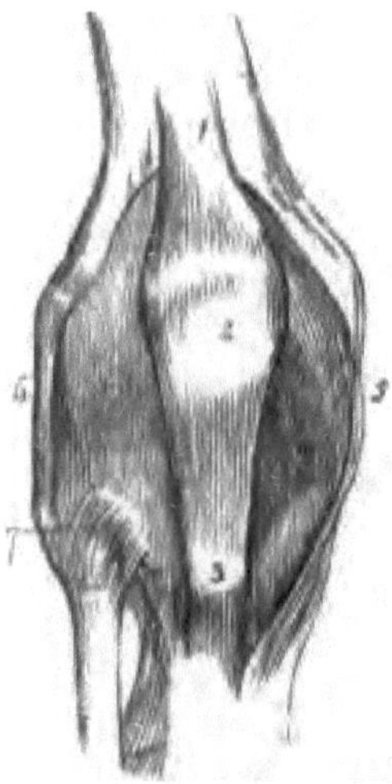

[To see the remaining ligaments, the popliteal vessels and nerves and the remains of the gastrocnemius and plantaris must be removed. The tendons of the biceps, semi-membranosus, and popliteus are to be traced out fully, and should be carefully preserved, but cut short.]

The **Posterior Ligament** (Fig. 66, 1) (ligamentum posticum Winslowii) is closely connected with the tendon of the semi-membranosus, from which a large band of fibres passes obliquely upwards and outwards across the back of the joint superficially to the fibres of the capsular ligament.

The **External Lateral Ligament** (Fig. 66, 6 & 7) is divided into two portions, the *long* and *short*, the long being the anterior one of the two. The divisions of the tendon of the biceps embrace the long ligament, and must be divided to expose it thoroughly. It is a round cord attached to the side of the external condyle immediately above the depression for the popliteus, and to the outer side of the head of

---

Fig. 65.—Anterior view of the ligaments of the knee joint (from Wilson).

1. Tendon of rectus.
2. Patella.
3. Ligamentum patellæ.
4, 4. Synovial membrane distended, the expansion of the vasti muscles having been removed.
5. Internal lateral ligament.
6. Long external lateral ligament.
7. Anterior superior tibio-fibular ligament.

the fibula. The tendon of the popliteus and the external inferior articular artery have been traced beneath it. The short ligament is placed behind the long, and reaches from the condyle to the head of the fibula, but it is very indistinct.

Fig. 66.

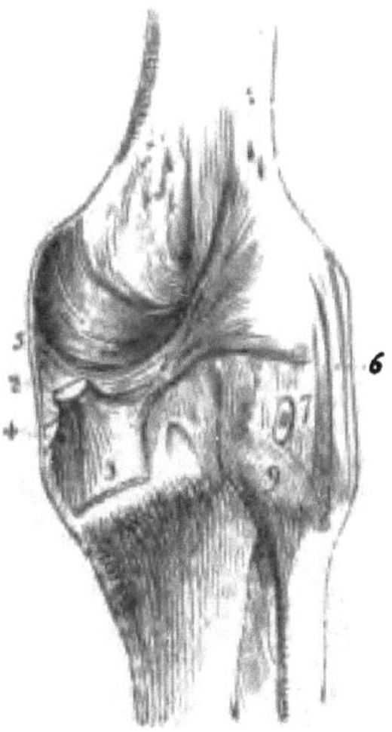

The **Internal Lateral Ligament** (Fig. 66, 5) is a broad band closely connected with the capsular ligament. It is attached to the side of the internal condyle, and to the inner side of the head of the tibia, where it covers in the tendon of the semimembranosus; and is then continued to the upper part of the shaft of the tibia, leaving an interval through which the inferior articular vessels of the knee pass.

The **Capsular Ligament** is a fibrous layer closely connected with the exterior of the synovial membrane and the inter-articular cartilages, and attached to the margins of the articular surfaces of the bones. It passes beneath the other ligaments and is strengthened by expansions from the several tendons, but is pierced by the tendon of the popliteus.

[The tendon of the rectus is to be divided and drawn down in order to see the extent of the synovial membrane

---

Fig. 66.—Posterior view of the ligaments of the knee joint (from Wilson).

1. Fasciculus of the ligamentum posticum Winslowii, derived from—
2. The tendon of the semi-membranosus muscle.
3. The process of the tendon which spreads out in the fascia of the popliteus muscle.
4. The process which is sent forwards beneath the internal lateral ligament.
5. Posterior border of the internal lateral ligament.
6. Long external lateral ligament.
7. Short external lateral ligament.
8. Tendon of the popliteus muscle cut through.
9. Posterior-superior tibio-fibular ligament.

above the patella, since it reaches for a couple of inches above the articular surface of the femur and gives attachment to the subcrureus. The synovial membrane is then to be divided on each side as near the condyles as possible and the patella turned down, when the ligamenta mucosum et alaria will be seen.]

The **Ligamentum Mucosum** (Fig. 68, 10) is a process of synovial membrane attached to the notch between the condyles, from which it stretches to immediately below the patella.

The **Ligamenta Alaria** (Fig. 68, 9) are two folds of synovial membrane with fringed borders, which are seen on each side of the ligamentum mucosum.

[The ligamentum mucosum is to be divided and the patella turned down completely, when the bursa of the ligamentum patellæ can be opened and seen. The capsular and posterior ligaments are to be removed, but the lateral ligaments are to be preserved and the crucial ligaments dissected.]

The **Intra-articular Ligaments** are the two crucial, the coronary, and the transverse ligaments. Within the joint also are the two semilunar cartilages.

The **Anterior Crucial Ligament** (Fig. 67, 2) is attached to the external condyle of the femur at its inner and back part, and passes obliquely across the joint to be inserted into the top of the tibia in front of the spine, between the attachments of the internal and external semilunar cartilages, with the latter of which it is closely connected.

The **Posterior Crucial Ligament** (Fig. 67, 3) is broader and stouter than the anterior. It is attached to the internal condyle at its anterior and outer part, and passes nearly vertically to the posterior margin of the top of the tibia and popliteal notch, being closely connected with the posterior extremity of the internal semilunar cartilage.

The crucial ligaments would alternately check extreme flexion and extension, the anterior ligament becoming tightened when the knee-joint is *extended*, and the posterior when the knee is *flexed*.

If the lateral ligaments are now divided, it will be found that the femur and tibia become more separated than before, showing that the object of the crucial ligaments is not simply to maintain the articular surfaces in apposition. It will also be seen that whilst rotation of the tibia outwards can now be carried to an extreme degree (so as to bring the crucial ligaments almost parallel), rotation inwards is immediately checked by the crossing of the ligaments, and it is thus that the tendency of the popliteus muscle to rotate the leg inwards is counterbalanced. The proof of this is that when, in extensive disease of the knee, the crucial ligaments are destroyed, there is a constant tendency for the leg to become inverted.

Fig. 67.

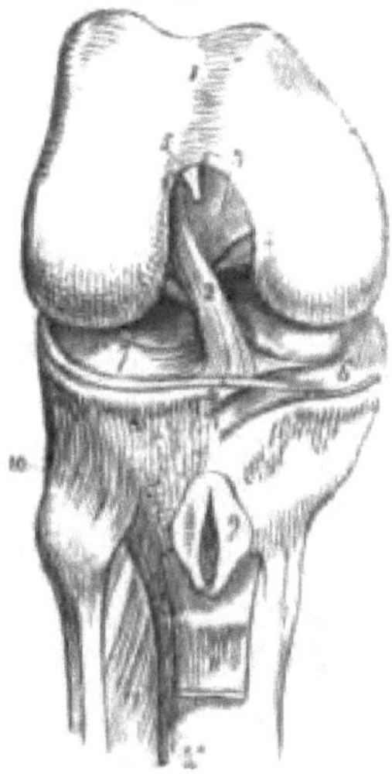

The **Transverse Ligament** (Fig. 67, 4) is a small band connecting the anterior extremities of the semilunar cartilages, and often not distinct.

The **Semilunar Cartilages** (Fig. 67) can be partly seen now, but will be fully exposed by dividing the crucial ligaments. They are fibro-cartilages which are thicker at the outer than at the inner margins, and which serve to deepen the condyloid cavities on the top of the tibia, being attached to the margin of the bone by short fibres called the *coronary ligaments*. The extremities of each cartilage

---

Fig. 67.—The right knee joint laid open from the front, and dissected to show the internal ligaments (from Wilson).

1. Cartilaginous surface of the lower extremity of the femur.
2. Anterior crucial ligament.
3. Posterior crucial ligament.
4. Transverse ligament.
5. Attachment of the ligamentum mucosum; the rest has been removed.
6. Internal semilunar cartilage.
7. External semilunar cartilage.
8. Part of the ligamentum patellæ turned down.
9. Bursa situated between the ligamentum patellæ and head of the tibia laid open.
10. Anterior superior tibio-fibular ligament.
11. Interosseous membrane.

are fixed respectively in front of and behind the spine of the tibia by short strong bands of fibrous tissue.

The **Internal Semilunar Cartilage** (Fig. 67, 6) is oval in form and embraces the extremities of the external cartilage. Its anterior extremity is attached to the tibia in front of the anterior crucial ligament, the posterior extremity is attached in front of the posterior crucial ligament.

Fig. 68.

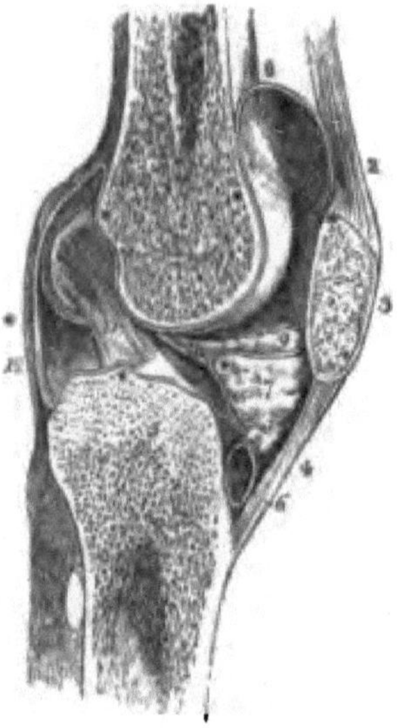

The **External Semilunar Cartilage** (Fig. 67, 7) is nearly circular; its anterior extremity is attached to the tibia immediately in front of the spine, and is connected with the insertion of the anterior crucial ligament. Its posterior extremity is attached to the back of the spine of the tibia, and is connected with the posterior crucial ligament by a distinct slip.

The **Attachments to the top of the Tibia** from before and backwards will be seen to be as follows:—

1, Transverse ligament; 2, Anterior extremity of internal semilunar cartilage; 3, Anterior crucial ligament;

Fig. 68.—Vertical section of the left knee joint, showing the reflections of its synovial membrane (from Wilson).

1. The lower part of the femur.
2. Tendon of the quadriceps extensor.
3. Patella.
4. Ligamentum patellæ.
5. The head of the tibia.
6. A bursa situated between the ligamentum patellæ and the head of the tibia.
7. Mass of fat projecting into cavity of the joint below the patella.

* * The synovial membrane.

8. The pouch of synovial membrane which ascends between the tendon of the extensor muscles of the leg and the front of the lower extremity of the femur.
9. One of the alar ligaments; the other has been removed with the opposite section.
10. Ligamentum mucosum left entire; the section being made to its inner side.
11. Anterior or external crucial ligament.
12. Posterior ligament.

4, Anterior extremity of external semilunar cartilage; 5, Posterior extremity of external semilunar cartilage; 6, Posterior extremity of internal semilunar cartilage; 7, Posterior crucial ligament.

The **Synovial Membrane** (Fig. 68) extends for at least two inches above the articular surface of the femur, forming a pouch beneath the extensor muscles. It is reflected from the articular surface of the femur over the crucial ligaments to the articular surface of the tibia, covering both aspects of the semilunar cartilages, and then lines the capsular ligament. It gives a tube around the tendon of the popliteus where it is within the capsule, and has already been seen to form the ligamentum mucosum and ligamenta alaria.

Fig. 69.

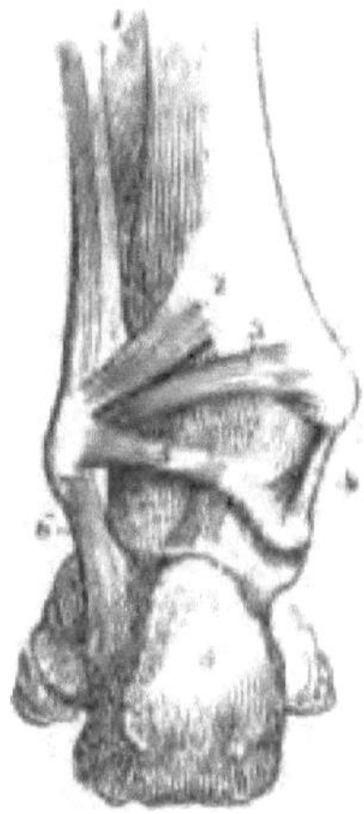

Tibio-Fibular Articulations.

[The whole of the muscular fibres connected with the bones of the leg and foot must be removed, but the tendinous insertions about the foot should be kept to be examined with the ligaments.]

The **Superior** tibio-fibular articulation (Fig. 67) is a simple arthrodial joint, the synovial membrane of which is occasionally continuous with that of the knee. The *anterior* and *posterior* ligaments are short bands connecting the anterior and posterior surfaces of the head of the fibula with the outer tuberosity of the tibia.

The **Middle** tibio-fibular articulation is formed by the

Fig. 69.—Posterior view of the ankle joint (from Wilson).

1. Lower part of the interosseous membrane.
2. Posterior inferior tibio-fibular ligament.
3. Transverse ligament.
4. Internal lateral ligament.
5. Posterior fasciculus of the external lateral ligament.
6. Middle fasciculus of the external lateral ligament.
7. Posterior ligament of the ankle joint.
8. Posterior tuberosity of the os calcis.

interosseous ligament or membrane, which is the great bond between the shafts of the bones of the leg, its fibres running downwards from the tibia to the fibula. It is attached to the sharp interosseous borders of both bones in all their lower part, a space being left above through which the anterior tibial vessels pass. The membrane has an opening in it near the lower part through which the anterior peroneal artery reaches the front of the leg.

Below the interosseous membrane is the *inferior interosseous ligament*, which consists of very short fibres passing between the triangular rough impressions on the adjacent surfaces of the tibia and fibula immediately above their inferior articulation.

The **Inferior** tibio-fibular articulation (Fig. 69) is a simple arthrodial joint continuous with the ankle-joint. It has *anterior* and *posterior* ligaments between the two bones resembling those above, and in addition a *transverse* (or *inferior*) *ligament* (3) placed below the posterior and distinct from it, which reaches from the external malleolus to the posterior surface and malleolus of the tibia, forming part of the true ankle joint.

Ankle Joint and Articulations of the Foot.

The ankle joint is a hinge joint capable of limited lateral motion when the toes are pointed. It has *in front* the tendons of the tibialis anticus, the tibial vessels and nerve, the extensor proprius pollicis, and the extensor longus digitorum with the peroneus tertius. *Behind*, to the inner sides are the tendons of the tibialis posticus, the posterior tibial vessels and nerve (or it may be the plantar vessels and nerves) ; the flexor longus pollicis near the middle ; and to the outer side, the tendons of the peroneus longus and brevis. The ankle joint has anterior, posterior, and two lateral ligaments.

The **Anterior Ligament** (Fig. 70, 7) is a broad membrane which is seldom seen entire. It is attached to the lower margin of the tibia, and to the anterior margin of the superior articular surface of the astragalus, joining the lateral ligament on each side.

The **Posterior Ligament** (Fig. 69, 7) is not described by many authors, and is little more than a few fibres stretching over the synovial membrane between the back of the tibia and the astragalus.

The **Internal Lateral Ligament** (Fig. 70, 6) is triangular in shape, and hence has been called *deltoid*. The apex is attached to the extremity of the internal malleolus, and the base to the posterior part of the astragalus, to the lesser process of the os calcis, and the tuberosity of the scaphoid bone. A strong *deep portion* passes from the apex of the malleolus to the side of the astragalus.

Fig. 70.

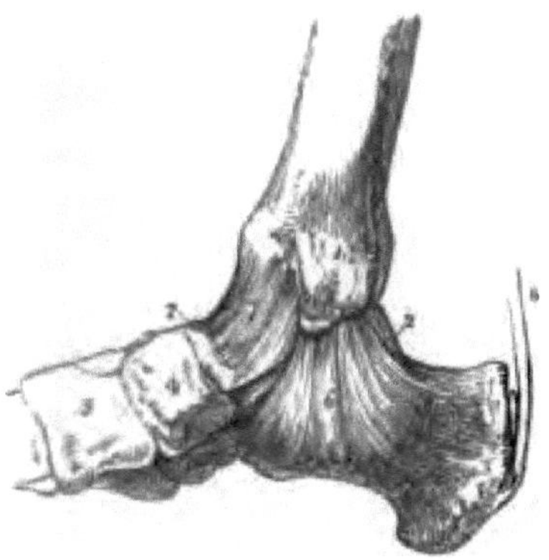

Fig. 71.

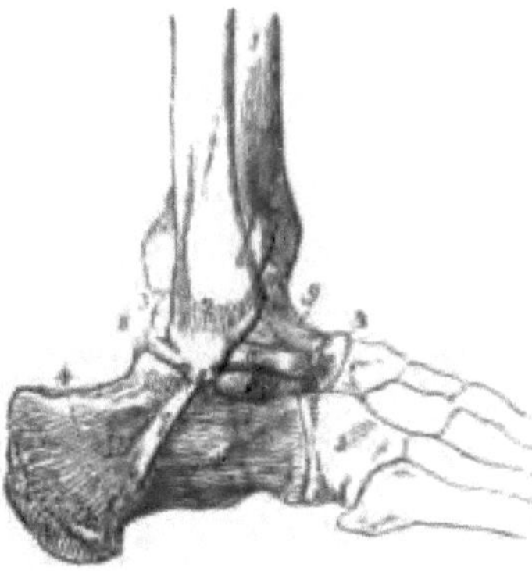

The **External Lateral Ligament** (Fig. 71) consists of three slips; two horizontal attached to the astragalus, and

---

Fig. 70.—Internal view of the ankle joint (from Wilson).

1. Internal malleolus of the tibia.
2, 2. Astragalus.
3. Os calcis.
4. Scaphoid bone.
5. Internal cuneiform bone.
6. Internal lateral or deltoid ligament.
7. Anterior ligament.
8. Tendo Achillis; a small bursa is seen interposed between this tendon and the posterior tuberosity of the os calcis.

Fig. 71.—External view of the ankle joint (from Wilson).

1. Tibia.
2. External malleolus of the fibula.
3, 3. Astragalus.
4. Os calcis.
5. Cuboid bone.
6. Anterior fasciculus of the external lateral ligament attached to the astragalus.
7. Its middle fasciculus, attached to the os calcis.
8. Its posterior fasciculus attached to the astragalus.
9. Anterior ligament of the ankle.

one more or less vertical, to the os calcis. The *anterior* slip (6) is attached to the front of the external malleolus and to the side of the astragalus in front of the superior articular surface; the *middle* (7) is a round cord which reaches backwards from the tip of the malleolus to the outer side of the os calcis about its middle; the *posterior* (8) is attached to the deep groove behind the articular surface of the external malleolus, and passes horizontally to the posterior surface of the astragalus behind the articulation with the tibia (Fig. 69, 5).

Fig. 72.

The **Calcaneo-astragaloid Articulation** has *external*, *posterior*, and *interrosseous* ligaments. The *external* is a short band passing obliquely backwards from the side of the astragalus to the upper margin of the greater process of the os calcis. The *posterior* is placed between the posterior parts of the two bones, close to the groove in the astragalus through which the tendon of the flexor longus pollicis passes. The *interosseous* ligament will be seen when the joint is opened. It is a strong thick band passing from

---

Fig. 72.—Ligaments of the sole of the foot (from Wilson).

1. Os calcis.
2. Astragalus.
3. Tuberosity of the scaphoid bone.
4. Long calcaneo-cuboid ligament.
5. Part of the short calcaneo-cuboid ligament.
6. Calcaneo-scaphoid ligament.
7. Plantar tarsal ligaments.
8, 8. Tendon of the peroneus longus muscle.
9, 9. Plantar tarso-metatarsal ligaments.
10. Plantar ligament of the metatarso-phalangeal articulation of the great toe; the same ligament is seen on the other toes.
11. Lateral ligaments of the metatarso-phalangeal articulation.
12. Transverse ligament.
13. Lateral ligaments of the phalanges of the great toe; the same ligaments are seen on the other toes.

the groove between the two concave articular surfaces of the astragalus and the corresponding groove between the two convex articular surfaces on the superior aspect of the calcaneum.

The *tendon of the tibialis posticus* is to be fully traced out now, and will be found to be principally attached to the tuberosity of the scaphoid bone and the adjacent internal cuneiform bone, giving slips also to all the adjacent bones of the tarsus, with the exception of the astragalus. These must be removed to see the true ligaments of the joints.

Ligaments of the Tarsus. On the dorsal aspect of the remaining tarsal bones will be found a number of short bands or fibres which pass between adjacent bones, and have received names as ligaments accordingly.

On the plantar aspect there are similar ligaments, three only of which are worthy of special notice, viz., the long and short calcaneo-cuboid ligaments, and the calcaneo-scaphoid ligament.

The **Long Calcaneo-cuboid Ligament** (Fig. 72, 4)—*ligamentum longum plantæ*—is a broad ligament which has been already seen in the dissection of the sole of the foot. It is extensively attached to the under surface of the os calcis in front of the tuberosities, and passes forward to be attached to the posterior margin of the peroneal groove in the cuboid bone. Some of the fibres pass on over the tendon of the peroneus (to which they form a sheath), and are attached to the bases of the second, third, and fourth metatarsal bones.

The **Short Calcaneo-cuboid Ligament** (Fig. 72, 5) lies to the inner side of the long ligament, and is more deeply placed. It reaches from the tubercle on the under surface of the greater process of the os calcis to the under surface of the cuboid behind the peroneal groove.

The **Calcaneo-scaphoid Ligament** is a broad band stretching between the lesser process of the calcaneum (*sustentaculum tali*) and the tuberosity of the scaphoid bone. It performs the important function of supporting the head of the astragalus, which rests upon it, and in this is assisted by the tendon of the tibialis posticus, which passes immediately beneath it. The upper surface of the ligament is covered with the synovial membrane of the astragalo-scaphoid articulation.

Numerous *interosseous ligaments* bind the bones of the tarsus firmly together and will be seen when the joints are opened.

The **Metatarsal Bones** (Fig. 72, 9) are connected with the tarsus by strong dorsal, plantar, and interrosseous ligaments.

The *dorsal* ligaments are one to each metatarsal bone from the tarsal bone with which it articulates; but the second metatarsal, in addition to its ligament from the middle cuneiform, has extra slips from the internal and external cuneiform bones, between which its base is wedged.

The *plantar* ligaments are less regular, and are united with the slips from the calcaneo-cuboid ligament.

The *interosseous ligaments* are short strong bands, and vary slightly in different subjects. There is always one between the internal cuneiform and the second metatarsal bone; a second may exist between the external cuneiform and the same metatarsal bone; and a third between the external cuneiform and the fourth metatarsal bone.

Transverse dorsal, plantar, and interosseous ligaments connect the bases of the metatarsal bones.

The **Synovial Membranes** (Fig. 63) of the foot, including the ankle joint, are seven in number, and will be seen by opening the several joints in succession.

The 1st is the synovial membrane of the ankle, reflected over the upper surface of the astragalus and upon the ligaments to the articular surfaces of the tibia and fibula: it passes also between the tibia and fibula into the inferior tibio-fibular articulation.

The 2nd synovial membrane lines the posterior calcaneo-astragaloid articulation.

The 3rd synovial membrane covers the head of the astragalus and the concave surface of the scaphoid, being prolonged backwards into the anterior calcaneo-astragaloid articulation.

The strong *interosseous ligament* connecting the under surface of the astragalus and the upper surface of the os calcis will now be seen. It is attached to the deep groove found between the two articulations in each bone. An *interosseous ligament* will also be seen, when the astragalus is displaced, passing from the upper border of the greater process of the os calcis to the scaphoid bone.

The 4th synovial membrane is placed between the greater process of the os calcis and the back of the cuboid bone, and when it is opened the peculiar manner in which the two bones are locked together will be seen. An *interosseous ligament* will be found between the os calcis and cuboid bone to the inner side of the synovial cavity.

The 5th synovial membrane is between the scaphoid and the three cuneiform bones, and will be found to pass between the middle and external cuneiform bones to their anterior surfaces and the bases of the second and third metatarsal bones. It is also prolonged into the articulations between the second, third, and fourth metatarsal bones, and sometimes into that between the external cuneiform and cuboid bones, in which case an interosseous ligament often existing between the external cuneiform and the fourth metatarsal bone is wanting.

Interosseous ligaments will be found between the scaphoid and cuboid bones; between the internal and middle cuneiform bones; between the middle and external cuneiform bones; and between the external cuneiform and cuboid bones.

The 6th synovial membrane will be found between the front of the cuboid bone and the bases of the fourth and fifth metatarsal bones, and entering the articulation between them.

The 7th is a separate synovial membrane for the first metatarsal bone, being placed between it and the internal cuneiform bone.

Surgery.—The shape of the articulations should be

particularly noticed with reference to the several amputations about the foot. At the ankle the astragalus is locked between the projecting malleoli, and it is only by dividing the lateral ligaments by the side of the astragalus and outside the joint altogether, that the bones can be separated, as in Syme's or Pirogoff's amputations.

The astragalo-scaphoid and calcaneo-cuboid articulations will be seen to be nearly parallel, and are easily opened in Chopart's amputation. If, in performing this operation, the articulation in front of the scaphoid should be opened by mistake, the operator would at once see the three articular surfaces for the cuneiform bones, instead of the single globular surface of the astragalus.

The operation of disarticulating the metatarsus from the tarsus (Hey's or Lisfranc's amputation) is rendered very difficult by the fact that the base of the second metatarsal bone is firmly wedged between the internal and external cuneiform bones, projecting farther into the tarsus than the other bones, and being also attached to the internal cuneiform by a strong interosseous ligament. In amputating at this point therefore, after opening the articulation of the first and three outer metatarsal bones, it is necessary to thrust the point of the knife between the internal cuneiform and second metatarsal bones to divide this ligament, or there will be danger of dragging away the internal cuneiform bone.

The articulations between the metatarsal bones and the phalanges, and between the phalanges themselves, are similar to those in the hand, but on a smaller scale, and therefore more difficult of dissection. Reference may be made to the hand for their description. (*Vide* p. 86.)

PART III.

DISSECTION OF THE ABDOMEN.

[*The Student is requested to read the 'Introduction' before commencing the dissection, unless he has done so previously.*]

Surgery.—Before the subject is tied up for the dissection of the perinæum, the student should practise the operation of introducing the catheter.

In the case of the **Male Subject** the operator should stand on the left side of the body, and having oiled a perfectly smooth and clean staff or sound, should grasp the end of the penis with the left hand and draw it upwards so as to stretch the urethra whilst introducing the instrument. The sound is to be held lightly in the right hand, and is to be passed at first along the fold of the groin, but when it has entered the urethra for a few inches it is to be brought parallel to the median line of the body. The introduction of an instrument is more difficult on the dead than on the living body, owing to the want of the mucous secretion of the urethra and the relaxed condition of the canal; and it is advisable therefore to withdraw the instrument and oil it afresh if any obstruction should occur.

When an obstruction to entering the bladder occurs, the left hand should be placed upon the perinæum at the point to which the staff has reached, and may be made to grasp the urethra and guide the instrument in its right course.

When the point of the staff has passed the bulb, the handle is to be depressed between the thighs, and the point of the instrument will probably enter the bladder. If it is found impossible to depress the handle, the student may be certain that he has failed to pass the triangular ligament, and should be very gentle in his manipulations, or he will very probably perforate the bulb. By a series of very cautious efforts, aided by the left hand on the perinæum, the staff may at last be passed, or if an organic stricture should exist, it will be necessary to use a smaller instrument.

When the handle of the instrument can be depressed between the thighs, the student may judge of its having entered the bladder by the readiness with which it passed, and by the ease with which the end of the staff can be rotated in the bladder. In all cases, however, it is advisable for the student to introduce his left fore-finger into the rectum, in order to ascertain that the instrument is not in a false passage, and also in order to appreciate the thickness of the coats of the bladder, the size of the prostate, and the course the catheter would take on the living body.

In the **Female Subject** it will be advisable to examine the position of the urethra by separating the labia before attempting to introduce an elastic catheter. It will be seen that the orifice of the urethra is separated from the clitoris by the space known as the vestibule, and that it is close above the entrance to the vagina. To pass the catheter, the student should stand on the right side of the subject, and having passed the left fore-finger between the thighs and labia, should place it at the anterior edge of the orifice of the vagina; the flexible catheter being then passed along the fore-finger can be slightly raised so as to enter the urethra, and will be felt in its passage through that canal by the finger at the orifice of the vagina. It will be well to practise the operation once or twice with the parts exposed to view, and then to repeat it with a cloth thrown over the pubes, as would be done in actual practice.

The Perinæum.

[The perinæum is now to be fully exposed by bringing the buttocks of the subject to the edge of the table, where they are to be raised upon a block. The thighs are then to be flexed upon the abdomen, with the legs bent; and the body is to be secured upon the table with a cord passing beneath it from one knee to the other, a block being placed lengthways between the knees to keep them sufficiently apart. The staff is to be retained in the male urethra, and secured in its place by tying a loop of string round the penis and fastening it to the handle of the staff.]

External Appearances.—*In the male* the skin is of a darker tint than that of the rest of the body, and is more or less covered with hair. In front will be found the *scrotum*, containing the two testes, and obscuring at present the root of the penis and deeper portion of the urethra. Behind the scrotum is the *anus* or orifice of the bowel, and between the two is the true *perinæum* which will be seen to be marked by a median raphé. Around the anus the skin is thrown into folds, which, if enlarged, form one variety of external piles. By drawing the anus slightly open, a white line will be seen, marking the junction of the skin with the mucous membrane, and corresponding to the separation between the internal and external sphincter muscles (Hilton).

In the female (Fig. 76) the scrotum may be considered to be split in the median line, to form the two *labia majora*, which however consist simply of cellular tissue and fat, and are continuous above with the *mons Veneris*, a thickened portion of the integument over the pubes (only partially seen in this view). The labia are united in front by the *anterior commissure*; they bound the *vulva* and are continued backwards to the anterior boundary of the perinæum, where they are united by the *posterior commissure*. In a female who has not borne children, a small transverse fold of mucous membrane may be seen just within the posterior commissure, which is called the *fourchette*, and between the two is the *fossa navicularis*.

position) from the tip of the coccyx. The fat filling the space between the anus and the edge of the muscle is to be carefully removed piecemeal, care being taken to preserve the inferior hæmorrhoidal vessels and nerves which run nearly transversely towards the bowel, and also a little branch of the fourth sacral nerve near the posterior part of the space, or ischio-rectal fossa.]

The **Ischio-rectal Fossa** (Fig. 73) is the space on each side of the rectum bounded by the tuber ischii and the gluteus maximus muscle. Its deep boundaries are, on the

Fig. 73.

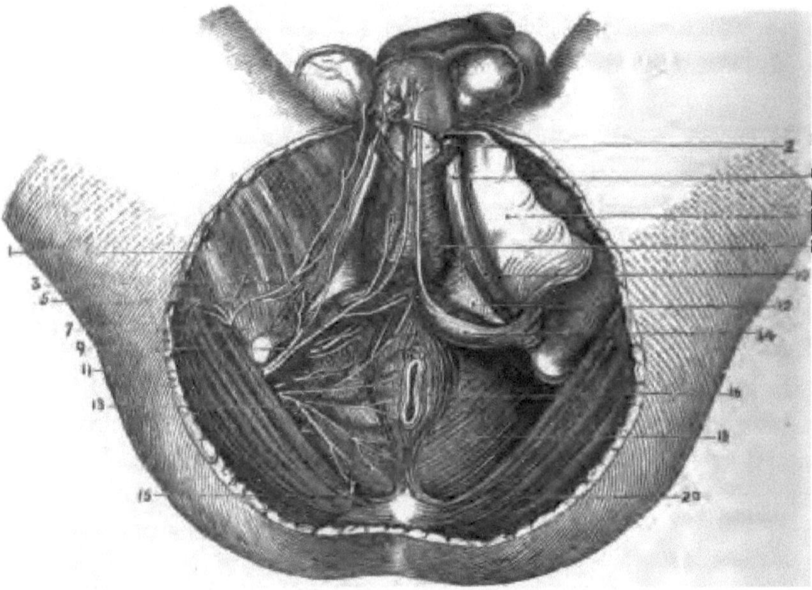

Fig. 73.—Superficial dissection of male perinæum (drawn by J. T. Gray).

1. Inferior pudendal nerve.
2. Urethra.
3. External or posterior superficial perinæal nerve.
4. Crus penis.
5. Superficial perineal artery.
6. Deep layer of superficial fascia (reflected).
7. Internal or anterior superficial perinæal nerve.
8. Accelerator urinæ.
9. Muscular branches of pudic nerv
10. Erector penis.
11. Pudic artery and nerve.
12. Deep perinæal fascia or triangular ligament.
13. Inferior hæmorrhoidal artery and nerve.
14. Reflection of the deep layer of superficial perinæal fascia round transversus perinæi.
15. Branch of fourth sacral nerve.
16. Sphincter ani.
18. Levator ani.
20. Gluteus maximus.

inner side the levator ani muscle, which slings the rectum, and has an oblique direction towards the middle line; and on the *outer* side the obturator internus lying on the internal surface of the thyroid foramen. Both these muscles are covered by processes of fascia derived from the pelvic fascia; that covering the obturator muscle being called the *obturator fascia*, and that upon the levator ani being the *anal fascia* (*see* diagram of pelvic fascia). When the finger is pushed into this space it is prevented from entering the pelvis by the reflection of this fascia upon the two muscles.

On the outer side of the space and behind the margin of the pubes the pudic vessels and nerves can be felt enclosed in a tube of fascia, and it is here that the artery can be effectually compressed in cases of hæmorrhage after lithotomy.

Surgery.—The tissues in the ischio-rectal fossa are more or less involved in the several forms of fistula in ano, and the sphincter ani, together with some of the fat and a small branch of artery, will be divided in the ordinary operation for the cure of that affection. In the operation of lateral lithotomy the left ischio-rectal fossa is cut into, and care has to be taken not to injure the rectum on the inner, or the pudic artery on the outer, side of the space.

The **Levator Ani Muscle** (Fig. 73, 18) *arises* within the pelvis, being attached to the back of the os pubis, to the spine of the ischium, and to the pelvic fascia between those points. It is now seen at its *insertion* into the central line of the perinæum, where it meets its fellow of the opposite side both in front of and behind the rectum; it is inserted also into the side of the rectum and into the anterior surface of the extremity of the coccyx with the opposite muscle.

The levator ani supports the pelvic viscera, and when in contraction, tends to prevent the expulsion of their contents. It may also be considered as an extraordir

By separating the labia the *clitoris* will be seen at the upper part of the vulva, resembling a diminutive penis in appearance and structure, and having a small and ill-defined *glans* and *prepuce*.

The *nymphæ* or *labia minora* extend obliquely downwards from each side of the clitoris, being connected both with the organ itself and its prepuce, and being lost in the labia majora and wall of the vagina below.

The *vestibule* is a triangular interval with the apex at the clitoris and the base at the orifice of the vagina, the sides being formed by the nymphæ. An inch below the clitoris is the *meatus urinarius* which is slightly prominent, and is placed immediately above the orifice of the vagina.

The entrance of the vagina may be more or less occluded by a *hymen*, which is a reduplication of mucous membrane, usually of a crescentic form with the concave border upwards. When ruptured it forms *carunculæ myrtiformes*, three or four little projections from the vaginal wall. On each side of the vagina immediately in front of the hymen, or its remains, are the orifices of the *vulvo-vaginal glands* or glands of Bartholin. The orifices of numerous sebaceous follicles will also be found scattered over the vulva.

The true *perinæum* of the female is between the posterior commissure of the labia and the anal orifice, and is from an inch to an inch-and-a-half long, but becomes considerably elongated during labour. The finger should be introduced into the vagina to ascertain the position and condition of the cervix and os uteri.

Boundaries of the Perinæum.—These are the same in both sexes, and can be best ascertained by placing a pelvis, on which the ligaments are preserved, in the same position as the subject, when the space under examination will be seen to correspond with the inferior aperture of the pelvis. In front is the symphysis pubis, with the divergent rami extending to the tuberosities of the ischia, which form the lateral boundaries. Posteriorly the great sacro-sciatic ligaments extend to the sacrum, but in the recent subject the margin of the glutei maximi muscles overlap these and reach to the coccyx, which is the posterior limit of the space.

The perinæum has been generally considered to resemble the heraldic lozenge, but has more aptly been compared to an inverted ace of hearts (Thompson). The inferior aperture of the pelvis is larger in the female than in the male, but considerable variation in the space between the rami of the pubes will be found in different individuals.

The perinæal space may be conveniently divided into halves by a line passing from the anterior border of one tuberosity of the ischium to the other; the anterior or urethral half containing the urino-genital organs, and the posterior or rectal half the lower extremity of the bowel.

POSTERIOR SPACE IN BOTH SEXES.

[A little cotton-wool is to be introduced into the rectum, and the anus is then to be carefully sewn up, the stitches being placed as near the white line mentioned above as possible. A transverse incision is then to be made in front of the anus connecting the two tuberosities, and a similar one across the lower extremity of the coccyx. These are to be joined by incisions carried round the anus as near the sutures as possible, and the skin is to be reflected on each side.]

The **Sphincter Ani** (Fig. 73, 16) (external) surrounds the anus and is close beneath the skin. It consists of two symmetrical halves, which are attached to the tip of the coccyx behind, and meet in front of the anus at the 'central point of the perinæum.' The external sphincter controls the lower end of the bowel and is *supplied* by the inferior hæmorrhoidal nerve, and by a branch of the fourth sacral nerve.

The **Internal Sphincter** cannot be seen at present. It is merely a thickened band of the unstriped circular fibres of the rectum, and differs therefore in structure from the external sphincter, which is a voluntary muscle.

[The edge of the gluteus maximus is now to be defined since it forms the boundary of the dissection; it will found extending obliquely upwards and outwards (in t

Veins corresponding to the arteries open into the internal pudic vein.

The **Superficial Perinæal Nerves** (Fig. 73) are two in number, and are branches of the internal pudic nerve.

The *external* or *posterior* (3) nerve has a short course in the ischio-rectal fossa, where it gives a branch to the anus. It then pierces the reflection of the deep layer of superficial fascia, and gives branches to the scrotum, forming a junction with the inferior pudendal nerve of Soemmering.

The *internal* or *anterior* (7) nerve accompanies the superficial perinæal artery either over or under the transversus perinæi muscle, and is distributed to the scrotum near the median line, where it unites with its fellow of the opposite side.

The **Inferior Pudendal Nerve** (Fig. 73, 1) (Soemmering) is a branch of the small sciatic nerve which pierces the deep fascia of the thigh about an inch in front of the tuberosity of the ischium. It varies a good deal in size and distribution, but generally runs inwards and forwards to the scrotum to join the external superficial perinæal nerve.

[The superficial vessels and nerves are to be turned aside and the muscles dissected out. The accelerator urinæ is in the median line, the erector penis parallel to the ramus of the pubes, and the transversus perinæi crosses between them; the muscles thus bounding a triangular space in which a part of the triangular ligament is seen.]

Superficial Muscles of the Perinæum.

The **Accelerator Urinæ** (Fig. 73, 8) (ejaculator seminis) is a single muscle composed of two symmetrical halves united in the middle line by a delicate fibrous raphé. The fibres *arise* from the central point of the perinæum and from the fibrous raphé in the middle line, and are thus *inserted*:—the posterior fibres which are nearly transverse in their direction, overlie the bulb and are lost upon the triangular ligament; the middle fibres enclose the urethra and corpus spongiosum; and the anterior fibres enclose the entire penis including the dorsal vessels.

The **Erector Penis** (Fig. 73, 10) covers the crus penis on the ramus of the os pubis. It *arises* from the anterior and inner surface of the tuberosity of the ischium, and also from the ramus of the ischium, and ends in an aponeurosis which is *inserted* into the sides of the crus penis.

The **Transversus Perinæi** (Fig. 73, 14) varies in size, and may be divided into two slips. It *arises* from the inner side of the ramus of the ischium, and meets its fellow muscle of the opposite side in the central point of the perinæum, where it is also blended with the fibres of the accelerator urinæ and sphincter ani.

The action of the accelerator is to expel the contents of the urethra by its sudden and spasmodic contraction. It is not put in action during the greater part of the act of micturition; but when the flow of urine has ceased, it serves to eject the small remaining portion, or may be used to arrest the flow of urine suddenly, when its action gives rise to considerable pain. Its action during coition is expressed by the name ejaculator seminis, and its anterior fibres assist materially in producing erection of the penis by compressing the dorsal vein. The erector penis assists somewhat in the production of erection, by compressing the corpus cavernosum.

Muscular Nerves (Fig. 73, 9). Passing either over or through the transversus perinæi muscle will be found branches of the pudic nerve, which arise from it either separately or in common with the anterior superficial perinæal nerve. These branches are distributed to the muscles of the perinæum, and a good-sized branch may be followed through the fibres of the accelerator urinæ to the bulb of the urethra (bulbo-urethral nerve of Cruveilhier).

Surgery.—The triangular space enclosed by the three superficial muscles, and of which the floor is formed by the triangular ligament or deep perinæal fascia, is important surgically as being the point to which the knife reaches in the incision for lateral lithotomy. In the first incision the surgeon cuts freely through the superficial structures, dividing the superficial perinæal and inferior hæmorrhoidal vessels and nerves, and aiming at the lower part of this space, at which point the finger will feel somewhat indistinctly the staff in the urethra. The deeper incision necessarily divides the transverse muscle and artery, together with the lower part of the triangular ligament, and enables the operator to reach the staff in the membranous portion of the urethra.

It should be noticed how the small space between the anus and the bulb of the urethra can be increased by drawing the whole penis upwards; this fact is of importance in relation to median lithotomy, which is performed in this space, and in which operation there is some risk of wounding the bulb. It is obvious that any cutting operation in the median line of the perinæum would be accompanied by but little hæmorrhage until the vascular structure of the bulb or corpus spongiosum urethræ was reached.

[The accelerator urinæ is to be divided in the middle line and dissected from the bulb beneath and from the triangular ligament, but its anterior fibres must not be interfered with. The transversus perinæi is to be removed, and the erector penis detached from the bone with the crus penis, which will be seen to be a vascular body attached to the ramus of the pubes for nearly an inch. This is to be drawn forward, a branch of artery going to it being carefully preserved; the triangular ligament will then be exposed.]

The posterior part of the **Corpus Spongiosum Urethræ** (Fig. 74, 6) will be seen to be a dark, vascular body, which expands to form the bulb immediately in front of the triangular ligament.

The **Bulb** is slightly expanded from side to side, and consists of two halves united by a delicate septum of fibrous tissue. It is closely connected with the triangular ligament with which its delicate fibrous covering is continuous.

— **Triangular Ligament** of the urethra or **Deep**

Perinæal Fascia (Fig. 74, 8) is a strong process of fibrous tissue stretching across the pubic arch, being attached to the rami of the pubes and ischium on each side behind the crura penis.

The apex of the ligament is connected with the subpubic ligament, and the base has been seen to be continuous with the deep layer of superficial fascia which turns around the transversus perinæi; but it is also connected with the central point of the perinæum and is continuous with the anal fascia over the levator ani muscle (p. 185).

About an inch below the pubes the urethra perforates the triangular ligament, with which it is inseparably united, and above this point are the openings for the dorsal vessels and nerves of the penis, the vein being in the median line with the artery and nerve on each side.

[It is advisable not to proceed any further with the dissection of the *left* side of the perinæum, in order that a good side view of this region may be obtained in a subsequent dissection. On the right side the triangular ligament is to be carefully detached from the bone and from its attachments below, and turned towards the median line; the bulb should also be drawn over to the left side with the hooks, so as to increase the space brought into view as much as possible.]

Parts behind the Triangular Ligament.*—These are, the deep muscles of the perinæum; the membranous portion of the urethra; Cowper's glands; the terminations of the pudic artery and nerve; the artery to the bulb; and the dorsal vein of the penis. Posterior to these is the pelvic fascia continuous with the capsule of the prostate.

Deep Muscles.—Muscular fibres having a transverse direction but varying considerably in size and distinctness,

* The triangular ligament of the urethra is by some authors said to consist of two layers, the anterior being the deep perinæal fascia, and the posterior the pelvic fascia. In that case the structures here enumerated must be considered as lying *between* the layers of the triangular ligament.

will be seen on the removal of the triangular ligament. These form the compressor urethræ, but the lower part when very distinct may be called the deep transverse muscle.

Fig. 74.

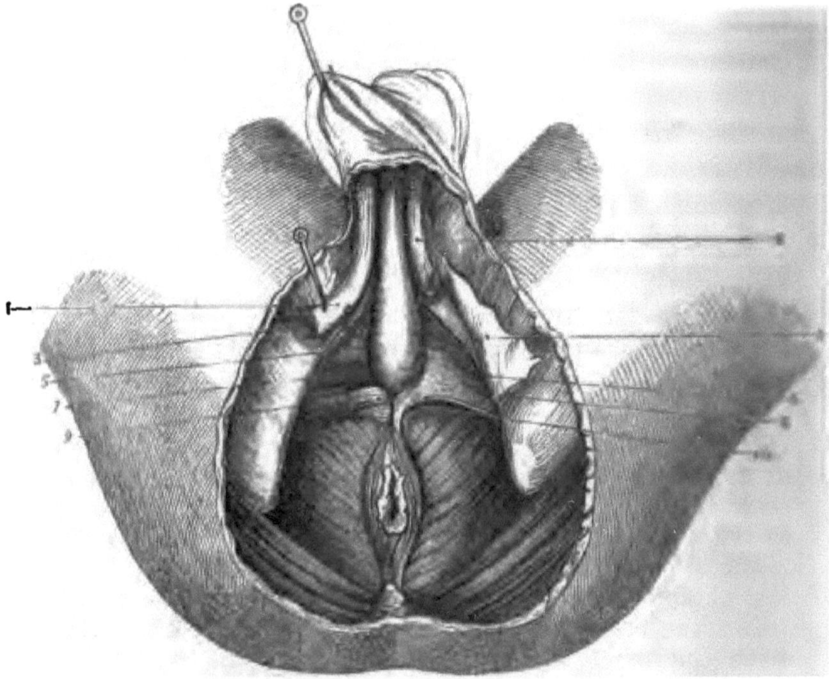

The **Compressor vel Constrictor Urethræ** (Fig. 74, 9) (Guthrie's muscle) *arises* from the posterior aspect of the ramus of the pubes, and divides into two slips which run inwards transversely to enclose the membranous portion of the urethra, meeting the opposite muscle in a median raphé, and being expanded upon the whole length of this portion of the canal. Circular muscular fibres surround

---

Fig. 74.—Deep dissection of the male perinæum, the triangular ligament being removed on the right side (drawn by J. T. Gray).

1. Right crus penis (cut).
2. Left crus penis (cut).
3. Pudic nerve.
4. Deep layer of superficial fascia (reflected).
5. Pudic artery.
6. Bulb of urethra.
7. Artery of the bulb.
8. Triangular ligament.
9. Compressor urethræ muscle.
10. Position of Cowper's gland (right).

the membranous urethra beneath the compressor, but these are involuntary, and continuous with the muscular fibres of the prostate (Ellis).

Behind the lower portion of this muscle near the middle line is situate one of Cowper's glands (10), and the muscle may conceal the artery to the bulb if it should arise lower than usual.

The **Deep Transverse Muscle** is generally merely the lower portion of the preceding muscle, which meets its fellow of the opposite side in the median line below the urethra.

[The muscular fibres are to be divided near the urethra in order to see one of Cowper's glands, and the pudic artery and nerve with their branches are to be followed out.]

Cowper's Glands are close to the membranous portion of the urethra, one being on each side of the median line. Each gland resembles a dried pea in size and shape, but consists of numerous minute lobules. Although the gland is so close to the membranous urethra, the duct opens into the bulbous portion of the canal, being nearly an inch long and perforating the floor of the urethra very obliquely.

The membranous portion of the urethra will be more satisfactorily examined afterwards than in the present dissection, but opportunity should be taken to feel the staff in the urethra at this point, since it is where that tube would be opened, on the left side, in the operation of lateral lithotomy, or in the middle line in the median operation.

The **Pudic Artery** (Fig. 75, 9) will be found under cover of the pubic arch. It ascends behind the triangular ligament and, after giving off the artery to the bulb and the artery to the corpus cavernosum, ends in the dorsal artery of the penis.

The **Artery to the Bulb** (Fig. 74, 7) is a good-sized branch, which is generally above the level of the compressor urethræ muscle. It passes transversly inwards to supply the bulb, to reach which it has to pierce the triangular

with the fascia of the body generally, and usually contains much fat, forming the labium.

The *deep layer* of superficial fascia is not so strongly marked as in the male, but has the same connections. It

Fig. 76.

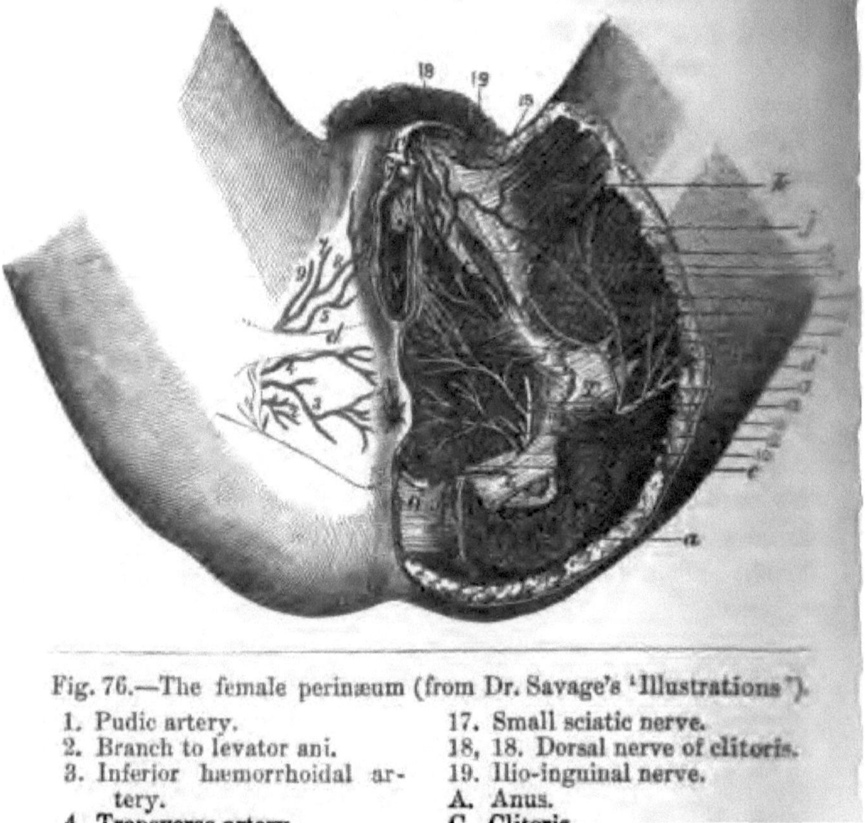

Fig. 76.—The female perinæum (from Dr. Savage's 'Illustrations').

1. Pudic artery.
2. Branch to levator ani.
3. Inferior hæmorrhoidal artery.
4. Transverse artery.
5. Great labial (superficial perineal artery).
7. Dorsal artery of clitoris.
8. Artery of bulb.
9. Artery to crus clitoridis.
10. Inferior hæmorrhoidal nerve.
11. Pudic nerve.
12. Muscular branch.
13. Internal superficial perineal nerve.
14. External superficial perineal nerve.
15. Its junction with—
16. Inferior pudendal nerve.
17. Small sciatic nerve.
18, 18. Dorsal nerve of clitoris.
19. Ilio-inguinal nerve.

A. Anus.
C. Clitoris.
M. Meatus urinarius.
L. Great sacro-sciatic ligament.
V. Vagina.
O. Coccyx.
T. Tuberosity of ischium.
a. Gluteus maximus.
c. Levator ani.
d. Superficial transverse muscle.
e. Compressor bulbi.
g. Erector clitoridis.
h. Triangular ligament (cut).
i. Biceps and semi-tendinosus.
j. Adductor magnus.
k. Gracilis.

s continuous with the deep fascia of the groin, and is attached on each side to the front of the rami of the pubes and ischium, nearly to the tuberosity. It then makes a turn around the transversus perinæi muscle on each side, and joins the deep perinæal fascia or triangular ligament, forming at the border of the transversus muscle the ischio-perinæal ligament of Savage. (Fig. 77, 15.)

Owing to the position of the vulva, the deep layer of superficial fascia is divided in the middle line and becomes continuous with the sheath of the vagina; and thus also the pouch between the deep layer of superficial fascia and the triangular ligament (p. 187) is divided in the female into two parts, which have been named *pudendal* or *vulvo-scrotal sacs* of Broca.

[The deep layer of superficial fascia is to be removed, and the superficial perinæal vessels and nerves dissected out.]

The **Superficial Perinæal Artery** (artery of the labium) (Fig. 76, 5) is larger than in the male. It is a branch of the internal pudic artery, and arises under cover of the ramus of the ischium to run forwards and inwards. It enters the pouch of fascia by piercing the lower border of the triangular ligament, and then passes over or under the transversus perinæi muscle to supply the superficial muscles of the perinæum and the labium.

The **Transverse Artery** (Fig. 76, 4) is a branch of the internal pudic arising immediately in front of or with the preceding vessel. It runs towards the median line close to the transversus muscle, and anastomoses with the opposite artery and with the neighbouring branches. *Veins* corresponding to the arteries open into the internal pudic vein.

The **Superficial Perinæal Nerves** (Fig. 76) are two in number and are branches of the internal pudic nerve.

The *internal* (13) nerve runs upwards and inwards, entering the pouch of fascia and being distributed to the posterior part of the vulva.

The *external* (14) nerve is larger, and is distributed over the superficial muscles of the perinæum, forming a junction with the inferior pudendal nerve of Soemmering.

The **Inferior Pudendal Nerve** (Soemmering) (Fig. 76, 16) is a branch of the small sciatic nerve which pierces the deep fascia of the thigh about an inch in front of the tuberosity of the ischium. It is usually of small size in the female, and runs upwards and inwards to join the external superficial perinæal nerve.

[The superficial vessels and nerves are to be turned aside and the muscles dissected out. The muscles are much less distinct than in the male perinæum, and it is impossible to trace all their fibres without maceration and preparation, which it is impracticable for the student to undertake; he will be able to make out sufficient however to assist materially in following the description.]

The **Superficial Muscles** (Figs. 76 and 77) of the female perinæum closely resemble those of the male, with the exception that the two portions of the central muscle (corresponding to the accelerator urinæ) are separated by the vulva. On each side therefore of the vulva will be found the compressor bulbi; lying on the ramus of the pubes, the erector clitoridis; and, passing from the tuber ischii to the central point of the perinæum, the transversus perinæi.

The **Compressor Bulbi Vaginæ** (sphincter vaginæ) (Fig. 77, 6) *arises* from the central point of the perinæum (where two or three of its superficial fibres are prolonged into the sphincter ani); also from the reflection of the deep layer of superficial fascia (ischio-perinæal ligament of Savage) for nearly an inch, thus overlapping a portion of the transversus. The fibres of the muscle are divisible into three sets corresponding to those of the accelerator urinæ of the male. The anterior fibres pass forward on each side of the vulva to meet those of the opposite muscle upon the clitoris, and to be partially *inserted* into the sheath of the clitoris; the middle fibres pass beneath the clitoris and over the great veins upon the urethra, which are analogous to the corpus spongiosum of the male, meeting those of the opposite side; the posterior fibres pass on to the triangular ligament (Savage).

The **Erector Clitoridis** (Fig. 77, 2) resembles the erector penis in the male, and though actually smaller in size, bears a much larger proportion to its organ than that muscle. It *arises* from the front of the tuber ischii, and is *inserted* on the sides of the crus clitoridis.

The **Transversus Perinæi** (Fig. 77, 10) *arises* broadly from the inner side of the ramus of the ischium, and meets its fellow muscle of the opposite side in the central point

Fig. 77.

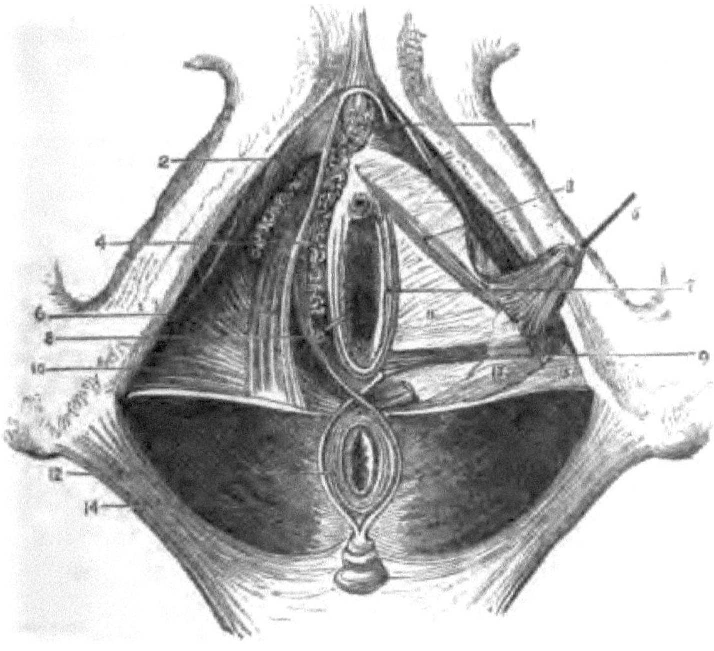

Fig. 77.—The muscles of the female perinæum, superficial on right, and deep on left side (from Savage and Luschka).

1. Clitoris.
2. Erector clitoridis.
3. Jarjavay's muscle.
4. Bulb of vagina.
5. Transversus perinæi (reflected).
6. Compressor bulbi (sphincter).
7. Constrictor vaginæ.
8. Gland of Bartholin.
9. Deep transversus perinæi.
10. Superficial transversus.
11. Posterior layer of triangular ligament.
12. Sphincter ani.
13. Triangular ligament.
14. Levator ani.
15. Ischio-perinæal ligament.

of the perinæum, where it is blended with the fibres of the compressor bulbi and sphincter ani, and the deep sphincter vaginæ.

[The compressor bulbi is to be divided at its origin and turned up, being dissected from the bulb of the vagina beneath and from the triangular ligament. The transversus perinæi is to be removed, and the erector clitoridis detached from the bone with the crus clitoridis, which will be seen to be a vascular body attached to the ramus of the pubes for half an inch. This is to be drawn forward, a branch of artery going to it being preserved; and the triangular ligament will be exposed.]

The **Bulb of the Vagina** (Fig. 77, 4) (bulbus vestibuli of Kobelt) corresponds precisely to one half of the bulb of the male. It is a vascular body placed on the side of the orifice of the vagina, and enclosed in a sheath of fascia derived from the triangular ligament or deep perinæal fascia. Each bulb is about an inch long, and is composed of a plexus of veins which communicate with the great veins of the clitoris.

The **Triangular Ligament** (Fig. 77, 13) or Deep Perinæal fascia resembles that of the male, and is a strong process of fibrous tissue stretching across the pubic arch, being attached to the rami of the pubes and ischium on each side behind the crura clitoridis. It is perforated by the urethra and by the vagina, with the coats of which it is firmly incorporated; it gives passage also to the terminal branches of the pudic arteries and nerves, as in the male, and to the large veins of the clitoris and bulbs which pass to the plexus on the neck of the bladder.

[The triangular ligament is to be carefully detached from the bone and removed, when the deep muscles and the terminal branches of the pudic artery and nerve are to be dissected.]

The **Deep Muscles*** are the Deep Transverse, the Constrictor of the Vagina, and Jarjavay's muscle.

* For the preparations from which these descriptions are made, I am indebted to Dr. Savage, who has most liberally placed at my disposal all his researches on the female perinæum, by which he has

The **Deep Transverse Muscle** (Fig. 77, 9) *arises* from the ramus of the ischium and passes to the central point of the perinæum, where it meets its fellow of the opposite side.

The **Constrictor of the Vagina** (Sphincter ?) (Fig. 77, 7) consists of a few fibres which *arise* from the central point of the perinæum, and ascend to meet the muscle of the opposite side over the urethra. It is joined by the following muscle.

Jarjavay's Muscle (depressor urethræ, Santorini) (Fig. 77, 3) *arises* from the ramus of the ischium close to the deep transverse muscle, and sometimes united with it. It runs upwards and inwards to join the fibres of the constrictor, and also to join its fellow-muscle over the urethra.

Involuntary circular fibres surround the whole length of the female urethra.

The **Glands of Bartholin** (Fig. 77, 8) correspond to Cowper's glands in the male. Each gland is of the size of a small bean, and is situated behind the triangular ligament opposite the lower border of the bulb. The minute duct of each gland opens on the mucous surface of the vagina, about half an inch from the orifice and in front of the hymen.

Pudic Artery and Nerve (Fig. 76). The terminations of the pudic artery and nerve in the female correspond closely to those of the male, but are much smaller. The artery lies under cover of the ramus of the pubes with the nerve, and gives off an artery to the bulb, to the crus clitoridis, and the dorsal artery of the clitoris.

The *artery of the bulb* (8) runs transversely inwards, and terminates in the bulb of the vagina.

---

been enabled to confirm in great measure the observations of Luschka. The accompanying drawing of the deep muscles was made under Dr. Savage's own superintendence, and I believe accurately represents the muscles as they will be found in a favourable subject after maceration and careful manipulation with needles, etc., but it will be impossible for the student to find more than indications of their existence in his ordinary dissection.

The *artery to the crus clitoridis* (9) is very small and is lost in that organ.

The *dorsal artery of the clitoris* (7) reaches the dorsum of the organ with the corresponding nerve.

The Dorsal Vein of the clitoris is of large size, and passes backwards to open into the great veins around the neck of the bladder.

The Pudic Nerve (11) gives off muscular branches (12), and ends in the dorsal nerve of the clitoris (18), which bears a much larger proportion to that organ than the corresponding nerve of the penis.

[The flaps of skin should be carefully fastened together with stitches before the body is moved from its position.]

Dissection of the Abdominal Wall.

The dissection of the Perinæum having been completed, the student should proceed to examine the abdominal wall, the condition of which will vary considerably in different subjects. In a fat subject the wall of the abdomen will be smooth and uniform, but in a thin muscular subject the prominences of the muscles will be readily recognized. In all bodies the umbilicus will be seen in the median line about midway between the pubes and the ensiform cartilage of the sternum, and it should be noted whether there is any protusion at this point constituting an *exomphalos* or umbilical hernia. The median line will be found to be slightly depressed, corresponding, as will be afterwards seen, to the linea alba, and it is here, below the umbilicus, that the operation of paracentesis or tapping the abdomen is performed.

On each side of the median line and parallel to it is the prominence caused by the rectus abdominis muscle, and occasionally the markings in that muscle known as the *lineæ transversæ* can be seen through the skin. Immediately above the pubes and close to the median line is occasionally seen the prominence of the pyramidalis muscle on each side. The pubic bone will be recognized with the finger, which should be carried outwards to the spine of the pubes, to which Poupart's ligament is attached. The ligament being traced out will be followed along the fold of the groin to the anterior superior spinous process of the ilium, from which the iliac crest passes backwards. On the sides of the lower part of the thorax the interdigitations of the obliquus externus with the serratus magnus may be seen through the skin in a well-developed subject.

The condition of the abdominal rings and inguinal canal should be carefully examined before the skin is removed. If the spermatic cord be traced up from the testis it may be felt to disappear through an opening known as the external abdominal ring, and even in a perfectly well-formed subject the forefinger can be passed into the ring with tolerable facility, if pushed up from below, carrying the scrotum before it. In a subject with a tendency to hernia the finger can be carried along the inguinal canal to the internal abdominal ring, or, if a rupture actually exists, may be passed into the abdominal cavity, pushing before it the loose skin. If any tumor of the scrotum should happen to be present it should be carefully examined and its nature diagnosed, with a view to confirmation by subsequent examination.

In the female, the abdominal ring is usually of small size, and generally too contracted to admit the finger, but may be enlarged by a hernia. The projection above the pubes which is covered with hair and known as the *mons Veneris* is due simply to a development of fat in the superficial fascia at this point. The skin of the abdomen of women who have borne children is marked by numerous small white scars, the result of its over-distension during pregnancy.

Percussion and palpation should be carefully practised in order to make out the limits of the principal organs of the abdomen before the dissection is begun, and the student is referred to Fig. 90, on which the position of the several viscera is shown in health, but from which considerable variations may be expected. The presence of fluid in the peritoneal cavity may be ascertained by the

Fig. 78.

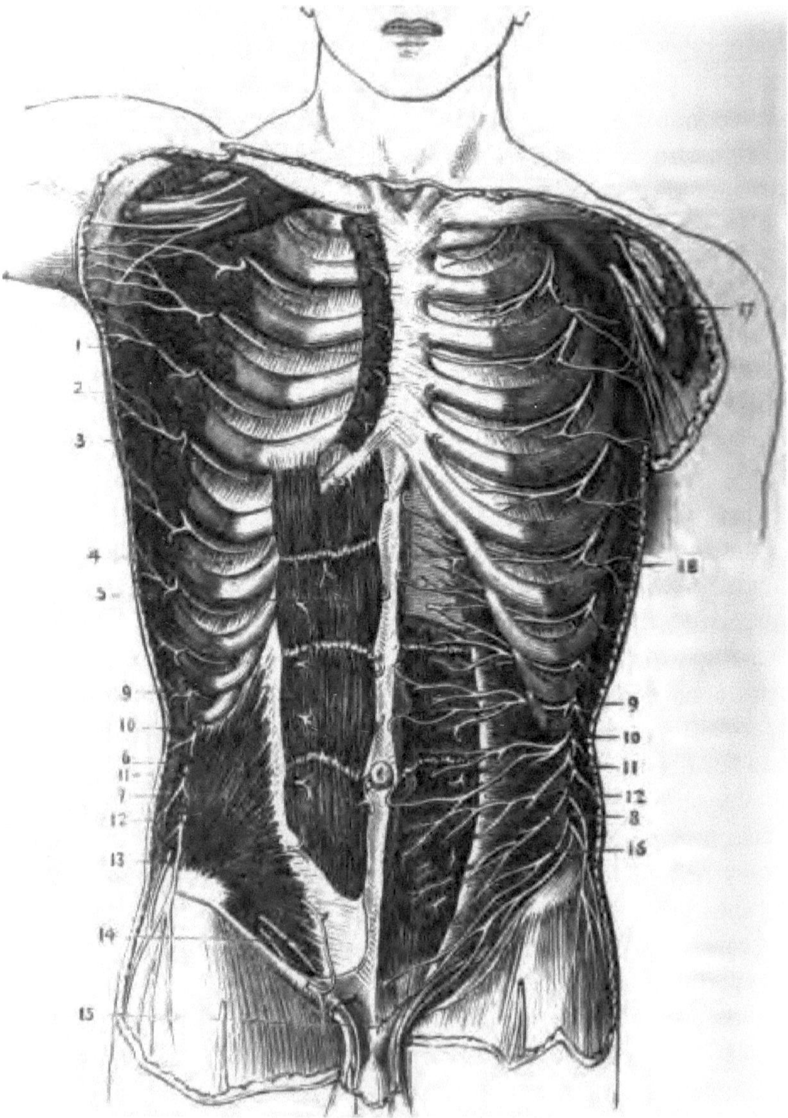

Fig. 78.—The nerves of the abdominal wall (from Hirschfeld and Leveillé).

1. Pectoralis major (cut).
2. Serratus magnus.
3. Latissimus dorsi.
4. Intercostal muscles.
5. Rectus abdominis.
6. Section of obliquus externus.
7. Obliquus internus.
8. Transversalis abdominis.

9, 9. Ninth dorsal nerve.

10, 10. Tenth dorsal nerve.

production of fluctuation between the hands applied on opposite sides of the abdomen.

[In order to dissect the abdominal wall it is necessary to distend the abdomen fully with air. To do this, a narrow knife should be thrust through the umbilicus into the peritoneal cavity, and a blowpipe be inserted. A circular incision is then to be made through the skin round the umbilicus, and a piece of whipcord tied round the blowpipe and in the groove cut in the skin. When the abdomen has been fully distended, the blowpipe is to be withdrawn and the opening carefully secured.

One incision is to be carried from the ensiform cartilage to the pubes (avoiding the umbilicus), and another at right angles to it along the crest of the ilium and from the anterior superior spine of the ilium to the middle line. A third at the level of the ensiform cartilage is to reach well back to the loins. The skin is then to be reflected from the median line by these incisions.]

The **Superficial Fascia** in this region resembles that throughout the body, but is sometimes much loaded with fat; it is to be reflected by the same incisions as the skin, and in doing this there will be found in it the terminations of the cutaneous nerves.

The **Cutaneous Nerves** (Fig. 78) consist of two sets, anterior and lateral. The *anterior* nerves, which are very small, will be found in two rows, one close to the linea alba, and the other piercing the tendon of the external oblique two inches from the median line; the *lateral* branches will be found of greater length, running along the fibres of the same muscle and also giving a few small twigs backwards. Both sets are derived from the dorsal spinal nerves. About two inches behind the anterior spine of the ilium and half an inch above the crest will be

11, 11. Eleventh dorsal nerve.
12, 12. Twelfth dorsal nerve.
13. Lateral cutaneous branch of first lumbar (ilio-hypogastric).
14. Anterior cutaneous branch of ilio-hypogastric.
15. Anterior cutaneous branch of ilio-inguinal.
16. Ilio-hypogastric and ilio-inguinal nerves.
17. Lateral cutaneous branch of second intercostal nerve.
18. Lateral cutaneous branch of intercostal nerve.

found the lateral cutaneous branch of the last dorsal nerve, and close upon the crest of the bone a little posterior to the preceding nerve will be seen the iliac branch of the ilio-hypogastric nerve (1st lumbar), both of them going to the buttock.

Accompanying the nerves will be found (in a well-injected body only) small branches of arteries derived from the intercostal, internal mammary, and epigastric arteries.

[The external oblique muscle is to be cleaned, the dissector beginning from below on the right side, and from above on the left, and taking care not to remove its tendon near the median line by mistake. The interdigitations with the serratus magnus and latissimus dorsi are to be carefully defined.]

The **Obliquus Externus** (Fig. 1, 12) has its *origin* from the outer surfaces of the eight lower ribs, and thus interdigitates with the serratus magnus by five points above, and with the latissimus dorsi by three points below. Its fibres pass obliquely downwards and forwards towards the middle line, except the posterior ones which pass vertically to the crest of the ilium.

[In order to see the whole of the insertion of this muscle it will be necessary to continue the dissection by reflecting the triangle of skin left upon the groin.]

The **Superficial Fascia of the Groin** (Fig. 79) differs from that over the abdomen in being divisible into two layers—superficial and deep.

The *superficial layer of superficial fascia* (*a*) resembles that found elsewhere, but the *deep layer* (*b*) (Scarpa's fascia) is more membranous, and is attached firmly to the fascia lata of the thigh a little below Poupart's ligament, thus preventing any urine which may have been infiltrated into the tissues from passing down the thigh.

Superficial Vessels and Nerves (Fig. 79).—Between the layers of superficial fascia will be found three small branches of the femoral artery with their accompanying veins.

Fig. 79.

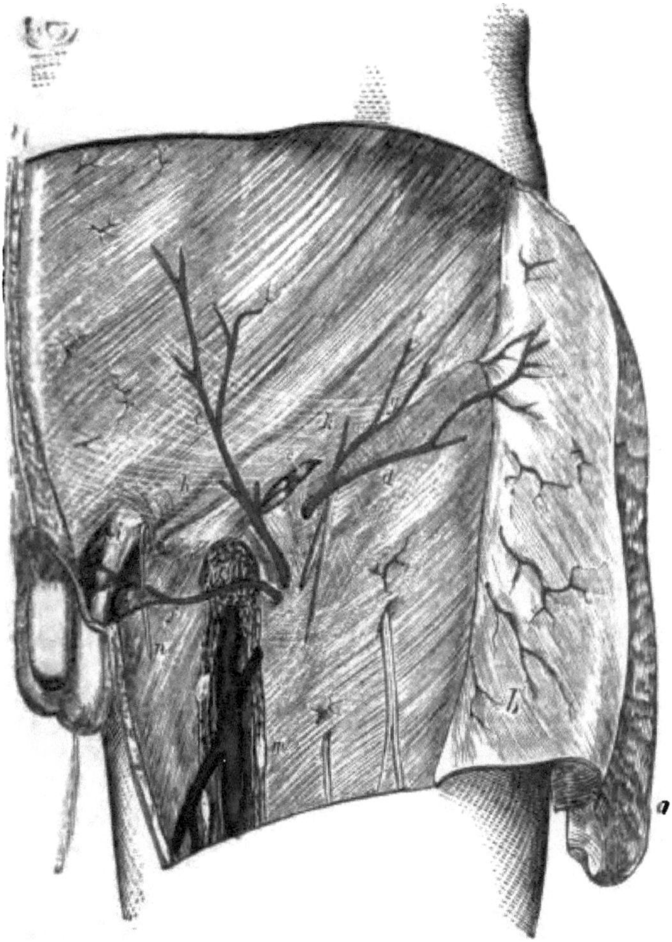

Fig. 79.—Superficial dissection of the inguinal and femoral regions (from Wood 'On Rupture').

a. Superficial layer of fascia (reflected).
b. Deeper layer of fascia (reflected) (the superficial vessels being left attached to the external oblique).
c. Inguinal lymphatic glands.
d. Superficial circumflex iliac artery.
e. Superficial epigastric artery.
f. Superior external pudic artery.
g. Poupart's ligament.
h. Intercolumnar fascia.
i. External abdominal ring.
k. Arciform fibres of external oblique.
l. Internal saphena vein.
m. Femoral lymphatic glands.
n. Ilio-inguinal nerve.
o. Saphenous opening.

The *superficial circumflexa ilii* (*d*) runs outwards about the level of Poupart's ligament to the anterior superior spine; the *superficial epigastric* (*e*) runs upwards for a short distance on the abdominal wall; and the *superior external pudic* (*f*) runs over the spermatic cord to the scrotum. Upon, or a little below, Poupart's ligament will be found two or three glands which receive the lymphatics from the penis and scrotum.

The tendon of the obliquus externus being exposed, a nerve will be found piercing it about two inches above the pubes—the *ilio-hypogastric*, from the 1st lumbar; and another may be traced upon the spermatic cord—the *ilio-inguinal* (*n*) also from the 1st lumbar (Fig. 78).

Insertion of Obliquus Externus (Fig. 79).—1st, the anterior half of the outer lip of the crest of the ilium; 2nd, the whole length of Poupart's ligament, *i.e*, from the anterior spine of the ilium to the spine of the pubes; 3rd, the front of the crest of the pubes; and 4th, the whole length of the *linea alba*, *i.e.*, from the ensiform cartilage to the pubes. In addition to these larger insertions there are two small ones which cannot be defined in the present stage, viz., into the ilio-pectineal line (forming Gimbernat's ligament) and into the crest of the pubes of the opposite side forming the triangular fascia.

Immediately above the pubes the spermatic cord (or the round ligament of the uterus in the female) will be seen to emerge from the tendon, the opening being obscured by the *intercolumnar* or *spermatic fascia* which must be removed to expose the abdominal ring.

The **External Abdominal Ring** (Fig. 79, *i*) is merely a slit in the external oblique tendon placed obliquely immediately above the spine of the pubes. In the healthy subject it is triangular with the base downwards, being about an inch in length and half an inch across, and its boundaries have been named the *pillars* or *columns of the ring*. It will be observed that the outer pillar is triangular at the upper part, but becomes round and thick below, and corresponds to Poupart's ligament; whilst the internal pillar is flat, and is continuous with the rest of the tendon. At the upper part of the ring may be seen some curved fibres running in a transverse direction and called the intercolumnar fibres, the continuation of which formed the intercolumnar or external spermatic fascia. It should be noticed that the spermatic cord does not lie in the centre of the ring, but upon the outer pillar or Poupart's ligament, which is slightly drawn down by the weight of the testis.

[The external oblique is to be reflected by cutting, both vertically and horizontally, through its muscular fibres where they join the tendon, and by detaching them from the ribs immediately below their origin. The greater part of the muscle can then be turned back towards the loin, and the corresponding tendon may be dissected up as far as possible towards the median line. The handle of the knife should be passed under the triangular portion of tendon left in the groin parallel to and as near the median line as possible, after which the tendon may be readily divided by a vertical incision, and reflected towards the thigh without endangering the subjacent parts. When this is done it will be found that the tendinous fibres of one side decussate or interlace with those of the other immediately above the pubes, and cross to the opposite side of the median line, where they become attached to the spine and crest of the pubes, and sometimes as far as the ilio-pectineal line, thus forming the *triangular fascia.** The internal oblique muscle will then be exposed, and will be most readily cleaned by beginning at the lower border of the muscle on the right side, and at the upper border on the left side of the subject.]

The **Obliquus Internus** (Fig. 78, 7).—The fibres of this muscle take the opposite direction to those of the obliquus externus, and therefore have their origin below and their insertion chiefly above. The *origin* is from the anterior two-thirds of the middle of the crest of the ilium; from the outer half of Poupart's ligament; and slightly from the fascia lumborum. The upper fibres pass to be *inserted*

---

* This tissue, which can be defined in well-developed subjects only, is better-named triangular *fascia* than triangular *ligament*, which is apt to be confounded with the deep fascia of the perinæum.

into the cartilages of the last seven ribs and into the linea alba, some directly and others by an aponeurosis, which is split to enclose the rectus and form its sheath, as will be afterwards seen; while the lower fibres, becoming gradually more and more horizontal in their course to the middle line, in the end arch downwards to help to form the *conjoined tendon*, and to be inserted into the crest of the pubes and the ilio-pectineal line.

Fig. 80.

In the male subject, some pale muscular fibres will be found to be continuous with the lower border of the internal oblique, and to overlie the spermatic cord, forming the *cremaster muscle.*

The Cremaster (Fig. 84, *h*).—To explain the formation of this muscle it will be necessary to refer to the early stage of fœtal life when the testis was situated within the abdomen and immediately below the kidney. A muscular band consisting of three sets of fibres and called the *gubernaculum testis* then passed from the pubes to the testis, and its divisions had the following attachments—one to the bottom of the scrotum, and one on each side of the external abdominal ring to Poupart's ligament and the crest of the pubes (Fig. 80). By the action of these bands the testicle was drawn down as far as the external abdominal ring, but it is obvious that when it had reached this point the lateral bands must have become horizontal, and thus have lost all power of traction; and the further progress of the testis depended therefore upon the middle band, which continued to draw the gland to the bottom of the scrotum (Fig. 81). The effect of this was to stretch the lateral bands attached

Fig. 81.

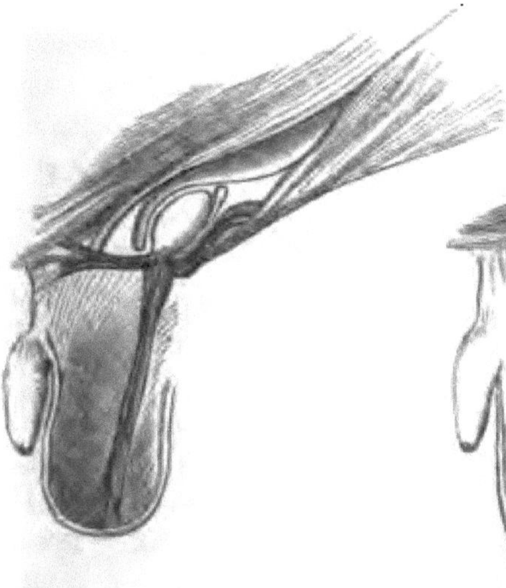

Fig. 82.

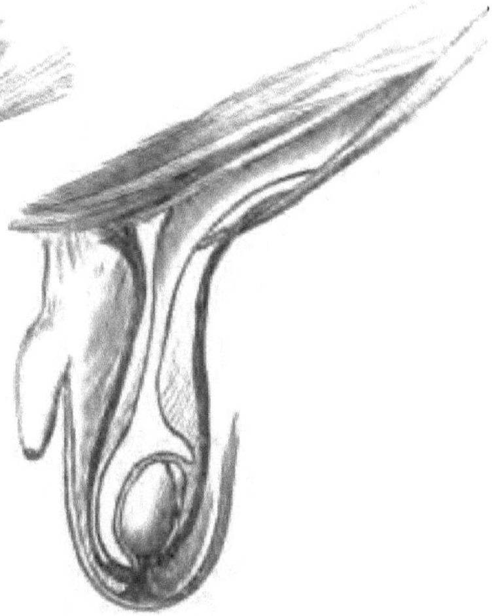

to the pubes on each side, and thus they formed more or less complete loops upon the cord, and constituted the cremaster muscle, being aided by the lower fibres of the internal oblique which were drawn down by the testicle in its progress (Fig. 82).

[The internal oblique is to be divided by a vertical incision from the last rib to the middle of the crest of the ilium, and this is to be carried along the crest and Poupart's ligament to near the lower border of the muscle. The muscular fibres must be carefully divided until a cellular interval is reached, in which, near the crista ilii, will be found the branches of an artery (circumflexa ilii). The muscle is then to be turned backwards and forwards, care being taken not to injure the exposed nerves or the subjacent transversalis muscle.]

The **Nerves** (Fig. 78) five or six in number, are the

Figs. 80, 81, 82.—Diagrams to illustrate the descent of the testis and the formation of the tunica vaginalis (drawn by J. T. Gray).

branches of the lower dorsal nerves, which run horizontally forwards to pierce the rectus muscle and emerge on the front of the abdomen. At this stage they may be seen giving off their *lateral cutaneous* branches, which pierce the obliquus internus at once. Close above the crista ilii may be found the branches of the first lumbar nerve (16) (ilio-hypogastric and ilio-inguinal) running downwards to the pubes.

The **Transversalis Muscle** (Fig. 83, 6) has its *origin*—1, from the inner surfaces of the cartilages of the lower six ribs, where it interdigitates with the diaphragm; 2, by three aponeurotic slips attached respectively to the roots of the transverse processes, to the

Fig. 83.

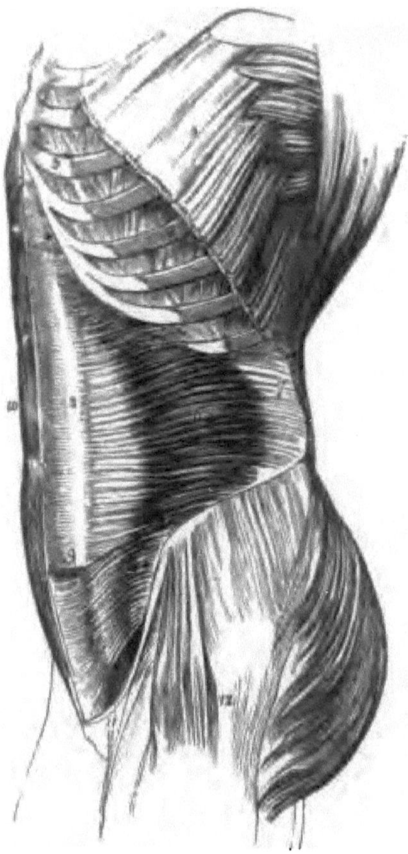

Fig. 83.—Lateral view of the transversalis abdominis muscle (from Wilson).

1. Latissimus dorsi.
2. Serratus magnus.
3. Upper part of the external oblique.
4. Two of the external intercostal muscles.
5. Two of the internal intercostals.
6. Transversalis.
7. Its posterior aponeurosis.
8. Its anterior aponeurosis passing behind the sheath of the rectus.
9. Lower part of the left rectus with the aponeurosis of the internal oblique and transversalis passing in front.
10. Right rectus muscle.
11. Poupart's ligament.
12. Tensor vaginæ femoris, gluteus medius and maximus invested by fascia lata.

tips of the transverse processes, and to the spinous processes of all the lumbar vertebræ (together forming what is commonly called the *fascia lumborum*) (Fig. 92); 3, from the anterior two-thirds of the inner lip of the crest of the ilium, and 4, from the outer third of Poupart's ligament. Its *insertion* is into the middle line from the ensiform cartilage to the pubes, being inseparably united with the internal oblique; and into the ilio-pectineal line, forming with the internal oblique the *conjoined tendon*.

Anatomy of Inguinal Hernia.

This will be the best time for examining the parts concerned in inguinal hernia, which is the protrusion of some part of the abdominal contents through certain openings in the region of the groin. To do this the muscles should be restored as nearly as possible to their natural positions, and the student will then see again the opening in the tendon of the external oblique called the external abdominal ring. Through this opening (the intercolumnar fascia having been previously removed) the spermatic cord with its vessels will be seen to pass, and on tracing it up by turning the muscles aside in their order, it will be found to disappear beneath the sharp arched border of the transversalis muscle. The tissue into which it disappears is the *fascia transversalis*, lining the abdomen externally to the peritoneum, and with a little care a hole may be artificially made in this fascia and around the cord, to which the name of *internal abdominal ring* is given. The opening of the internal ring is necessarily artificial because the fascia transversalis was prolonged upon the cord when the testis was drawn out of the abdomen, and thus took the shape of a funnel, being hence called the *infundibular fascia*.* (Fig. 84, *i*).

The **Internal Abdominal Ring** (Fig. 84) is an opening in the fascia transversalis placed midway between the symphysis pubis and the anterior superior iliac spine, and about half an inch above Poupart's ligament.

* The infundibular fascia has also been called the internal spermatic fascia.

The **Inguinal Canal** (Fig. 85) is the space between the external and internal abdominal rings, an inch and a half in length, and is occupied by the spermatic cord. The

Fig. 84.

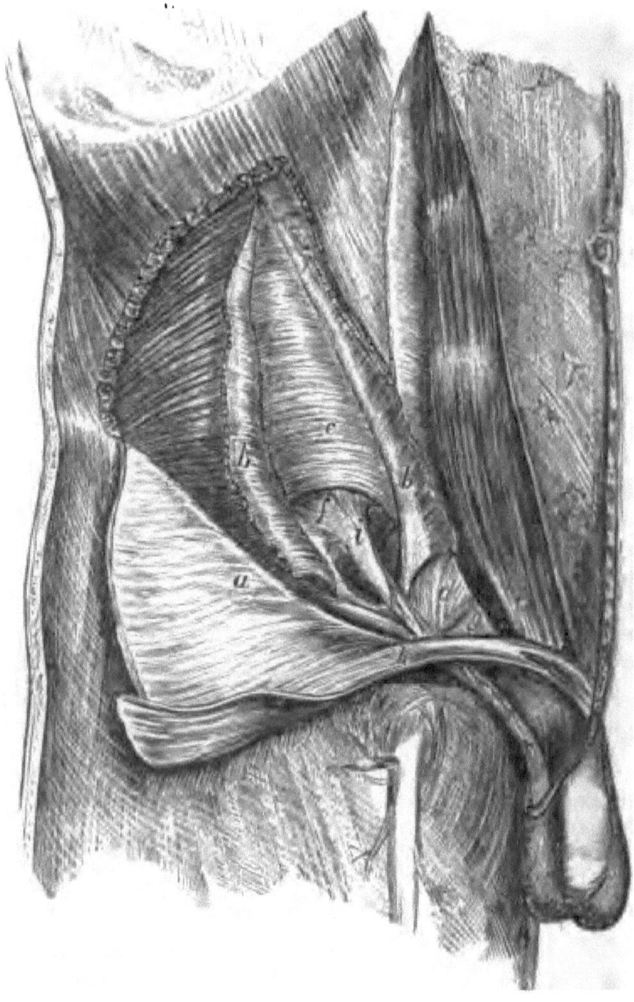

Fig. 84.—Dissection of the inguinal canal (from Wood, 'On Rupture').

a. External oblique (turned down).
b, *b*. Internal oblique.
c Transversalis.
d. Conjoined tendon.
e. Rectus abdominis with sheath opened.
f. Fascia transversalis.
g. Triangular fascia.
h. Cremaster.
i. Infundibular fascia.

simplest way therefore to study the boundaries of the canal will be to take them in their relation to the cord which lies in it.

In front are (1) the tendon of the external oblique, (2) the lowest part of the muscular fibres of the internal oblique, and (3) a small portion of the cremaster. *Behind* are—beginning nearest the internal ring—(1) the fascia transversalis, (2) the conjoined tendon, (3) the triangular fascia (if it exist). *Below* is Poupart's ligament. *Above* are the arched border of the transversalis muscle and the space between that muscle and the internal oblique.

Fig. 85.

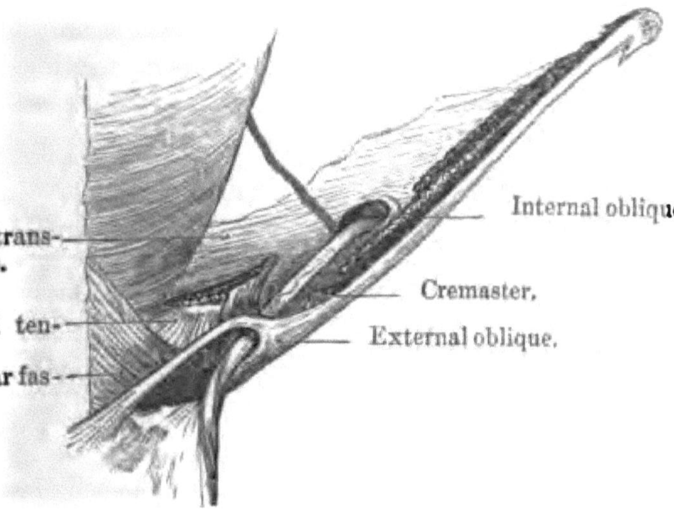

Inguinal Hernia is of two kinds, *oblique* and *direct.* Oblique passes through the internal abdominal ring, along the inguinal canal, and out of the external ring. Direct breaks through or pushes before it the posterior wall of the inguinal canal at its lower part, and emerges at the external ring, thus taking a more *direct* course than the *oblique* variety.

The *coverings* of each hernia will be best seen by tracing them from within outwards; thus oblique hernia would

Fig. 85.—Diagram of the inguinal canal, showing its anterior and posterior boundaries (drawn by J. T. Gray).

have (1) peritoneal sac; (2) in coming through the internal ring it would have the fascia transversalis, or infundibular fascia; (3) in passing beneath the internal oblique it would have the cremaster muscle, the fibres of which are united by cellular tissue and are sometimes known as the cremasteric fascia; in emerging from the external ring it would have (4) the intercolumnar or spermatic fascia; and (5) the superficial fascia and skin.

Direct hernia would have, (1) the peritoneum, and (2) the fascia transversalis, (for both line the abdomen); then pushing into the inguinal canal it would have a covering (3) from the conjoined tendon; and lastly emerging from the external ring, it would have (4) the intercolumnar or spermatic fascia, and (5) the superficial fascia and skin. Thus the number of coverings of both varieties is the same, the difference consisting in the substitution of the conjoined tendon for the cremaster muscle.

Coverings of Inguinal Hernia.

Oblique.	*Direct.*
Peritoneum or sac.	Peritoneum or sac.
Fascia transversalis (or infundibular fascia).	Fascia transversalis.
Cremaster muscle (or fascia).	Conjoined tendon.
Intercolumnar or spermatic fascia.	Intercolumnar or spermatic fascia.
Superficial fascia.	Superficial fascia.
Skin.	Skin.

By dividing the fascia transversalis about half an inch to the inner side of the internal abdominal ring, there will be found an artery called the epigastric, and the varieties of inguinal hernia have been named *external* or *internal* according to their relation to this vessel. It is obvious that a hernia passing through the internal ring must be *external* to this artery, and hence oblique hernia and external hernia are synonymous terms. In the same way direct hernia protrudes through the abdominal wall to the *inner* side of the epigastric artery, and thus internal is the same as direct hernia.

The anatomy of the parts concerned in hernia being fully comprehended, it will be readily seen that the direction in which force is to be applied to return the protruded bowel into the abdomen—commonly called the *taxis*—must differ in the two cases; and the student should notice the effect which the position of the body and limbs has upon the tissues near the groin. Thus, with the body and legs extended at full length, the lower part of the abdomen is rendered tense, whilst if the thigh is flexed upon the pelvis, and the body bowed forward, the structures become relaxed, and would more readily admit of the return of the protruded bowel.

Varieties of Hernia.—There are two varieties of oblique inguinal hernia, termed *congenital hernia* and *infantile* or *encysted hernia*, the anatomy of which requires explanation.

To understand these fully it will be necessary to refer for a moment to the descent of the testis (page 210). When the testicle is in the abdomen it has the peritoneum attached to its surface, and during its descent into the scrotum it accordingly drags a fold of peritoneum down with it, so that at first a tube extends along the whole length of the inguinal canal (Fig. 82). By a natural process during the later months of fœtal life however, an obliteration of the portion of tube within the canal takes place, and thus a small sac is formed around the testis which is called the *tunica vaginalis*, while a mere dimple is left in the peritoneum at the internal ring. Thus when in the ordinary condition of the parts a hernia descends, its sac is above and quite distinct from the tunica vaginalis (Fig. 86).

Occasionally however this obliteration does not take place, and the tunica vaginalis remains continuous with the peritoneum. In that case the hernia descends at once into the tunica vaginalis and lies over the testis, and being the consequence of a congenital defect it is termed *congenital hernia* (Fig. 87).

In what is termed *congenital hydrocele* the anatomical arrangement is the same, and the fluid runs from the peritoneal cavity into the scrotum and back again according to the position of the patient.

Infantile or *encysted hernia* exists when only a partial obliteration of the peritoneal tube has taken place, and the tunica vaginalis is left larger than usual and prolonged into the inguinal canal to a slight extent. Then when a hernia descends, it pushes down behind this portion of

Fig. 86.

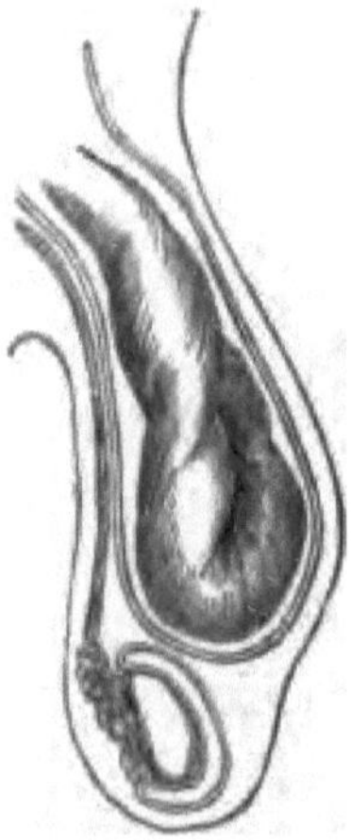

Fig. 87.

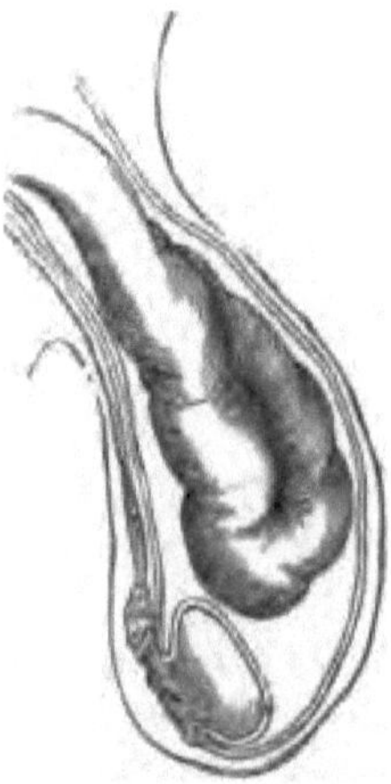

Fig. 88.

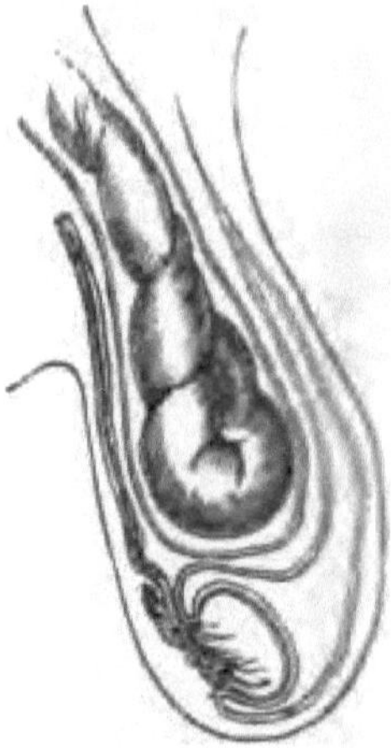

the tunica vaginalis, and there are therefore three layers of serous membrane in front of that covering the intestine, two being formed by the tunica vaginalis and one by the peritoneal sac (Fig. 88).

Surgery.—*Operation for Strangulated Hernia.*—In this operation the object is to reach and divide any tissue which presses upon the bowel and prevents its return within the abdomen. An incision is made over the external ring, and the surgeon dissects carefully down through the several layers (which are never distinctly separable in practice) until he reaches the peritoneal sac, which is recognized by its bluish appearance. Passing the forefinger carefully around this he feels for the stricture,

Fig. 86.—Diagram of a common scrotal hernia, showing the relation of the sac to the tunica vaginalis testis (J. T. Gray).

Fig. 87.—Diagram of a congenital hernia, the sac being continuous with the tunica vaginalis testis (J. T. Gray).

Fig. 88.—Diagram of an infantile hernia showing the tunica vaginalis prolonged in front of the sac (J. T. Gray).

which may possibly be at the external ring (where it is easily divided), or more probably within the inguinal canal. Having discovered the stricture, he next passes the hernia-knife along the finger, and, using it or a director as a guide, insinuates the knife, held flat, beneath the stricture, and divides it by bringing the edge of the knife upwards and pressing it against the tissues with the finger.

The direction in which the superficial incision is made, is of little consequence, but in the deep incision the rule is to cut directly *upwards*, so as to be parallel to the epigastric artery; for since a mistake in the diagnosis between an external and an internal hernia may occur, owing to the parts becoming distorted and the abdominal rings drawn together in old-standing herniæ, it is well to be careful lest the artery should be divided.

Radical Cure of Hernia.—Various operations have been suggested for the cure of rupture by obliterating the inguinal canal, of which the principal are Wurtzer's and Wood's. Wurtzer's operation consisted in the invagination of the skin of the scrotum, which was fastened into the inguinal canal by a plug of wood, from which a needle passed transfixing the front wall of the canal. Inflammation was thus excited, and the portion of skin remained *in situ* for a time, effectually plugging up the canal, but was found to descend almost invariably after any exertion, and the operation has therefore been generally abandoned.

Wood's operation consists essentially in closing the lower part of the inguinal canal and external abdominal ring, either with or without the invagination of a piece of fascia. The left forefinger being passed beneath the margin of the conjoined tendon, a needle is thrust through it and afterwards through the internal and external pillars of the ring, and these three points are drawn together, thus occluding the canal.*

The **Spermatic Cord** may now be examined, and will

* For full particulars of the modifications of this operation the student is referred to Mr. Wood's valuable work on Ru

be found to consist of the *vas deferens* or duct of the testicle together with the spermatic artery and veins, bound together with loose areolar tissue and covered more or less by the cremaster muscle. In a well-injected subject a small branch may be traced from the epigastric artery to the cord, which is the *cremasteric* branch; and the genital branch of the genito-crural nerve may be found in the cremaster muscle, which it supplies. The position of the vas deferens at the back of the cord should be noticed, and may always be recognized by its feeling like a piece of whipcord. Upon the vas may be found a small branch of the vesical artery—the *deferential* branch.

Tunica Vaginalis Testis (Fig. 86).—By drawing the testicle out of the scrotum and making an incision over the anterior part of it, the *tunica vaginalis* will be opened, and will be seen to consist of two parts, the tunica vaginalis *propria* upon the testis, and the *reflexa* or part around. The tunica vaginalis propria converts the subjacent tunica albuginea into a *fibro-serous* membrane, the other examples of which are the pericardium and the dura-mater.

The **Penis** should now be examined, and the student may advantageously practise the operation of circumcision. It should be noticed that the skin becomes continuous with the mucous membrane at the margin of the fore-skin or *prepuce*, which is lined by the membrane continued from the *glans penis*. Below the orifice of the urethra will be found the fold called the *frænum præputii*, and around the base of the glans penis is a circular ridge, the *corona glandis*, upon which open a number of sebaceous glands, the *glandulæ odoriferæ*, that secrete the *smegma præputii*. An incision is to be carried along the upper surface of the penis, when there will be found the *dorsal vein* with the *dorsal artery* and *nerve* on each side of it. The artery and nerve are the terminations of the pudic artery and nerve, and the vein passes through the triangular ligament to open into the prostatic plexus.

The *Suspensory Ligament* of the penis is a triangular piece of strong fibrous tissue, attached to the front of the symphysis pubis and to the upper surface of the penis, where it is divided into two layers to give passage to the dorsal vessels and nerves.

[Returning to the abdominal wall the dissector will find the **Rectus** muscle close to the median line and still enclosed in its sheath, through which it is indistinctly visible. The line between the recti is known as the *linea alba*, that at the outer border of each muscle as the *linea semilunaris*, while the transverse markings, three or four in number, are the *lineæ transversæ*.]

The **Sheath of the Rectus** (Fig. 92) having been opened by a vertical incision, will be found to be formed by the tendon of the internal oblique, which splits at the linea semilunaris to enclose it. Thus in front of the muscle there is the tendon of the obliquus externus and one-half of the tendon of the obliquus internus, whilst behind is the other half of the tendon of the obliquus internus and the tendon of the transversalis. This arrangement does not hold good throughout, for upon lifting up the lower part of the rectus it will be found that a little below the umbilicus (seldom so low as midway between the umbilicus and pubes) the sheath is wanting behind, *i.e.* it passes entirely in front, and there is nothing between the muscle and the peritoneum but the fascia transversalis.

The point where the sheath ceases to pass behind the rectus is marked by a sharp curved margin (fold of Douglas), beneath which the epigastric vessels will be seen to pass, and which will be afterwards seen from behind (Fig. 89, *c*).

[In many subjects immediately above the pubes, in front of the rectus and enclosed in its sheath, will be found a little muscular slip, the Pyramidalis. Both muscles are to be cleaned in the direction of their fibres.]

The **Pyramidalis** (Fig. 78) *arises* from the crest of the pubes close to the median line, and passes up for about two inches to be *inserted* into the linea alba.

The **Rectus Abdominis** (Fig. 78, 5) *arises* by two heads one from the crest and the other from the pubis. It expands into a broad thin muscle, and is *inserted* into the cartilages of the last three true ribs (5th. 6th, and 7th).

The lineæ transversæ are tendinous intersections corresponding to abdominal ribs; one is placed nearly opposite the umbilicus, a second at the level of the ensiform cartilage, and a third between them. A fourth is occasionally found between the umbilicus and the pubes.

The **Epigastric Artery** (Fig. 84) is to be traced into the rectus muscle, which it will be found to enter at the part where the sheath is deficient, passing beneath the fold of Douglas. The muscle should be divided and carefully dissected off the artery, which may be traced up to the sternum, where it anastomoses with the internal mammary artery. This is a most important anastomosis, forming a direct communication between the subclavian and the external iliac arteries, which comes into play after ligature of the external or common iliacs, or obstruction of the descending aorta.

The nerves piercing the rectus are the terminations of the lower intercostal nerves which pass through it and the sheath to reach the skin (Fig. 78).

Action of the Abdominal Muscles.—The abdominal muscles compress the viscera when put in action, and thus act as muscles of *expiration*, and in direct opposition to the diaphragm. The attachment of the muscles to the lower ribs enables them at the same time to depress the wall of the thorax, thus assisting the internal intercostal muscles. The abdominal muscles are called into play chiefly in violent expiratory efforts, such as coughing, sneezing, etc., and have a most important action upon the viscera during the acts of micturition, defæcation, vomiting, and parturition. The rectus abdominis has the power of approximating the pelvis and thorax, as seen in rising from the supine position without the aid of the hands, or in climbing. When spasmodically contracted in tetanus it produces the bowed condition of the trunk known as *emprosthotonos.*

The abdominal muscles are *supplied* by the lower six

intercostal nerves, and by the ilio-hypogastric and ilio-inguinal branches of the first lumbar nerve.

The abdominal wall having been finished on both sides, the dissectors should cut through the remains of the

Fig. 89.

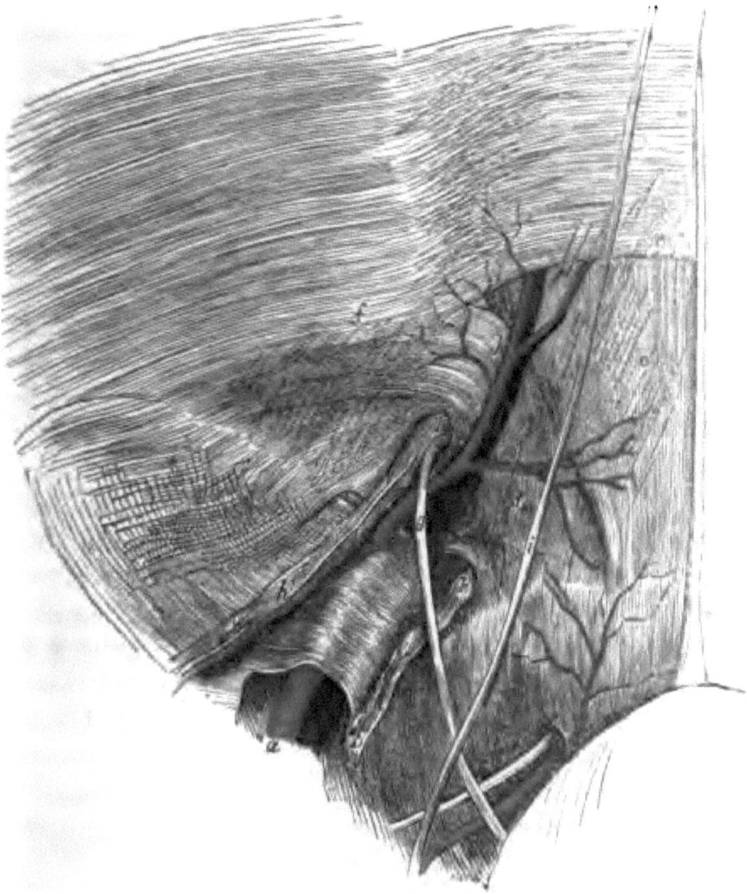

Fig. 89.—Dissection of the lower part of the abdominal wall from within, the peritoneum having been removed (from Wood, 'On Rupture').

a. External iliac artery.
b. Epigastric artery.
c. Border of the posterior part of the sheath of the rectus (fold of Douglas).
d. Conjoined tendon in the triangle of Hesselbach.
e. Posterior surface of rectus.
f. Fascia transversalis.
g. Vas deferens.
h. Spermatic vessels.
i. Obliterated hypogastric artery.
k. Lymphatics in crural ring.
l. Internal abdominal ring.

abdominal muscles in two or three places until the peritoneum is exposed but without injuring the latter, in order to study the appearance of the membrane which forms the sac of a hernia. They should then open the abdomen and dissect the viscera together.

Cavity of the Abdomen.

[The abdomen is to be opened by a transverse incision at the level of the umbilicus, and another cut is to be carried to the sternum a little to the left of the median line.]

By holding up the lower portion of the abdominal wall before dividing it in the median line, and viewing it from behind, the dissectors will perceive through the peritoneum five cords taking a course towards the umbilicus (Fig. 89). The central cord is the obliterated urachus, those on each side of it are the obliterated hypogastric arteries (*i*), and the outside ones are the two epigastric arteries (*b*) seen through the wall. The peritoneum is raised into distinct folds by the obliterated hypogastric arteries, and external to the epigastric artery on each side may be seen a small dimple in the membrane marking the position of the internal abdominal rings. The term "Hesselbach's triangle" (*d*) has been applied to the triangular space bounded by the border of the rectus internally, the epigastric artery externally and Poupart's ligament below, and it is through this that direct inguinal hernia forces its way. The fold of Douglas (*c*) or sharp lower margin of the posterior sheath of the rectus will be readily perceived through the peritoneum.

[By dividing the lower flap in the median line it can be reflected on each side, and the abdominal contents will be exposed.]

Abdominal Cavity (Fig. 90).—The parts first seen on opening the abdomen are the stomach and right lobe of the liver with a small portion of the gall-bladder, whilst the great omentum passing down from the stomach conceals the rest of the intestines. Should, however, the great omentum be turned up, as often happens, the mass of the small intestines will be visible, and immediately below the stomach the transverse colon. The ascending and descending colon will be more or less visible on each side, according to the amount of their distension. A very much distended bladder, or the pregnant uterus, would be visible at once, and the amount of liver and stomach which will appear will depend upon the healthy condition of those organs, and upon their not having been displaced by tight-lacing.

For convenience of description the abdomen is divided

Fig. 90.

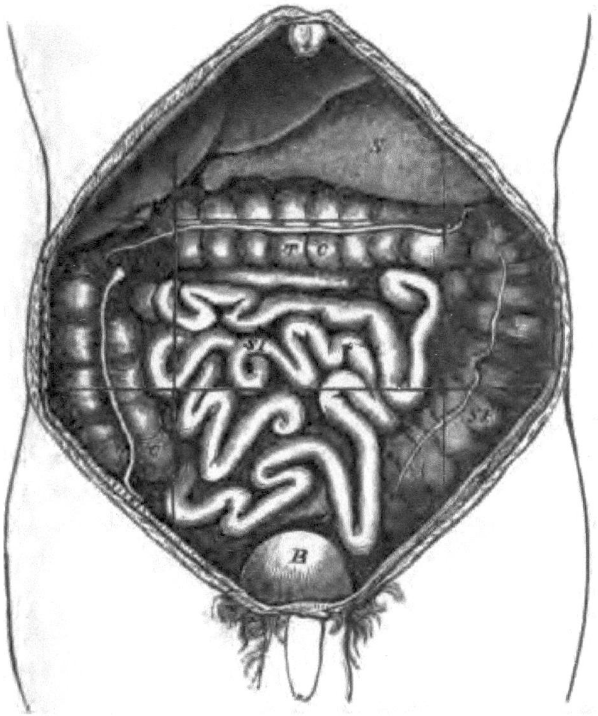

Fig. 90.—The abdominal viscera *in situ*, the great omentum having been removed in order to show the parts more distinctly (drawn by J. T. Gray). The lines mark the regions of the abdomen

L. Liver.
S. Stomach.
S I. Small intestine.
C. Cæcum.
T C. Transverse colon.
S F. Sigmoid flexure.
B. Bladder.

into the following regions by imaginary lines passing vertically from the seventh costal cartilage to the middle of Poupart's ligament on each side; and horizontally at the level of the ninth costal cartilages, and the spines of the ilium respectively.

REGIONS OF THE ABDOMEN.

Right Hypochondriac.	Epigastric.	Left Hypochondriac.
Right Lumbar.	Umbilical.	Left Lumbar.
Right Iliac.	Hypogastric.	Left Iliac.

The position of the several viscera should now be noted and the different organs identified.

Beginning at the upper part, the liver will be found to occupy the right hypochondriac, the epigastric, and a small part of the left hypochondriac regions; passing to it from the umbilicus will be seen the obliterated umbilical vein. The stomach will be seen in the epigastric region, with its greater or left end reaching into the left hypochondrium. By drawing this end out from beneath the ribs, the spleen will be found attached to it, and will also frequently be found to be bound firmly to the diaphragm by old inflammatory adhesions. Tracing the stomach to the right side it will be found to terminate in the small intestine (duodenum), and the point of demarcation between the organs is marked by a thickened band of fibres, the pylorus. The duodenum can be traced for a short distance only at present. By drawing up the great omentum the transverse colon will be exposed crossing the umbilical region, and continuous on the right side with the ascending colon, and on the left with the descending colon, which should be traced out. Occupying the hypogastric region are the coils of the small intestine, two-fifths of which constitute the jejunum, and the remainder the ileum; this may be seen to end in the large intestine in the right iliac fossa, where the commencement of the colon (caput cæcum coli, or cæcum) will be recognized by its little appendage, the appendix vermiformis. In the left iliac region will be seen the tortuous portion of the large intestine, called the sigmoid flexure, which may be traced to its termination in the rectum at the brim of the true pelvis.

TABLE OF ABDOMINAL CONTENTS.

Right Hypochondriac Region.	*Epigastric Region.*	*Left Hypochondriac Region.*
Right lobe of liver and gall-bladder, 1st part of duodenum, hepatic flexure of colon, right supra-renal capsule, and part of right kidney.	Stomach (centre and pylorus), left lobe of liver, cæliac axis, abdominal aorta, vena cava, semi-lunar ganglia, receptaculum chyli, and vena azygos.	Stomach (cardiac end), spleen and tail of pancreas, splenic flexure of colon, left supra-renal capsule, and part of left kidney.
Right Lumbar Region.	*Umbilical Region.*	*Left Lumbar Region.*
Ascending colon, small intestine, 2nd part of duodenum, head of pancreas, right kidney.	Great omentum, transverse colon, 3rd portion of duodenum, body of pancreas.	Descending colon, small intestine, left kidney.
Right Iliac Region.	*Hypogastric Region.*	*Left Iliac Region.*
Cæcum coli, ureter, spermatic vessels.	Small intestines, apex of bladder in distension and in children. Pregnant uterus.	Sigmoid colon, ureter, spermatic vessels.

The above table of necessity includes organs whose positions cannot be seen at present, and will be subsequently studied.

It will be well for the student to notice the distinctive appearances of the several parts of the intestinal canal, and particularly the differences between the large and small intestines. The condition of the intestines will vary so much according to the mode of death, that mere size is no criterion, the small intestines being occasionally distended to a much greater size than the large; but attention to the following points will prevent all possibility of error. The cylinder of the small intestine is uniform throughout, and the surface is perfectly smooth; the large intestine on the contrary will be found to be pouched or slightly constricted at short intervals throughout, except the rectum. Moreover it will be seen to be marked by distinct bands of longitudinal muscular fibres, three in number, and in most subjects small pieces of fat (appendices epiploicæ) will be found attached along the free border. In the healthy condition the large intestine is about as large again as the small intestine.

The **Peritoneum** (Fig. 91).—The serous membrane covering the intestines is now to be traced, in doing which the student must bear in mind that he has already opened the cavity of the peritoneum, which was previously a shut sac. The best way to learn the peritoneum is to trace it as simply as possibly at first, and afterwards to study any complications.

Beginning at the umbilicus the membrane will be seen to line the abdominal wall, and so to reach the diaphragm; from this it is reflected on the upper surface of the liver, round its anterior margin, and on the under surface of the organ as far back as the transverse fissure. It then passes down to the stomach forming the anterior layer of the *gastro-hepatic* or *lesser omentum*, covers the anterior surface of the stomach, and is prolonged downwards over the intestines to form the anterior layer of the *great omentum* or *epiploon*. By turning up the great omentum, the peritoneum will be seen to be reflected upon itself to form the posterior layer of the great omentum, which is continued upwards until it meets with the transverse colon, under which it is prolonged to the spine forming the under layer of the *transverse meso-colon*. It is now carried over the mesenteric vessels to the small intestine and back to the spine, thus forming the two layers of the *mesentery;* and it can then be seen to pass over the abdominal aorta into the pelvis, where it invests the rectum (*meso-rectum*) and passes from that intestine to the bladder, forming in the male the *recto-vesical pouch*. In the female the peritoneum is reflected from the rectum to the upper part of the vagina, forming the *recto-vaginal pouch*, then over the uterus and between the uterus and bladder, forming the *utero-vesical pouch.* It is lastly carried over the bladder to the lower part of the abdominal wall and so to the umbilicus.

Fig. 91.

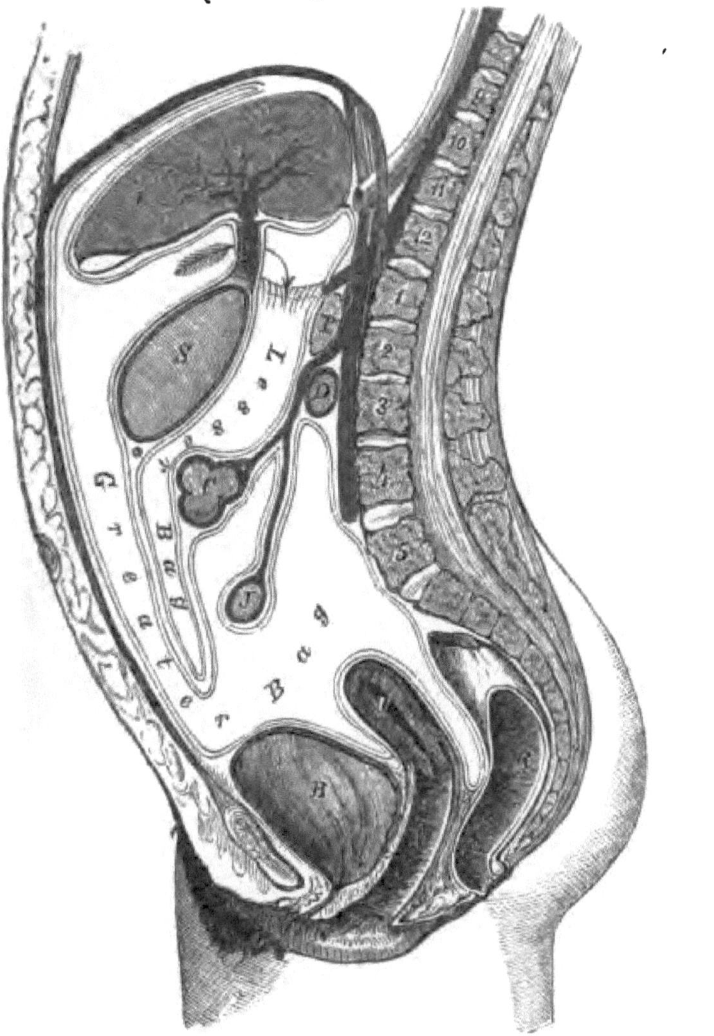

Fig. 91.—Diagram of vertical section of the peritoneum (drawn by J. T. Gray).

The continuation of the greater with the lesser bag through the foramen of Winslow is marked by an arrow.

L. Liver.	J. Jejunum.	B. Bladder.
S. Stomach.	P. Pancreas.	U. Uterus.
C. Colon.	D. Duodenum.	R. Rectum.

slig
the
dis
mb
e
la

Foramen of Winslow

parts and produced by the
wards and upwards to the
foramen it will feel the
the lesser omentum containing
duct and portal vein; behind, the
vein and the inferior vena cava;
passes forwards from the
of the liver.

preferred to trace the
together, which can be readily
Beginning at the liver, one layer
and the other the back of the under surface of
the two meet at the transverse fissure to
the lesser omentum. They then separate to enclose
the stomach, uniting at its lower border to form the an-
terior two layers of the great omentum; being reflected
upon themselves they next form the posterior two layers,
which separate to enclose the transverse colon, form

erus ar

·on, as they are prolonged to the spine. The w leave one another, and the upper one (lesser ·nged over the pancreas to the under surface ·r, where it commenced. The under layer ;) forms the mesentery around the small in- ·e recto-vesical pouch between the bladder and ·d passes over the abdominal wall to the dia- ·om which it is reflected on to the liver, where ·tion commenced.

tracing the peritoneum vertically as has been . should be traced horizontally as follows. Below of the transverse colon the circle will be found ·eedingly simple. Beginning at the median line

Fig. 92.

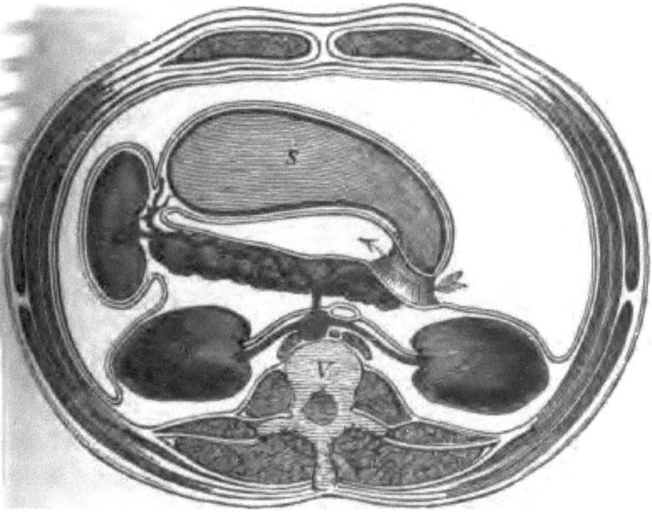

of the abdominal wall, the peritoneum may be traced to the right iliac region, where it will be found to cover the front of the cæcum and ascending colon (forming the *meso-cæcum* and *ascending meso-colon*); it then forms the *mesentery* around the small intestine, any portion of which

---

Fig. 92.—Transverse section of the abdomen at the level of the foramen of Winslow (drawn by J. T. Gray).

The arrow passes from the greater bag to the lesser bag of the peritoneum through the foramen of Winslow, which is seen in section.

S. Stomach. Sp. Spleen. V. Vertebra.
P. Pancreas. K. Kidney. A. Aorta.

will serve to show it; and lastly, covering the sigmoid flexure of the colon (*sigmoid meso-colon*) it is brought round again to the abdominal wall.

A fold attaching the top of the descending colon to the under surface of the diaphragm (*costo-colic*, Jenner) should be noticed, since it passes below the spleen and influences the movements of that organ.

Above the colon the arrangement is a little complicated by the existence of the two sacs, the continuity of which however may thus be traced (Fig. 92). Beginning at the median line of the abdominal wall the peritoneum may be traced into the right hypochondrium and over the right kidney; it then passes through the loop of the hepatic artery (foramen Winslowii), and across the body in front of the pancreas, then to the posterior surface of the stomach and back to the foramen again, where it forms the posterior layer of the lesser omentum. Reflected at this point upon itself, the membrane will be seen to form the anterior layer of the lesser omentum and to be continued over the front of the stomach; thence to the spleen which it encloses (forming the *gastro-splenic omentum*), and so to the left hypochondrium and round the abdominal wall. The continuity of the greater with the lesser bag is thus evident, and the foramen of Winslow is seen to be merely the narrowed tube of communication between the two cavities.

Besides the folds of peritoneum which have been named in tracing the membrane, there are others which form certain ligaments of the liver and bladder, and of the uterus in the female.

Passing from the umbilicus to the Liver the obliterated umbilical vein or *round ligament* is seen, and around it is reflected a double fold of peritoneum, the *suspensory* or *falciform ligament* of the liver. This is prolonged on each side over the diaphragm and on to the upper surface of the liver, where it forms the upper layer of the *coronary ligament;* the under layer being formed partly by the lesser bag of the peritoneum, and the triangular surface between the two being attached by firm areolar tissue to the diaphragm. The doubled edges of the upper layer of the coronary ligament on each side are called the right and left *lateral ligaments.*

The *false ligaments* of the Bladder are formed by the peritoneum; they are five in number, two posterior, two lateral, and one superior. The *posterior* false ligaments are the margins of the recto-vesical pouch, and are formed by the membrane being reflected over the obliterated hypogastric arteries. The *lateral* false ligaments are the pieces of peritoneum reaching from the bladder to the sides of the pelvis, and the *superior* is that passing over the obliterated hypogastric arteries and urachus to the umbilicus.

Fig. 93.

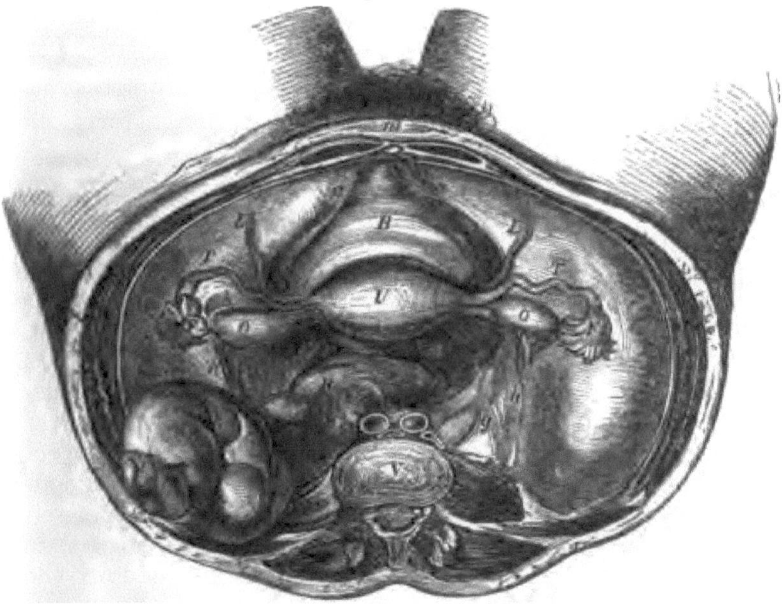

Fig. 93.—Female pelvic organs, seen from above (from Savage).

a, a. Obliterated hypogastric arteries.
b. Broad ligament of uterus.
c. Posterior ligaments of uterus.
g. Ureter.
h, h. Ovarian vessels.
m. Linea alba and section of urachus.
B. Bladder.
L. Round ligament of uterus.
F. Fimbriated extremity of Fallopian tube.
T. Fallopian Tube.
O. Ovary.
R. Rectum.
U. Uterus.
V. Fifth lumbar vertebra.

In the female (Fig. 93), the pelvis is divided transversely by a fold of peritoneum reaching from each side of the uterus to the side of the pelvis, and called the *broad ligament* of the uterus, in which will be found the ovary, the Fallopian tube, and the round ligament. It is by the open mouth of the Fallopian tube that the peritoneum has a communication with the mucous membrane of the uterus. The Uterus and Vagina intervene between the rectum and bladder of the female, and the peritoneum is prolonged from the rectum to the posterior wall of the vagina, and then over the uterus to the bladder, thus forming the *recto-vaginal* and *utero-vesical* pouches, the edges of which are the *anterior* and *posterior ligaments* of the uterus.

In the child an arrangement may occasionally be found which is extremely rare in the adult—the posterior layer of the great omentum ascending directly to the pancreas, and then being again reflected to the colon to form its meso-colon.

MESENTERIC VESSELS.

[In order to dissect the mesenteric arteries, the great omentum and transverse colon must be turned up over the cartilages of the ribs and the small intestines drawn down. Beginning at the upper end of the jejunum (*i.e.*, at the point where the small intestine can first be distinctly seen) the dissector should remove the whole of the anterior layer of the mesentery as far as the edge of the bowel, where it is to be cut off. This must be continued along the whole length of the small intestine, and the mesentery will be found to pass obliquely downwards from left to right. After the vessels contained in the mesentery are cleaned, the under layer of the meso-colon must be removed in the same way, to expose the vessels going to the large intestine.]

The **Superior Mesenteric Artery** (Fig. 94, 10) is the vessel which supplies the small and half the large intestines. A branch of the abdominal aorta, it comes off immediately beneath the pancreas and passes forward between it and the duodenum, giving off the *inferior* *pancreatico duodenal* branch (13) which takes a curved course to the right and between the head of the pancreas

Fig. 94.

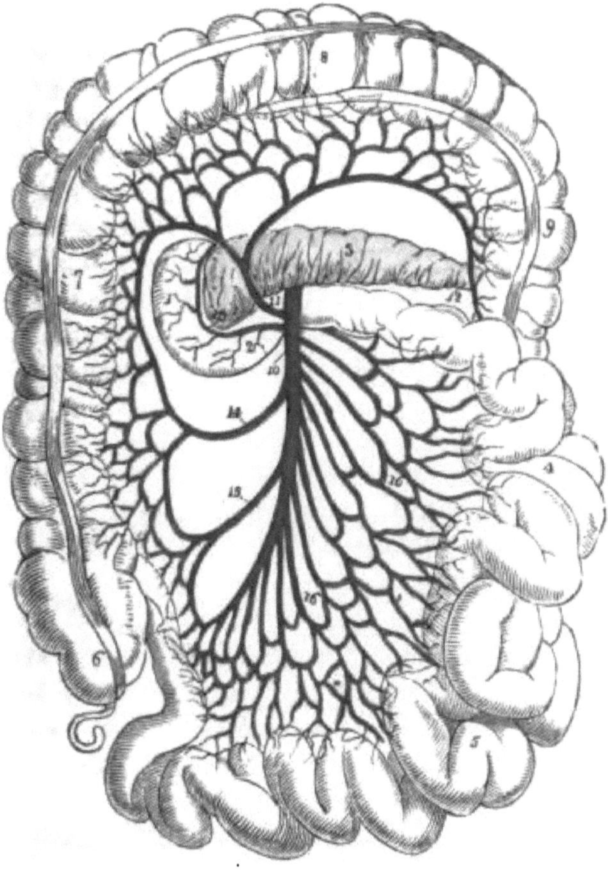

Fig. 94.—Course and distribution of the superior mesenteric artery (from Wilson).

1. Decending portion of the duodenum.
2. Transverse portion.
3. Pancreas.
4. Jejunum.
5. Ileum.
6. Cæcum and appendix vermiformis.
7. Ascending colon.
8. Transverse colon.
9. Descending colon.
10. Superior mesenteric artery.
11. Colica media.
12. Its anastomosis with the colica sinistra.
13. Pancreatico-duodenalis inferior.
14. Colica dextra.
15. Ileo-colica.

16, 16. Vasa intestini tenuis.

and the duodenum to anastomose with the *pancreatico-duodenalis superior* artery. The trunk of the superior mesenteric artery will be seen to take a slightly curved direction from left to right, and from its convexity or left side are given the branches to the small intestine (16) (*vasa intestini tenuis*), while from its concavity or right side are given three branches to the large intestine, *ileo-colic* (15), *right colic* (14) and *middle colic* (11) arteries. The branches to the small intestine form a series of primary and secondary loops which inosculate freely with one another, and at last ramify in the submucous areolar tissue of the intestine. The branches to the large intestine take the course indicated by their names; thus the ileo-colic goes to the end of the ileum and commencement of the colon (cæcum), the right colic to the ascending, and the middle colic to the transverse colon. Each of these arteries divides into primary loops, which spread out on each side, and inosculate with one another and with the branches to the small intestine; and thus a free anastomosis is established from one part to another of the intestinal canal.

Corresponding branches of veins will be found running with the arteries, all of which unite in the **Superior Mesenteric Vein** (Fig. 98, 6) which helps to form the vena portæ. In addition there may be found ramifying upon the arteries the branches of the superior mesenteric plexus, derived from the solar plexus of the sympathetic; and interspersed are the lymphatic glands which receive the lacteals from the intestines.

[The whole of the small intestines are to be turned over to the right side, and the remainder of the meso-colon is to be removed to expose the inferior mesenteric vessels.]

The **Inferior Mesenteric Artery** (Fig. 95, 9) will be seen to arise from the abdominal aorta about an inch and a half above the bifurcation, and to give off the *left colic* (10) to the descending colon; the *sigmoid* (12) (three or four) to the sigmoid flexure; and lastly, the *superior hæmorrhoidal artery* (13) which descends into the pelvis to supply the rectum. These arteries form loops and an-

Fig. 95.

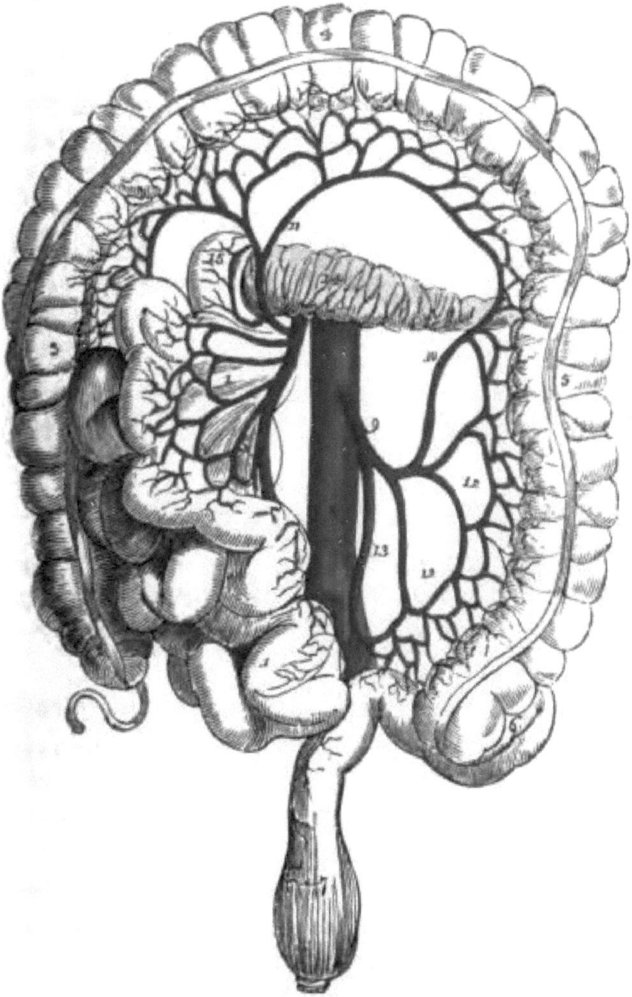

Fig. 95.—Branches of the inferior mesenteric artery (from Wilson).

1, 2. The superior mesenteric and small intestines turned over to the right side.
2. Cæcum and appendix cœci.
3. Ascending colon.
4. Transverse colon turned upwards.
5. Descending colon.
6. Sigmoid flexure.
7. Rectum.
8. Abdominal aorta.
9. Inferior mesenteric artery.
10. Colica sinistra, inosculating with—
11. Colica media.
12, 12. Sigmoid branches.
13. Superior hæmorrhoidal artery.
14. Pancreas.
15. Descending portion of the duodenum.

astomose freely with one another, and moreover, the left colic inosculates with the middle colic of the superior mesenteric, and the superior hæmorrhoidal with the middle hæmorrhoidal of the internal iliac artery.

The **Inferior Mesenteric Vein** (Fig. 98, 1) will be seen to have branches corresponding to those of the artery, and anastomoses by its hæmorrhoidal branch with the middle hæmorrhoidal branch of the internal iliac vein. It passes upwards beneath the pancreas to join the splenic vein, and so into the vena portæ, thus forming a communication between the two great venous systems. Branches of the inferior mesenteric plexus of the sympathetic may be traced upon the vessels.

It may be noticed that the mesenteric veins never present dilatations, which is owing to their having no valves.

The Cœliac Axis.

[The intestines are now to be replaced, and the liver fastened up to the ribs, either with hooks or, better, by a few stitches; the stomach being drawn down, the lesser omentum is then to be carefully removed.]

Between the layers of the lesser omentum will be found three important structures, and in the following positions (Fig. 97). To the right and most in front is the bile-duct (7), to the left the hepatic artery (5), and behind and between the two the large vena portæ (9).

Behind the omentum in the middle line will be found the short trunk of the **Cœliac Axis** and its branches, viz., the gastric (coronaria ventriculi), the hepatic, and the splenic arteries.

The **Gastric Artery** (Fig. 96, 13), the smallest of the three, passes along the lesser curvature of the stomach from left to right, giving branches to both surfaces of that organ and to the œsophagus; it anastomoses at the pylorus with the pyloric branch of the hepatic artery. Branches of the sympathetic (gastric plexus) may be traced upon it. Its accompanying vein opens into the vena portæ.

The opportunity may be taken to notice the distribution of the pneumo-gastric nerves to the stomach. The left pneumo-gastric gives numerous branches to the front of the stomach and the right to the back of the viscus, and these can be generally seen through the peritoneal coat without further dissection.

The **Hepatic Artery** (Fig. 96, 14) passes upwards and to the right side (thus forming the inferior and anterior boundaries of the foramen of Winslow), and will be seen to divide into the right and left hepatic branches, which enter the transverse fissure of the liver. The artery gives off the *pyloric* branch (15) which has been seen to anastomose with the gastric, and a trunk which immediately disappears behind the duodenum and is called the *gastro-duodenalis* (16). From the right hepatic branch the small *cystic* artery is given off, which is distributed to the gall-bladder by two branches, one passing between it and the liver and the other along its free surface.

The **Splenic Artery** (Fig. 96, 20) can now be seen only in part, taking a tortuous course behind the stomach and along the upper border of the pancreas to the spleen.

[To continue the dissection it will be necessary to remove the intestines which have been examined, by putting a couple of ligatures upon the upper part of the jejunum, and also upon the upper part of the rectum, and dividing the intestine between the two ligatures at each point. By cutting the superior mesenteric artery the small intestine will be at once set free, but the large intestine must be carefully removed by dividing the meso-colon close to the intestine throughout, and lastly by cutting the inferior mesenteric artery.]

The **Gastro-duodenal Artery** (Fig. 95, 16) can be traced out by turning up the stomach. After passing behind the duodenum it divides into the gastro-epiploica dextra and the pancreatico-duodenalis arteries.

The *Gastro-epiploica Dextra* (17) runs from right to left along the great curvature of the stomach and between the folds of the great omentum; it anastomoses with the gastro-epiploica sinistra of the splenic artery.

The *Pancreatico-duodenalis superior* (18) will be found between the duodenum and the head of the pancreas, the latter of which it surrounds; it anastomoses with the small branch from the commencement of the superior

Fig. 96.

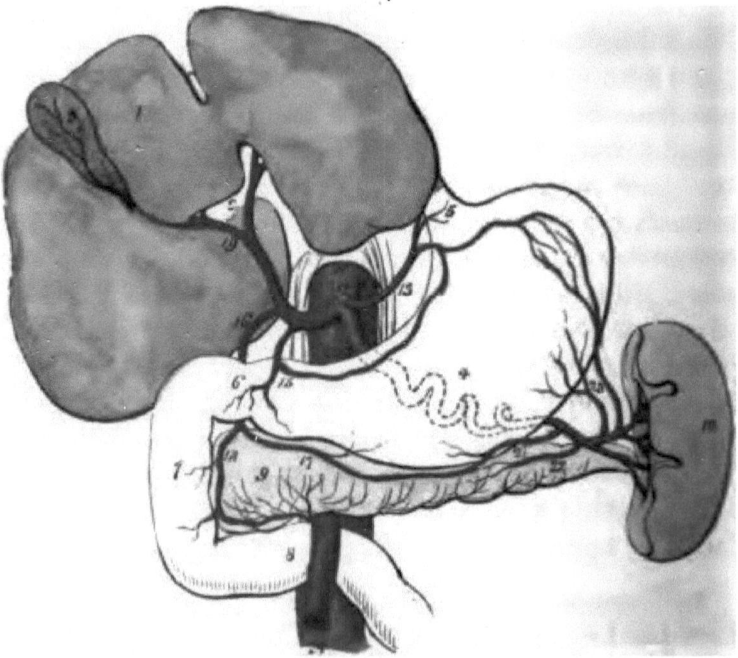

Fig. 96.—Branches of the cœliac axis (from Wilson).

1. Liver.
2. Its transverse fissure.
3. Gall-bladder.
4. Stomach.
5. Œsophagus.
6. Pylorus.
7. Duodenum, descending portion.
8. Transverse portion of the duodenum.
9. Pancreas.
10. Spleen.
11. Abdominal aorta.
12. Cœliac axis.
13. Gastric artery.
14. Hepatic artery.
15. Its pyloric branch.
16. Gastro-duodenalis.
17. Gastro-epiploica dextra.
18. Pancreatico-duodenalis, inosculating with the inferior pancreatico-duodenalis.
19. Division of the hepatic artery into right and left branches: the right giving off the cystic branch.
20. Splenic artery, traced by dotted lines behind the stomach to the spleen.
21. Gastro-epiploica sinistra.
22. Pancreatica magna.
23. Vasa brevia to the great end of the stomach.
24. Superior mesenteric artery, emerging from between the pancreas and transverse portion of the duodenum.

mesenteric artery called the *inferior pancreatico-duodenal*, and gives branches to both pancreas and duodenum.

The **Splenic Artery** (Fig. 96, 20) can now be seen in the whole of its course to the spleen, near which it breaks up into four or five branches which enter the hilum or fissure on the concave surface of that organ. As it passes along the upper border of the pancreas it gives off small branches to it (*pancreaticæ parvæ*), and one (*pancreatica magna*) which runs along the whole length of the gland with the duct. It also gives small branches to the great end of the stomach (*vasa brevia*), and one large one (*gastro-epiploica sinistra*), which runs along the great curvature to anastomose with the gastro-epiploica dextra of the hepatic. Upon the hepatic and splenic arteries are corresponding plexuses of the sympathetic.

The **Splenic Vein** (Fig. 98, 3) is placed below its artery and lies behind the pancreas. It receives branches corresponding to those of the artery, and also at about the middle of the pancreas, the inferior mesenteric vein. It is joined by the superior mesenteric vein behind the head of the pancreas, the two forming the vena portæ.

[The pancreas and duodenum should now be thoroughly defined, and the bile and pancreatic ducts traced to their entrance into the intestine. If desired, the stomach may be removed by tying and cutting it near the pylorus, and at the œsophagus.]

The **Duodenum** (Fig. 97, 11) is from eight to ten inches long (twelve finger-breadths), and makes a peculiar bend upon itself before it passes beneath the mesenteric artery to become the jejunum; thus entering successively the right hypochondriac, the right lumbar, and the umbilical regions.

The *first* or *ascending portion* reaches from the pylorus to the under surface of the right lobe of the liver, where it is in close proximity to the gall-bladder. It is covered both in front and behind by a prolongation from the lesser omentum, and is therefore moveable with the stomach. Behind it are the vena portæ, the common bile duct, and the gastro-duodenal artery.

The *second* or *descending portion* is covered in front by the peritoneum, which binds it to the surface of the right kidney, in front of which it descends for about three inches. It lies close to the outer side of the head of the pancreas with the pancreatico-duodenal artery intervening, and into its posterior surface the biliary and pancreatic ducts open.

The *third* or *transverse portion* is between the layers of

Fig. 97.

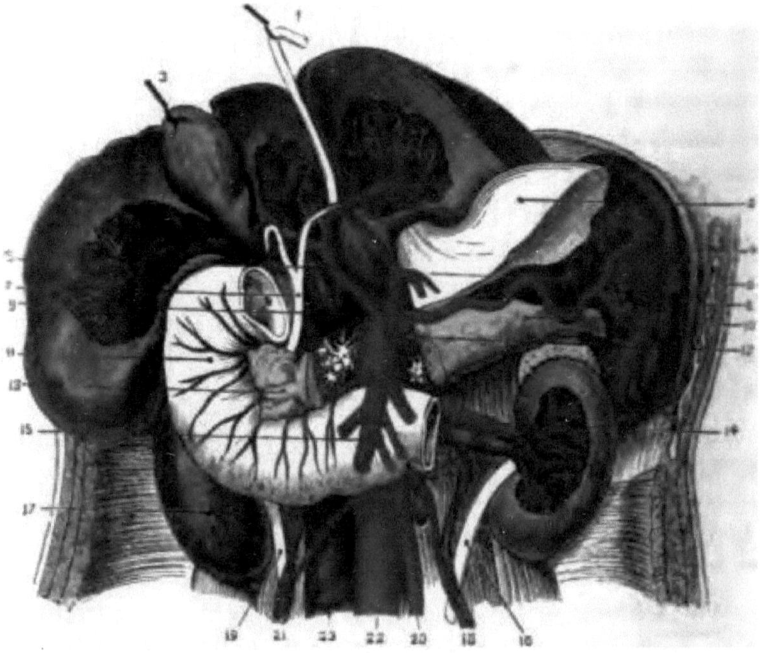

Fig. 97.—Cœliac axis (altered from Hirschfeld and Leveillé).

1. Round ligament of liver.
2. Great end of stomach (cut).
3. Gall-bladder.
4. Spleen.
5. Hepatic artery.
6. Coronary artery.
7. Common bile duct.
8. Splenic artery and vein.
9. Vena portæ.
10. Tail of pancreas.
11. Duodenum.
12. Cœliac axis.
13. Head of pancreas.
14. Left renal vessels.
15. Superior mesenteric artery.
16. Left ureter.
17. Right kidney.
18. Left spermatic vessels.
19. Right ureter.
20. Inferior mesenteric artery.
21. Right spermatic vessels.
22. Aorta.
23. Vena cava.

the meso-colon, and therefore has no proper peritoneal covering. It is crossed by the superior mesenteric artery and vein, and lies upon the vena cava, and the aorta

Fig. 98.

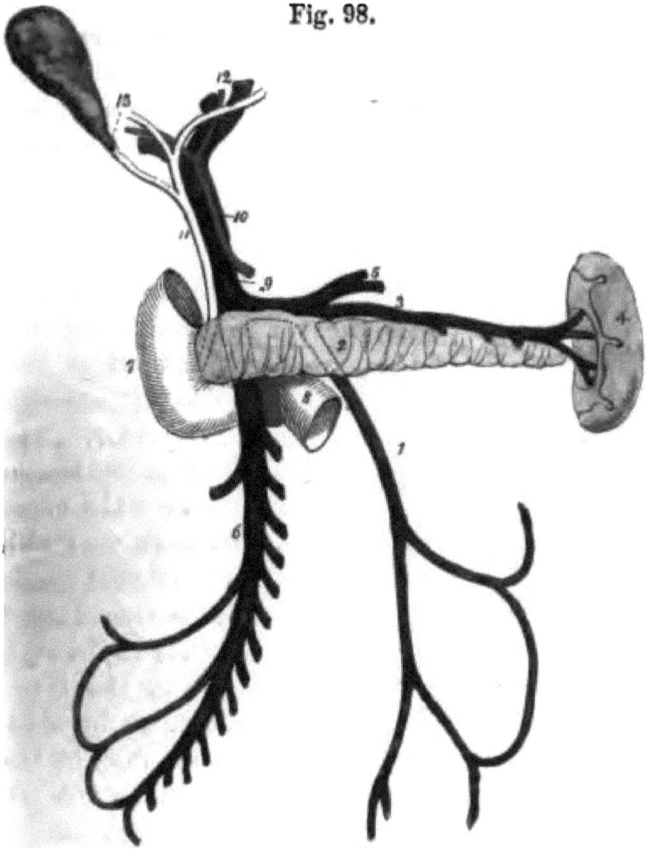

Fig. 98.—The portal vein (from Wilson). The pancreas drawn down to show the splenic vein behind it.

1. Inferior mesenteric vein; it is traced by means of dotted lines behind—
2. The pancreas to terminate in the splenic vein 3.
4. Spleen.
5. Gastric veins, opening into the splenic vein.
6. Superior mesenteric vein.
7. Descending portion of the duodenum.
8. Its transverse portion, crossed by the superior mesenteric vein and part of the trunk of the superior mesenteric artery.
9. Portal vein.
10. Hepatic artery.
11. Ductus communis choledochus.
12. Division of the duct and vessels at the transverse fissure of the liver.
13. Cystic duct leading to the gall-bladder.

placed between the crura of the diaphragm, gradually ascending from the level of the third to that of the second lumbar vertebra, at the left side of which it becomes the jejunum. Immediately above it is the body of the pancreas with the pancreatico-duodenal arteries.

The **Pancreas** (Fig. 97, 13) lies transversely in the abdomen, occupying the right lumbar, epigastric, and left hypochondriac regions; the *head* being embraced by the duodenum, the *body* crossing the aorta between the cœliac axis and the superior mesenteric artery, and the *tail* touching the spleen. It was covered by the lesser bag of the peritoneum, and from right to left it crosses the vena portæ, the vena cava, and the aorta between the crura of the diaphragm; and then lies over nearly the whole length of the splenic vein, crossing the two mesenteric veins as they ascend to join it. The duct runs along the whole length of the gland to join the common bile duct, which lies in the groove between the duodenum and pancreas; and the united ducts open very obliquely into the back of the second part of the duodenum.

The **Bile Duct** (Fig. 97, 7) should be traced up to the liver, where it will be found to commence in the right and left hepatic ducts, which emerge from the transverse fissure; these unite to form the common hepatic duct, which is joined at an acute angle by the duct of the gall-bladder (cystic), thus forming the common bile duct or *ductus communis choledochus*. The bile duct lies between the layers of the lesser omentum to the right of the hepatic artery and portal vein, and passes behind the first portion of the duodenum to join the pancreatic duct and open into the duodenum as described above.

The **Vena Portæ** (Fig. 98, 9) is formed by the junction of the splenic and superior mesenteric veins behind the pancreas and to the right of the median line. It ascends between the layers of the lesser omentum to the transverse fissure of the liver, between but posterior to the bile duct and the hepatic artery. It receives the small gastric and cystic veins and is remarkable for having no valves.

The **Spleen** (Fig. 97, 4) occupies the left hypochondrium and is attached to the greater end of the stomach by the gastro-splenic omentum, which is reflected from it to the diaphragm. It lies against the left supra-renal capsule and the diaphragm, and the tail of the pancreas touches its lower margin. The splenic vessels will be seen to enter the *hilum* or slit on its concave surface.

Deep Dissection of the Abdomen.

[The stomach, duodenum, pancreas, and spleen should now be taken away by dividing the œsophagus, bile duct, and vessels, and the liver should then be carefully removed in order to see the arrangement of the coronary ligament. The round and falciform ligaments should first be cut and the liver pulled down; by carrying the knife along the attachment to the diaphragm the upper layer of the coronary ligament will then be divided, as well as the lateral ligaments on each side which are formed from it. The vena cava and a part of the liver uncovered by peritoneum will now be exposed, the latter being attached to the diaphragm by areolar tissue, and when these are carefully divided, the right supra-renal capsule and the top of the right kidney will be brought into view. The vena cava must be again cut through at the lower margin of the liver, when it will only be necessary to divide the under layer of the coronary ligament (derived chiefly from the lesser bag of the peritoneum) to remove the organ.

The viscera should all be placed in a covered pan for examination during the time that the subject is turned.

The Diaphragm is to be cleaned by stripping off the peritoneum, as far as possible in the direction of the muscular fibres; and at the same time the inner surface of the flap of abdominal muscles still attached should be cleaned, in order to follow out the fibres of the transversalis. Care must be taken not to puncture the diaphragm, and to leave the branches of artery and nerve ramifying on its surface. The crura or muscular slips on each side of the aorta must be carefully cleaned, and two tendinous arches over the psoas and quadratus lumborum muscles defined, for which purpose it will be well to detach the kidney and turn it over to the opposite side temporarily.]

The **Diaphragm** (Fig. 99) is found to be arched after death, being concave towards the abdomen (particularly on the right side), owing to the fact that *expiration* is the last act of life. It *arises* from the back of the ensiform cartilage and from the inner surfaces of the six lower ribs, where it interdigitates with the transversalis abdominis;

Fig. 99.

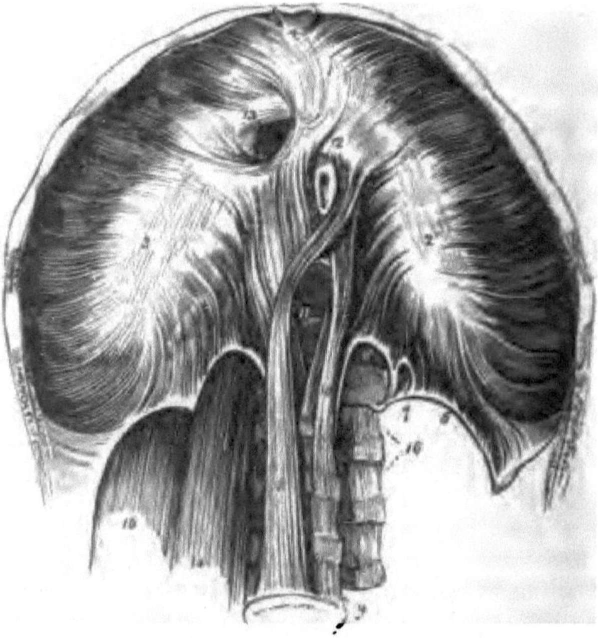

also from the *ligamenta arcuata, externa* and *interna*, of which the internal one passes from the body to the trans-

---

Fig. 99.—Under, or abdominal surface of the diaphragm (from Wilson).

1. The central leaflet of the tendinous centre.
2. The left or smallest leaflet.
3. The right leaflet.
4. Fasciculus from the ensiform cartilage.
5. Ligamentum arcuatum externum of the left side.
6. Ligamentum arcuatum internum.
7. A small arched opening occasionally found, through which the least splanchnic nerve passes.
8. Right crus.
9. Fourth lumbar vertebra.
10. Left crus.
11. Aortic opening.
12. Œsophageal opening.
13. Opening for the inferior vena cava.
14. Psoas magnus passing beneath the ligamentum arcuatum internum.
15. Quadratus lumborum passing beneath the ligamentum arcuatum externum.

verse process of the first lumbar vertebra, and the external one from the same transverse process to the last rib; the two crossing the psoas and quadratus lumborum muscles respectively.

The *ligamentum arcuatum externum* is only the artificially isolated border of the fascia covering the quadratus lumborum, which is the anterior lamella of the fascia lumborum derived from the posterior tendon of the transversalis abdominis muscle (Fig. 92).

The *ligamentum arcuatum internum* is formed solely by the posterior tendinous border of the diaphragm, and to it the fascia of the psoas is attached.

The *crura* or pillars of the diaphragm *arise* from the front of the bodies of the upper lumbar vertebræ; the right crus is the larger, and reaches as low as the third vertebra and intervertebral substance, while the left only reaches the second lumbar vertebra.

All the fibres of the diaphragm are *inserted* into the central cordiform or trefoil tendon, but the crura decussate with one another and form a figure-of-8 before they pass into the tendon, thus surrounding the aorta and the œsophagus.

Openings.—There are three large openings in the diaphragm, viz. the Aortic, the Œsophageal and the Vena Caval openings, and two or three small orifices in the crura for the splanchnic nerves and vena azygos minor, and also for the vena azygos major if it does not pass through the aortic opening.

The *Aortic* and *Œsophageal* openings are in the loops formed by the crura, and the front or œsophageal one, which is also to the left side, is entirely muscular; but that for the aorta will be found upon dissection to become tendinous, owing to some deep fibres which pass transversely over the vessel. The aortic opening is strictly speaking *behind* the diaphragm, being bounded posteriorly by the vertebral column.

The opening for the *Vena Cava* is more or less quadrilateral *(foramen quadratum)*, is situated in the right division of the cordiform tendon, and is entirely tendinous.

Through the aortic orifice pass the aorta, the vena azygos major, and the thoracic duct; through the œsophageal opening, the œsophagus with the two pneumogastric nerves; through the quadrilateral opening only the vena cava inferior. The greater and lesser splanchnic nerves pass through a slit in each crus, and the vena azygos minor (if it exists) passes through the left crus. The gangliated cord of the sympathetic is continued into the abdomen from the thorax beneath the ligamentum arcuatum internum.

The position of the diaphragm alters considerably during inspiration, for the muscle then becomes flattened, and slopes downwards and backwards so as to deepen the posterior part of the chest. The diaphragm is *supplied* by the phrenic nerves from the cervical plexus, which, after traversing the thorax, pierce the fibres of the muscle and are distributed principally on its under surface.

[The vessels and muscles of the abdominal cavity are now to be cleaned. Around the aorta and its branches will be found a quantity of tough tissue, which is the sympathetic nerve giving off plexuses to the several vessels. The greater part of this may be removed, but the great splanchnic nerves should be traced to their termination in the semilunar ganglia, situated on each side of the aorta close to the diaphragm and closely resembling lymphatic glands in appearance; and the gangliated cord of the sympathetic on each side of the vertebræ is to be preserved. The vena cava inferior (a piece of which it will be remembered has been removed with the liver) is to be cleaned, and care must be taken not to cut off the right spermatic vein on its anterior surface. The vein may be secured in its proper position with a pin if necessary. The kidneys and supra-renal capsules are to be dissected out and the vessels going to each defined, and the spermatic vessels should be traced out.]

The **Abdominal Aorta** (Fig. 100) is the direct continuation of the thoracic aorta, and reaches from the last dorsal vertebra to the left side of the fourth lumbar vertebra, where it divides into the two common iliacs. It was covered originally and quite superficially by the stomach, transverse colon, and small intestines, and was crossed more closely (from above downwards) by the pancreas and splenic vein between the cœliac axis and superior mesenteric artery; by the duodenum, immediately below the mesenteric artery; and by the left renal vein. Immediately below the renal vein the spermatic arteries lie in front of the aorta for a short distance, and lower down is the inferior mesenteric artery, from which the peritoneum was reflected over the aorta to the bifurcation.

The artery lies against the bodies of the vertebræ and the anterior common ligament, and crosses the left lumbar veins. To its right is the vena cava, in close connection below but separated at the upper part by the right crus of the diaphragm, and between the upper part of the aorta and the right crus are the receptaculum chyli, with the thoracic duct, and the vena azygos major. To its left is the left gangliated cord of the sympathetic.

Surgery (Fig. 101).—The abdominal aorta has been tied immediately above the bifurcation for iliac aneurism, but without success. It may be reached by an incision prolonged above and below the umbilicus, by which Sir Astley Cooper originally tied it, the peritoneum being necessarily opened in front, and again at the point of ligature; or by an incision at the side similar to that for ligature of the common iliac artery, but larger, by which the peritoneum is turned forward and left unopened. Mr. Lane has proposed to apply the ligature above instead of below the origin of the inferior mesenteric artery in order to preserve the anastomoses of that vessel.

The **Branches** of the abdominal aorta are—1. Phrenic or diaphragmatic. 2. Cœliac axis (dividing into gastric, hepatic, and splenic). 3. Superior mesenteric. 4. Suprarenal or capsular. 5. Renal. 6. Spermatic (ovarian in female). 7. Inferior mesenteric. 8. Sacra media. 9. Lumbar arteries. 10. Common iliacs.

Fig. 100.

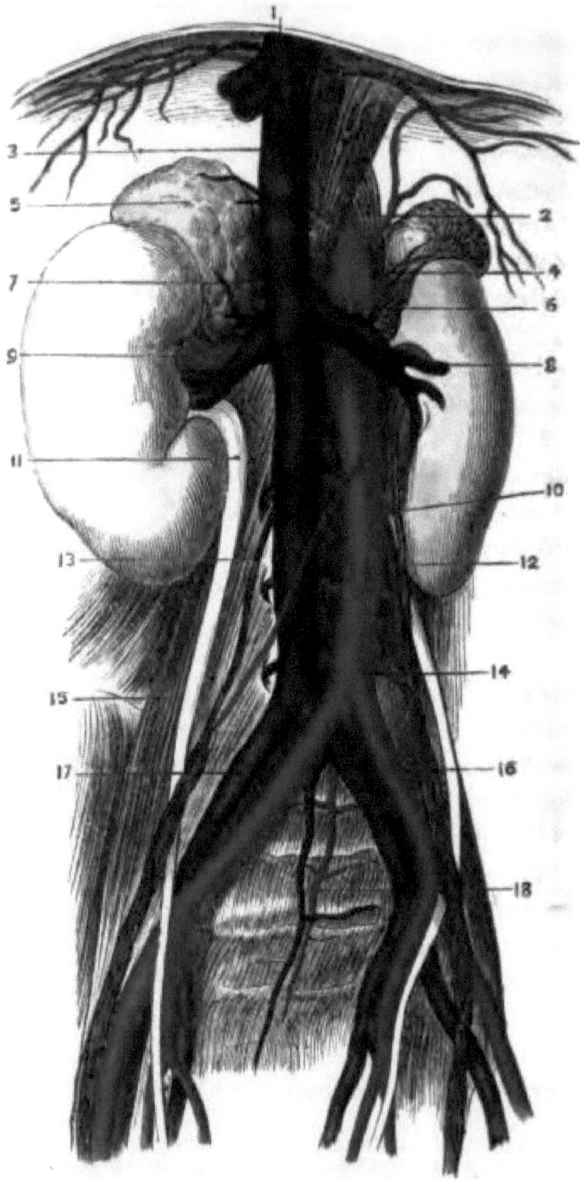

Fig. 100.—Abdominal aorta and vena cava (from Henle).

1. Hepatic veins (cut).
2. Phrenic arteries.
3. Vena cava.
4. Cœliac axis (cut).
5. Supra-renal capsule.
6. Superior mesenteric artery (cut).
7. Supra-renal artery.

The **Phrenic** arteries (Fig. 100, 2) are two small branches which arise close below the diaphragm and ramify on its under surface, where they form a circle and anastomose with one another and with the internal mammary artery, giving also a branch to the supra-renal capsule. The veins open into the vena cava.

The **Cœliac Axis** (Fig. 100, 4) is only half an inch in length and divides at once into the gastric, hepatic, and splenic arteries (*v.* p. 238).

The gastric and splenic veins open into the vena portæ, but the hepatic veins into the vena cava.

The **Superior Mesenteric** artery (Fig. 100, 6) supplies all the small and half the large intestine (*v.* p. 234). Its vein opens into the vena portæ.

The **Supra-renal** arteries (Fig. 100, 7) pass, one on each side, transversely to the supra-renal capsule, the right going beneath the vena cava. In addition to this artery (middle capsular) the capsule is supplied by a branch from the phrenic (superior) and one from the renal artery (inferior). The right vein enters the vena cava, the left the renal or phrenic vein.

The **Renal** arteries (Fig. 100, 9) arising on each side of the aorta immediately below the capsular arteries, pass transversely to the concave border of the kidney, the right going beneath the vena cava. The aorta being a little to the left of the vertebræ at this point, the right renal artery is necessarily a little longer than the left, and the veins are the reverse, owing to the position of the vena cava on the right of the column. The renal artery breaks up into three or four branches just as it enters the kidney, and one or more of them may pass in front of the vein, but as a rule the relation of the parts at the hilum of the kidney is (1) Vein (2) Artery (3) Ureter,

---

8. Renal vein.
9. Renal artery.
10. Left spermatic vessels.
11. Right ureter.
12. Inferior mesenteric artery.
13. Right spermatic vein.
14. A lumbar artery.
15. Psoas magnus.
16. Common iliac artery.
17. Common iliac vein.
18. Middle sacral artery.

from before backwards. The renal veins open into the vena cava, the left crossing the aorta.

The **Spermatic** arteries (Fig. 100, 10) are two long slender branches which arise from the front of the aorta about an inch below the renal, and pass downwards to the internal abdominal ring and through the inguinal canal to the testes. The right crosses obliquely over the vena cava, and both rest upon the external iliac arteries near Poupart's ligament, and are close to the epigastric arteries at the ring. In the female the spermatic arteries become the *ovarian*, and pass into the pelvis to ramify in the broad ligament of the uterus (Fig. 93, *h*).

The spermatic *veins* take a different course on the two sides, the left opening at a right angle into the left renal vein, and the right opening directly into the vena cava; which arrangement is said to account in part for the more frequent occurrence of 'varicocele' on the left side.

The great length of the spermatic arteries is accounted for by the fact that the testicle was originally in the abdomen opposite the point where the arteries arise, and that as the testicles descended the vessels necessarily became elongated.

The **Inferior Mesenteric** artery (Fig. 100, 12) supplies the lower half of the large intestine (*v.* p. 236) and its vein opens into the splenic vein.

The **Middle Sacral** (Fig. 100, 18) is a small branch arising at the bifurcation of the aorta and passing down into the pelvis along the middle of the sacrum. It anastomoses with the lateral sacral branches of the iliac arteries, and its corresponding vein opens into the left common iliac vein.

The **Lumbar** arteries (Fig. 100, 14), four or five in number, arise in a double row from the back of the aorta and pass transversely, close upon the vertebræ, beneath the gangliated cord of the sympathetic, and on the right side beneath the vena cava. Only small portions of these arteries can be seen at present, but when followed out each will be found to divide into an anterior and posterior branch. Of these the anterior runs between the abdominal muscles, and the posterior passes between the transverse processes to the muscles of the back, giving also a branch into the vertebral canal through the intervertebral foramen, which supplies the membranes of the cord, and ends in the body of the vertebra. The lumbar veins open into the vena cava.

The **Common Iliac Arteries** (Fig. 100, 16) are the direct continuations of the abdominal aorta, and extend from the left side of the fourth lumbar vertebra to the sacro-iliac synchondrosis, where each divides into the external and internal iliac arteries. The right common iliac is a little longer than the left, and the relations differ on the two sides owing, principally, to the fact that the two common iliac veins unite to form the vena cava on the right side of the vertebral column. Both arteries are covered in front by the peritoneum, and are crossed by the ureter near their termination, but the *left* is crossed in addition by the rectum and inferior mesenteric vessels The *left* artery lies against the psoas magnus, but the *right* artery lies upon its own vein at the lower part and then crosses the vein of the opposite side. The *left* common iliac vein will be seen to lie to the inner side of its own artery, and then to pass beneath the right artery to reach the vena cava, whilst the *right* vein is at first behind and then to the outer side of the right artery until it unites with the left. Thus the right common iliac artery crosses both the veins.

The only branches from the trunks of the vessels are small twigs to the iliacus muscle and to the ureter.

The **Vena Cava Inferior** (Fig. 100, 3) lies to the right of the aorta and in close relation to it except at the upper part. It is formed by the junction of the two common iliac veins on the right of the fifth lumbar vertebra, and lies on the right of the vertebral column as high as the liver; there it passes forwards to enter the posterior border of that gland, through which it reaches the quadrilateral opening in the diaphragm. It receives the lumbar, right spermatic, renal, right supra-renal, and phrenic veins, and as it passes through the liver the large hepatic veins open into it;

and thus the whole of the blood of the abdomen and lower extremities is returned to the heart.

The **External Iliac Artery** (Fig. 125, 11) reaches from the point of bifurcation of the common iliac to Poupart's ligament, where it becomes the femoral. It passes along the brim of the true pelvis and is bound down to the psoas muscle, on the outer side, by a process of fascia which includes both it and the vein and must be removed in order to expose them. The crural branch of the genito-crural nerve lies close to the outer side of the artery, and the spermatic vessels lie upon it near Poupart's ligament. The *vein* lies to the inner side of the artery in the whole of its course upon the left side, but is beneath its upper part on the right side. Close to Poupart's ligament one of the branches of the vein (circumflexa ilii) will be seen to cross the artery, and just above it the *vas deferens* hooks round the epigastric artery and descends into the pelvis to the inner side of the external iliac artery.

Branches.—The epigastric and circumflexa ilii arteries are given off just above Poupart's ligament.

The *Epigastric* artery passes upwards immediately outside the peritoneum and enters the rectus muscle at its posterior aspect, to which it has been already traced (p. 222). The branches of the epigastric are *pubic* to the back of the pubes, *cremasteric* to the cord, and *muscular*. An abnormal branch of large size is frequently found coming off from the epigastric close to its origin, and descending into the pelvis; this is the obturator artery, and its important relations to femoral hernia have been already described (p. 97).

The *Circumflexa Ilii* artery takes a course outwards along Poupart's ligament and is at first immediately outside the peritoneum. It soon, however, pierces the transversalis, and a branch of it serves as the guide between that muscle and the internal oblique. The artery runs along the crest of the ilium between the muscles, and anastomoses with the ilio-lumbar branch of the internal iliac and with the last lumbar artery. The vein has been seen to cross the external iliac artery close to Poupart's ligament to open into the external iliac vein.

Surgery.—The operations of tying the iliac arteries cannot be conveniently practised on the subject, as the muscles of the abdomen would be necessarily damaged. The following is the usual mode of proceeding :—

Fig. 101.

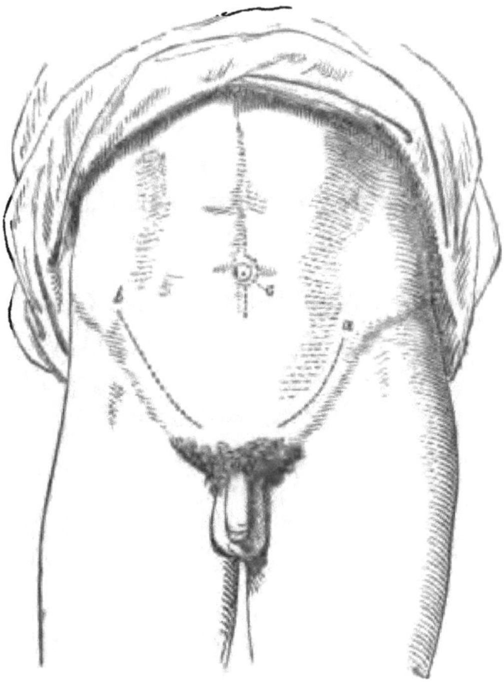

To tie the External Iliac artery, a lunated incision three and a half inches long is to be made about an inch above and parallel to Poupart's ligament (Fig. 101, *a*). The tendon of the external oblique and the fleshy fibres of the internal oblique and transversalis muscles are to be divided in the same direction and to the same extent, and it is usual to divide the transversalis on a director to avoid the possibility of injuring the peritoneum. Care must also be taken not to divide the epigastric artery. The fascia transversalis, now exposed, is to be divided on a director, and the peritoneum is then to be carefully stripped up from the iliac fossa and turned forward with the viscera,

Fig. 101.—(From Fergusson's 'Surgery.')

a. Line of incision for tying the external iliac artery.
b. Line of incision for tying the common iliac artery.
c. Sir Astley Cooper's incision for tying the abdominal aorta.

the ureter and spermatic vessels generally adhering to the membrane and being therefore drawn out of the way. The finger can now reach the inner border of the psoas, at which the artery can be felt, but the sheath must be carefully opened before the needle can be passed between the vein and the artery. The vessel should be tied as nearly as possible midway between the origins of the internal iliac and epigastric arteries.

The Internal and Common Iliac arteries may be tied through a similar incision prolonged to a length of about five inches (Fig. 101, *b*), or by one of similar length placed so that its centre is opposite the anterior superior iliac spine. The steps of the proceeding are the same as those for tying the external iliac, but the operations are more arduous owing to the greater depth of the arteries and the consequent difficulty in reaching them.

The **Kidney** (Fig. 97, 14) may be roughly said to lie between the last rib and the crest of the ilium. It generally however reaches as high as the eleventh rib, and therefore lies on a small portion of the diaphragm and on the quadratus lumborum muscle, but the right kidney is always a little lower than the left. The upper extremities of the organs are nearer the spine than the lower, and upon them are fixed the supra-renal capsules. It will be remembered that in front of the kidney was the peritoneum, with the ascending colon on the right, and the descending colon on the left side. The outer border of the kidney is convex, the inner concave, and the vessels entering at this border are the renal vein, renal artery, and ureter, which lie in that relation to one another from before backwards.

The *Ureter* (Fig. 100, 11) should be traced out, and will be seen to lie over the psoas muscle, to pass beneath the spermatic vessels, and, having crossed the common or external iliac artery, to disappear in the pelvis, where it will be afterwards traced to the bladder.

The **Supra-renal Capsule** (Fig. 100, 5) is a small triangular yellow body, somewhat resembling a cocked-hat, placed immediately above each kidney and lying against the crus of the diaphragm. The vessels to it should be traced out, and one kidney and capsule are then to be removed and put aside with the viscera for after-examination.

[The kidney being removed or turned out of the way, the psoas and quadratus lumborum muscles are to be cleaned, and all nerves passing out of them are to be carefully preserved.]

Fascia of the quadratus lumborum.—The anterior lamella of the fascia lumborum, which forms part of the origin of the transversalis abdominis muscle (*v.* p. 212), will be found to lie in front of the quadratus muscle; what has been described as the ligamentum arcuatum externum of the diaphragm being merely the border of this, artificially isolated.

The *fascia over the psoas* is a distinct structure, although connected partially with the fascia of the quadratus. It is continuous with the iliac fascia in the pelvis, and passes over the psoas, to be attached to the lower edge of the ligamentum arcuatum internum of the diaphragm, and to the sides of the bodies of the vertebræ from which the muscle takes its origin. It is owing to this fascia that "psoas abscess," the result of diseased spine, is conducted along the psoas muscle to the inner side of the thigh.

The *Iliac Fascia* covers the iliacus internus muscle and is continuous with the fascia over the psoas magnus, being strengthened at the lower part by an expansion from the psoas parvus when that muscle is present. The iliac fascia should be traced to Poupart's ligament, beneath which it passes to unite with the fascia transversalis in forming the sheath of the femoral vessels (*v.* p. 93).

The **Psoas Magnus** muscle (Fig. 102, 6) *arises* from the sides of the bodies and roots of the transverse processes of the last dorsal and all the lumbar vertebræ, and from their intervertebral substances. A series of tendinous arches from which muscular fibres arise, connected with the intervertebral substance and upper border of each vertebra, will be seen to spring across the hollow in the side of the body to become attached to the lower border of the same vertebra and to the next in[illegible] substance; and thus the lumbar vessels are protected from injury. The muscle passes along the border of the true pelvis in close relation to the iliac vessels, and disappears beneath Poupart's ligament to be *inserted* into the back of the lesser trochanter of the femur.

The **Psoas Parvus** (Fig. 102, 4) is a small muscle occasionally found upon the front of the psoas magnus. It *arises* from the bodies of the last dorsal and first lumbar vertebræ, and ends in a long tendon which is *inserted* into the ilio-pectineal eminence and part of the ilio-pectineal line, giving an expansion to the iliac fascia.

The **Iliacus Internus** (Fig. 102, 8) lies in the iliac fossa, from the greater part of which it *arises*, as well as from the ilio-lumbar ligament and the base of the sacrum. A few fibres also are attached to the capsular ligament of the hip-joint. The iliacus is united with the psoas at its *insertion* into the femur.

The psoas and the iliacus are flexors and rotators outwards of the thigh, but taking their fixed point below are flexors of the trunk upon the thigh. This is seen in dislocation of the femur downwards into the thyroid foramen, when the body is always bowed forwards owing to the stretching of the muscles. The psoas is *supplied* by the lumbar nerves, the iliacus by the anterior crural nerve.

The **Quadratus Lumborum** (Fig. 102, 5) is the short muscle filling the space between the last rib and the crista ilii. It has its origin below, from the posterior portion of the inner lip of the crest of the ilium and from the adjacent ilio-lumbar ligament; and ascends to be *inserted* into the lower border of the last rib, and by three or four slips into the tips of the transverse processes of the lumbar vertebræ. It is sometimes described as consisting of two distinct parts.

Twelfth dorsal nerve (Fig. 102, 3).—Crossing the quadus, and immediately below the ligamentum arcuatum ernum will be found a large nerve, the twelfth dorsal, ich will be seen to disappear through the transversalis scle, and must not be confounded with the branches of lumbar plexus.

The **Lumbar Plexus** (Fig. 102).—At the outer border of the psoas and about midway between the last rib and the crest of the ilium will be found two small nerves (often united), crossing the quadratus obliquely and piercing the abdominal muscles immediately above the crest, and generally close together. These are the *ilio-hypogastric* (7) and *ilio-inguinal* (9) nerves, and their size varies considerably in different subjects, one being often larger than the other, or there being one trunk only which subdivides in the abdominal wall. The after course of the nerves has been given in the dissection of the abdominal wall, the ilio-hypogastric being distributed to the skin of the lower part of the abdomen, and the ilio-inguinal emerging from the external abdominal ring to supply the scrotum.

The *Genito-crural nerve* (11) appears on the front of the psoas at its upper part and is of small size. It descends on the muscle, and divides into a *genital* (29) branch which accompanies the spermatic cord through the inguinal canal to the cremaster, and a *crural* (27) branch which runs under Poupart's ligament to the outer side of the external iliac artery, and is lost in the skin of the thigh.

The *External cutaneous nerve* (15) appears at the outer border of the psoas just below the level of the crista ilii. It crosses obliquely to the anterior superior spinous process, close to which it passes under Poupart's ligament to the outer side of the thigh. (This frequently comes from the anterior crural nerve and is nearly transverse in its direction.)

The *Anterior crural* (23) is the large nerve lying close to the outer border of the psoas at its lower part, and imbedded in the iliacus. It passes under Poupart's ligament into the thigh, but in the abdomen gives branches to the iliacus internus muscle.

The *Obturator nerve* (25) will be better dissected with the pelvis, but if desired may be seen passing below the brim of the true pelvis to disappear through the upper part of the thyroid foramen.

An *Accessory Obturator* nerve is occasionally found

Fig. 102.

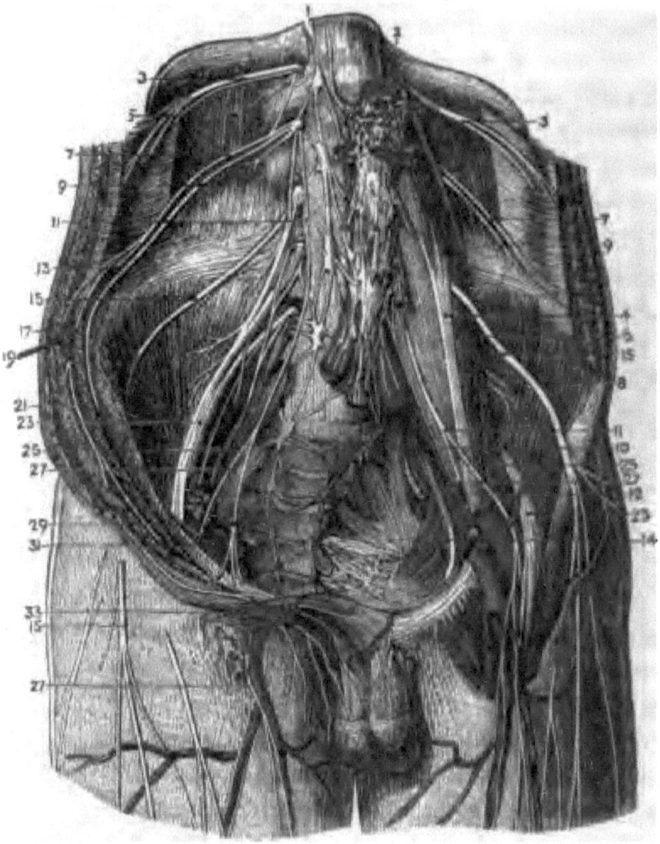

Fig. 102.—Lumbar plexus (from Hirschfeld and Leveillé).

1. Right gangliated cord of sympathetic.
2. Abdominal aorta.
3, 3. Last dorsal nerves.
4. Psoas parvus.
5. Quadratus lumborum.
6. Psoas magnus.
7, 7. Ilio-hypogastric nerves.
8. Iliacus internus.
9, 9. Ilio-inguinal nerve.
10. Lumbo-sacral nerve.
11, 11. Genito-crural nerves.
12. Gluteal nerve.
13. Iliac branch of ilio-hypogastric nerve.
14. Sacral plexus.
15, 15, 15. External cutaneous nerves.
17. Transversalis abdominis.
19. Obliquus internus.
21. Obliquus externus.
23, 23. Anterior crural nerves.
25, 25. Obturator nerves.
27, 27. Crural branch of genito-crural nerve.
29. Genital branch of genito-crural nerve.
31. External iliac artery.
33. External abdominal ring.

arising with the obturator nerve, which is then diminished in size. The accessory nerve passes beneath the psoas over the brim of the pelvis, and joins the obturator nerve on the inner side of the thigh.

[The psoas should be carefully removed piecemeal on one side, in order to dissect the lumbar nerves to their origins, and to trace out the plexus completely.]

The *Lumbar plexus* will now be seen to be formed by the 1st, 2nd, 3rd, and part of the 4th lumbar nerves, with a branch from the 12th dorsal, the remainder of the 4th joining with the 5th to form the lumbo-sacral cord. The several nervous trunks unite more or less, and from them the nerves are given off thus:—from the 1st the ilio-hypogastric and ilio-inguinal nerves; from the 1st and 2nd the genito-crural nerve; from the 2nd and 3rd the external cutaneous nerve; from the 2nd, 3rd, and 4th the anterior crural nerve; from the 3rd and 4th the obturator nerve. Thus it will be seen that the origins of the nerves are very simple, a nerve being added or subtracted in regular sequence, as follows:—

Lumbar nerves	1 . .	Ilio-hypogastric and ilio-inguinal.
	1 2 . .	Genito-crural.
	2 3 .	External cutaneous.
	2 3 4	Anterior crural.
	3 4	Obturator.

Sympathetic Nerve (Fig. 97).—On each side of the aorta and close to the crura of the diaphragm will be seen an irregular mass of a somewhat crescentic form called the semi-lunar ganglion, and into it the great splanchnic nerve from the thoracic ganglia may be traced.

The *Solar plexus* is formed by fibres crossing from one semi-lunar ganglion to the other, and receives a branch from the right pneumo-gastric. Its continuation, the aortic plexus, will have been more or less seen in cleaning the aorta, and so also the secondary plexuses derived from it and distributed upon the several branches, viz., the phrenic plexus, gastric plexus, hepatic plexus

plexus, superior mesenteric plexus, supra-renal plexus, renal plexus, spermatic plexus, and inferior mesenteric plexus. In addition to the great splanchnic nerve, the lesser splanchnic nerve may, in a favourable subject, be traced into the cœliac plexus, and the third or least splanchnic nerve into the renal plexus.

The *Gangliated Cord of the sympathetic* (Fig. 102, I) will be found on each side of the spinal column and behind the vena cava upon the right side. A ganglion is situated opposite each vertebra, and branches of communication may be traced to the several lumbar nerves and to the neighbouring plexuses.

Some lymphatic glands may also be found by the side of the aorta at the lower part, which receive the lymphatics from the testis and are therefore affected in malignant disease of that organ.

The **Vena Azygos Major** may be seen lying close to the right side of the aorta and communicating with one or more of the lumbar veins. It passes through the aortic opening.

The **Vena Azygos Minor** (if it exists) may be seen to commence in the left lumbar veins, and to pass through the left crus of the diaphragm.

Receptaculum Chyli.—By detaching the right crus of the diaphragm from the vertebræ, the *receptaculum chyli*, an irregular tube about two inches long, will be seen lying to the right side of the aorta, opposite the second lumbar vertebra. It receives the lymphatics and lacteals of the abdomen, and is continued upwards as the thoracic duct.

The subject will now be turned for the dissection of the back, and the dissectors of the abdomen should occupy the time with the examination of the viscera.

Viscera of the Abdomen.

[The stomach and intestines should be cleaned by allowing a stream of water to pass through them; and the stomach, as well as a piece of small and large intestine, should be moderately distended with air by means of a blowpipe.]

The stomach (Fig. 103) presents a lesser and a greater curvature at its upper and lower borders, and a greater and lesser end. The great end is the dilatation on the left

Fig. 103.

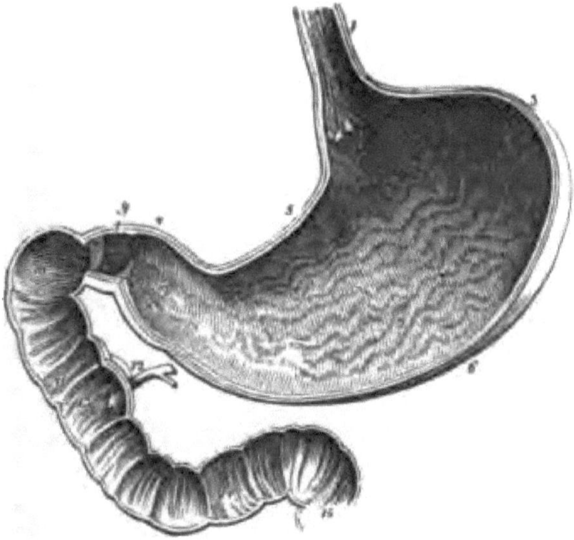

side of the body near the entrance of the œsophagus, and is called the cardiac or splenic end, or is sometimes known as the fundus. The small end terminates in the duodenum and is called the pyloric end, the junction of

---

Fig. 103.—Vertical and longitudinal section of the stomach and duodenum (from Wilson).

1. Œsophagus.
8. Cardiac orifice of the stomach.
3. Great end of the stomach.
4. Lesser or pyloric end.
5. Lesser curve.
6. Greater curve.
7. Dilatation at the lesser end of the stomach, the antrum of the pylorus.
8. Rugæ of the stomach formed by the mucous membrane; their longitudinal direction is shown.
9. Pylorus.
10. Ascending portion of the duodenum.
11. Descending portion.
12. Pancreatic duct and ductus communis choledochus close to their termination.
13. Papilla upon which the ducts open.
14. Transverse portion of duodenum.
15. Commencement of jejunum. In the interior of the duodenum and jejunum, the valvulæ conniventes are seen.

the two viscera being marked by a thickened band of fibres which can be both seen and felt, and is called the *pylorus*.

The coats of the stomach are four—peritoneal, muscular (comprising longitudinal, circular and oblique fibres), areolar and mucous.

On laying open the stomach along its lesser curvature, the mucous membrane will be seen to be arranged in

Fig. 104.

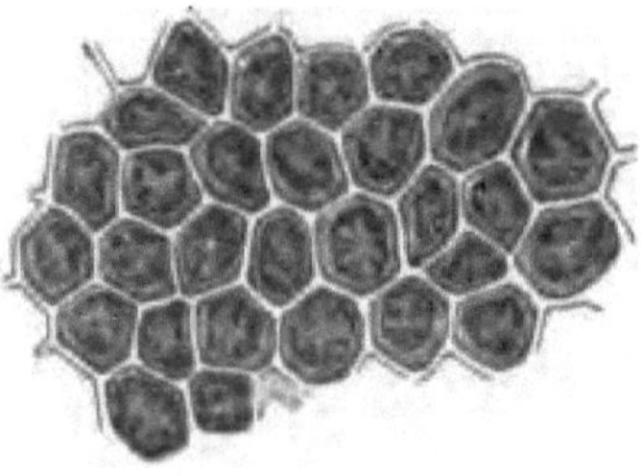

Fig. 105.

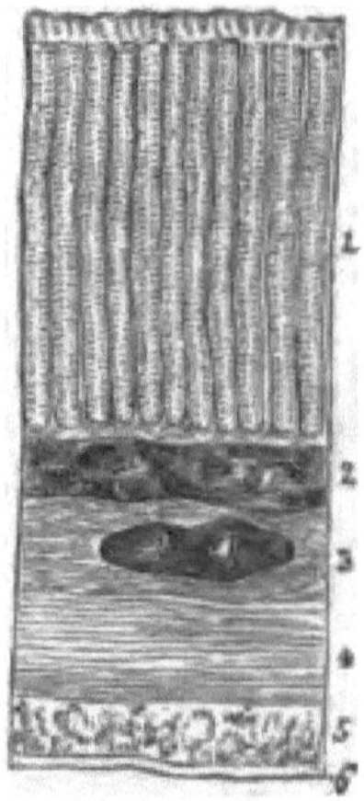

Fig. 106.

longitudinal folds or ridges called *rugæ*, which disappear when the viscus is distended and are more evident in the

---

Fig. 104.—Portion of the mucous membrane of the stomach, magnified 75 times. The trifid or quadrifid division of a small artery is seen at the bottom of each alveolus, and in the depressions between the divisions of the artery, the apertures of the gastric follicles; two, three, or four in each depression (from Wilson).

Fig. 105.—Perpendicular section of the stomach in the pyloric region; showing its coats and the gastric glands; from the pig (from Wilson, after Kölliker).

1. Gastric glands.
2. Muscular layer of the mucous membrane.
3. Submucous coat, in which a vessel has been divided.
4. Transverse fibres of the muscular coat.
5. Longitudinal fibres of the muscular coat.
6. Serous coat.

Fig. 106.—A tubular gastric gland (from Wilson, after Kölliker).

1. Columnar epithelium occupying its upper third.
2. Flexuous and twisted deeper portion, filled with nucleated cells.
3. One of the nucleated cells.

child than in the adult. By inverting the pyloric end the *pylorus* or *pyloric valve* will be seen to consist of a set of strong circular fibres, surrounding the intestine and acting as a sphincter muscle.

Minute Anatomy.—The mucous membrane is divided into a series of shallow cavities more or less hexagonal in form (stomach cells), at the bottom of which two or more tubes (stomach tubes) open. These latter secrete the gastric juice and are imbedded in the submucous areolar layer. The epithelium is columnar.

The **Small Intestines** extend from the pylorus to the ileo-cœcal valve, averaging twenty feet in length, and have certain general characteristics in common; thus they are of an uniform calibre throughout, and their external surface is smooth and regular. They have four coats, viz. peritoneal, muscular (longitudinal and circular), areolar, and mucous. The peritoneal coat is not found throughout, thus the first part of the duodenum is completely invested except where the vessels enter, the second part only in front, and the third part not at all; the jejunum and ileum are completely covered by peritoneum. The longitudinal muscular fibres are external, and the circular internal,* and they are both complete and continuous layers. The mucous membrane has the following characteristics throughout—valvulæ conniventes, villi, Lieberkühn's follicles, solitary glands, and a columnar epithelium; but the duodenum and ileum have certain special structures in addition.

The **Duodenum** (Fig. 103) should be laid open along its free border, in order to see the opening of the bile and pancreatic ducts, which is marked by a papilla situated at the back of the vertical portion of the intestine, and about or below its middle. A probe should be passed into the orifice and along the pancreatic duct. The duodenum is from eight to ten inches long, and has a special structure (Brunner's glands) in addition to the ordinary characteristics of the small intestine.

* An artificial aid to the remembrance of this fact is afforded by noting that lOngitudinal=Outer, and cIrcular=Inner [illegible]

Valvulæ conniventes are transverse folds of mucous membrane, which will be seen to commence a little beyond the pylorus, and to increase in size at the lower part of the duodenum.

Villi (Fig. 107, 1) are minute projections from the surface of the mucous membrane, each containing a little capillary loop and a lacteal vessel, and being covered with columnar epithelium. They may be best seen by floating a piece of intestine in water,

Lieberkühn's follicles (Fig. 107, 2) are minute tubes which are placed between the villi; they dip into the submucous areolar tissue and are lined with columnar epithelium.

Fig. 107.

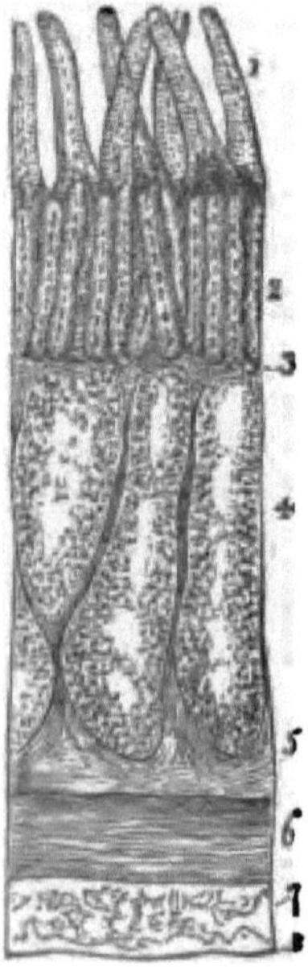

Solitary glands are minute white bodies consisting of closed vesicles containing milky fluid and covered with villi.

Brunner's glands, peculiar to the duodenum, are small lobulated glands situated in the submucous areolar tissue and having small ducts which open into the intestine. To see these it is necessary to pin out a piece of duodenum with the mucous membrane downwards, and then carefully to dissect away the muscular coat.

[The jejunum and ileum (except a few inches to be left attached to the cæcum) should be laid open along the line of attachment of the mesentery, so as to avoid damaging Peyer's patches which are on the free border.]

Fig. 107.—Section of the lower part of ileum of the calf, through a Peyer's gland (from Wilson, after Kölliker).

1. Villi.
2. Tubular glands or Lieberkühn's follicles.
3. Muscular stratum of the mucous membrane.
4. Sacculi of Peyer's gland.
5. Deep portion of the submucous coat.
6. Circular layer of the muscular coat.
7. Longitudinal layer of the muscular coat.
8. Serous coat.

The **Jejunum** includes about two-fifths of the remaining small intestine, and is remarkable for the large size of the valvulæ conniventes and the number of solitary glands, but has no special characteristic.

In the **Ileum** the valvulæ conniventes will be found to diminish rapidly in size and to be wanting at the lower part.

Peyer's patches (Fig. 107, 4), the special characteristic of this part of the intestine, vary very much in size and are occasionally invisible. They will be recognised by their dotted appearance and by being of a lighter colour than the rest of the intestine, and are always found on the part of the intestine farthest from the attachment of the mesentery. Each patch is formed of a collection of the solitary glands, and may be as long as a couple of inches. Peyer's glands will be found strongly marked, and sometimes ulcerated in subjects which have died of typhoid fever.

The **Large Intestine** (Fig. 95) is between five and six feet in length, and is remarkable for the pouched appearance consequent upon its longitudinal muscular fibres being shorter than the intestine itself. These longitudinal fibres are arranged in three distinct bands, two of which can be seen through the peritoneum, and the third between the layers of the meso-colon. (In the rectum which is not pouched, these fibres spread over the whole gut as in the small intestine.) The *appendices epiploicæ* are small processes containing fat attached along the free border of the intestine, which will be found to vary considerably in size in different subjects.

The **Cæcum** (Fig. 108, 1) (caput cæcum coli) is distinguished by its rounded shape and by the opening of the ileum into it. It has attached to its lower and back part the *appendix vermiformis*, which is a little tubular prolongation terminating in a blind extremity.

[The cæcum with a few inches of small and large intestine should be detached, and the large intestine turned inside out to see the ileo-cæcal valve. After this has been done and the intestine reinverted, the cut ends may be tied and the piece inflated and dried, that the valve may be again examined in the dry condition.]

The *Ileo-cæcal valve* (Fig. 108, 5) is formed by a prolongation of a piece of ileum through the wall of the cæcum, to which it is firmly attached; and being stretched from side to side, the small intestine forms two folds which project into the cæcum and form the valve. The upper fold, which is horizontal in direction, is sometimes known as the *ileo-colic*, and the lower or more vertical one as the *ileo-cæcal* fold.* The ridge on each side, formed by the prolongation of the two folds, is called the *frænum*. It is obvious that these folds permit the ready passage of fluid from the small into the large intestine, but would prevent regurgitation unless the parts were very much over-distended.

Fig. 108.

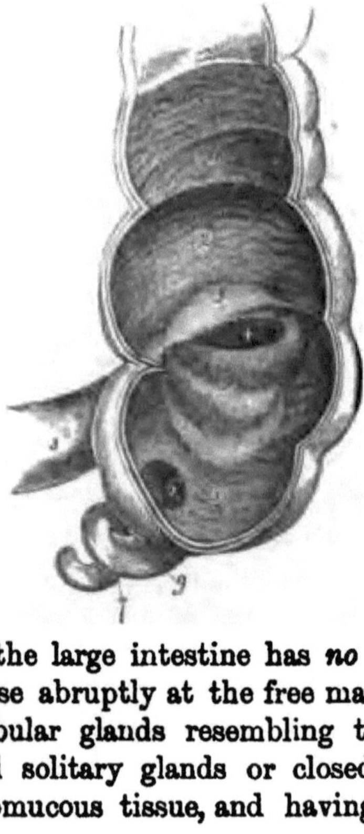

The mucous membrane of the large intestine has *no* villi, for they are found to cease abruptly at the free margin of the ileo-cæcal valve. Tubular glands resembling the follicles of Lieberkühn, and solitary glands or closed follicles embedded in the submucous tissue, and having over them small depres-

* Artificial memory, S H I P—Superior Horizontal, Inferior Perpendicular.

Fig. 108.—The cæcum laid open, showing the entrance of the ileum, and ileo-cæcal valve (from Wilson).

1. Cæcum.
2. Commencement of colon.
3. Ileum.
4. Aperture of entrance of the ileum into the large intestine.
5, 5. Ileo-cæcal valve.
6. Aperture of appendix vermiformis cæci.
7. Appendix vermiformis.
8, 8. Sacculi of the colon.
9. Mesentery of appendix vermiformis.

ions on the mucous surface, are found throughout the large intestine (Figs. 109 and 110).

The **Pancreas** (Fig. 97) should be cleaned but be left attached to the duodenum. It is about seven inches long, and the larger end, where it touches the bowel, is called the head, the central portion the body, and the small extremity the tail. Its average weight is three ounces. A probe being passed into the duct from the duodenum, it should be dissected out, and will be found to pass along the whole length of the gland, and nearer the posterior than the anterior surface. The pancreas consists of lobules of a

Fig. 109.

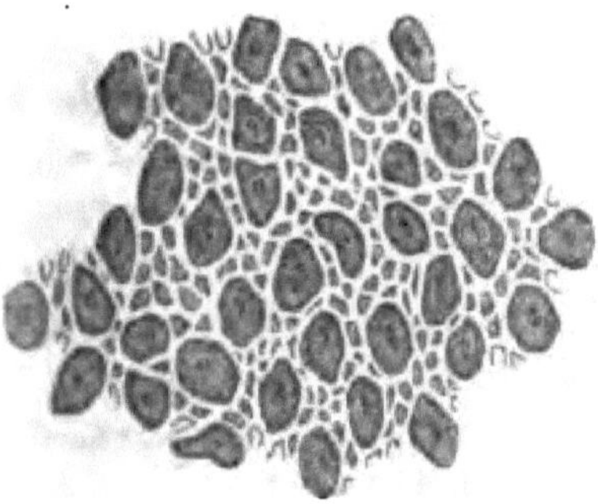

Fig. 110.

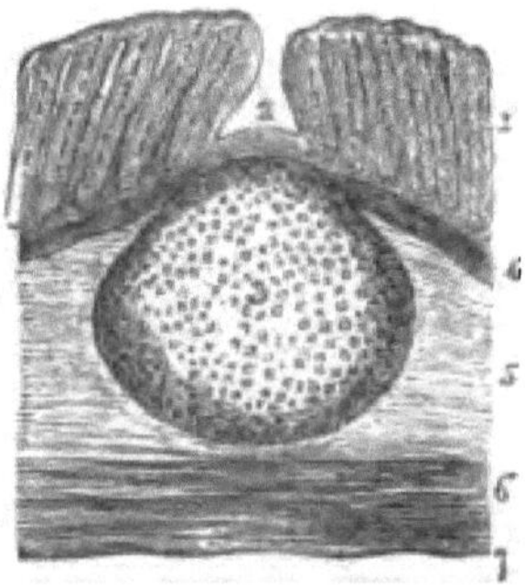

yellowish colour held together by loose fibrous tissue, a small duct passing from each lobule into the main trunk.

The **Spleen** (Fig. 98, 4) is purple in colour and somewhat oval in form, and has a convex and a concave surface, with an anterior border which is sharp and slightly notched, and a posterior border which is thicker. On the concave surface is a vertical fissure, the *hilum*, into which the splenic vessels should be traced. Beneath

---

Fig. 109.—Portion of the mucous membrane of the large intestine, magnified 75 times (from Wilson).

Fig. 110.—Solitary gland from the colon of a child, as shown in a section of the intestine (from Wilson, after Kölliker).

1. Tubular glands.
2. Depression on the surface of the membrane corresponding with the central prominence of the gland.
3. The gland.
4. Muscular layer of the mucous coat.
5. Submucous coat.
6. Muscular coat.
7. Serous coat.

the peritoneal coat which invests the organ will be found a fibrous coat, which is also continued into the splenic substance, forming the *trabeculæ* or meshes in which the splenic *pulp* is contained. In this pulp, which consists principally of blood-vessels, are found the *Malpighian corpuscles* which are special to the spleen, and are small vesicular bodies containing white semi-fluid matter attached to the minute divisions of the splenic artery (Fig. 111). The minute anatomy and function of this organ are not fully understood.

Fig. 111.

The **Liver** (Fig. 112) presents an upper and an under surface united by a thick posterior and a thin anterior margin. Its ordinary measurements are, twelve inches across, six from before backwards, and three inches in thickness; its weight is about fifty ounces avoirdupois. The upper surface is divided into two unequal parts, of which the right is the larger, by the attachment of the falciform ligament; and on tracing this back, its two layers will be found to diverge, forming the upper layer of the coronary and also the two lateral ligaments. If the left lateral ligament be traced it will be found to consist of a double fold of the peritoneum which may be followed to the under surface of the left lobe. The upper layer of the coronary ligament if traced outwards in the same way will be found to be reflected upon itself to form the right lateral ligament, and then will be followed to the under surface of the right lobe of the liver behind the transverse fissure, to form the under layer of the coronary ligament; thus leaving a more or less triangular interval between the two, in which the liver is attached to the diaphragm by cellular tissue, and through which the vena cava passes to the diaphragm. In this

* Fig. 111. Arterial twig of the splenic artery, showing the connexion of the splenic corpuscles with the small vessels. From the spleen of the dog (from Wilson, after Kölliker).

manner four ligaments of the liver are formed by peritoneum, and the fifth or round ligament is the obliterated umbilical vein.

The **under surface** of the liver (Fig. 113) presents five *fissures* and five *lobes* for examination.

Fig. 112.

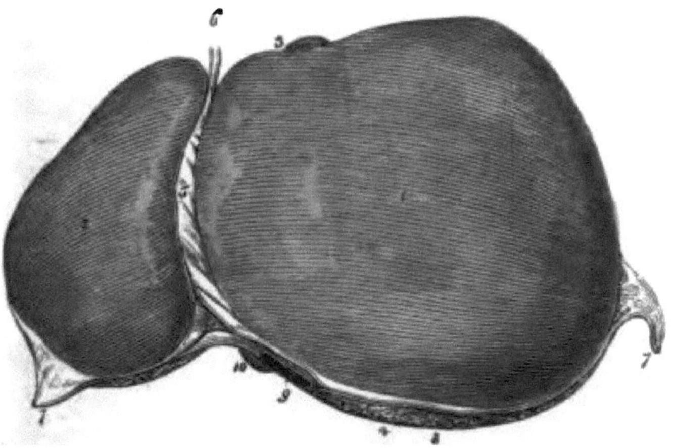

Fissures.—The *longitudinal fissure* divides the under surface into a right and a left lobe and is occupied by the round ligament or umbilical vein (1). Its continuation behind the transverse fissure is the *fissure for the ductus venosus* (8) and contains that obliterated vessel. The *transverse fissure* is at right angles to the longitudinal, and gives passage to the hepatic duct (3), hepatic artery (4), and portal vein (5), which lie in that order from before backwards.* Parallel to the longitudinal fissure is the *fissure*

* It will be observed that the order of vessels entering the liver is the reverse of those entering the kidney, *q. v.* p. 251.

Fig. 112.—Upper surface of the liver (from Wilson).

1. Right lobe.
2. Left lobe.
3. Anterior or free border.
4. Posterior or rounded border.
5. Falciform or suspensory ligament.
6. Round ligament.
7, 7. The two lateral ligaments.
8. The space left uncovered by the peritoneum and surrounded by the coronary ligament.
9. Inferior vena cava.
10. Point of the lobus Spigelii.
3. Fundus of the gall-bladder seen projecting beyond the anterior border of the right lobe.

for the gall-bladder (2) which last should be distended with air by a blowpipe placed in the duct. Placed obliquely at the posterior margin of the liver and behind the gall-bladder, is the short groove or *fissure for the vena cava* (7) a portion of which vessel has been left *in situ*.

Lobes.—The whole of the liver substance on the right of the longitudinal fissure is strictly the *right lobe* (R) but

Fig. 113.

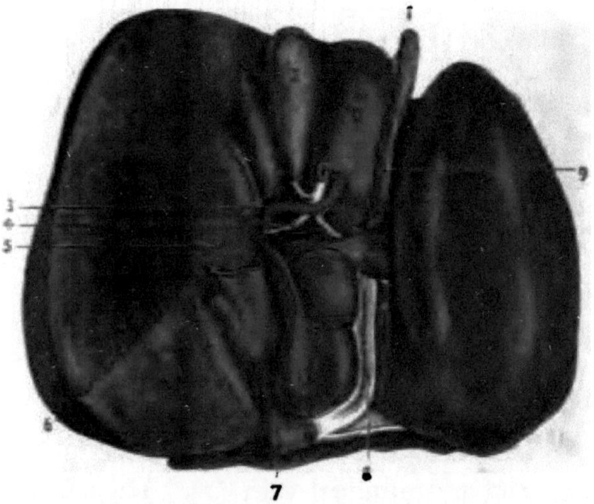

certain parts of it have received special names, viz. lobus quadratus, lobus Spigelii, and lobus caudatus. The right lobe has two shallow depressions on its under surface, the anterior one being where the ascending colon touches the liver, and the posterior one corresponding to the anterior surface of the right kidney, over which it lies in the erect posture.

Fig. 113.—Under surface of the liver (from Bonamy).

R. Right lobe.
L. Left lobe.
Q. Lobus quadratus.
S. Lobus Spigelii.
C. Lobus caudatus.
1. Umbilical vein in longitudinal fissure.
2. Gall-bladder in its fissure.
3. Hepatic duct in transverse fissure.
4. Hepatic artery in transverse fissure.
5. Portal vein in transverse fissure.
6. Line of reflexion of peritoneum.
7. Vena cava.
8. Obliterated ductus venosus.
9. Ductus communis choledochus.

The *Lobus quadratus* (Q) is the square lobe bounded by the longitudinal and transverse fissures and the gall-bladder. The *Lobus Spigelii* (S) is the projecting portion behind the transverse fissure, and between the fissures for the vena cava and the ductus venosus. The *Lobus caudatus* (C) is the small "*tail*" of liver substance which connects the lobulus Spigelii with the right lobe. The *left lobe* (L) is on the opposite side of the longitudinal fissure, across which there is often a bridge of liver substance (pons hepatis).

The vessels entering the transverse fissure should be defined, and the fibrous tissue around them (capsule of Glisson) removed. The *right* and *left hepatic ducts* (4) will be found to emerge from the corresponding lobes and to unite in the *common hepatic duct*, which is about two inches long; this is afterwards joined by the *cystic duct* from the gall-bladder, to form the *common bile duct* (ductus communis choledochus) (9) which is nearly three inches long, and should be traced into the duodenum.

The *gall-bladder* (2) is a pear-shaped bag attached by cellular tissue to the liver, and covered superficially by the peritoneum. When distended, its large end or *fundus* projects beyond the anterior border of the liver; the neck is curved upon itself twice and ends in the cystic duct, which is about an inch and a half long and joins the hepatic duct.

The hepatic artery and the portal vein each divide into right and left, and from the right artery the cystic branch should be traced to the gall-bladder.

The piece of the vena cava is to be laid open, when the large hepatic veins will be seen passing into it.

In order to learn the arrangement of the vessels in the liver, the vena portæ should be carefully laid open for some distance with a pair of scissors. Through the thin wall of the vein a branch of injected hepatic artery will then be seen, and accompanying it is a branch of hepatic duct, the three vessels thus lying side by side in a *portal canal*. One of the hepatic veins being laid open in a similar way will be seen to be totally unaccompanied

An incision into any part of the liver will show on its cut surfaces a number of openings, some wide open,—the hepatic veins; and others more or less collapsed and having by their sides the section of a small injected artery. These last are the portal veins lying in the portal canals, and they collapse on account of the loose attachment of the fibrous tissue around them.

Minute Anatomy (Fig. 114).—The blood from the chylopoietic viscera is brought to the liver by the vena portæ, and from it the bile is secreted. The divisions of the vein have been seen to pass through the portal canals, in which they receive the vaginal branches from the fibrous tissue of the organ, and subsequently divide until their branches run *between* the minute lobules, and are therefore called *interlobular veins* (*b*). The lobules are surrounded by these interlobular veins, from which the lobular plexus of capillaries is derived, which converges to the centre of each lobule and pours its blood into the *intralobular vein* (*a*). Each intralobular vein passes out of the lobule at right angles to the portal vessels, and unites with other intralobular veins to form the *sublobular* veins; these open into the *hepatic veins* which have been traced into the vena cava. Thus then the blood is traced from the portal vein into the vena cava.

Fig. 114.

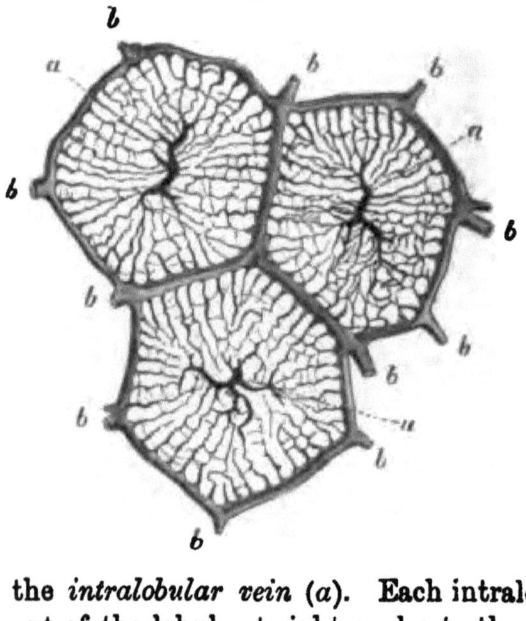

Fig. 114.—Diagram of the circulation in the lobules of the liver (after Kiernan).

a, *a*. Intralobular veins. *b*, *b*. Interlobular veins.

The commencement of each primary *hepatic duct* is in e centre of each minute lobule, being continuous, ac-rding to Beale, with the basement membrane upon which e hepatic cells are placed. The minute ducts run to-ards the circumference of the lobule, and the radicles of djacent lobules unite, the resulting ducts running along he portal canals by the side of the portal vein to emerge ventually at the transverse fissure.

The *Hepatic artery* is principally for the nourishment f the tissue of the organ, and has little if anything to lo with its function. It gives off *vaginal* and *capsular* branches which run in the portal canals and supply the fibrous tissues, the blood being eventually returned into the portal vein.

The *interlobular* branches accompany the interlobular veins, and according to the latest researches their blood enters the plexus within the lobules from which the bile is secreted.

The **Kidney** (Fig. 97, 17) is convex on its anterior and flattened on its posterior surface, and its upper extremity is somewhat larger than the lower. Its outer border is convex and regular, while the internal is notched at the point where the vessels enter. An average kidney measures about four inches in length, two-and-a-half in breadth, and rather over an inch in thickness. Its weight in the male is about 4½ ounces, and rather less in the female. The renal artery enters, and the renal vein and ureter emerge at a slit on the inner border called the *hilum*, and the *trunks* of the vessels have the following relation from before backwards, viz., vein, artery, and ureter (compare vessels entering transverse fissure of liver in the reverse order, p. 271). Very generally, however, one of the branches of the renal artery will enter the hilum in front of the vein.

The ureter, if traced upwards, will be found to expand, forming the *pelvis* of the kidney, and it should be noticed that the pelvis has a direction downwards, so that by referring to this, to the position of the vessels, and to the flattening of the posterior surface, the side to which the kidney belongs can be readily ascertained. The kidney is

Fig. 115.

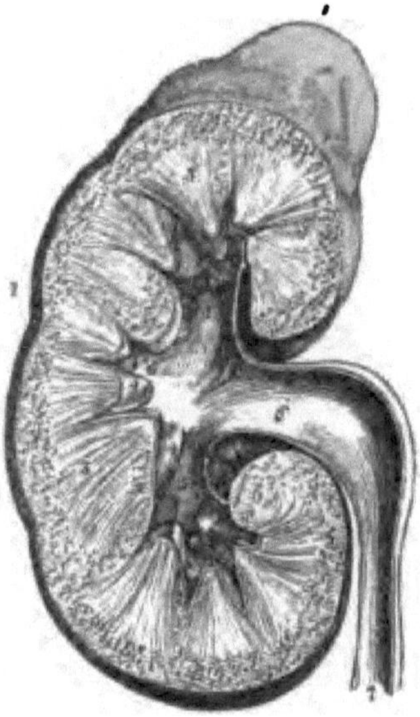

Fig. 116.

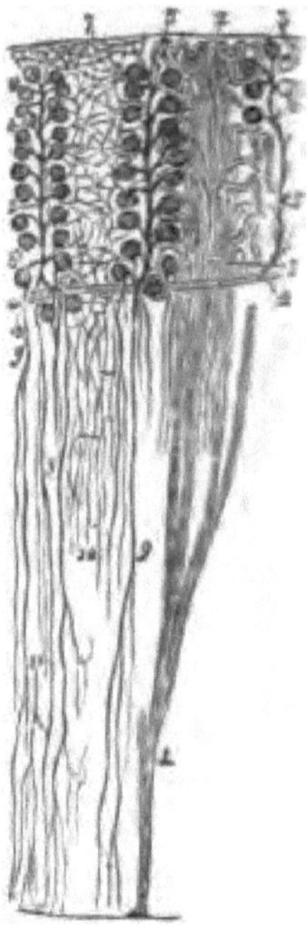

Fig. 115. Section of the kidney, surmounted by the suprarenal capsule (from Wilson).

1. Suprarenal capsule.
2. Cortical portion of the kidney.
3, 3. Tubular portion, consisting of cones.
4, 4. Two of the papillæ projecting into their corresponding calices.
5, 5, 5. The three infundibula; the middle 5 is situated in the mouth of a calyx.
6. Pelvis.
7. Ureter.

Fig. 116.—Plan of the structure of the kidney (from Wilson, after Kölliker).

1, 1. Tubulus rectus, dividing into branches as it ascends.
2, 2. Pyramid of Ferrein; with the tubuli contorti.
3. Origin of two tubules from their respective Malpighian bodies.
4, 4. Branch of an artery.
5, 5. Three interlobular arteries giving off afferent twigs to the Malpighian bodies, 6, 6.
7. Capillary plexus formed by the efferent vessels, of which one is seen at 8.
9, 9. Efferent vessels proceeding to the tubular structure of the pyramids and constituting the arteriolæ rectæ.
10. Capillary plexus of the tubular structure of the pyramid.
11. One of the venulæ rectæ.

enclosed in a tough fibrous capsule, which may be peeled off to see the kidney substance. To see the internal structure of the organ it should be opened by a vertical incision through the convex border, which is to be carried into the pelvis.

The *Pelvis* (Fig. 115, 6) will be found to be subdivided at the upper, middle, and lower part of the kidney into three portions, which are called the *Infundibula* (5) and are again subdivided into the *Calices*. Projecting into the calices are small nodules of kidney substance termed the *papillæ* (4), which will be seen to be the apices of certain triangular dark-coloured bodies called the *pyramids* or *cones* (3). These pyramids, which are arranged more or less regularly side by side, constitute what is called the *tubular* or *medullary* substance of the kidney, while the lighter granular portion at the margin of the section is the *cortical* substance (2).

The renal artery breaks up into four or five branches, which again subdivide and pass between the pyramids into the cortical substance (Fig. 116). In the cortical substance the *Malpighian bodies* or *glomeruli* are formed, each of which consists of a capillary plexus with an arterial twig (or afferent vessel) entering, and a venous radicle (or efferent vessel) leaving it, and these bodies are arranged along the arteries "like bunches of currants." Each Malpighian body or plexus is enclosed in a *capsule*, which is continuous with a minute *uriniferous tube* (Fig. 117). The tube is convoluted in the cortical substance, and a secondary *intertubular plexus* is formed upon it by the venous radicle, after it has emerged from the Malpighian tuft. The uriniferous tubes enter the bases of the pyramids and converge at their apices, where they empty themselves into the calices; from these the urine passes into the infundibula, and thence into the pelvis and ureter. Henlé described a second series of small tubes, passing from one capsule to another and unconnected with the uriniferous tubes; but the latest researches have shown that these smaller tubes are continuous with the convoluted tubes of the cortical substance, and are reflected for some distance in the pyramids between the straight tubuli uriniferi, which they eventually join.

The venous radicles, after forming the secondary plexus on the tubes, unite at the base of the pyramids, between which they pass to emerge at the hilum and form the renal vein.

The **Supra-Renal Capsule** (Fig 115, 1) is irregularly triangular in shape, or is commonly said to resemble a cocked hat. If a vertical section be made, its external or cortical layer will be seen to be yellow in colour and indistinctly striated, owing to its being composed of minute tubes. The internal portion is darker and softer and surrounds a small cavity in the centre, which, however, is only the result of decomposition (Wilks). The use of the capsule is unknown.

Fig. 117.

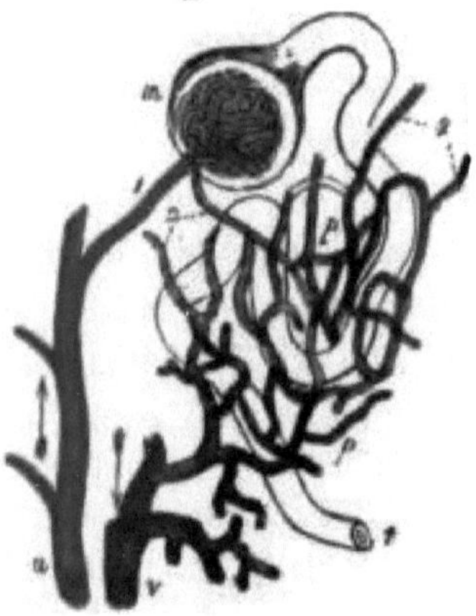

The subject being replaced on its back, the dissectors should proceed with the examination of the pelvis, unless they were unable to finish the aorta, lumbar plexus, and iliac arteries before the body was turned, in which case they must revert to the dissection of them at once (*v.* p. 248). In any case the dissector is strongly advised to read through the dissection of the deep parts of the abdomen again, before proceeding any further.

Fig. 117.—Plan of the renal circulation (from Wilson, after Bowman).

a. Branch of the renal artery giving off several Malpighian twigs.

1. An afferent twig to the capillary tuft contained in the Malpighian body, *m;* from the Malpighian capsule the uriniferous tube is seen taking its tortuous course to *t*.

2, 2. Efferent vessels; that which proceeds from the Malpighian body is smaller than the corresponding artery.

p, p. The capillary plexus, ramifying on the uriniferous tube. This plexus receives its blood from the efferent vessels, 2, 2, and transmits it to the branch of the renal vein, *v.*

THE PELVIS.

[The pelvis with two lumbar vertebræ is to be separated from the trunk with the saw, when, by placing it on a table with the sacrum towards himself, the dissector will be able to get a better view of the contents than he has yet done. The cavity of the pelvis should be carefully sponged out, and having removed the tow or cotton-wool from the anus, the dissector should let a stream of water run through the rectum. The bladder should be emptied of any urine it may contain by pressure with the hand, and be moderately distended with air.]

The **Peritoneum** (Fig. 119) should be examined first and will be found to pass over the rectum, binding the upper part to the front of the sacrum (*meso-rectum*); from the rectum to the bladder forming the *recto-vesical pouch*, and thence over the back of the bladder to the abdominal wall. On each side of the recto-vesical pouch is a distinct fold of peritoneum, formed by the reflection of the membrane over the obliterated hypogastric arteries and the ureters; these are the *posterior false ligaments* of the bladder. The peritoneum passing from the side of the bladder to the pelvis forms the *lateral false ligaments*, and the portion reaching to the back of the abdominal wall over the urachus is called the *superior false ligament* of the bladder. Thus the false ligaments of the bladder, five in number, are all formed by peritoneum.

In the **Female** (Fig. 93) the peritoneum passes from the rectum to the posterior wall of the vagina, forming the *recto-vaginal pouch*, and then over the uterus to the bladder, forming the *utero-vesical pouch*. On each side of the uterus it is stretched across the pelvis, forming the *broad ligament* of the uterus, which contains the round ligament, the ovary, and the Fallopian tube, the fimbriated extremity of which is to be noticed. The false ligaments of the bladder are the same as in the male, but are less distinctly marked; the margins of the recto-vaginal and utero-vesical pouches have been named the *anterior* and *posterior ligaments of the uterus*.

[The peritoneum is to be stripped off the upper part and sides of the bladder, but the recto-vesical pouch is not to be interfered with at present. By scraping away a little fat with the handle of the knife the pelvic fascia will be brought into view, but to examine it thoroughly the fol-

Fig. 118.

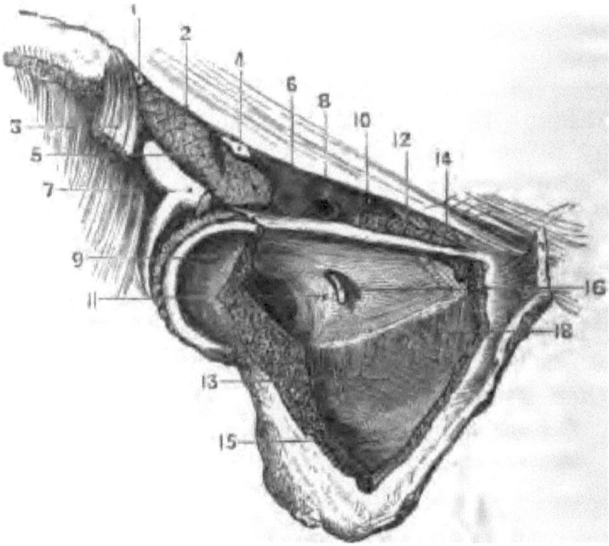

lowing dissection is to be made, in order to obtain a view from the outside. One dissector holding the pelvis firmly, the other is to clear away the remains of the adductor muscles on the right side of the pubes so as to expose the obturator externus muscle, which must then be carefully removed. Beneath the muscle will be found the branches of the obturator artery, forming a circle around the foramen and lying upon the obturator membrane or ligament, which gives passage to both obturator artery and nerve at

---

Fig. 118.—Pelvic fascia seen from the outside (drawn by J. T. Gray).

1. External cutaneous nerve.
2. Poupart's ligament.
3. Sartorius.
4. Anterior crural nerve.
5. Psoas and iliacus muscles.
6. Femoral artery.
7. Origins of rectus femoris.
8. Femoral vein.
9. Acetabulum, partly removed.
Crural ring.
11. Pelvic fascia.
12. Pectineus muscle.
13. Obturator fascia.
14. Gimbernat's ligament.
15. Pudic vessels and nerve in a sheath of fascia.
16. Obturator vessels and nerve.
18. Fascial origin of levator ani (white line).

its upper part. On removing the obturator ligament the fibres of the obturator internus muscle will be brought into view. With the saw a horizontal cut is now to be made from the upper margin of the obturator foramen into the cotyloid cavity, and a similar one at the lower margin of the obturator foramen, the extremities of the two cuts being about an inch apart in the bottom of the cavity. These are to be joined by a vertical cut with the chisel, and the piece of bone having been loosened with that instrument can be removed with the bone-forceps. With the chisel and bone-forceps the margins of the obturator foramen may then be cut away so as to leave only a ring of bone. The obturator internus being now fully exposed, should be carefully detached from the adjacent structures, and may then be readily removed by grasping the tendon with the bone forceps and drawing the whole muscle out through the lesser sacro-sciatic foramen. The outer surface of the pelvic fascia will then be exposed.]*

The **Pelvic Fascia** (Fig. 118) is continuous with the fascia over the psoas and iliacus muscles, and lines the pelvis, being very thin behind the obturator foramen, where it covers the pyriformis muscle and sacral plexus. In front it is a single layer, down to the level of a white line (18) which can be seen more or less distinctly stretching in a curve across the obturator foramen. This marks the point where the pelvic fascia divides into two parts, the *recto-vesical* and the *obturator* layers, and also forms a part of the origin of the levator ani muscle, the fibres of which can be indistinctly seen through the obturator fascia. The *obturator* layer (13) can be seen passing downwards on the inner side of the obturator internus muscle to be attached to the spine and tuberosity of the ischium and great sacro-sciatic ligament, and to the ramus of the ischium, where it gives a covering to the pudic vessels and nerve. In addition, it gives a thin layer to the under surface of the levator ani, called the *anal* fascia (Fig. 120), which is lost on the rectum and external sphincter, and was seen in the dissection of the perinæum.

* The credit of the invention of this dissection is due to Mr Wood, of King's College, London.

(The relation of the pelvic and recto-vesical fasciæ has been compared to that of the wall-paper and carpet of a room.)

Fig. 119.

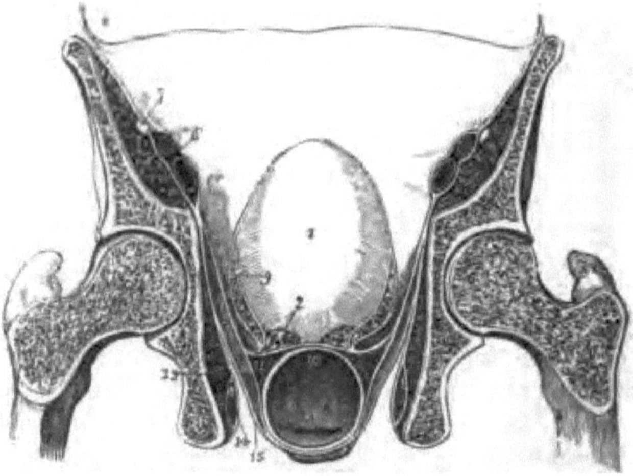

The *recto-vesical* layer is seen within the pelvis, and is continued over the internal surface of the levator ani to

Fig. 119.—Transverse section of the pelvis, seen from behind, showing the distribution of the pelvic fascia (from Wilson).

1. Bladder.
2. Vesicula seminalis of one side, divided.
3. Rectum.
4. Iliac fascia, covering in the iliacus and psoas (5); and forming a sheath for the external iliac vessels (6).
7. Anterior crural nerve excluded from the sheath.
8. Pelvic fascia, splitting into the recto-vesical and obturator layers.
9. Recto-vesical layer, forming the lateral ligament of the bladder of one side, and a sheath to the vesical plexus of veins.
10. A layer of fascia passing between the bladder and rectum.
11. A layer passing around the rectum.
12. Levator ani.
13. Obturator internus, covered in by the obturator fascia, which also forms a sheath for the internal pudic vessels and nerve (14).
15. Anal fascia investing the under surface of the levator ani.

Figures 14, 15 are placed in the ischio-rectal fossa.

Note.—This diagram is copied (with acknowledgement) in Gray's Anatomy, but has been confused by the addition of the vertebræ drawn so as to make the view from the *front*, which alteration renders the relations impossible.

the bladder, forming the *lateral true ligaments*, and from the back of the pubes, forming the *anterior true ligaments of the bladder*; whilst between them it is seen to dip down to the prostate, of which it forms the capsule, and is thus continued from one side to the other of the pelvis.

Thus, the levator ani muscle is seen to be enclosed by the recto-vesical and anal layers of fascia; the ischio-rectal fossa to be lined by the obturator and anal layers; and four of the true ligaments of the bladder to be formed by the recto-vesical layer,—the fifth being the obliterated urachus.

Fig. 120.

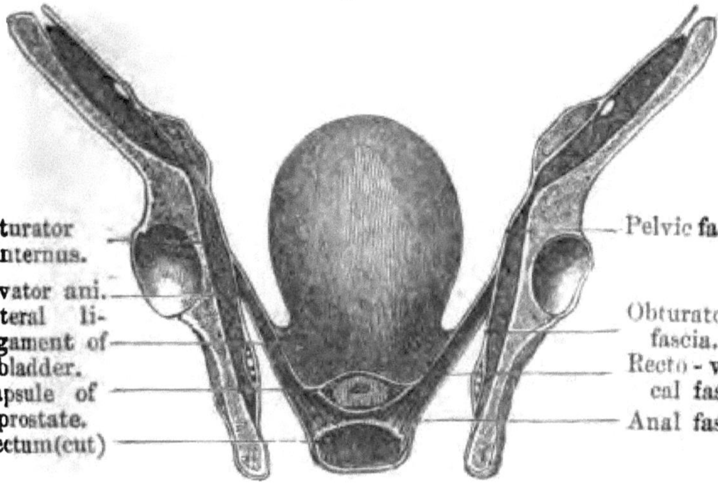

In the **Female** the recto-vesical layer forms the ligaments of the bladder in the same way as in the male, but, in addition, gives a covering to the vagina.

[A side-view of the contents of the pelvis is now to be obtained in the following way:—

The left obliterated hypogastric artery with the vesical and hæmorrhoidal arteries (and uterine and vaginal in the female) are to be severed, and the bladder and rectum turned as much as possible to the right side. The left lateral true ligament of the bladder is to be divided, and

Fig. 120.—Section of pelvis to show the pelvic fascia from the front (drawn by J. T. Gray).

the knife carried through the subjacent levator ani muscle. A cut is then to be made with the saw through the pubes just external to the left anterior true ligament of the bladder and another is to be made through the

Fig. 121.

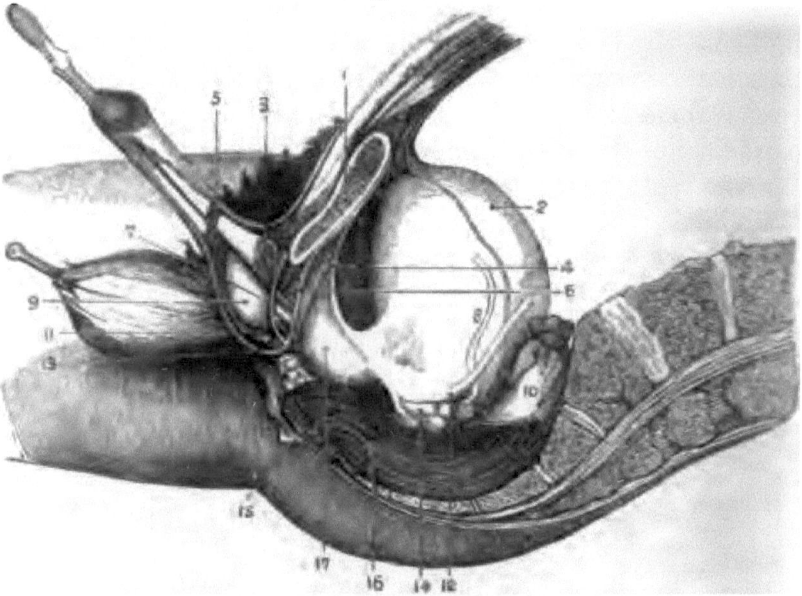

whole length of the sacrum and coccyx just to the left of the middle line, which will necessarily divide the right common iliac artery and left common iliac vein.

Fig. 121.—Section of pelvis to the left of the median line at the pubes, and through the middle line of the sacrum (drawn by J. T. Gray).

1. Section of left pubic bone.
2. Peritoneum on bladder.
3. Left crus penis (cut).
4. Pelvic fascia forming anterior ligaments of bladder.
5. Part of accelerator urinæ.
6. Posterior layer of triangular ligament, or pelvic fascia forming the capsule of the prostate.
7. Anterior layer of triangular ligament, or deep perinæal fascia.

Between 6 and 7 are seen the following: — membranous urethra, deep muscles of urethra (insertion), and Cowper's gland of the left side.

8. Vas deferens.
9. Bulb of urethra.
10. Rectum.
11. Cut edges of accelerator urinæ and transversus perinæi.
12. Left ureter.
13. Reflection of deep layer of superficial fascia round transversus perinæi.
14. Left vesicula seminalis.
15. Cut edge of levator ani.
16. Rectum.
17. Prostate.

The sacrum may be conveniently sawn partly on its anterior and partly on its posterior surface, and the two cuts united with the chisel, by which time will be saved.

The knife is now to be carried closely along the ramus of the pubes of the left side so as to detach the urethral muscles and triangular ligament, when by using a little force the pelvis will be divided into two parts, the right containing all the viscera, and the left the internal iliac vessels and sacral plexus of that side. A staff should be passed through the urethra into the bladder, which is to be moderately distended with air, and the rectum is to be carefully distended with tow or cotton-wool, and secured by a string to the top of the sacrum or lumbar vertebræ.]

Side View of the Pelvis (Fig. 121).—Beginning in front and supposing the dissection of the perinæum to have been made as directed, the bulb of the urethra (9) will be plainly seen lying in front of the triangular ligament or deep perinæal fascia (7), which has been purposely left untouched on the left side. The edge of the triangular ligament which has just been detached from the ramus of the pubes will be recognised, and immediately behind will be seen some pale muscular fibres, the cut edges of the *compressor urethræ* (Guthrie). The staff may be felt through these as it lies in the membranous portion of the urethra, which by a little dissection will be exposed, and may be seen in a favourable subject to be surrounded by circular muscular fibres. One of Cowper's glands, which resembles a pea in size and appearance, may be found immediately below the urethra. Behind the membranous portion of the urethra will be found another edge of fascia, the recto-vesical layer of the pelvic fascia passing on to the prostate to form its capsule (6).

The **Membranous portion of the Urethra** is a very important part, and should be specially noticed in the present view. It is seen to extend from the deep perinæal fascia or triangular ligament to the pelvic fascia or capsule of the prostate, or it is sometimes described as lying "between the two layers of the triangular ligament." It is nearly an inch in length and is not quite horizontal in its position, the anterior extremity being about an inch below the symphysis pubis, and the posterior a little lower. This is the portion of the urethra opened by the deep incision in lithotomy, and it should be noticed how much its position will vary according to the movements of the staff,—whether that instrument be hooked under the pubes or depressed towards the rectum.

The *Pelvic Fascia* should be traced on to the prostate, bladder, and rectum, and the relations of those organs noticed.

The **Rectum** (Fig. 121, 10) is about eight inches long, and extends from the left sacro-iliac synchondrosis to the anus, lying a little to the left of the median line of the pelvis. It is divided into three portions, a superior or vertical, a middle or horizontal, and an inferior or vertical portion; and the upper two parts of the intestine are closely attached to the front of the sacrum and coccyx, while the third part passes through the levator ani and sphincter muscles to the anus, immediately in front of the tip of the coccyx.

The peritoneum surrounds the first vertical portion and binds it to the sacrum, forming the meso-rectum; the membrane only covers the front of the upper part of the second portion as it forms the recto-vesical pouch; and has no connection with the third part at all. (Thus the relations of the rectum correspond to those of the duodenum as respects the peritoneum, *v.* p. 241).

The bladder lies immediately over the second portion of the rectum, and between the two organs are the vesiculæ seminales and vasa deferentia. The prostate is close above the commencement of the third portion of the rectum, but as the bowel turns down to the coccyx, an interval is left between it and the urethra, corresponding to the perinæum proper or surface between the back of the scrotum and the anus.

[The left vesicula seminalis should be dissected out that its position may be seen, and the vas deferens and ureter of the left side are also to be traced out; the recto vesical pouch of peritoneum is to be laid open along the side so that its extent may be fully seen.]

The **Bladder** (Fig. 121, 2) varies in position according to its state of distension, being in the anterior part of the pelvis when empty, but filling the pelvis and rising into the abdomen when distended. The *neck* of the bladder is the narrowed portion attached to the prostate, and the remainder of the viscus is termed the *body*. The upper part of the body is the *apex* or *summit*, to which the obliterated urachus is attached, and over which the obliterated hypogastric arteries pass to the abdominal wall. The *base* or *fundus* of the bladder is the lowest part of the organ, which is seen to rest on the horizontal portion of the rectum, with the vesiculæ seminales and vasa deferentia intervening.

The peritoneum is reflected from the second portion of the rectum on to the fundus of the bladder at the level of the entrance of the ureters, and is then continued over the back of the bladder until it is reflected over the urachus to the abdominal wall, leaving the anterior surface uncovered.

It should be noticed how much the extent of bladder uncovered by peritoneum, both above and below, depends upon the distension of the viscus; since in the contracted state its anterior surface is in contact with the pubes, whilst in the distended condition it rises above the bone and is in contact with the posterior surface of the abdominal muscles, and thus tapping above the pubes without injury to the serous membrane is possible. The depth of the recto-vesical pouch will similarly be found to vary, the peritoneum reaching nearer to the prostate when the bladder is empty than when it is full; consequently the operation of tapping by the rectum can be only safely undertaken when the bladder is distended.

The left **Ureter** (Fig. 121, 12) descending from the kidney is contained in the posterior false ligament of the bladder, and can now be traced beneath the peritoneum to its point of entrance into the posterior part of the fundus of the bladder, at a point nearly parallel to the posterior extremity of the vesicula seminalis.

The left **Vas Deferens** (Fig. 121, 8) can be traced from the testicle, and has been seen to turn down into the pelvis to the inner side of the external iliac artery, after leaving the inguinal canal. It is now seen to wind over the back of the bladder, crossing the obliterated hypogastric artery, and to be continued beneath the peritoneum to the fundus of the bladder, where it hooks round the ureter to become internal to it; the vas will be afterwards traced to the prostate.

The **Prostate Gland** (Fig. 121, 17) is now seen in front of the bladder, and enclosed by a process of recto-vesical fascia which forms the *capsule* of the gland. It is placed over the commencement of the third part of the rectum, and in front of, but below the level of, the bladder. In old age the prostate is frequently enlarged, in which case it presses upon the rectum as well as upon the urethra, which passes through its substance nearer the upper than the lower surface.

[A transverse cut is now to be made at the bottom of the recto-vesical pouch, which will allow the bladder to be drawn forward, when a little dissection will expose the surface of the viscus uncovered by peritoneum and lying over the rectum.]

The portion of the base of the bladder uncovered by peritoneum is triangular in shape, and is bounded on each side by the *vasa deferentia* and *vesiculæ seminales;* the base of the triangle is formed by the reflection of peritoneum at the line of the entrance of the ureters, and the apex is at the back of the prostate. It is here that the bladder is opened when punctured from the rectum.

A process of the recto-vesical layer of pelvic fascia can now be traced between the rectum and bladder, giving a covering to those organs, to the prostate, and to the *vesiculæ seminales.*

The **Vesiculæ Seminales** (Fig. 121, 14) are two convoluted sacs situated on the under surface of the bladder, and converging to enter the prostate close to the median line. They are somewhat expanded at their posterior extremities, which are about two inches apart and reach nearly to the recto-vesical pouch. They have the vasa deferentia lying close to their inner sides, and together they form the sides of the triangle mentioned above.

The vasa deferentia become somewhat enlarged as they pass beneath the bladder; they lie close to the inner sides of the vesiculæ, and the duct of each vesicula joins the corresponding vas deferens at the prostate, thus forming the *common ejaculatory duct.*

In the Female (Fig. 122), the short urethra will be seen to pass obliquely upwards to the bladder; and immediately below it is the vagina passing up to the uterus, which must be held in position by a string passed through its fundus. The rectum occupies the same position as in the male, but is rather larger; and the peritoneum will be seen to pass from it to the posterior surface of the upper part of the vagina, thus forming the recto-vaginal pouch, being then reflected over the uterus to form the utero-vesical pouch between it and the bladder.

The pelvic fascia gives coverings to the rectum and vagina which should be defined, and the latter canal may be laid open along the side to study the position of the *os uteri*, when the uterus is in its proper position and when displaced.

The **Bladder** (Fig. 122, 4) closely resembles that viscus in the male, but is rather larger, and its fundus is in contact with the vagina instead of with the rectum.

The **Urethra** (Fig. 122, 6) in the female is about an inch and a half long, placed immediately beneath the pubes and curved upwards slightly to the bladder. It may be said to correspond to the membranous urethra of the male.

The **Uterus** (Fig. 122, 18) is situated behind the bladder, and, if in its proper position and healthy, its upper end or *fundus* should not reach above a line drawn from the upper edge of the pubic symphysis to the lumbo-sacral articulation (Savage). The uterus is tilted slightly for-

Fig. 122.

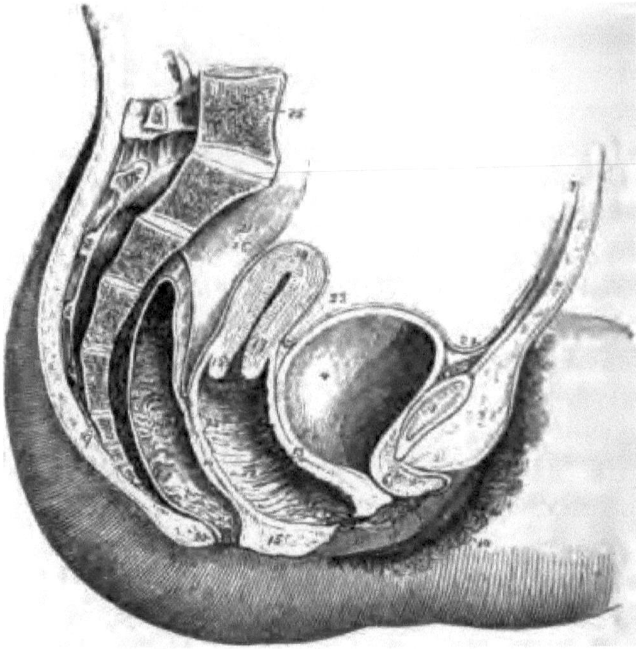

Fig. 122.—Side view of the viscera of the female pelvis (from Wilson).

1. Symphysis pubis.
2. Abdominal parietes.
3. Collection of fat, forming the prominence of the mons veneris.
4. Urinary bladder.
5. Entrance of left ureter.
6. Canal of the urethra.
7. Meatus urinarius.
8. Clitoris, with its præputium, divided through the middle.
9. Left nympha.
10. Left labium majus.
11. Meatus of the vagina, narrowed by the contraction of its sphincter.
12, 22. Canal of the vagina, on which the transverse rugæ are apparent.
13. Thick wall of separation between the base of the bladder and the vagina.
14. Wall of separation between vagina and rectum.
15. Perinæum.
16. Os uteri.
17. Cervix uteri.
18. Fundus uteri.
19. Rectum.
20. Anus.
21. Upper part of rectum, invested by peritoneum.
23. Utero-vesical fold of peritoneum. The recto-vaginal fold is seen between the rectum and posterior wall of the vagina.
24. Superior false ligament of the bladder.
25. Last lumbar vertebra.
26. Sacrum.
27. Coccyx.

wards so that its axis corresponds to that of the pelvis, and its lower end or *cervix* is received into the upper end of the vagina.

The **Broad Ligament** (Fig. 93) of the uterus is the fold of peritoneum attached to each side of the fundus, in which are three partial subdivisions, viz., a superior in which is the Fallopian tube, an anterior in which is the round ligament of the uterus, and a posterior covering the ovary.

The right round ligament of the uterus is now to be traced from the internal abdominal ring to the fundus of the uterus, but the Fallopian tube and ovary may be more conveniently examined with the uterus at a later period.

The **Vagina** (Fig. 122, 12) is a curved tube of which the anterior wall is considerably shorter than the posterior, the former in the virgin rarely exceeding two inches in length, but the latter being nearly twice as long (Savage). It is in contact with the bladder in front, and with the rectum behind at the lower part, but the recto-vaginal pouch of peritoneum intervenes between the two organs above. The neck of the uterus projects backwards into the upper extremity of the vagina, and its lower extremity passes through the levator ani muscle and is surrounded by the sphincter vaginæ.

[The viscera of the pelvis being drawn down to the left, the internal iliac vessels and sacral plexus of the right side can be examined, and they should also be dissected on the left half of the pelvis, which has been preserved for the purpose. Opportunity may be taken to trace branches of the sympathetic nerve to the sides of the pelvic organs.]

The **Internal Iliac Artery** (Fig. 123, 4) is a branch of the common iliac, from which it comes off at the sacro-iliac synchondrosis, and immediately passes into the pelvis. In the fœtus the hypogastric artery, the direct continuation of the internal iliac in size and direction, passed forwards over the bladder to the umbilicus, but it is now converted into a fibrous cord, which however is pervious for a short distance, and thus gives branches to the bladder. The internal iliac is crossed by the ureter and closely

Fig. 123.

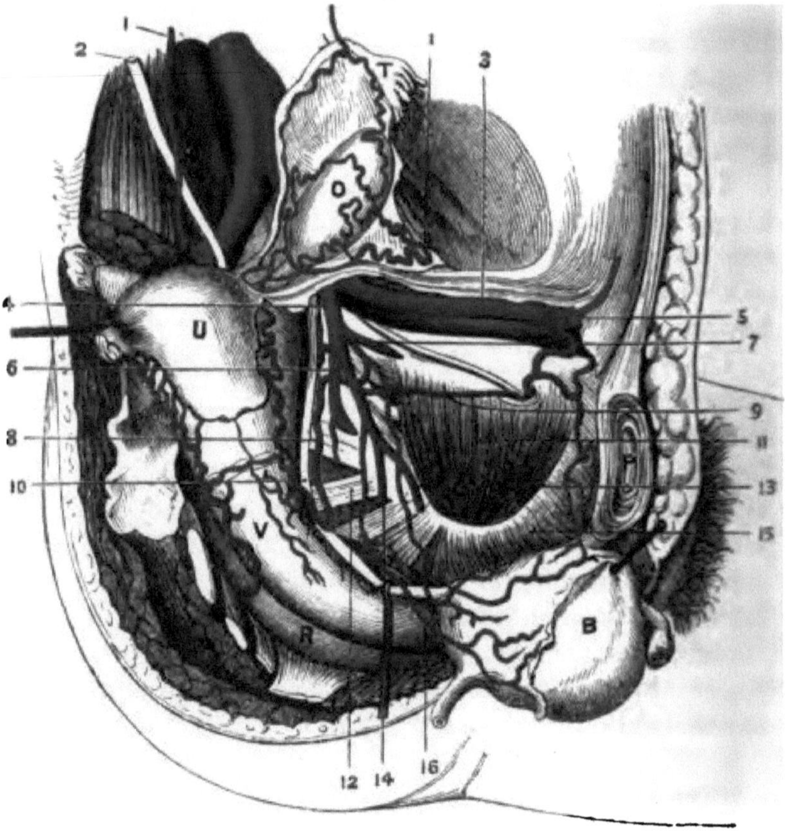

Fig. 123.—Side view of female pelvis with internal iliac artery (altered from Savage).

1, 1. Spermatic or ovarian artery.
2. Right ureter (cut).
3. External iliac vessels.
4. Internal iliac artery.
5. Epigastric artery giving an obturator branch.
6. Posterior trunk of internal iliac dividing into gluteal, ilio-lumbar, and lateral sacral branches.
7. Anterior trunk of internal iliac artery giving off superior vesical and uterine arteries, and ending in the obliterated hypogastric (cut).
8. Uterine artery anastomosing with the ovarian in the broad ligament, and giving a vaginal branch.
9. Obturator artery.
10. Left ureter.
11. Obturator internus.

covered by the peritoneum in front; behind it are the internal iliac vein and the lumbo-sacral cord, with a portion of the pyriformis muscle. At the upper border of the great sacro-sciatic notch it divides into anterior and posterior trunks, which give branches to the viscera and to the inside and outside of the pelvis.

Branches (Fig. 123).—From the anterior division are given off three visceral and three parietal branches; from the posterior trunk three parietal branches.

Anterior Division.

Visceral branches.	Superior vesical. Inferior vesical. Middle hæmorrhoidal. Uterine } additional Vaginal } in female.	Parietal branches.	Obturator. Sciatic. Pudic.

1. The *Superior vesical* (7) arteries are small branches from the unobliterated portion of the hypogastric artery, which are distributed to the upper part of the bladder. A middle vesical branch is sometimes derived from one of these.

2. The *Inferior vesical*, and 3. *Middle hæmorrhoidal* (16) arteries generally come off together, and are distributed to the under surface of the bladder and vesiculæ seminales, and to the adjacent portion of the rectum. The middle hæmorrhoidal artery anastomoses with the superior hæmorrhoidal branch of the inferior mesenteric above, and with the inferior hæmorrhoidal branch of the pudic artery below.

4. The *Uterine* (8) and *Vaginal* arteries supply those organs, the uterine passing to the cervix and then giving branches to the body of the uterus.

5. The *Obturator* (9) artery passes directly forwards below the level of the nerve to the obturator foramen,

---

12. Sacral plexus.
13. Pudic artery.
14. Sciatic artery.
15. Levator ani.
16. Inferior vesical and middle hæmorrhoidal arteries.

B. Bladder.
U. Uterus.
O. Ovary.
F. Fallopian Tube.
V. Vagina.
R. Rectum.
P. Pubes.
S. Sacrum.

through the upper part of which it disappears with the nerve, after giving off a small *pubic* branch to the back of the bone, and also an *iliac* branch to the iliacus internus and ilium. The distribution of the obturator artery outside the pelvis has been already seen on the right side (p. 280) but may now be followed out on the left side.

When the obturator arises from the epigastric artery there is generally a small branch running in the proper position with the obturator nerve (*v.* p. 97).

6. The *Sciatic* (14) and 7. the *Pudic* (13) arteries can only be seen for a short distance within the pelvis, as they lie on the pyriformis to pass through the lower part of the greater sacro-sciatic foramen below that muscle. The pudic is generally smaller than the sciatic, and a little in front of it, but the size and relation of the vessels vary. Supposing the perinæum and buttock to have been dissected, the opportunity may be taken to trace the sciatic artery through the great sacro-sciatic foramen to its distribution outside the pelvis, and to follow the pudic in its course around the spine of the ischium, and through the lesser foramen to the perinæum, where it will be seen running close to the margin of the pubes.

The three parietal branches from the posterior division of the internal iliac artery are, the gluteal, ilio-lumbar, and lateral sacral arteries.

1. The *Gluteal* (6) artery is a thick trunk, disappearing at once through the upper part of the great sacro-sciatic foramen between the lumbo-sacral and first sacral nerves and above the pyriformis. It is distributed to the buttock (p. 119).

2. The *Ilio-lumbar* (16) artery passes into the iliac fossa beneath the psoas and iliacus muscles ; and then divides into a *lumbar* branch which anastomoses with the last lumbar artery, and an *iliac* branch which supplies the iliacus and the bone, anastomosing with the circumflexa ilii branch of the external iliac artery.

3. The *Lateral sacral* (18) artery (often double) descends on the front of the sacrum internally to the sacral foramina

to the coccyx, where it anastomoses with the artery of the opposite side and with the sacra media from the aorta. In its course it gives branches to the sacral nerves and the pyriformis muscle.

The *Veins* corresponding to the branches of the internal iliac artery open into the internal iliac vein, which passes beneath its artery to the common iliac vein.

Fig. 124.

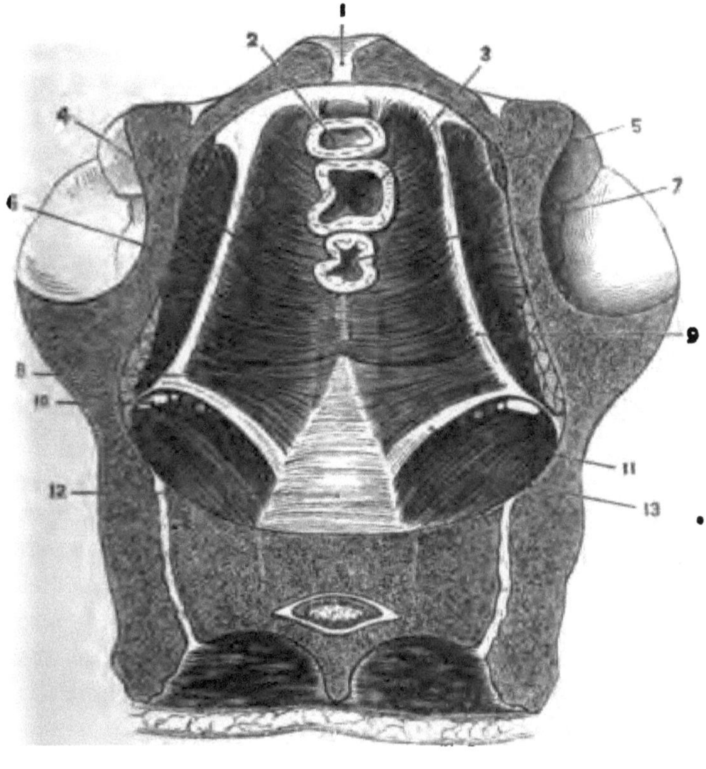

Fig. 124.—Floor of female pelvis (from Savage).

1. Symphysis pubis.
2. Section of bladder.
3. Section of vagina.
4. Levator ani muscle.
5. Obturator vessels.
6. Obturator internus muscle.
7. Section of rectum.
8. Coccygeus muscle.
9. 'White line' of pelvic fascia.
10. Sciatic and pudic vessels and nerves.
11. Lesser sacro-sciatic ligament.
12. Front of sacrum.
13. Pyriformis muscle.

[The bladder and rectum are now to be drawn out of the pelvis as much as possible, to do which it will be necessary to divide the ligaments of the bladder and the recto-vesical layer of pelvic fascia; this will bring into view the structures which close the lower aperture or outlet of the pelvis.]

The **Floor of the Pelvis** (Fig. 124) is formed by the following structures. In front the levatores ani of the two sides blend in the middle line, supporting and being closely connected with the pelvic viscera. Immediately behind these and separated only by a very narrow space of cellular tissue, are the two coccygei muscles; and posterior to these the two smaller sacro-sciatic ligaments, which bound the great sacro-sciatic foramen. This foramen is closed by the pyriformis muscle, above and below which vessels and nerves leave the pelvis.

The **Levator Ani Muscle** (Fig. 125, 23) takes its *origin* from the back of the pubes, and from the front of the spine of the ischium; between those points its fibres are attached to the white line marking the division of the pelvic fascia, and seen on the dissection of the fascia from without (Fig. 118). The anterior fibres pass beneath the bladder and prostate to unite with those of the opposite side in the median raphé of the perinæum, the name *levator prostatæ* being sometimes given to those fibres which pass beneath the prostate [in the female they are connected with the vagina]; the middle fibres are inserted into the rectum and blend with the sphincter; whilst the posterior fibres unite with those of the opposite side in the median raphé behind the anus, and are attached to the apex of the coccyx.

The **Coccygeus** (Fig. 125, 34) is the little triangle of pale muscular fibres having its *insertion* into the side of the coccyx, and its *origin* (the apex of the triangle) from the spine of the ischium immediately behind the attachment of the levator ani, from which the muscle is separated by a narrow cellular interval. The posterior border of the coccygeus is intimately connected with the lesser sacro-sciatic ligament, which passes from the sacrum to the spine of the ischium.

The **Obturator Internus Muscle** (Fig. 125, 19) has been destroyed on the right side, but on the left side of the pelvis may be seen both above and below the level of the fascial origin of the levator ani. It *arises* from the posterior aspect of the rami of the pubes and ischium immediately in front of the thyroid foramen; from the inner surface of the obturator ligament; and from the surface of bone behind the foramen as far as the great sacro-sciatic notch. The fibres end in a tendon, which passes out of the lesser sacro-sciatic foramen to be *inserted* on the upper margin of the great trochanter of the femur, its deep surface being subdivided into four or five small tendons as it winds over the edge of the sacro-sciatic notch, which is encrusted with cartilage and lubricated by a bursa.

[The pelvic viscera are now to be removed by carefully detaching the urethra and bladder from the pubes, dividing the levator ani, and severing the slight connections between the rectum and coccyx. The whole of the urinary and genital organs should be carefully preserved for after examination. The sacral nerves can now be dissected on both sides of the pelvis, and in order to see them clearly it will be well to remove the remaining branches of the internal iliac artery. All the branches of the sacral nerves should be preserved, and care be taken not to remove the gangliated cord and hypogastric plexus of the sympathetic. The upper sacral nerves will be readily seen emerging from the foramina, but the fifth sacral and the coccygeal nerve will be found piercing the coccygeus muscle close to the side of the coccyx.]

The **Sacral Plexus** (Fig. 125) is formed by the upper four sacral nerves and the lumbo-sacral cord derived from the fourth and fifth lumbar nerves, which is seen descending into the pelvis. The several nerves receive branches from the cord of the sympathetic as they leave the sacral foramina, and then lying on the pyriformis muscle they unite to form one large flat band, which is the plexus. This is seen to disappear through the great sacro-sciatic foramen immediately above the lesser sacro-sciatic ligament, and by looking at the outside of the pelvis, the dissector will be able to see the three nerves into which

Fig. 125.

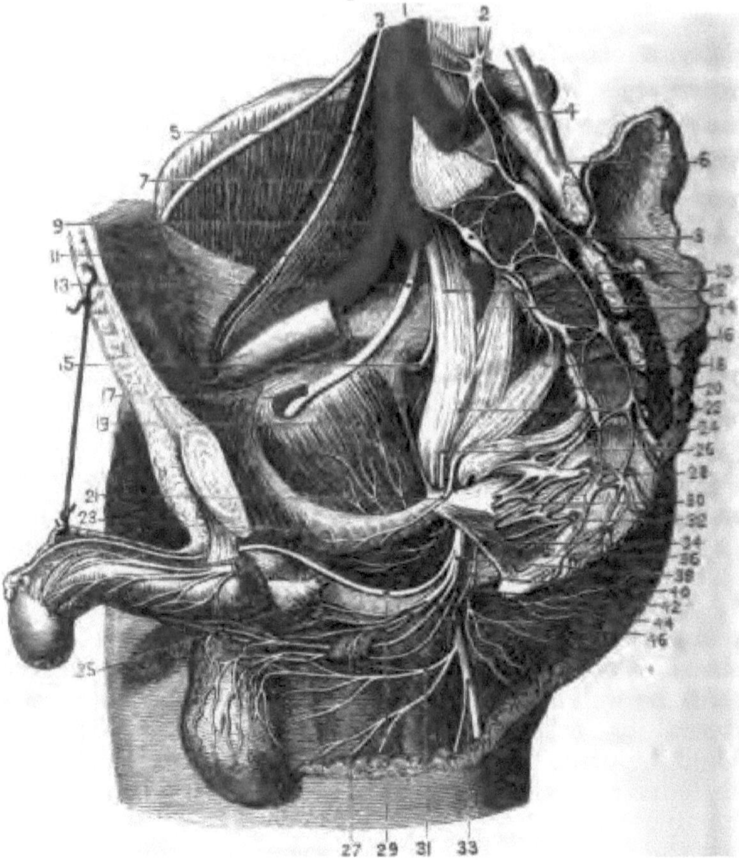

Fig. 125.—Side view of the nerves of the pelvis, the viscera having been removed (from Hirschfeld and Leveillé).

1. Abdominal aorta.
2. Gangliated cord of sympathetic (left side).
3. Genito-crural nerve.
4. Left common iliac artery.
5. Right common iliac artery.
6. Left lumbo-sacral cord.
7. Psoas muscle.
8. Gangliated cord of sympathetic (right side).
). Internal iliac artery (cut).
). Gluteal artery (cut).
l. Right external iliac artery.
12. Right lumbo-sacral nerve.
13. Obturator nerve.
14. First sacral nerve.
15. Gluteal nerve.
16. Pyriformis of left side (cut).
17. Obturator artery.
18. Second sacral nerve.
19. Obturator internus.
20. Pyriformis of right side.
21. Nerve to levator ani.
22. Sacral plexus.
23. Levator ani (cut).
24. Third sacral nerve.

it divides, viz., Great Sciatic, Small Sciatic, and Internal Pudic; and to trace the pudic nerve around the spine of the ischium and through the lesser sacro-sciatic foramen to the perinæum.

Branches within the Pelvis. A couple of branches are given to the *pyriformis* either from the plexus or from the sacral nerves before they unite. The nerve to the *obturator internus* (26) arises at the point of union of the lumbo-sacral with the first sacral nerve and, lying upon the plexus, passes with the pudic nerve through the sacro-sciatic foramina to the inner surface of the muscle, supplying in its course the gemellus superior. The nerve to the *quadratus* from the lower part of the plexus, or sometimes from the great sciatic nerve, passes out beneath that nerve close upon the hip-joint, and supplies the quadratus and the gemellus inferior (p. 122).

At the upper margin of the great sacro-sciatic foramen will be seen the *superior gluteal nerve* (15), derived from the lumbo-sacral cord and passing out above the pyriformis with the gluteal artery.

Below the plexus are the branches of the other sacral nerves. These last can seldom be made much of, since the nerves themselves are very small and a quantity of fibrous tissue is mixed with them.

From the *4th Sacral nerve* a branch may be traced to its junction with the 5th nerve, and from the 4th also are derived *communicating branches* to the sympathetic on the bladder and rectum, and *muscular branches* to the levator ani and coccygeus; as well as a branch which passes through the great sacro-sciatic ligament to the perinæum to supply the external sphincter.

---

25. Bulb of urethra covered by accelerator urinæ.
26. Nerve to obturator internus.
27. Transversus perinæi (cut).
28. Fourth sacral nerve.
29. Dorsal nerve of penis.
30. Visceral branches (cut).
31. Inferior pudendal nerve.
32. Fifth sacral nerve.
33. Small sciatic nerve.
34. Coccygeus muscle.
36. Sixth or coccygeal nerve.
38. Internal pudic nerve.
40. Inferior hæmorrhoidal nerve.
42. Posterior superficial perinæal nerve.
44. Anterior superficial perinæal nerve.
46. Deep perinæal nerves to bulb and muscles.

The *5th Sacral nerve* communicates with the 4th and the coccygeal nerve, and terminates by piercing the coccygeus muscle, which it supplies.

The *Coccygeal nerve* (6th) also pierces the coccygeus muscle to end on the back of the coccyx.

The **Gangliated Cord of the Sympathetic** (Fig. 125, 8) may be traced on each side of the sacrum internally to the foramina, and has upon it four or five ganglia; the two cords uniting on the coccyx in the *ganglion impar*. Communicating branches are given to the sacral nerves and to the hypogastric plexus.

The *Hypogastric plexus* is placed on the front of the sacrum, and is continuous with the sympathetic plexus upon the aorta. It is continued forward upon the two internal iliac arteries to form the *pelvic plexuses*, which are placed on each side of the bladder and rectum.

In addition there are upon the several hollow viscera plexuses from the sympathetic, parts of which are visible in the various stages of the dissection. They are derived from the pelvic plexuses, and are distributed over the branches of arteries supplied to the several viscera, receiving corresponding names.

The **Pyriformis Muscle** (Fig. 125, 10) if not sufficiently well seen may be more fully exposed by drawing the sacral plexus out of the sacro-sciatic foramen. It *arises* from the front of the sacrum between the 1st, 2nd, 3rd, and 4th sacral foramina, and leaves the pelvis through the great sacro-sciatic foramen, having the gluteal vessels and nerve at its upper, and the sciatic and pudic vessels and nerves at its lower border. It is *inserted* into the upper border of the great trochanter of the femur, and is *supplied* by branches from the upper sacral nerves.

Pelvic Viscera in the Male.

[The bladder with the urethra and the rectum having been removed as directed, should be laid on the table with the rectum uppermost, and that viscus should be cleaned, and its muscular fibres exposed.]

The **Rectum** is about eight inches long and is generally largest just above the anus, which will be seen to be surrounded by the sphincter muscle. The arrangement of the peritoneum upon the rectum has been already described (p. 286), and the longitudinal muscular fibres will be now seen to differ from those of the rest of the large intestine, in being spread all around it instead of being collected in three bands. The internal or circular muscular fibres become thickened near the anus to form the internal sphincter muscle, which is best seen when the bowel is opened. The numerous arteries entering the rectum are derived from three sources—the superior hæmorrhoidal artery from the inferior mesenteric; the middle hæmorrhoidal from the internal iliac; and the inferior hæmorrhoidal from the internal pudic. The veins open both into the *vena portæ* and the internal iliac vein.

The intestine being opened along its unattached border, the mucous membrane will be seen to be smooth, except near the anus, where it is thrown into longitudinal folds by the sphincter. Here also it frequently happens that hæmorrhoids exist. Three or four transverse ridges of mucous membrane (valves or folds of Houston) will be found from three to six inches above the anus, which occasionally cause difficulty in the introduction of a bougie or rectum tube.

[The rectum is now to be carefully dissected from the bladder, which is then to be moderately distended with air, and the under surface of the penis and bladder being then placed before the student, he is to dissect out thoroughly the whole of the parts exposed.]

The **Corpus Spongiosum Urethræ** (Fig. 126, 20) is seen between the two *corpora cavernosa penis*, and behind the point where these bodies cease will be seen the *bulb* (21), from which any remains of the accelerator urinæ muscle should be removed. Behind the bulb is the slender *membranous portion* (10) of the urethra which terminates at the *prostate* (7), the under surface of which is now exposed with the *vesiculæ seminales* (9), *vasa deferentia*, and base of the bladder. On each side of the membranou portion of the urethra may be found, by careful dissectio a little yellow body of the size of a pea, *Cowper's glan* (11), the duct of which opens into the bulbous portion o the urethra. By turning the preparation over, the dors aspect of the penis will be brought into view and the dorsal vessels and nerves will be seen. The dorsal vein

Fig. 126.

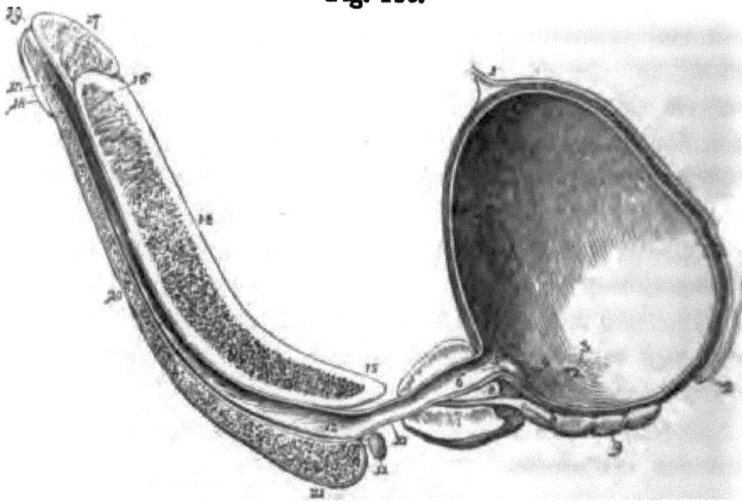

Fig. 126.—Longitudinal section of the bladder, prostate gland, and penis (from Wilson).

1. Urachus.
2. Recto-vesical fold of peritoneum.
3. Opening of the right ureter.
4. A slight ridge formed by the muscle of the ureter.
5. The neck of the bladder.
6. Prostatic portion of the urethra.
7. Prostate gland.
8. The common ejaculatory duct.
9. Right vesicula seminalis; the vas deferens is cut short.
10. Membranous portion of the urethra. Its direction is the reverse of this when *in situ*.
11. Cowper's gland of the right side, with its duct.
12. Bulbous portion of the urethra.
13. Fossa navicularis.
14. Corpus cavernosum.
15. Right crus penis.
16. A portion of the septum pectiniforme.
17. The glans penis.
18. Corona glandis.
19. Meatus urinarius.
20. Corpus spongiosum.
21. Bulb of the corpus spongiosum.

may now be traced to its junction with the plexus of veins around the prostate and neck of the bladder.

The **Prostate** (Fig. 126, 7) is shaped like a chestnut, having the small end or apex in front. The length of the organ from base to apex is in health from 1¼ to 1½ inch, and the greatest transverse diameter about 1⅝ inch (H. Thompson); but these measurements undergo great variation in old persons, the subjects of "enlarged prostate." The prostate consists of two lateral lobes, the division between which is marked by a slight groove on the under surface, the third or middle lobe described by some authors being the result of morbid changes. A little hollow in the centre of the posterior aspect, or base, receives the common ejaculatory duct. The structure of the prostate is partly glandular and partly muscular, the fibres (of the unstriped variety) being continuous with those of the bladder.*

The **Vesiculæ Seminales** (Fig. 126, 9) are two membranous receptacles for the semen, attached to the under surface of the bladder immediately behind the prostate, from which they diverge. Each vesicle consists of a convoluted tube, which becomes straight at the base of the prostate, and joins the corresponding *vas deferens* to form the *common ejaculatory duct*, to be afterwards seen opening into the urethra.

The **Bladder** (Fig. 126, 4).—The *peritoneal coat* of the bladder has been already dissected (p. 287).

The *muscular coat* consists of three layers, external or longitudinal, middle or circular, and internal (submucous of Ellis), also longitudinal.

The *external layer*, sometimes known as the *detrusor urinæ*, is attached to the anterior true ligaments of the bladder at the back of the pubes. It is continued over the bladder and joins the posterior part of the prostate.

The *middle layer* consists of circular fibres, which form

* For full particulars of the minute structure of the prostate, bladder, etc., the student may refer to Kölliker's 'Microscopic Anatomy,' or Professor Ellis's paper in Medico-Chirurgical Trans., vol. xxxix.

the *sphincter vesicæ* around the neck of the bladder, and are then continued over the prostate. They become more or less oblique in their distribution over the bladder.

The *internal layer* is continuous with the longitudinal fibres of the urethra, and is joined by the fibres of the ureters, which meet in the middle line and are also spread over the trigonum (Ellis).

Sir Charles Bell described special muscular bands continued from the ureters to the back of the prostate, the existence of which as separate structures is doubtful. According to Dr. Pettigrew, whose beautiful preparations are in the museum of the College of Surgeons of England, the arrangement of the muscular fibres of the bladder is reducible to a series of figure-of-8 loops, the so called circular fibres depending upon compression of the two halves of the figure,—thus 8 becomes 8.

Beneath the muscular is the *fibrous coat* of the bladder, composed of loose connective tissue.

[The bladder is to be opened along its upper surface by a longitudinal cut, which is to be continued with a strong pair of scissors through the upper surface of the prostate and the whole length of the urethra, in the latter portion being a little to one side of the middle line.]

Interior of Bladder and Urethra (Figs. 126 and 127). —The mucous membrane of the bladder if healthy will be pale and more or less in folds throughout, except near the urethra, where is the triangular space called the *trigone* or *trigonum vesicæ*. This space corresponds to that between the vesiculæ seminales externally, and the sides of the triangle are about an inch and a half long, being bounded by the orifices of the ureters, which will be recognised as little slits in the mucous membrane. If a probe be introduced into one of the ureters, it will be seen to pass very obliquely through the coats of the bladder, with which it is firmly united. The epithelium is tesselated.

The *Uvula vesicæ* is a small elevation of the mucous membrane at the orifice of the urethra, which is not often seen.

The **Urethra** (*Fig. 126*) averages eight inches in length and is divided into the prostatic, membranous, bulbous, and spongy portions.

Fig. 127.

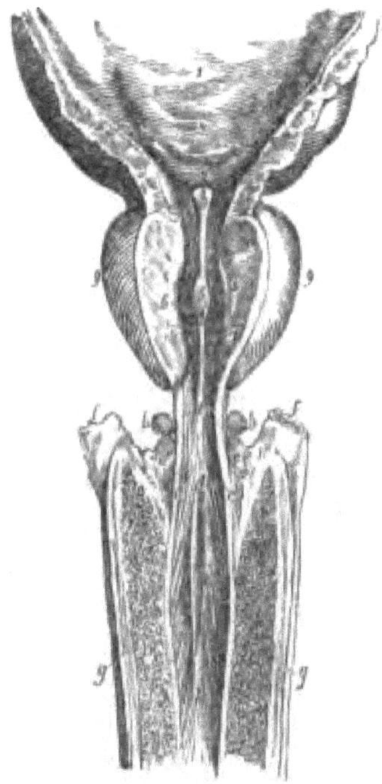

The *Prostatic portion* (Fig. 127) lies in the prostate and is about an inch and a quarter long. On each side of it will be seen the cut substance of the prostate, which consists partly of glandular and partly of muscular tissue. At the bottom of the prostatic portion of the urethra is a ridge called the *veru montanum* or *caput gallinaginis* (5) dividing the canal into two hollows called the *prostatic sinuses* (8) into which the prostatic ducts open. On the veru montanum and near its posterior part will be found the *sinus pocularis* (6), a little blind pouch directed backwards, just within which are placed the orifices of the common ejaculatory ducts.*

* The position of the sinus pocularis is given differently by various authors; thus Quain, Gray and Wilson place it on the anterior, and Cruveilhier and Ellis on the posterior part of the veru montanum. The discrepancy arises from slight variations in the shape and extent of the ridge.

Fig. 127.—The bulbous, membranous, and prostatic urethra, with part of the bladder, laid open from above (from Wilson).

1. Part of the urinary bladder: its internal surface.
2. Trigonum vesicæ.
3. Openings of the ureters.
4. Uvula vesicæ.
5. Veru montanum.
6. Opening of the sinus pocularis.

7, 7. Apertures of the ejaculatory ducts.

8, 8. Openings of the prostatic ducts in the prostatic sinus.

9, 9. Lateral lobes of the prostate gland.

a. Membranous portion of the urethra.

b, b. Cowper's glands.

c, c. Apertures of the excretory ducts of Cowper's glands.

d. Commencement of the bulbous portion of the urethra.

e, e. Upper surface of the bulb of the corpus spongiosum.

f. f. Crura penis.

g, g. Corpus cavernosum.

h. Spongy portion of the urethra.

The *Membranous portion* (*a*) is the shortest and narrowest division of the urethra, and being placed between the two layers of the triangular ligament which meet below, its upper wall is somewhat longer than its floor, the former measuring about ¾ inch, and the latter ½ inch (Thompson). The wall of this portion is thin and membranous, but it will be seen to be enclosed by circular involuntary fibres continuous with the circular fibres of the bladder and prostate. The mucous membrane is smooth, and presents no orifices; it is frequently stained of a red colour.

The *Bulbous portion* (*d*) is the part corresponding to the bulb externally, and is about an inch long, but there is no line of demarcation between it and the spongy portion. The canal is dilated at this spot, and in the floor of it are the two minute orifices of the ducts of Cowper's glands, which run obliquely through the wall of the urethra for some distance. The cut edge of the bulb will be seen to be continuous with the corpus spongiosum which surrounds the spongy or anterior portion of the urethra, and expands again to form the *glans penis*.

The *Spongy portion* is the longest part of the canal, and averages five inches in length. Its calibre is somewhat smaller than that of the bulbous portion, but it expands in the glans penis to form the *fossa navicularis*, again becoming contracted at the *meatus* or orifice, at which point the urethra is as small as in the membranous portion, and occasionally smaller. Along the floor of the spongy portion are numerous mucous follicles or *lacunæ*, the orifices of which are directed towards the meatus; a few similar follicles are situated on the upper surface of the urethra, one of which opposite the fossa navicularis is the *lacuna magna*. This can seldom be made out when the urethra is opened from above as here directed.

The mucous membrane is smooth and pale in health, but the student should notice carefully any traces of stricture, which will probably be found, if present, in the bulbous portion. The epithelium of the posterior part of the urethra is columnar, but becomes tesselated near the glans penis.

Beneath the mucous membrane of the urethra there is a layer of longitudinal involuntary muscular fibre continuous with the submucous layer of the bladder (Ellis), and intermixed with fibrous tissue. According to Hancock, circular muscular fibres also invest the whole length of the urethra at the same level.

The Corpora Cavernosa Penis (Fig. 126, 14) are now seen to be two vascular bodies enclosed in fibrous capsules, which unite to form a median septum between them, known as the *septum pectiniforme* (16) from the comb-like appearance produced by apertures in it. The elastic capsule consists of two layers of fibres, longitudinal and circular, from which *trabeculæ* or processes are distributed through the vascular tissue of the organ. The attachment of the corpora cavernosa or crura penis has been seen in the dissection of the perinæum.

Fig. 128.

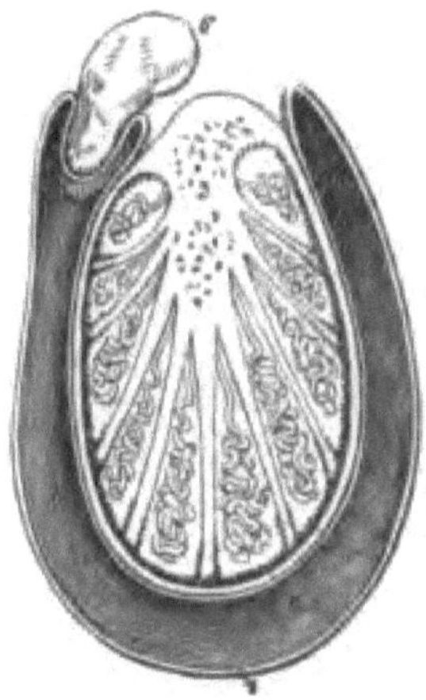

The Testicle (Fig. 128) has already been seen in its position in the scrotum, and its structure should now be investigated as far as possible, but cannot be thoroughly seen without mercurial injection.

The testis consists of two parts, an anterior or larger portion, the *body* of the testis, and the accessory and more slender posterior part, the *epididymis*, from the lower end of which the *vas deferens* or duct arises.

Fig. 128.—Transverse section of the testicle (from Wilson).

1. Cavity of the tunica vaginalis.
2. Tunica albuginea.
3. Mediastinum testis giving off numerous septa. The cut extremities of the vessels below the figure belong to the rete testis; and those above to the arteries and veins of the organ.
4. Tunica vasculosa, or pia mater testis.
5. One of the lobules, consisting of the convolutions of the tubuli seminiferi, and terminating by a single duct, the vas rectum.
6. Section of the epididymis.

The testicle has three tunics or coverings. The tu *vaginalis* (1) is a serous covering derived originally fr the peritoneum, and described as consisting of t portions, the *tunica vaginalis propria*, which covers t organ, and the *tunica vaginalis reflexa* or reflected porti The *tunica albuginea* (2) is a fibrous covering enclosi the body of the testis, and sending into its centre a ve tical process called the *mediastinum* (3) or corpus High morianum. From this secondary processes or *septa* ar derived, which separate the *lobules* of the testicle. The tunica albuginea is inseparably connected with the tunica vaginalis, thus forming a fibro-serous membrane. The *tunica vasculosa* (4) is beneath the tunica albuginea, and contains the ramifications of the blood-vessels, which cannot be seen except when finely injected.

Fig. 129.

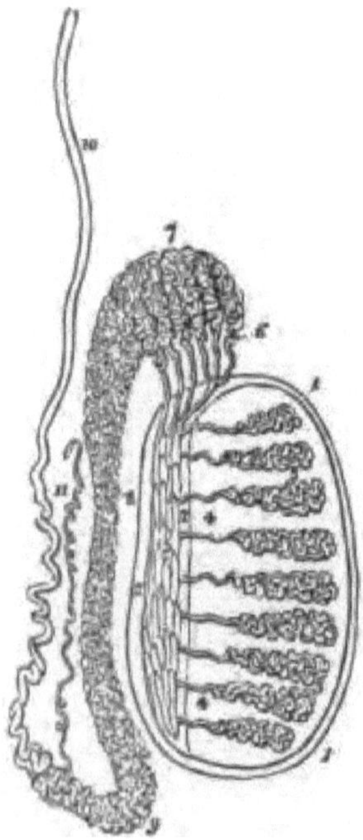

The *Epididymis* (Fig. 129) is divided into three parts. The upper extremity is the *head* or *globus major* (7) which is divisible into *coni vasculosi;* the central portion is the *body* (8), and the lower extremity the *tail* or *globus minor* (9), which is larger than the body, but not so large as the head. From this the *vas deferens* (10) or duct of the testicle arises, and close to its point of origin may

---

Fig. 129.—Anatomy of the testis (from Wilson).

1, 1. Tunica albuginea.
2, 2. Mediastinum testis.
3, 3. Lobuli testis composed of tubuli seminiferi.
4, 4. Vasa recta.
5. Rete testis.
6. Vasa efferentia, of which six only are represented in this diagram.
7. Coni vasculosi, constituting the globus major epididymis.
8. Body of the epididymis.
9. Globus minor epididymis.
10. Vas deferens.
11. Vas aberrans.

ound a single tube extending along the cord and d the *vas aberrans* (11).

n making a section of the uninjected testis, little more be seen than a mass of a drab colour, which may be wn out in threads, these being the minute seminal es. In an injected preparation the tubes may be ced as follows (Fig. 129):—

Each lobule is composed of convoluted *tubuli seminiferi*), which anastomose with one another, and these as they merge from the lobules become straight, and are called *asa recta* (4), about twenty in number. Entering the mediastinum, they anastomose to form the *rete testis* (5), which terminates in the *vasa efferentia* (6); and these, from twelve to fifteen in number, pierce the tunica albuginea to form the *coni vasculosi* (7) of the globus major of the epididymis. The ducts are now all collected into one, which by its convolutions forms the body and globus minor of the epididymis, and ends in the vas deferens.

Attached to the top of the testis immediately in front of the epididymis may generally be found a little vesicle called the *hydatid of Morgagni*, the remains of Müller's duct; and in the cellular tissue of the cord immediately above the epididymis may be found a small white body which has been supposed by M. Giraldès, who first described it, to be part of the remains of the Wolffian body of the fœtus.

Pelvic Viscera in the Female.

[The organs removed from the pelvis are to be laid out upon the table with the rectum upwards; this is to be cleaned, and after it has been examined is to be dissected from the uterus.]

The **Rectum** (Fig. 122, 19) corresponds in all particulars to that of the male, but is sometimes much distended. The description of the rectum will be found at page 301.

[The preparation being reversed, the bladder and urethra are next to be dissected and examined. The peritoneum is to be dissected up from the bladder and left attached to the uterus.]

The **Bladder** (Fig. 122, 4) is more capacious than in the male, and is directly continuous with the urethra, without the intervention of a prostate, the relation to the vesiculæ seminales and vasa deferentia being of course wanting. The bladder is firmly attached to the neck of the uterus and front wall of the vagina. On each side of the upper part of the vagina the ureters turn upwards to gain the sides of the bladder, running obliquely for a short distance in the vesico-vaginal septum, then forwards obliquely through its coats to open by two orifices an inch and a half apart, about where they would open in the male (Savage). The structure of the bladder in the female corresponds closely to that of the male, which is given at page 303.

The **Urethra** (Fig. 122, 6) of the female is a simple tube about an inch and a half long, the structure of which is best seen by laying it open with the bladder. The urethra is about a quarter of an inch in diameter, but is larger near the bladder, this being also its most dilatable portion. The mucous and muscular coats (circular and longitudinal) resemble those of the membranous urethra of the male.

[The bladder and urethra are to be carefully removed, and the vagina and uterus exposed. The broad ligament of the uterus should be pinned out for the examination of the Fallopian tube and round ligament, and the preparation must afterwards be reversed for the dissection of the ovary, which is placed posteriorly.]

The **Vagina** (Fig. 130, 5) has been seen in its position in the pelvis (p. 291) when the anterior wall was found to be shorter than the posterior, thus giving it a curved direction which is lost when the organs are removed from the body. The vagina is composed of longitudinal mus- ' - fibres enclosing erectile tissue, and lined by a mucous membrane, which when the tube is opened will be seen to form transverse *rugæ*. Two longitudinal ridges at the anterior and posterior surfaces of the vagina have been named the *columns* of the vagina. At the upper part of the vagina the cervix uteri projects into its cavity, and the vagina will be found to be attached higher on

Fig. 130.

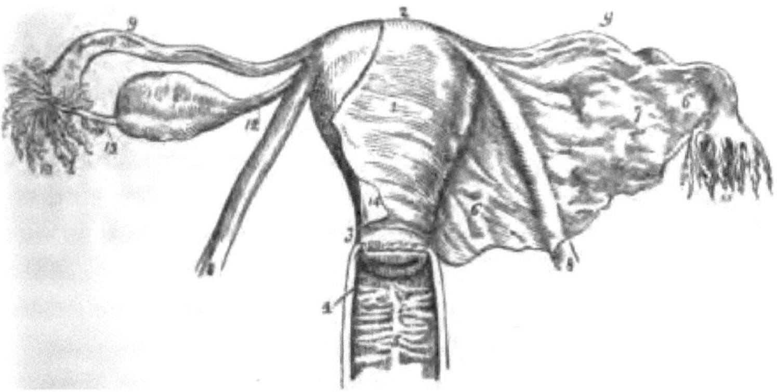

its posterior than its anterior surface. At the lower end will be found the *carunculæ myrtiformes* or remains of the *hymen*, and outside the vagina may be found the two small glands of Bartholin, the ducts of which open within the nymphæ.

The **Uterus** (Fig. 130, 1) is more or less pyriform in

Fig. 130.—Uterus with its appendages viewed from the front (from Wilson).

1. Body of the uterus.
2. Fundus.
3. Cervix.
4. Os uteri.
5. Vagina.
6, 6. Broad ligament of the uterus.
7. Convexity of the broad ligament formed by the ovary.
8, 8. Round ligaments of the uterus.
9, 9. Fallopian tubes.
10, 10. Fimbriated extremities of the Fallopian tubes; on the right side the mouth of the tube is turned forwards in order to show its *ostium abdominale*.
11. Ovary.
12. Utero-ovarian ligament.
13. Fallopio-ovarian ligament.
14. Peritoneum of anterior surface of uterus, continuous with the anterior layer of the broad ligament.

shape, the *fundus* or upper part of the body being broad and somewhat flattened from before backwards, and the lower end or *cervix* being cylindrical. The cervix is perforated by the *os uteri* (externum), a transverse opening which divides it into the two lips, anterior and posterior, of which the anterior is the thicker.

[The anterior wall is to be divided vertically with scissors, and the cut prolonged on each side towards the Fallopian tube.]

Immediately within the os uteri on the posterior aspect of the cervix is seen a vertical ridge with oblique ridges radiating from it, producing the appearance known as the *arbor vitæ.* A similar appearance in the mucous membrane of the anterior surface of the cervix has been necessarily destroyed by the incision. Above the canal of the cervix is the *os internum*, which opens into the general cavity of the uterus.

The *cavity* of the uterus is triangular in shape, the apex being downwards and the extremities of the base corresponding with the orifices of the Fallopian tubes. It is lined by a smooth mucous membrane, in which are the orifices of numerous small glands.

The uterus is composed of pale unstriped muscular fibre together with a large quantity of areolar tissue. In pregnancy the muscular fibres become much developed and are divisible into three layers.

The nerves of the uterus are derived from the hypogastric plexus of the sympathetic, and Dr. Robert Lee has described numerous ganglia on the surface of the organ, and believes that the nerves enlarge during pregnancy.

The **Fallopian Tube** (Fig. 130, 9) is found at the upper part of the broad ligament on each side of the uterus, into the angles of which the two tubes open. Each tube is about four inches long and of very small calibre at the uterus, but gradually expands to form the *fimbriated extremity*, a trumpet-shaped opening surrounded by fringe-like processes. One of the fimbriæ is attached to the ovary, and serves to direct the tube in grasping the ovule as it emerges from that organ. The tube is muscular in structure and is lined by a mucous membrane with ciliated epithelium.

The **Round Ligament** (Fig. 130, 8) placed in the anterior part of the broad ligament has been already traced (p. 291). It is composed of unstriped muscular and areolar tissue, and is covered by peritoneum, which in the child can be traced for a short distance into the inguinal canal, forming the *canal of Nuck*.

The **Ovary** (Fig. 130, 11) is an oval body situated in the posterior part of the broad ligament on each side of the uterus. In the young subject it is smooth and plump, but in the old subject it is shrivelled and scarred. It is attached to the fundus of the uterus by a slender cord, the *ligament of the ovary* (12), and has one of the fimbriæ of the Fallopian tube attached to its surface. It is composed of a fibrous capsule containing the Graafian vesicles, from which the ovules are discharged by a process of dehiscence or bursting, producing a cicatrix on the surface of the ovary, and within it a shrivelled clot called the *corpus luteum*.

Between the Fallopian tube and the ovary and contained in the broad ligament, are the remains of the Wolffian body, a fœtal structure which becomes the epididymis in the male, and has been called the *organ of Rosenmüller* in the female. It consists of a number of tubes with blind extremities. Connected with the lower surface of the ovary is a venous erectile tissue, to which the name of *Bulb of the Ovary* has been given by Mr. Reeves Traer, who first described it.* This is supposed to maintain the ovary in its position and to prepare it for the grasp of the Fallopian tube during the passage of the ovule.

LIGAMENTS OF THE PELVIS.

[The remaining soft tissues are to be removed from the halves of the pelvis and the several ligaments dissected out.]

* 'Sur l'Arrangement des Veines de l'Ovaire,' read before the Anatomical Society of Paris, 1857.

Articulation of the Vertebral Column with the Pelvis (Fig. 131).—The last lumbar vertebra is connected with the sacrum by the continuation of the several ligaments common to the vertebral column, and by an intervertebral substance, which is peculiar in being of greater thickness in front than behind. There are also two special ligaments, the lumbo-sacral and ilio-lumbar.

The *lumbo-sacral* or *sacro-vertebral ligament* (2) is short, thick, and triangular, being attached to the transverse process of the last lumbar vertebra above, and the upper border of the sacrum below.

The *ilio-lumbar ligament* (3) is triangular, and passes horizontally from the tip of the transverse process of the last lumbar vertebra to the crest of the ilium close above the sacro-iliac articulation.

Sacro-iliac Articulation (Fig. 131).—The *anterior sacro-iliac ligament* (4) consists of short fibres passing between the anterior surfaces of the two bones.

The *posterior sacro-iliac ligament* is composed of strong ligamentous bands, connecting the rough portion of the ilium behind the auricular surface of articulation with the posterior surface of the sacrum external to the posterior foramina.

The *oblique sacro-iliac ligament* is a superficial portion of the posterior ligament, and reaches from the posterior superior spine of the ilium to the third lateral tubercle on the sacrum.

When the ilium and sacrum are separated, after the dissection of the other ligaments of the pelvis, a distinct layer of cartilage will be found between the two bones, with occasionally an intervening space, the joint being an instance of 'amphiarthrosis.'

Sacro-sciatic Ligaments (Fig. 45).—These ligaments have been partially exposed in the dissection of the gluteal region.

The *Great sacro-sciatic ligament* (2) is broad at its upper attachment, which is to the posterior inferior spine of the ilium and to the sides of the sacrum and coccyx, and narrows to its insertion into the inner and back part of e tuberosity of the ischium. A portion of this ligament, prolonged to the ramus of the ischium, has been en to protect the pudic vessels and nerve.

The *Lesser sacrosciatic ligament* (3) is attached to the side of the sacrum and coccyx close to, but in front of, the great ligament, and reaches to the spine of the ischium; thus separating the greater from the lesser sacro-sciatic foramen.

Fig. 131.

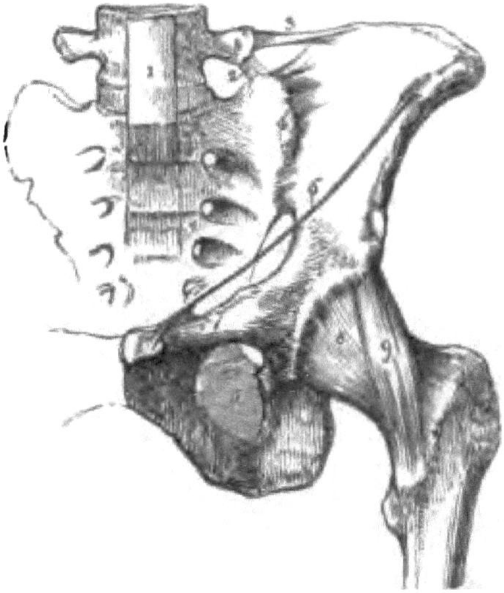

Articulation of the Sacrum and Coccyx.—The *anterior sacro-coccygeal ligament* is a thin fasciculus passing from the anterior surface of the sacrum to the front of the coccyx.

The *posterior sacro-coccygeal ligament* is attached to the lower opening of the sacral canal, and passes along the whole length of the posterior surface of the coccyx.

A small *inter-articular cartilage* intervenes between the sacrum and coccyx, and the cornua of the latter are connected to the former by short ligamentous bands.

Symphysis Pubis.—The *anterior pubic ligament* passes in front of the two bones, the superficial fibres interlacing obliquely, and the deeper ones running transversely.

Fig. 131.—Ligaments of the pelvis and hip-joint (from Wilson).

1. Lower part of the anterior common ligament of the vertebræ.
2. Lumbo-sacral ligament.
3. Ilio-lumbar ligament.
4. Anterior sacro-iliac ligament.
5. Obturator membrane.
6. Poupart's ligament.
7. Gimbernat's ligament.
8. Capsular ligament of the hip-joint.
9. Ilio-femoral or accessory ligament.

The *posterior pubic ligament* can hardly be said to exist, consisting merely of a few scattered fibres.

The *superior pubic ligament* connects the angles of the pubes.

The *sub-pubic ligament* is a strong band passing from one pubic bone to the other and arching beneath the symphysis.

[To see the arrangement of the inter-articular cartilages, a vertical transverse section of the symphysis should be made with the saw near its posterior part.]

An *inter-articular fibro-cartilage* covers the roughly-grooved surface of each pubic bone, the two cartilages being connected in front by strong elastic tissue, but separated posteriorly by a small cavity lined by a synovial membrane. This joint also is an example of "amphi-arthrosis."

PART IV.

DISSECTION OF THE HEAD AND NECK.

[*The Student is requested to read the 'Introduction' before commencing the dissection, unless he has done so previously.*]

BEFORE beginning the dissection the student should make himself fully acquainted with the external configuration of the part, and the relation of surface-markings to deeper structures. If he has already dissected this region, he should make the incisions necessary for exposing the common carotid artery and the third part of the subclavian artery, and may also advantageously perform the operations of laryngotomy and tracheotomy, and practise extraction of the teeth and the operation for plugging the posterior nares. The surface of the scalp presents nothing worthy of note, but in a well-injected body the temporal artery will be prominent and arteriotomy may be performed upon it.

The superficial appearances on the face have been described more particularly under that head, and reference may be made to that section now if time allows.

In the neck, the median line is the boundary of the dissection, and presents the following prominences and depressions, which should be fully recognised with the finger. First is the symphysis of the lower jaw, next a depression corresponding to the mylo-hyoid muscle, followed by the projection of the hyoid bone, the cornua of which can be felt distinctly on each side in some subjects. The depression corresponding to the thyro-hyoid membrane and the prominent *pomum Adami* of the thyroid cartilage are next felt, below which are the interval of the crico-thyroid membrane and the ring of the cricoid cartilage. Below this may occasionally be felt the isthmus of the thyroid gland or body, and even the trachea in a thin subject; and on each side there is (particularly in women) the slight projection of the thyroid gland.

The crico-thyroid membrane and the trachea should be especially noticed as the positions in which laryngotomy and tracheotomy are performed.

The clavicle and sternum bound the dissection below, and across the side of the neck the prominence of the sterno-mastoid marks the division into anterior and posterior triangles. In an injected body the carotid artery may be felt near the middle of the anterior border of the muscle. The external jugular vein, if full of blood, will probably be seen crossing the sterno-mastoid obliquely, and may be opened *secundum artem* by an incision parallel to the muscle.

It will be found to be impossible to dissect both sides of the head and neck simultaneously, and the students should therefore arrange to work together, and to take turns at dissecting and reading.

The dissection of both sides of the scalp should be completed on the first day, in order that the brain may be removed on the second morning.

THE SCALP.

[The head being shaved and raised on a block, one incision is to be carried from the root of the nose to the occiput, and another at right angles to it from the front of the ear across the head to the corresponding point on the opposite side, and the flaps thus marked out are to be dissected by beginning at the top of the head. Great care must be taken to remove only the skin and not the subcutaneous tissues, which are very thin; the best guides are the roots of the hair, which should be exposed on the under surface of the flaps. The flaps having been turned down to the level of the brow in front, and the occipital protuberance behind, the dissector may, if he choose, attempt to define the small muscles of the external ear, which are seldom well developed.]

The **Muscles of the Pinna** (Fig. 132) are the *Attollens aurem* (5) (superior), the *Attrahens aurem* (1) (anterior), and the *Retrahens aurem* (9) (posterior). The names sufficiently indicate their positions, and they are all inserted into the pinna, the anterior *arising* from the anterior part of the aponeurosis of the occipito-frontalis, and being *inserted* into the helix; the superior also *from* the middle part of the same aponeurosis and *inserted* into the back of the pinna; and the posterior *from* the mastoid process *to* the back of the concha.

[The muscular fibres of the occipito-frontalis are to be cleaned both on the forehead and on the occiput, care being taken not to damage the intervening aponeurosis, or the nerves which pierce the muscle at several points.]

The **Occipito-frontalis** (Fig. 132).—The *anterior belly* (2) has no bony attachment, but takes its *origin* from the fibres of the orbicularis palpebrarum and the pyramidalis nasi, and is also connected with the corrugator supercilii on the face. The *posterior belly arises* from the superior curved line of the occipital bone and very slightly from the mastoid process of the temporal bone. Both bellies are *inserted* into the broad tendinous aponeurosis which extends over the top of the skull. The aponeuroses of the two sides are continuous over the top of the head, and are freely moveable upon loose areolar tissue which intervenes between them and the periosteum of the skull.

Cutaneous Vessels and Nerves (Fig. 132).—In front of the ear will be found branches of the three divisions of the 5th nerve and branches of the 7th nerve, together with branches of the ophthalmic and temporal arteries. Behind the ear are, a branch of the 7th cranial nerve, two branches of the cervical plexus (great auricular and small occipital), and a branch of the second cervical nerve (great occipital) together with the posterior auricular and occipital arteries.

The *Supra-orbital nerve* (8) [first div. of 5th] will b most readily found by feeling for the supra-orbital notch

Fig. 132.

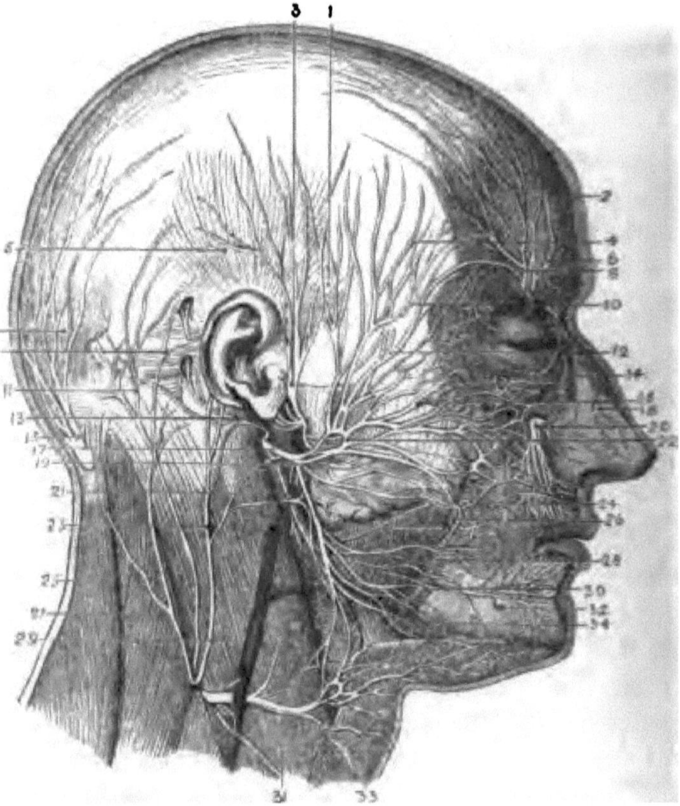

Fig. 132.—Nerves of the face and scalp (from Hirschfeld and Leveillé).

1. Attrahens aurem muscle.
2. Anterior belly of occipito-frontalis.
3. Auriculo-temporal nerve.
4. Temporal branches of facial nerve (7th).
5. Attollens aurem muscle.
6. Supra-trochlear nerve (5th).
7. Posterior belly of occipito-frontalis.
8. Supra-orbital nerve.
9. Retrahens aurem muscle.
10. Temporal branch of temporo-orbital nerve (5th).
11. Small occipital nerve.
12. Malar branches of facial nerve.
13. Posterior auricular nerve (7th).
14. Malar branch of temporo-malar nerve (5th) (subcutaneus malæ).
15. Great occipital nerve.
16. Infra-orbital branches of facial nerve (7th).
17. Facial nerve (7th).

ıen cutting through the occipito-frontalis at that It turns round the margin of the orbit at this point, .ivides into two branches (inner and outer), which pierce the occipito-frontalis, and supply the skin.

ə *Supra-trochlear nerve* (6) [first div. of 5th] is a ł nerve which leaves the orbit at its inner angle, and supplies the forehead by piercing the occipito-talis.

ne *Supra-orbital artery* [ophthalmic] accompanies the ra-orbital nerve, and also gives superficial branches to surface of the occipito-frontalis muscle.

'he *Frontal artery* [ophthalmic] is of small size and ompanies the supra-trochlear nerve.

The *Supra-orbital* and *Frontal veins* unite to form the gular vein, which is the commencement of the facial ein.

The *Temporal branch* of the *orbital nerve* (10) [second ıv. of 5th] will be found with difficulty, as it is of small ze. It pierces the temporal fascia immediately above ıe zygoma, and is distributed to the neighbouring skin.

The *Auriculo-temporal nerve* (3) [third div. of 5th] lies nmediately in front of the ear and close to the temporal tery. Emerging from the parotid gland it passes up the de of the head with the temporal artery, and, after ving a branch to the pinna, is distributed to the skin of ıe scalp.

Temporal branches of the Facial nerve (4) [7th] will be und running obliquely over the zygoma to the temple, here they join with the several branches of the 5th,

---

. Nasal nerve (5th).
Cervico-facial division of 7th.
. Infra-orbital nerve (5th).
Branches to digastric and stylo-hyoid (7th).
Temporo-facial division of 7th.
Great auricular nerve.
Buccal branches of facial nerve.
Trapezius.
26. Buccinator nerve (5th).
27. Splenius capitis.
28. Masseter.
29. Sterno-mastoideus.
30. Supra-maxillary branches of facial nerve (7th).
31. Superficial cervical nerve.
32. Mental nerve (5th).
33. Platysma.
34. Infra-maxillary branches of facial nerve (7th).

already described, and supply the orbicularis palpeb the anterior belly of the occipito-frontalis, and the hens aurem muscles.

The *Temporal artery* [external carotid] is found i of the ear, and after giving small anterior aur branches to the pinna and the middle temporal (which pierces the deep fascia above the zygom divides into anterior and posterior branches. The *an* lies upon the temporal fascia and is subcutaneou reaches as far forward as the forehead to anast with the supra-orbital artery. The *posterior* turns above the ear and anastomoses with the occipital art

The *Temporal vein* is formed by branches correspon to those of the artery, and disappears in the parotid g to join the internal maxillary vein, and form the *exte* jugular vein.

Surgery.—It is on the anterior branch of the temp artery that the operation of arteriotomy is performed w it is desired to abstract blood from the head. In operation it is necessary to make only a puncture in artery and, when sufficient blood has flowed, to divide vessel, in order that the ends may be able to retract a prevent the formation of a false aneurism.

The *Posterior Auricular nerve* (13) [7th] will be foun immediately behind the ear and running over the mastoi process with an accompanying artery. It is distribute to the posterior belly of the occipito-frontalis and to th attollens and retrahens aurem muscles.

The *Posterior Auricular artery* [external carotid] accompanies the nerve in this part of its course, and *is* distributed to the pinna and to the skin behind it. Its *vein* opens into the external jugular.

The *Great Auricular nerve* (23) [superficial cervical plexus] ascends to the pinna, to which it gives numerous branches, principally on its posterior surface, forming a junction with the posterior auricular nerve.

The *Small Occipital nerve* (11) [superficial cervical

will be found about an inch behind the mastoid and just at the posterior margin of the sterno-d. It is a small nerve, and passes directly upwards scalp, which it supplies—giving also a branch to nna.

Great Occipital nerve (15) [2nd cervical nerve] is a nerve, and will be found piercing the trapezius close to the occipital protuberance. Its branches er the posterior part of the occipito-frontalis and istributed to the scalp, communicating with the occipital nerve and with the fellow nerve of the ite side.

e *Occipital Artery* [external carotid] lies close to the occipital nerve, but its relation to it is inconstant. amifies on the back of the scalp, anastomosing with artery of the opposite side and with the posterior ricular and temporal arteries of the same side.

The **Temporal Fascia** will be fully exposed by removing the temporal artery and the branches of nerves over . It is attached above to the temporal ridge on the ontal, parietal, and temporal bones, and below to the ygoma, where it is split so as to be attached to both its orders; between the layers are the small orbital branch of the temporal artery and the temporal branch of the orbital nerve of the superior maxillary. It is pierced by the middle temporal artery and the temporal branch of the orbital nerve (*v.* p 321), and gives origin to fibres of the temporal muscle by its deep surface.

The **Temporal Muscle** will be exposed in part by removing the fascia covering it. It *arises* from the temporal fossa and from the under surface of the temporal fascia, and its fibres are seen to converge to a tendon, which will be found to be *inserted* into the coronoid process of the lower jaw in a subsequent dissection.

Removal of the Brain.

[Both sides of the scalp having been dissected, the knife should be drawn round the skull, marking out a line an inch above the orbit in front, and half an inch above the occipital protuberance behind, and passing through the temporal muscle between those points. The outer plate of the skull is to be carefully sawn through, particularly at the anterior and posterior extremities of the temporal ridge, and at the occiput. With the chisel the remaining thickness of the skull can then be readily divided without injuring the brain, and the calvaria removed.]

Upon Removing the Skull-cap the rough outer surface of the dura mater will be seen, and it should be noticed that it formed the internal periosteum of the calvaria, which will be found to be bare. The inner surface of the skull-cap presents a longitudinal groove in the median line, which corresponds to the superior longitudinal sinus, and on each side may frequently be seen depressions of variable size corresponding to the Pacchionian glands. The groove for the middle meningeal artery will be seen on the inner surface of the parietal bone. The dura mater is cream coloured, but often has a bluish appearance, resulting from its translucency, which allows the veins of the surface of the brain to be partially seen. On each side of the middle line will be seen the *glandulæ Pacchionii*, which are growths from the pia mater, and frequently project through the dura mater and indent the skull. Ramifying upon the dura mater will be found the large middle meningeal artery [internal maxillary] and in a well-injected body the anterior meningeal [ethmoidal] may sometimes be seen.

The **Superior Longitudinal Sinus** (Fig. 133, 1) should be opened at once. It runs along the middle line of the head, and will be seen to be triangular in shape, being formed by a splitting of the dura mater and lined with a serous coat. It commences in a small vein which enters the skull through the foramen cæcum, and it ends in the torcular Herophili, running along the whole length of the upper border of the falx cerebri. Within the sinus may be seen some small tendinous cords—*chordæ Willisii*, and also the orifices of the veins opening into it, which all pass from behind forwards, *i.e.*, contrary to the current of blood in the sinus itself.

[The dura mater is to be divided longitudinally on each side of the sinus and turned down. By this the cavity of the arachnoid will be opened, and one layer of that serous membrane will be seen to line the dura mater, whilst the other is reflected upon the pia mater covering the brain. By slightly separating the two hemispheres of the brain with the handle of the knife, the falx cerebri will be seen, and must be detached from the crista galli of the ethmoid bone.]

Removal of the Brain.—The head being tilted a little backwards, the anterior lobes of the cerebrum are to be raised, when the Olfactory nerves and bulbs (1st pair) will probably be detached from the cribriform plate of the ethmoid bone with them, or if not, they can be readily displaced with the handle of the knife. The Optic nerves (2nd pair) of large size, will be seen close to the anterior clinoid processes, and should be divided, when the Internal Carotid Arteries will be brought into view, and must also be cut, together with the *infundibulum* attached to the *pituitary body* in the middle line. The 3rd pair (motores oculorum) will be seen piercing the dura mater behind, and a little external to the carotid arteries, and after dividing these, the *tentorium cerebelli* will be brought into view with the 4th nerve (patheticus vel trochlearis) just at its edge. The tentorium is to be divided on each side by carrying the knife parallel to the margin of the petrous bone, and the 4th pair will be cut at the same time. The tentorium having been turned aside, the remaining cranial nerves will be exposed, and must be divided, in their numerical order. The 5th pair (trifacial) will be found to be composed of two parts, the larger superior one being sensory, and the deeper one, which is anterior at its origin, motor in function. The 6th pair (abducentes) will be found nearer the median line. The 7th consists of two distinct nerves with a small intermediate portion. The anterior is the facial nerve, and the posterior the auditory, the intervening portion being called the *pars intermedia.* The entire nerve is seen to disappear through the *meatus auditorius internus* with an accompanying auditory artery. The 8th pair consists of three distinct nerves. The most anterior fibres form the small Glosso-pharyngeal nerve, which pierces the dura mater separately; the next and largest portion is the Pneumo-gastric or Vagus; and the part coming out of the foramen magnum is the Spinal-accessory. The whole nerve disappears

Fig. 133.

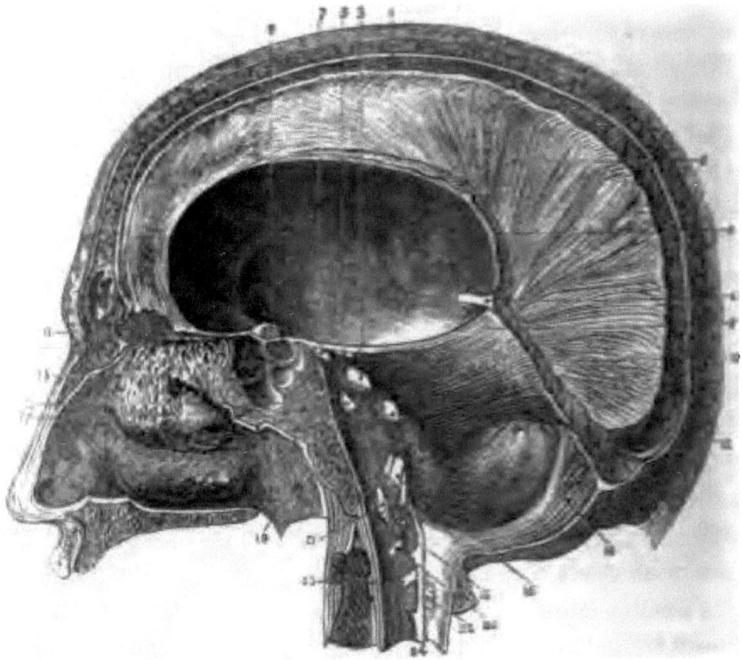

Fig. 133.—Sinuses of the skull (side view) (from Hirschfeld and Leveillé).

1. Superior longitudinal sinus.
2. Falx cerebri.
3. 5th nerve.
4. Inferior longitudinal sinus.
5. 4th nerve.
6. Venæ Galeni.
7. 3rd nerve.
8. Tentorium cerebelli.
9. 2nd nerve.
10. Straight sinus.
11. Crista galli of ethmoid.
12. Torcular Herophili.
13. 1st nerve.
14. Falx cerebelli.
15. 6th nerve.
16. 8th nerve.
17. 7th nerve.
18. Vertebral artery.
19. Eustachian tube.
20. 1st cervical nerve.
21. 9th nerve.
22. Posterior root of 2nd cervical nerve.
23. Anterior root of 2nd cervical nerve.
24. Ligamentum denticulatum.

rough the *foramen lacerum posterius* or jugular hole. he 9th (hypoglossal) nerve consists of two bundles placed ear the median line, which pierce the dura mater at eparate points to pass through the *anterior condyloid oramen.* The Veins of Galen, which pass from the brain o the anterior border of the tentorium, should be divided so that that fold of dura mater may be quite free.

The knife is now to be pushed through the foramen magnum so as to divide all the parts passing into the skull viz., the spinal cord and its membranes, the two vertebral arteries, the small spinal vessels, and the two spinal accessory nerves; and it should be noticed that the cord can be cut considerably lower than the point at which the vertebral arteries enter the spinal canal. The brain can now be lifted out of the skull by slipping the fingers beneath the cerebellum and allowing the head to fall back.

[Before placing the brain in methylated spirit, the dissectors should pick off the arachnoid and the pia mater from the base, and carefully lay out the nerves in their proper position, taking care not to interfere with the arteries. The brain should be placed in a pan with the base upwards, and with a piece of calico bandage beneath it, so that it may be easily lifted out.]

THE DURA MATER AND SINUSES OF THE SKULL.

The **Dura Mater** is firmly attached to the base of the skull, and sends processes through the several foramina to form sheaths for the nerves. It is continuous with the dura mater of the spinal cord through the foramen magnum, but is closely attached to its margin.

The **Falx Major** (Fig. 133, 2) or *falx cerebri* is a vertical process of dura mater placed between the two hemispheres of the cerebrum. It is attached in front to the crista galli of the ethmoid bone, and behind, where it is much deeper, it is attached to the internal occipital protuberance and to the tentorium cerebelli.

The **Tentorium Cerebelli** (Fig. 133, 8) is the horizontal

process of dura mater placed between the cerebrum and cerebellum, and when perfect it leaves only an oval opening of small size, through which the crura cerebri pass. In front it is attached to the anterior and posterior clinoid processes of the sphenoid bone and the superior border of the petrous bone, and behind it is attached to the upper margin of the lateral sinus where it grooves the occipital, parietal, and temporal bones.

The **Falx Minor** (Fig. 133, 14) or *falx cerebelli* is below the tentorium in the median line, and reaches to the foramen magnum, being attached to the median ridge of the occipital bone. It generally divides below, sending a slip to each side of the foramen magnum.

The **Sinuses** are the venous canals of the skull into which the veins of the brain, orbit, and diploë empty themselves, and differ from the ordinary veins, in being (1) enclosed in sheaths of dura mater, to which they are firmly attached, so as to remain patent when divided; (2) in having no valves; and (3) that the veins open into them in a direction contrary to that of the blood within them.

The **Superior Longitudinal Sinus** (Fig. 133, 1) runs along the upper border of the falx major (*v.* p. 324), and along its lower border is a much smaller canal, the **Inferior Longitudinal Sinus** (4). This begins about the middle of the falx and opens into the straight sinus.

The **Straight Sinus** (Fig. 133, 10) must be opened along the line of attachment of the falx to the tentorium. It is formed by the *venæ Galeni* of the brain and the inferior longitudinal sinus, and opens into the torcular Herophili, thus connecting the two longitudinal sinuses together.

The **Occipital Sinuses** (Fig. 134, 18) are two in number, but of smaller size. They pass forwards one on each side from the torcular Herophili, and along the bifurcated attachment of the falx cerebelli to the jugular vein.

The **Torcular Herophili** (Fig. 134, 21) is the point where the superior longitudinal, straight, occipital, and two lateral sinuses meet. It is usually situated on the right of the internal occipital protuberance, but is sometimes to its left.

The **Lateral Sinuses** (Fig. 134, 20) are the largest and, commencing at the torcular Herophili, take a curved course to the foramen jugulare on each side. It will be

Fig. 134.

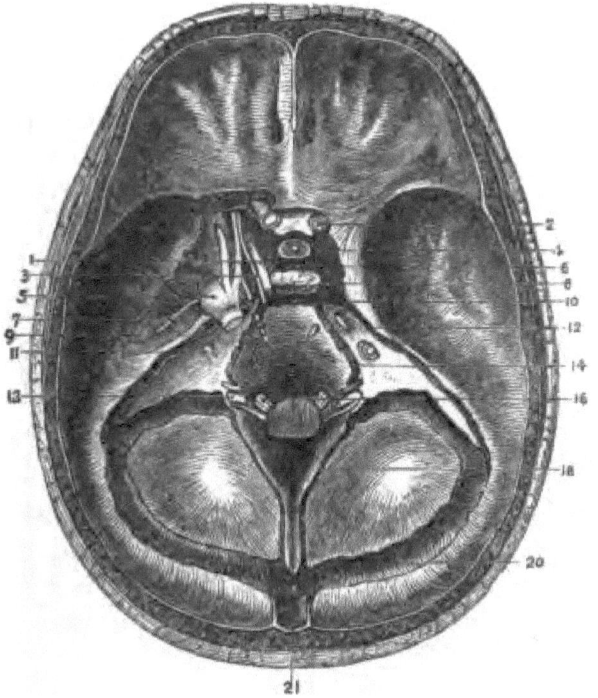

necessary to cut through the tentorium, in order to expose the cavity of the sinus, which grooves the following bones,—occipital, parietal, temporal, and the occipital

---

Fig. 134.—Dissection of the sinuses of the skull and cranial nerves; the cavernous sinus dissected on the left side (drawn by J. T. Gray).

1. 3rd nerve.
2. Optic nerve.
3. 4th nerve.
4. Internal carotid artery.
5. Gasserian ganglion of 5th nerve with its three divisions.
6. Circular sinus.
7. Superficial petrosal nerve.
8. Cavernous sinus.
9. 6th nerve.
10. Transverse or basilar sinus.
11. 7th pair.
12. Superior petrosal sinus.
13. 8th pair.
14. Inferior petrosal sinus.
16. 9th nerve of left side.
18. Occipital sinus.
20. Lateral sinus.
21. Torcular Herophili.

bone again close to the jugular process.* The lateral sinuses receive the mastoid veins through the mastoid foramina, and end in the jugular veins, returning nearly the whole of the blood from the brain.

The **Superior Petrosal Sinus** (Fig. 134, 12) will be exposed by running the knife along the upper border of the petrous bone, from which the tentorium was detached. It is of small size, and runs from the cavernous sinus back to the lateral sinus, just where the latter turns downwards in the temporal bone.

The **Inferior Petrosal Sinus** (Fig. 134, 14) will be opened by carrying the knife from the foramen jugulare directly forwards along the lower border of the petrous bone to the clinoid process. It communicates with the cavernous sinus, and with the jugular vein outside the skull, passing through the foramen jugulare in front of the 8th pair.

The **Cavernous Sinus** (Fig. 134, 8) is placed by the side of the pituitary body, and requires careful dissection on account of the nerves in relation with it. They are the 3rd, 4th, 5th, and 6th, and their positions should be at once ascertained. The 3rd, of good size (1), will be recognised close behind the anterior clinoid process; the 4th, very small (3), in the cut edge of the tentorium; the 5th, the largest and somewhat flattened out (5), will be seen immediately beneath the tentorium, and the 6th, piercing the dura mater nearer the median line, upon the basilar process (9).

[The left sinus will be found the most convenient for dissection, and should therefore be taken first. The 4th nerve is to be followed out carefully to the sphenoidal fissure by dividing the dura mater with a sharp knife, and the 3rd should be taken next. The dura mater over the 5th, having been freely divided, should be torn up from the bone, by which means and a very little dissection, the Gasserian ganglion and middle meningeal artery will be brought fully into view.]

The **Gasserian Ganglion** (Fig. 134, 5) (Ganglion Semi-

* Artificial memory, O P T O

unare) is the largest of the cranial ganglia, is of a pinkish colour, and is placed upon the tip of the petrous portion of the temporal bone. It is formed upon the larger or sensory portion of the 5th, and the smaller or motor portion passes beneath it. From the ganglion the three great divisions of the 5th are to be traced as follows:—The first or ophthalmic division to the sphenoidal fissure; the second or superior maxillary division to the foramen rotundum; and the third or inferior maxillary division to the foramen ovale. The motor trunk of the nerve joins this last division outside the skull. The ophthalmic division is seen to pass by the side of the cavernous sinus and to give off a branch (nasal), before it enters the sphenoidal fissure. It also sends a minute recurrent branch (Arnold) to the tentorium cerebelli.

[The 6th nerve is to be followed by dividing the dura mater over it, and will be found to lie against the outer side of the carotid artery, where it makes its sigmoid turn at the side of the sella turcica. This vessel should be fully defined, and an attempt made to see the carotid plexus of the sympathetic upon it, from which a branch goes to join the 6th. It will render the dissection more useful if a small piece of the lesser wing of the sphenoid is clipped off with the bone-forceps so as to open up the sphenoidal fissure, and the nerves can then be carefully followed to their entry into the orbit.]

The 3rd and 4th nerves, and the first division of the 5th nerve have been seen to lie in the outer wall of the cavernous sinus immediately beneath the dura mater, and between them and the 6th nerve, which is placed against the carotid artery, is the venous canal through which the blood is returned from the orbit. This is very delicate and can seldom be defined, but it receives the ophthalmic vein and opens into the two petrosal sinuses (Fig. 135).

The relation which the nerves hold to one another in the cavernous sinus is at first from above downwards according to their *numerical* order, *i.e.* 3, 4, 5, 6; and their relations to the sinus have been already described. Before they reach the orbit, however, they change thei- -relation positions, which at the sphenoidal fissure are roughly as follows, from above downwards—4, 5, 3, 6.

Fig. 135.

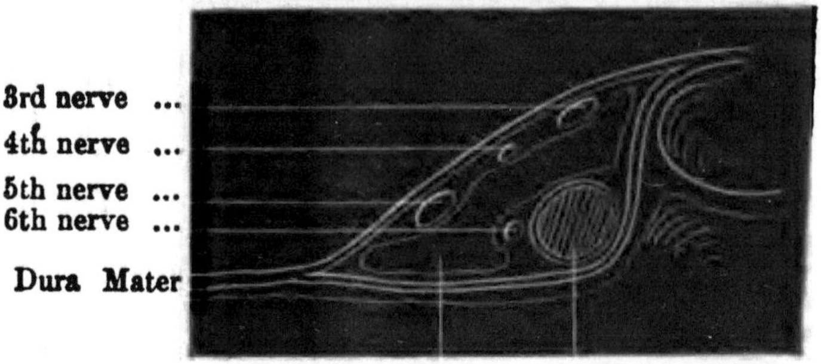

If however the fissure has been opened up as advised, and the dissection be carefully made, the 3rd nerve before entering the orbit will be found to divide into two parts

Fig. 136.

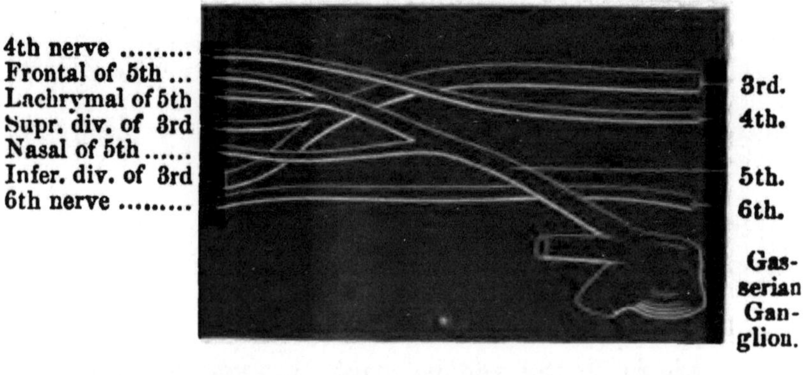

(between which the nasal branch of the 5th takes its course), and the 5th to have divided into the *frontal* and *lachrymal* branches. The order therefore of parts passing

---

Fig. 135 —Diagram of left cavernous sinus seen in section from behind (drawn by J. T. Gray).

Fig. 136.—Diagram of the nerves passing through the cavernous sinus (drawn by J. T. Gray).

through the sphenoidal fissure from above downwards would strictly be as follows :—

Parts passing through the sphenoidal fissure (Fig. 136):—

Fourth nerve.
Frontal and lachrymal nerves [5th].
Upper division of 3rd nerve.
Nasal branch of 5th nerve.
Lower division of 3rd nerve.
Sixth nerve.

Ophthalmic vein (most internal).

The **Internal Carotid Artery** (Fig. 134, 4) should be traced to the upper opening of the carotid canal in the petrous bone. It makes a sigmoid turn by the side of the sella turcica, and then winds close behind the anterior clinoid process, where it was cut off in removing the brain. In a well-injected body, a few small branches (*arteriæ receptaculi*) may be seen in the cavernous sinus ; and the ophthalmic artery will be seen entering the optic foramen with the optic nerve.

Sympathetic Nerve.—A few filaments may be traced upon the carotid artery, forming the *carotid plexus*, which gives a branch of communication to the sixth nerve ; and the *cavernous plexus* may be found close to the anterior clinoid process.

The **Circular** or **Coronary Sinus** (Fig. 134, 6) surrounds the pituitary body in the sella turcica, and forms a communication between the two cavernous sinuses.

The **Transverse** or **Basilar Sinus** (Fig. 134, 10) crosses the basilar process at a variable point, and communicates between the two inferior petrosal sinuses.

By displacing the Gasserian ganglion, a little groove in the upper surface of the petrous bone will be seen, in which are a small nerve and artery taking a course to the hiatus Fallopii, through which they disappear. They are the *superficial petrosal* nerve and artery, the nerve a branch of the Vidian going to join the 7th, and the artery a branch of the middle meningeal.

Meningeal Arteries.—The *Anterior Meningeal Arteries* from the ethmoidal arteries, if injected, will be seen to enter the cranium through the ethmoidal foramina.

The *Middle Meningeal Artery* can be seen entering at the foramen spinosum. It ramifies on the outside of the dura mater, grooving deeply the anterior inferior angle of the parietal bone.

The *Small Meningeal* is a branch of the middle meningeal artery, which enters the skull through the foramen ovale.

The *Meningeal Branch of the Ascending Pharyngeal* artery may be found perforating the foramen lacerum medium.

The *Posterior Meningeal Arteries* are small branches of the occipital and vertebral, which enter by the foramen jugulare and foramen magnum respectively.

POSTERIOR TRIANGLE OF THE NECK.

[In order to dissect the posterior triangle, the blocks beneath the head should be removed, and it should be drawn downwards and to the opposite side and secured with hooks. The shoulders are to be raised to a convenient height by blocks beneath the scapulæ, and the arm drawn down, and if possible secured in that position. The side of the neck is thus put fully on the stretch, and probably the external jugular vein distended with blood will be seen beneath the skin.

One incision is to be carried from the back of the pinna to the clavicle near its inner end, and another from this point along the whole length of that bone; a third incision is to be made (if necessary) transversely from the pinna to the occiput, and the flap of skin is then to be reflected backwards. The dissection should be begun at the inferior angle, the dissector of the *right* side standing by the shoulder, and the dissector of the *left* side at the head, of the subject.]

Beneath the skin is the *superficial fascia*, and between the layers of it at the lower part of the neck is the platysma muscle, the fibres of which may be cleaned at
removing the skin. Care must be taken as
pper border of the platysma is reached to

keep close to the skin, or the superficial nerves may be injured.

The **Platysma Myoides** (Fig. 137, 6) is only seen now in its lower part, the remainder of it will be found in the dissection of the anterior triangle and face. It is a subcutaneous muscle *arising* from the fascia over the pectoral and deltoid muscles, and very slightly from the front of the clavicle; and it passes obliquely across the neck to be *inserted* into the lower jaw and the side of the mouth, reaching as high as the zygoma.

Through the platysma will generally be seen the **External Jugular Vein** (Fig. 137, 13) which, it is to be noticed, runs parallel or nearly so to the fibres of the muscle; if therefore, in bleeding from the vein, an incision were made in the direction of the vein, there would be no opening by which the blood might escape, and hence the rule in practice, viz., to cut parallel to the sterno-mastoid, *i.e.*, across the fibres of the platysma, so that they may retract.

The platysma is to be detached from the clavicle and turned forward, and the branches of the **Superficial Cervical Plexus** are to be found. These are in two sets, —the ascending, comprising the Superficial Cervical, Great Auricular, and Small Occipital nerves, and the descending set comprising the Sternal, Clavicular, and Acromial branches.

a. The **Superficial Cervical Nerve** (Fig. 137, 17) [from the second and third nerves] will be found turning round the posterior border of the sterno-mastoid, and crossing it transversely beneath the platysma to ramify over the anterior triangle, where it communicates with the inframaxillary branches of the facial nerve.

b. The **Great Auricular Nerve** (Fig. 137, 11) [from the second and third nerves] is the largest branch of the set, and turning round the posterior border of the sterno-mastoid, passes obliquely to the pinna, which it supplies, communicating also with the 7th nerve.

A *mastoid branch*, often of large size, arises from this nerve and runs to the mastoid process, and may be mistaken for the following:

Fig. 137.

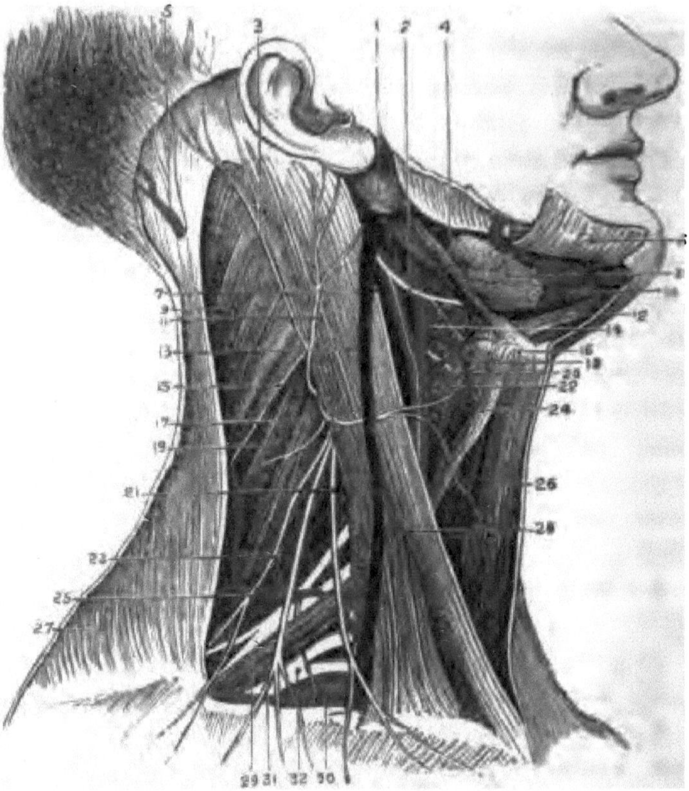

Fig. 137.—Superficial dissection of the triangles of the neck (drawn by J. T. Gray).

(In order to see the relations of the posterior triangle properly, the figure should be placed in the recumbent position, since it is impossible to see the subclavian as represented when the body is upright.

1. Parotid gland.
2. Hypoglossal nerve (9th).
3. Small occipital nerve.
4. Posterior belly of digastricus, and stylo-hyoideus.
5. Occipital artery and great occipital nerve.
6. Platysma (turned up).
7. Internal jugular vein.
8. Submaxillary gland.
9. Splenius capitis.
10. Mylo-hyoideus.
11. Great Auricular nerve.
12. Anterior belly of digastricus.
13. External jugular vein.
14. External carotid artery.
15. Spinal accessory nerve.
16. Hyoid bone.
17. Superficial cervical nerve.
18. Superior laryngeal nerve.
19. Levator anguli scapulæ.
20. Inferior constrictor of the pharynx.
21. Descending branches of the superficial cervical plexus.
22. Common carotid artery with descendens noni nerve.
23. Scalenus posticus and medius.

c. The **Small Occipital Nerve** (Fig. 137, 3) (from the second nerve) is always to be found at the posterior margin of the sterno-mastoid muscle, and runs directly upwards to the scalp, where it has been already seen (p. 322).

The **Descending Branches** (Fig. 137, 11) sternal, clavicular, and acromial, take the directions indicated by their names, and passing over the clavicle are distributed to the skin upon the pectoral muscle and shoulders, where they are seen by the dissector of the arm.

[The anterior and posterior boundaries of the triangle (the sterno-mastoid and trapezius) are now to be defined, but only the edges of the muscles need be cleaned. The descending nerves being detached and turned up, the branches of the deep cervical plexus and the spinal-accessory nerve passing obliquely between the borders of the sterno-mastoid and the trapezius are to be dissected out, and also the omo-hyoid muscle, which will be found at a variable distance above the clavicle; the external jugular vein is to be followed as far as convenient.]

The **Deep Cervical Fascia**, which is very dense, is now seen. It commences at the spinous processes of the vertebræ and encloses the trapezius muscle, then crosses the posterior triangle to split and enclose the sterno-mastoid, and so on to the median line of the neck. At the lower part of the triangle it gives a special covering to the omo-hyoid muscle, and is attached to the clavicle and first rib.

[To see the floor of the triangle the fascia is to be removed, and will be most readily detached by beginning above at the junction of the sterno-mastoid and trapezius, and carrying the knife obliquely so as to be parallel to the fibres of the muscles beneath. It should be removed

---

24. Anterior belly of omo-hyoid.
25. Transversalis colli artery.
26. Sterno-hyoideus.
27. Trapezius.
28. Sterno-mastoid.
29. Posterior belly of omo-hyoid.
30. Subclavian artery.
31. Transversalis humeri artery.
32. Brachial plexus partly covered by omo-hyoid.

only as low as the omo-hyoid at present, and all branches of nerve should be carefully preserved.]

The **Posterior Triangle** (Fig. 137) is the space bounded in *front* by the sterno-mastoid, *behind* by the trapezius; the *base* of the triangle being formed by the middle third of the clavicle, and its *apex* by the meeting of the anterior and posterior boundaries at the occiput. It is covered in by skin, superficial and deep fasciæ, and at the lower part by the platysma. Its *floor* is formed from above downwards by the splenius capitis, levator anguli scapulæ, scalenus posticus and scalenus medius, and the upper digitation of the serratus magnus, which cannot be seen at present.

The splenius colli does not form part of the floor because it is overlain by the levator anguli scapulæ. In some subjects, where the apex of the triangle is wider than usual, a portion of the complexus may be seen, and will be recognized by the vertical direction of its fibres. Not unfrequently if the attachment of the sterno-mastoid to the clavicle is not well developed, both the anterior scalenus and the phrenic nerve lying on it are seen in the posterior triangle.

The triangle is subdivided into two portions by the posterior belly of the omo-hyoid muscle, the height of which above the clavicle may vary, when the contents of the subdivision will be modified accordingly.

The *upper* or *occipital portion* contains the branches of the superficial cervical plexus derived from the first four cervical nerves; the spinal-accessory nerve (which may be traced into the trapezius, where it communicates with the deep cervical plexus); muscular branches of the deep cervical plexus; the transversalis colli vein; and the *glandulæ concatenatæ* or lymphatic glands placed along the posterior border of the sterno-mastoid.

The *lower* or *clavicular portion* contains the brachial plexus; the third part of the subclavian artery; the transversalis colli artery; the transversalis humeri vessels; and the external jugular vein, all of which must now be dissected.

The tendon of the **Omo-hyoid Muscle** (Fig. 137, 29) s bound down to the back of the clavicle and to the first rib by a process of the deep cervical fascia, so that either belly of the muscle may act on this fixed point. Upon the muscle may be traced a branch of the descendens noni nerve which supplies it. The origin and insertion of the muscle are found in other dissections.

The Transversalis Colli Artery (Fig. 137, 25) is generally found at the level of the tendon of the omo-hyoid, and the vein a little above it, so that their relation to the subdivisions of the triangle is different.

The artery, a branch of the thyroid axis [subclavian], passes transversely under the sterno-mastoid, crosses the posterior triangle, and divides under the

Fig. 138.

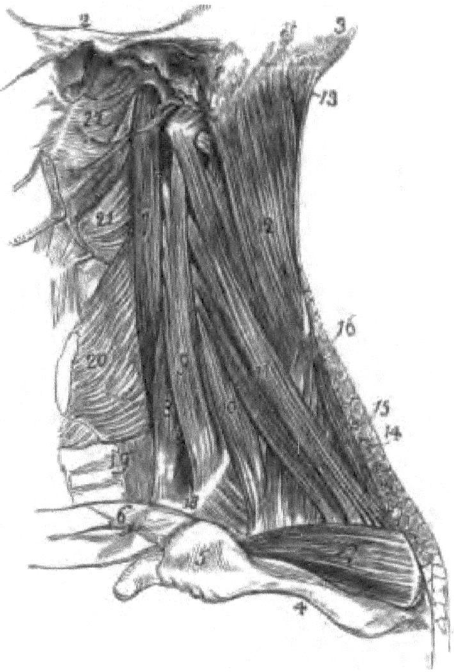

Fig. 138.—Lateral view of the muscles of the prævertebral region and side of the neck (from Wilson).

1. The mastoid process of the temporal bone.
2. The zygoma.
3. The occipital bone.
4. The spine of the scapula.
5. The acromion process.
6. The clavicle.
7. Longus colli muscle.
8. Scalenus anticus.
9. Scalenus medius.
10. Scalenus posticus.
11. Levator anguli scapulæ.
12. Splenius.
13. Complexus.
14. Cut edge of the trapezius.
15. Rhomboideus minor.
16. Serratus posticus superior.
17. Supra-spinatus.
18. The first rib.
19. Œsophagus and trachea.
20. Inferior constrictor of the pharynx.
21. Middle constrictor.
22. Superior constrictor.

border of the trapezius into the *superficial cervical* and the *posterior scapular* arteries. The vein lies above the level of the artery and usually joins the external jugular.

The **Transversalis Humeri Artery** and **Vein** (Fig. 137, 31) lie close behind the clavicle. The artery is a branch of the thyroid axis and becomes the suprascapular.* The vein opens into the external jugular just before it joins the subclavian vein.

[The great cords of the brachial plexus can only be roughly dissected at present, but the third part of the subclavian artery should be thoroughly observed before the clavicle is cut.]

The **Subclavian Artery** (Fig. 137, 30) (3rd portion) extends from the outer border of the anterior scalenus obliquely downwards and outwards beneath the clavicle to the lower border of the first rib, where it becomes the axillary artery. It has *in front* the skin, platysma, superficial nerves and deep fascia; the clavicle, subclavius muscle, and transversalis humeri vessels; and is crossed by the external jugular vein. It lies *against* the lowest cord of the brachial plexus, the scalenus medius, and the first rib. *Above* it are the remaining cords of the brachial plexus; and *below* (at some distance) the subclavian vein. Ordinarily this portion gives off no branch, but should there be one it will probably be the posterior scapular artery.

Surgery.—The position, relations, and direction of the third portion of the subclavian should be especially noticed, because it is that on which a ligature is most frequently applied for aneurism lower down. The student should particularly accustom his finger to feel for the scalene tubercle on the first rib, which is taken as the guide to the artery; and should notice how materially the position of the vessel is altered by raising or depressing the shoulder, the effect of an aneurism in the axilla being to raise the clavicle considerably, and thus complicate the operation.

* The whole length of this artery is sometimes called the Suprascapular.

The operation of tying the subclavian in its third part is thus performed. The head being thrown back and the shoulder depressed as much as possible, the skin of the lower part of the triangle is to be drawn down upon the clavicle, and a lunated incision carried from the edge of the trapezius to the sterno-mastoid, cutting at once on to the bone. The skin being then allowed to resume its natural position, the incision will be immediately above the clavicle, and may be deepened at the anterior part. The external jugular vein is to be carefully guarded, and held aside if necessary, and a cautious dissection made until the finger can feel the scalenus anticus and the tubercle on the first rib, to which it is attached. Immediately behind this the artery will be found, and even in the uninjected state will be readily recognized by its rolling on the bone beneath the finger. The aneurism needle should be passed from *above* so as to avoid the brachial nerves, which are more liable to be included in the ligature than the vein, since this is quite below the artery.

Fig. 139.

[Supposing the dissection of the axilla to be sufficiently advanced, the clavicle should now be divided at the outer border of the sterno-mastoid and again at the edge of the trapezius. The knife being then passed carefully beneath

Fig. 139.—Incisions for tying (*e*) the common carotid and (*b*) the subclavian artery (from Fergusson's 'Surgery').

and close to the bone to detach the subclavius, the piece may be removed. The parts will still be retained in position by the subclavius muscle, and the opportunity should be taken to trace a small branch of nerve to it from the brachial plexus. The transversalis humeri artery and vein will now also be better seen, and can be thoroughly cleaned. The subclavius muscle is to be divided close to the trapezius, when the scapula will fall back, and the brachial plexus will be fully exposed and should be carefully cleaned. The upper digitation of the serratus magnus which was mentioned as forming part of the floor of the triangle can now be seen.]

Fig. 140.

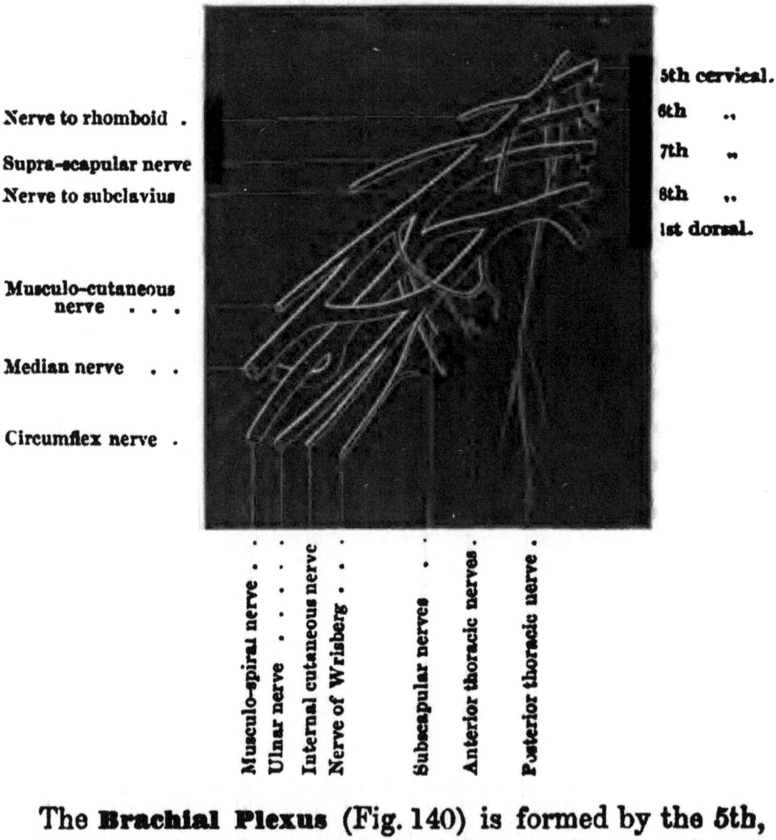

The **Brachial Plexus** (Fig. 140) is formed by the 5th, 6th, 7th, and 8th cervical nerves and the 1st dorsal nerve, a small branch of communication being given by

---

Fig. 140.—Diagram of the Brachial Plexus (drawn by J. T. Gray).

the lowest cord of the cervical plexus (4th). The nerves appear at the outer border of the anterior scalenus as four cords, because the 8th cervical and 1st dorsal unite close to the vertebral foramina; they lie against the scalenus medius and posticus, and all, with the exception of the lowest, above the level of the subclavian artery. The 5th and 6th next unite, and are joined by the 7th at the outer border of the scalenus posticus, and thus two cords are formed. At the level of the clavicle each of these cords gives off a large branch, and these unite to form a third cord; the three nerves then come into relation with the axillary artery, being at first external, but afterwards internal, external, and posterior to that vessel.

It sometimes happens that the formation of these cords does not take place higher than the level of the pectoralis minor, but the arrangement of these nerves is very inconstant. The 7th nerve very frequently bifurcates, one branch going to the upper and one to the lower cord, or it may itself form the posterior cord, being joined by branches from the other trunks.

The *branches above the clavicle* (Fig. 4, p. 18)—

a. The *nerve to the subclavius* is a small branch lying on the front of the plexus and derived from the 5th and 6th nerves.

b. The *nerve to the rhomboid and levator anguli scapulæ* is from the 5th nerve, and pierces the scalenus medius to reach the under surface of those muscles.

c. The *Supra-scapular nerve* is the largest branch above the clavicle and appears at the outer side of the plexus, being given off from the upper cord of the plexus (5th and 6th). It disappears beneath the trapezius to reach the supra-scapular notch.

d. The *nerve to the serratus magnus* (posterior thoracic, external respiratory of Bell) lies behind the plexus against the serratus magnus muscle, and is derived from the 5th and 6th (and often from the 7th) nerves before they unite.

e. A *branch to the phrenic nerve* of very variable size is given by the 5th nerve at the upper part, or sometimes the branch appears to pass from the phrenic to the 5th.

f. Small *branches to the longus colli and scaleni* muscles arise from the nerves close to the foramina.

[If the time for turning the subject has not arrived when both the posterior triangles of the neck are dissected, the orbits should be at once proceeded with; but if not now dissected they had better be postponed until after the dissection of the anterior triangle.]

The Orbit.

[The head being raised to a convenient height, the saw is to be applied to the edge of the skull close to the inner and outer angles of the orbit (the proper points being readily ascertained with the finger), and the cuts carried into the orbit through the frontal bone. With the chisel these incisions are to be prolonged backwards till they meet at the sphenoidal fissure, when the triangle of bone can be readily tilted forward by a blow with the hammer. Any small remnant of bone

Fig. 141.

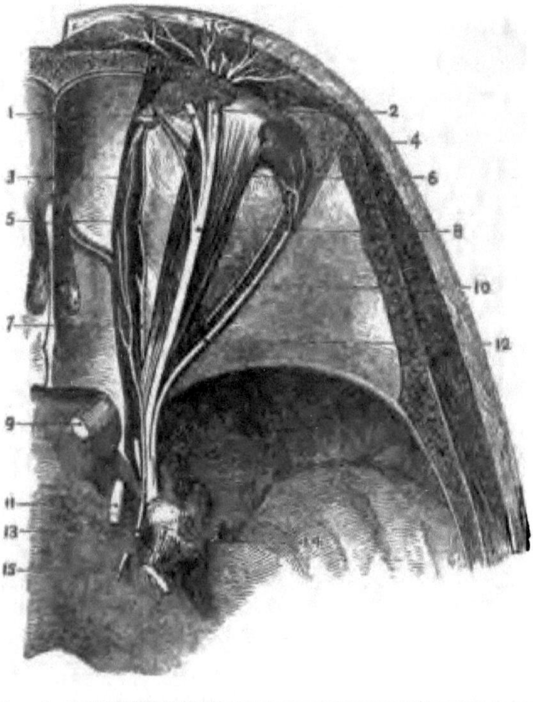

Fig. 141.—Superficial dissection of the nerves of the orbit (from Hirschfeld and Leveillé).

1. Supra-trochlear nerve.
2. Supra-orbital nerve.
3. Obliquus superior.
4. Lachrymal gland.
5. Infra-trochlear nerve.
6. Levator palpebræ superioris.
7. Nasal nerve (5th).
8. Frontal nerve (5th).
9. Second nerve.
10. Rectus superior.
11. Third nerve.
12. Lachrymal nerve.
13. Fourth nerve.
14. Gasserian ganglion of 5th.
15. Sixth nerve.

can be removed so as to expose fully the cavity, but the margin of the optic foramen should be left untouched.]

The first thing seen on opening the orbit is the **Periosteum**, which is detached from the triangle of bone and will be seen to be continuous with the dura mater of the skull through the sphenoidal fissure and the optic foramen. At the margin of the orbit the periosteum splits into two layers; one which is continuous around the margin with the external periosteum of the skull, and the other which forms the palpebral ligament of each eyelid.

The periosteum being divided in the centre and carefully turned aside, and some soft fat removed, three nerves and two arteries are brought into view—the *frontal nerve* [5th] with the *supra-orbital artery* in the centre, the *lachrymal nerve* and *artery* on the outer side, and the little *fourth nerve* on the inner side; the lachrymal gland is also seen.

The **Fourth Nerve** (Fig. 141, 13) (*trochlearis* or *patheticus*) has already been traced through the cavernous sinus to the sphenoidal fissure, where it occupies the highest position, and is now seen passing to the inner side of the orbit to enter the superior oblique muscle on its *orbital* surface; this being an exception to the other muscles of the orbit, which are supplied on their *ocular* surface.

The **Frontal Nerve** (Fig. 141, 8) [5th] the largest nerve entering through the sphenoidal fissure, lies in the centre of the orbit and divides at its anterior part into the *supra-orbital* and *supra-trochlear* branches; these turn round the margin of the orbit to the forehead, the supra-orbital through the supra-orbital notch, and the supra-trochlear close to the inner angle of the orbit. Frequently the supra-trochlear nerve, the more internal of the two, is double.

The **Lachrymal Nerve** (Fig. 141, 12) (5th) lies on the outer side of the orbit and passes to the lachrymal gland, which it supplies. It frequently sends a branch to the forehead close to the outer angle of the orbit.

The **Supra-orbital Artery** (Fig. 142, 8) is a branch of the ophthalmic artery and accompanies the supra-orbital nerve to the forehead.

Fig. 142.

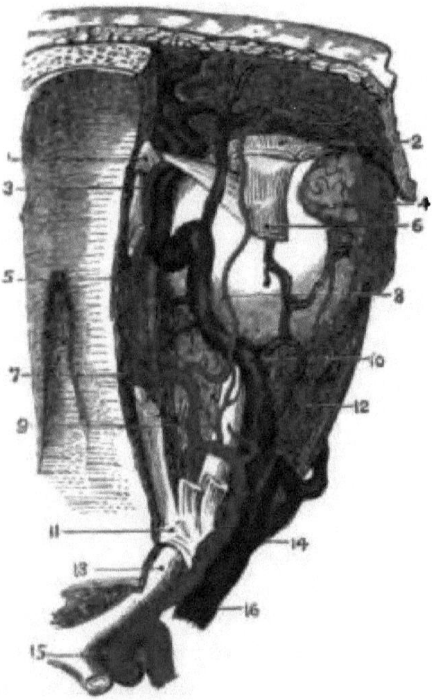

The **Lachrymal Artery** (Fig. 142, 12) is a smaller branch also from the ophthalmic, and supplies the lachrymal gland.

The **Lachrymal Gland** (Fig. 142, 4) is an almond-shaped body of a reddish-brown colour, situated in the anterior and outer part of the orbit. It is convex on its superior aspect, to fit against the orbital plate of the frontal bone; concave beneath, where it is placed over the eyeball. It secretes the tears, which its ducts, seven or eight in number, carry to the surface of the conjunctiva beneath the upper eyelid.

[The frontal nerve is to be divided, and a hook inserted into the upper eyelid so as to put the levator palpebræ in the centre of the orbit on the stretch; this muscle and the superior oblique are then to be cleaned, care being

Fig. 142.—Arteries and veins of orbit (from Hirschfeld and Leveillé).

1. Pulley of superior oblique tendon.
2. Levator palpebræ (cut).
3. Trunk of ophthalmic artery from which the frontal, nasal, and palpebral branches are derived.
4. Lachrymal gland.
5. Anterior ethmoidal artery.
6. Rectus superior (cut).
7. Posterior ethmoidal artery.
8. Supra-orbital artery.
9. Ciliary arteries.
10. Ophthalmic vein.
11. Origins of obliquus superior, levator palpebræ, and superior rectus.
12. Lachrymal artery.
13. Optic nerve.
14. Ophthalmic artery.
15. Carotid artery.
16. Cavernous sinus.

taken not to destroy the little pulley through which the tendon of the latter works.]

The **Obliquus Superior** (Fig. 141, 3) is the most superficial muscle of the orbit at its *origin*, which is from the upper margin of the optic foramen, above and a little to the inner side of the levator palpebræ. It lies superficially along the inner side of the orbit, and ends in a round tendon which, passing through the *trochlea* or pulley-like ring of fibrous tissue attached to the frontal bone, turns down abruptly to the eye-ball. The tendon becomes flattened near its insertion, and passes beneath the superior rectus to be *inserted* on the outer side of the eyeball between it and the external rectus. With a little dissection a delicate synovial membrane can be seen lubricating the tendon where it passes through the *trochlea*. The superior oblique is *supplied* by the 4th nerve on its orbital aspect.

The **Levator Palpebræ Superioris** (Fig. 141, 6) *arises* from the upper margin of the optic foramen, below the superior oblique muscle and partly overlapped by it; it is narrow at its origin, but expands anteriorly to be *inserted* into the anterior surface of the tarsal cartilage of the upper eyelid. It is *supplied* by the 3rd nerve.

[The hook is to be removed from the eyelid and fixed into the eyeball, which is to be drawn gently forwards; the levator palpebræ is to be cut and the little branch of the 3rd nerve traced to it, and the superior rectus which is then seen should be cleaned.]

The **Rectus Superior** (Fig. 142, 6) *arises* from the margin of the optic foramen, below and a little overlapped by the levator palpebræ; and is *inserted* into the sclerotic coat of the eye-ball at its upper and anterior part. It is *supplied* by the upper division of the 3rd nerve, which can be seen entering its under surface when the muscle is divided.

[The rectus having been divided, a quantity of

be brought into view, through which the optic nerve passes to the eyeball. The nasal branch of the 5th is at once to be looked for crossing the optic nerve from

Fig. 143.

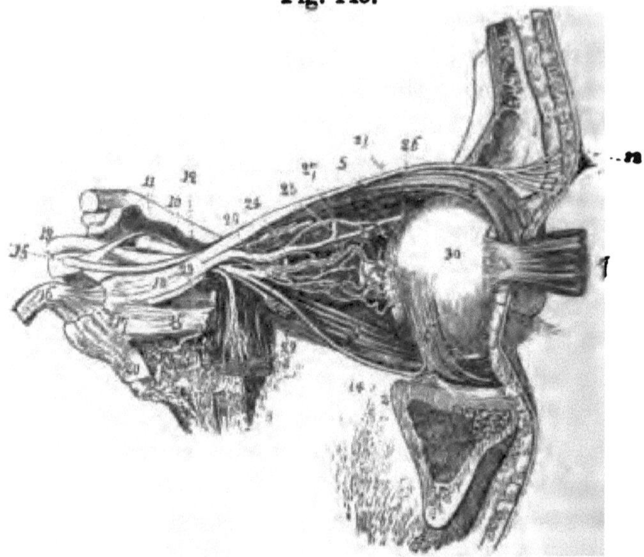

Fig. 143.—Nerves of the orbit, seen from the outer side (from Wilson, after Arnold).

1. Section of frontal bone.
2. Superior maxillary bone.
3. Part of sphenoid bone.
4. Levator palpebræ and superior rectus muscle.
5. Superior oblique muscle.
6. Inferior oblique.
7. External rectus drawn forwards.
8. Orbital origin of the external rectus turned downwards with the sixth nerve.
9. Inferior rectus.
10. Optic nerve.
11. Internal carotid artery emerging from the cavernous sinus.
12. Ophthalmic artery.
13. Third nerve.
14. Inferior division of the third nerve.
15. Fourth nerve.
16. Trunk of the fifth nerve.
17. Gasserian ganglion.
18. Ophthalmic division of fifth.
19. Superior maxillary division.
20. Inferior maxillary division.
21. Frontal nerve.
22. Supra-orbital and supratrochlear nerves.
23. Lachrymal nerve.
24. Nasal nerve.
25. Nasal nerve passing over the internal rectus muscle to the anterior ethmoidal foramen.
26. Infra-trochlear nerve.
27. A long ciliary branch of the nasal nerve.
28. Long root of the lenticular ganglion from the nasal nerve.
29. Lenticular ganglion, giving off from its fore part the short ciliary nerves.
30. Globe of the eye.

without inwards, and the lenticular ganglion, a minute pink body, is to be found on the outer side of the optic nerve. All small branches of nerve running from the ganglion along the optic nerve are to be preserved, and all branches of the ophthalmic artery.]

The **Nasal Nerve** (Fig. 141, 7, and 143, 24) (5th) enters the orbit between the heads of the external rectus, and then crosses the optic nerve from without inwards to pass through the anterior ethmoidal foramen. It thus re-enters the cranium, and next passes into the nose by a slit in front of the cribriform plate of the ethmoid bone, to appear ultimately on the face, between the nasal bone and the lateral nasal cartilage. It gives off, 1, the *long root* to the lenticular ganglion on the outer side of the optic nerve; 2, the *long ciliary nerves*, which run along the optic nerve to the eyeball; 3, the *infra-trochlear nerve*, which, arising just before the nerve enters the ethmoidal hole, passes forwards to the eyelid close to the inner angle of the orbit.

The **Lenticular Ganglion** (Fig. 144) (*ciliary* or *ophthalmic*) is a minute pink body placed on the outer side of the optic nerve near the back of the orbit, which has (like all the cranial ganglia) three roots, sensory, motor, and sympathetic. The *sensory* or long root is derived from the nasal branch of the 5th, and enters the posterior superior angle of the ganglion. The *motor* or short root is derived from the inferior division of the 3rd nerve, which can be seen passing to the bottom of the

Fig. 144.

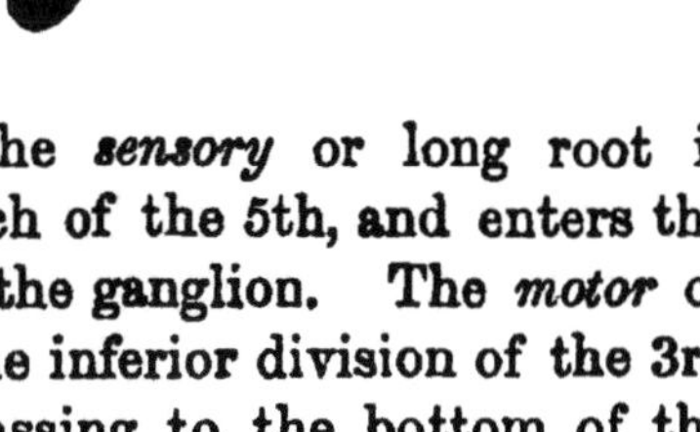

Fig. 144.—Diagram of the lenticular ganglion (from Hirschfeld and Leveillé).

1. Lenticular ganglion.
2. Long root from—
3. Nasal branch of 5th.
4. Short root from—
5. Inferior division of 3rd.
6. Sympathetic root.
7. Short ciliary nerves.
8. Long ciliary nerves.

orbit, and enters the ganglion at its posterior inferior angle. The *sympathetic* root is from the cavernous plexus which is derived from the plexus on the carotid artery, and enters the orbit through the sphenoidal fissure; it joins the ganglion between the other two roots, and can rarely be seen.

Fig. 145.

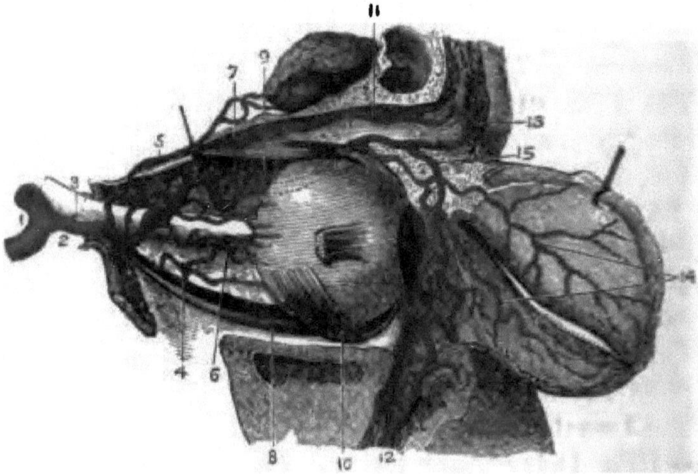

The *short ciliary branches* of the ganglion arise from its upper and lower angles in front, and are eight or ten in number. They run along the optic nerve to the *eyeball*, which they pierce to supply the iris.

The **Ophthalmic Artery** (Fig. 145, 2) arises from the internal carotid close to the anterior clinoid process, and enters the orbit through the optic foramen with the optic nerve, but to its outer side. Its branches are:—

a. The *Lachrymal artery* (5) accompanies the lachrymal

Fig. 145.—Arteries of the orbit from the outer side (from Hirschfeld and Leveillé).

1. Internal carotid.
2. Ophthalmic artery.
3. Arteria centralis retinæ.
4. Muscular branches.
5. Lachrymal artery.
6. Ciliary artery.
7. Posterior ethmoidal artery.
8. Rectus inferior.
9. Anterior ethmoidal artery.
10. Obliquus inferior.
11. Supra-orbital artery.
12. Facial artery.
13. Frontal artery.
14. Palpebral arteries.
15. Nasal artery.

nerve along the outer side of the orbit to the lachrymal gland, which it supplies.

b. The *Supra-orbital artery* (11) ascends to join the frontal nerve and afterwards the supra-orbital nerve, which it accompanies through the supra-orbital notch to the forehead.

c. *Muscular branches* (4) are given to all the muscles of the orbit, entering them on their ocular surfaces.

d. The *Ciliary branches* (6) are numerous small arteries which run by the side of the optic nerve to pierce the sclerotic, and have been divided into *long* and *short.* One of them pierces the optic nerve to run in it to the retina, and is called the *arteria centralis retinæ* (3).

e. The *Ethmoidal arteries*, anterior (9) and posterior (7), pass through the ethmoidal foramina in the inner wall of the orbit, the anterior accompanying the nasal nerve.

f. The *Papebral arteries* (14) leave the orbit at the inner side, to be distributed to the upper and lower eyelids.

g. The *Frontal artery* (13), one of the terminal branches, turns round the margin of the orbit at its inner angle to accompany the supra-trochlear nerve on the forehead.

h. The *Nasal artery* (15), the other terminal branch, leaves the orbit at the inner side above the tendo oculi, and anastomoses on the side of the nose with the angular branch of the facial artery (12).

The **Ophthalmic Vein** (Fig. 142, 10) has branches corresponding to those of the artery, which unite to form a single trunk, passing between the heads of the external rectus to the cavernous sinus.

The **Optic Nerve** (Fig. 142, 13) (2nd) enters the orbit by the optic foramen, and passes through its centre to the eyeball, to end in the retina. It pierces the back of the sclerotic about its own width to the inner side of the axis of the eyeball.

[The optic nerve is to be divided, and the globe turned forwards to bring into view the muscles beneath.]

The remaining **Recti Muscles** (Fig. 146) are seen

below, and to the inner and outer sides of, the optic nerve, and are named accordingly, *inferior*, *internus*, and *externus*. They *arise* by a common origin (ligament of Zinn) around the optic foramen, but the externus has an attachment to the margin of the sphenoidal fissure, completing its lower head, as well as a second head (the upper) arising from the upper part of the optic foramen close to the rectus superior. Between these two heads pass the 3rd nerve, the nasal branch of 5th nerve, the 6th nerve, and the ophthalmic vein.

The recti muscles are to be traced forward to their *insertion* into the sclerotic coat, immediately behind the cornea.

The **Sixth Nerve** (Fig. 143, 8) (abducens) is seen on the inner surface of the rectus externus, which it supplies.

. The **Third Nerve** (Fig. 143) (motor oculi) enters the orbit in two parts, one above and the other below the nasal nerve (*v.* p. 332).

The upper division has been traced to the levator palpebræ and superior rectus muscles ; the lower division (14) is now seen to give small branches to the internal and inferior recti, and the short root to the lenticular ganglion ; lastly a large long branch which runs on the inferior rectus to the inferior oblique muscle.

[To expose the obliquus inferior, it will be necessary to draw the eyeball to the upper and outer part of the orbit, and to remove the conjunctiva at the inner angle.]

The **Obliquus Inferior** (Fig. 146, 8) muscle lies transversely in the orbit beneath the rectus inferior. It *arises* from the superior maxilla behind the lachrymal groove, and passing below the inferior rectus, but between the external rectus and globe, is *inserted* into the sclerotic above the level of the rectus externus and close to the insertion of the obliquus superior.

The actions of the muscles of the orbit are not very readily appreciated, owing to the fact that the movements of the eyeball depend upon the combined influence of several of them.

The action of the Levator palpebræ is implied by its me; it raises the eyelid, and therefore when paralysed om some affection of the third nerve, the characteristic ·opping of the lid, ptosis, results.

The Recti muscles, when acting together, would draw ie eyeball towards the back of the orbit, but each when cting separately would draw the front of the eye in its wn direction; the uperior and inerior recti having n addition a tenlency to draw the eye inwards. Internal and external strabismus is produced by the direct action of the internal and external recti respectively.

Fig. 146.

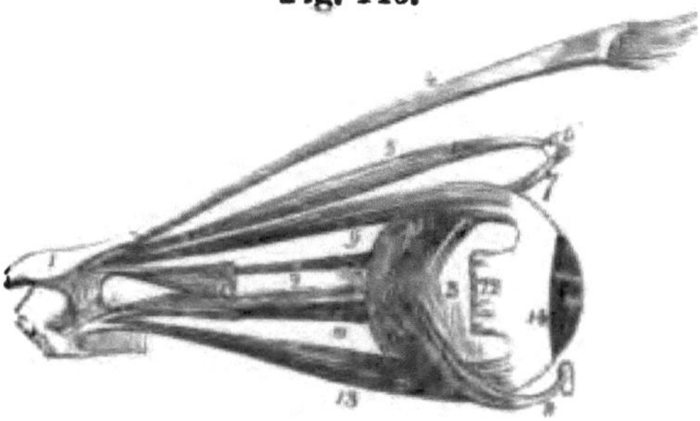

The Oblique muscles when acting alone are rotators of the eyeball on its own axis, the superior oblique rolling the eye downwards and outwards, the inferior oblique rolling it upwards and outwards; but these muscles always act in combination with the recti, and thus give the oblique movements to the eyeball. The obliqui are not used unless the spectator, or the object seen, is moving.

If all the contents of the orbit are removed, the orbital branch of the superior maxillary nerve may be seen passing through the spheno-maxillary fissure, and dividing

Fig. 146.—Muscles of the eyeball, the view is taken from the outer side of the right orbit (from Wilson).

1. Portion of the sphenoid bone.
2. Optic nerve.
3. Globe of the eye.
4. Levator palpebræ muscle.
5. Superior oblique.
6. Its fibrous pulley.
7. Its reflected tendon.
8. Inferior oblique muscle arising from the inner wall of the orbit.
9. Superior rectus.
10. Internal rectus partly concealed by the optic nerve.
11. Part of external rectus, showing its two heads of origin.
12. Insertion of external rectus.
13. Inferior rectus.
14. Sclerotic.

into two branches—the *subcutaneus malæ* which pierces the malar bone, and the *temporal branch* which pierces the outer wall of the orbit to reach the temporal fossa.

The Side of the Neck.

[The head is to be drawn as far back as possible by means of a hook placed in the chin, and the side to be dissected should be fully exposed by means of blocks placed beneath. Before beginning the dissection the student should pass his finger along the median line of the neck, and recognise the following points. The lower jaw, and from an inch and a half to two inches below it the slight projection of the hyoid bone; next a hollow, corresponding to the thyro-hyoid membrane, and below it the projecting angle of the thyroid cartilage (pomum Adami) which is very small in women. About an inch below the pomum Adami is a dip corresponding to the crico-thyroid space, and this is especially to be noticed as it is the space in which laryngotomy is performed. Below this will be felt the hard ring of the cricoid cartilage, and in a thin subject the rings of the trachea may be recognised lower down, sometimes also the isthmus of the thyroid body may be felt crossing the trachea at a variable point.

An incision is to be made from the chin to the sternum, and the flap of skin is to be dissected up over the face. The platysma may be cleaned in the upper part of the space at once, and it will facilitate the operation if the part detached from the clavicle is held down with hooks.]

The **Platysma Myoides** (Fig. 137, 6) is now seen to reach to the side of the lower jaw, where it is partially inserted. The muscle decussates with its fellow of the opposite side in the median line for a short distance close to the jaw, but at the lower part of the neck a large triangular interval exists between them. A few small branches of the superficial cervical nerve will be found piercing the muscle, and an anterior jugular vein may sometimes be seen through its fibres.

[The platysma is to be carefully reflected like the skin, the superficial nerves are to be dissected out and the sterno-mastoid cleaned.]

The **Superficial Cervical Nerve** (Fig. 137, 17), which is of small size, is seen to divide into two or three branches which supply the skin over the anterior triangle, the upper one communicating with branches of the facial nerve below the jaw.

The **Sterno-Cleido-Mastoid Muscle** (Fig. 137, 28) *arises* by an oval tendon from the top of the sternum, and by a broad tendinous origin from the inner third of the posterior border of the clavicle. This latter is of very variable extent, and between the two heads of origin is a cellular interval, which may extend for some distance up the neck. The muscle is *inserted* across the mastoid process of the temporal bone and into the outer half of the superior curved line of the occipital bone. The *action* of each sterno-mastoid is to twist the head and flex it, so as to throw the chin over the opposite shoulder; if both muscles act together they draw the head and upper part of the spine forwards, as in rising from a pillow. The sterno-mastoid is *supplied* by the spinal accessory nerve.

The **Deep Cervical Fascia** is now seen to cover in the parts included in the triangle, and to extend to the median line; and the anterior border of the sterno-mastoid should be dissected up a little and turned back to see the continuation of the fascia beneath it. The fascia gives sheaths to all the muscles of the front of the neck, and is continued into the thorax, where it is attached to the pericardium. It forms the sheath of the carotid artery, which is now to be seen beneath the sterno-mastoid.

[Opportunity should be taken, before the tissues are in any way disturbed, to notice the parts involved in the operation of tying the common carotid artery. The vessel may be felt and indistinctly seen enclosed in a sheath of fascia, and the point where the ligature would be applied is at the angle formed by the sterno-mastoid and omo-hyoid muscles, the latter of which can now be seen through the fascia.

It will be advisable to examine the ligaments of the inner end of the clavicle before detaching it, in order to reflect the sterno-mastoid muscle. The sternal origin of the sterno-mastoid must be cut, and any remains of the pectoralis major be removed, in order that the ligaments between the clavicle, sternum, and first rib, and also between the two clavicles, may be cleaned.]

Sterno-Clavicular Articulation (Fig. 147).

The *Inter-clavicular ligament* (2) is a strong band passing across the inter-clavicular notch of the sternum, and attached to that bone as well as to the inner extremities of the two clavicles.

The *Sterno-clavicular ligament* (1) may be considered to form a *capsule* to the joint, or may be divided into *anterior* and *posterior* ligaments. The fibres pass obliquely from the inner extremity of the clavicle to the margin of the articular surface of the sternum.

Fig. 147.

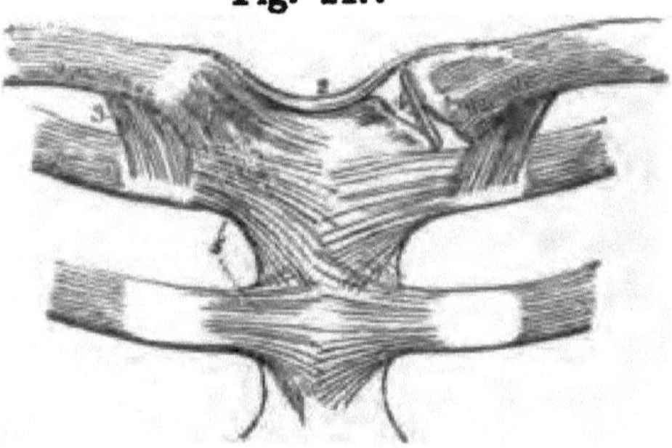

The *Costo-clavicular* or *Rhomboid ligament* (3) is a strong band of fibres of a rhomboidal shape, attached to the upper surface of the cartilage of the first rib, and to the rough impression on the under surface of the inner end of the clavicle, close to the articulation.

[It is supposed that the clavicle has been cut close to the attachment of the sterno-mastoid in the dissection of the posterior triangle of the neck, but if this has not been the case it should now be divided. The inner end of the clavicle being then drawn up, the costo-clavicular

Fig. 147.—Ligaments of the sterno-clavicular and costo-sternal articulations (from Wilson).

1. Anterior sterno-clavicular ligament.
2. Inter-clavicular ligament.
3. Costo-clavicular or rhomboid ligament; seen on both sides.
4. Interarticular fibro-cartilage, brought into view by the removal of the anterior and posterior ligaments.
5. Anterior costo-sternal ligaments of the first and second ribs.

ligament is to be divided and the knife passed into the sterno-clavicular articulation from below and close to the clavicle. By this one of the synovial membranes will be opened, and the other can be exposed by cutting from above close to to the sternum, thus leaving the interarticular cartilage uninjured.]

The *Interarticular fibro-cartilage* (4) is circular and flat, and is thinner in the centre than at the circumference, being occasionally perforated. It is attached to the clavicle above, and to the sternum and cartilage of the first rib below, and by its circumference to the ligaments of the articulation. A synovial membrane is placed on each side of it. The sterno-clavicular is an arthroidal joint admitting of extensive movements of the clavicle in all directions.

[The interarticular fibro-cartilage is to be divided and the inner end of the clavicle dislocated, the fibres of the sterno-hyoid which are attached to it being separated. The sterno-mastoid (with the portion of the clavicle) is then to be turned back, being carefully separated from the fascia beneath. The spinal accessory nerve will be found to pierce it at the upper part, and some branches of the deep cervical plexus enter the under surface of the muscle.]

The small **Descendens Noni Nerve** (149, 14) is now to be dissected out. It is either upon or within the sheath of the carotid vessels, and is to be traced upwards to the 9th nerve (which crosses just below the digastric muscle), and downwards to the muscles in the front of the neck which it supplies, viz., sterno-hyoid, sterno-thyroid, and omo-hyoid. A branch, which may be double, will be found to come forward from the cervical plexus to join the nerve and form a loop beneath the sterno-mastoid. This is the *communicans noni* nerve (7) and comes from the 2nd and 3rd cervical nerves.

The **Anterior Jugular Vein** will be found of very variable size and near the median line of the neck. It may receive any of the veins of the upper part of the neck when of good size, and opens into the subclavian or external jugular vein. Its size depends upon that of the external jugular, varying inversely with it.

[The fascia is now to be removed from the superficial muscles of the space, but the carotid sheath should be left untouched for the present.]

The **Sterno-hyoid** (Fig. 148, 14) is the most superficial muscle; it is next the median line and only separated by a small cellular interval from its fellow of the opposite side. It *arises* from the back of the first piece of the sternum, from the first costal cartilage, and from the back of the inner extremity of the clavicle; and is *inserted* into the body of the hyoid bone.

The **Sterno-thyroid** (Fig. 148, 15) is deeper and broader than the preceding muscle, by which it is partly covered. It *arises* from the back of the sternum immediately below the sterno-hyoid, and from the first costal cartilage; and is *inserted* into the oblique line on the side of the thyroid cartilage. It very generally has a transverse tendinous intersection in its fibres.

The **Thyro-hyoid** (Fig. 148, 16) is a direct continuation of the last muscle. It *arises* from the oblique line of the thyroid cartilage, and is *inserted* into the body and part of the great cornu of the hyoid bone beneath the omo-hyoid. A small special branch of nerve from the 9th should be traced to this muscle.

The **Omo-hyoid** (Fig. 148, 17) can now be seen in its whole length, crossing the neck beneath the sterno-mastoid, and consisting of two bellies united by a small tendon, which has been seen to be held down to the clavicle and first rib by a process of the deep cervical fascia (*v.* p. 339).

It *arises* from the upper margin of the scapula close to the notch, and from the transverse ligament which converts the suprascapular notch into a hole. It is *inserted* into the body of the hyoid bone external to the sterno-hyoid, and superficial to the thyro-hyoid muscle.

These four muscles are direct or indirect depressors of the hyoid bone. They are all *supplied* by the descendens noni nerve, except the thyro-hyoid, which receives a special branch from the 9th nerve.

Fig. 148.

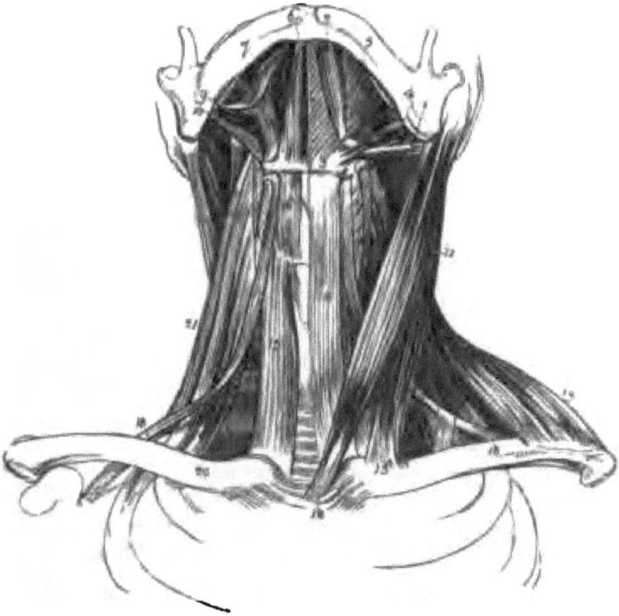

The **Digastric** (Fig. 148, 1) muscle consists of two

Fig. 148.—Muscles of the anterior aspect of the neck; on the left side of the figure the superficial muscles are seen, on the right the deep (from Wilson).

1. Posterior belly of digastricus.
2. Its anterior belly. Aponeurotic pulley, through which its tendon is seen passing, attached to the body of the os hyoides 3.
4. Stylo-hyoideus muscle.
5. Mylo-hyoideus.
6. Genio-hyoideus.
7. The tongue.
8. Hyo-glossus.
9. Stylo-glossus.
10. Stylo-pharyngeus.
11. Sterno-cleido-mastoideus.
12. Its sternal origin.
13. Its clavicular origin.
14. Sterno-hyoid.
15. Sterno-thyroid of the right side.
16. Thyro-hyoid.
17. Anterior belly of the omo-hyoid.
18, 18. Its posterior belly; on the left side, the tendon of the muscle is seen to be bound down by a portion of the deep cervical fascia.
19. Clavicular portion of the trapezius.
20. Scalenus anticus, of the right side.
21. Scalenus posticus; the scalenus medius is seen between the two.

portions, placed in front of and behind the submaxillary

Fig. 149.

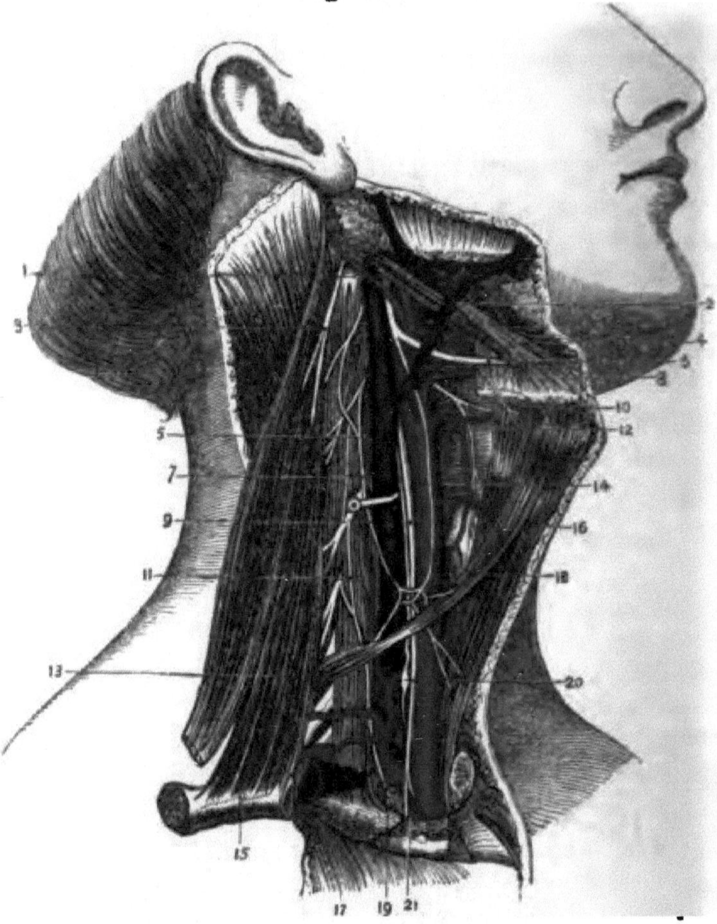

Fig. 149.—The side of the neck (drawn by J. T. Gray).

1. Occipital artery.
2. Facial vein.
3. Spinal accessory nerve.
4. Facial artery.
5. Internal jugular vein.
6. Hypoglossal nerve.
7. Communicans noni nerve.
8. Lingual artery.
9. Pneumo-gastric nerve.
10. Superior laryngeal nerve.
11. Phrenic nerve.
12. Superior thyroid artery.
13. Sterno - cleido - mastoideus (reflected).
14. Common carotid artery with descendens noni nerve.
15. Inner end of clavicle (reflected).
16. Sterno-hyoid.
17. Subclavian vein (cut).
18. Omo-hyoid.
19. Subclavian artery giving off the thyroid axis and the internal mammary artery.
20. Inferior cervical ganglion of sympathetic.
21. Apex of pleura.

gland above the hyoid bone, to which the intermediate tendon is attached. It *arises* from the groove on the inner side of the mastoid process of the temporal bone; and is *inserted* into the digastric fossa at the lower border of the inferior maxilla close to the median line. Its action is to open the mouth, or if the jaw is fixed, to raise the hyoid bone and larynx. Its posterior belly is *supplied* by a branch of the 7th nerve, and the anterior by the mylo-hyoid branch of the inferior dental nerve (5th).

The **Stylo-hyoid** (Fig. 148, 4) is the muscle in immediate connection with the posterior belly of the digastric, which passes through its fibres close to the hyoid bone. It *arises* from the outside of the styloid process of the temporal bone between the stylo-pharyngeus and stylo-glossus muscles; and is *inserted* into the upper surface of the body of the hyoid bone. It is *supplied* by a branch of the 7th nerve.

[The sheath of the carotid artery derived from the cervical fascia is now to be opened. The descendens noni nerve has already been traced upon it, and within will now be found the common carotid artery nearest the median line; external to that the pneumo-gastric (or vagus) nerve, and still more externally the internal jugular vein. Behind the sheath will be found the trunk of the sympathetic nerve lying parallel with the vessels, and crossing transversely behind them at the lower part will be seen the inferior thyroid artery. The sheath is to be carefully dissected away, and the branches of the artery and vein followed out and cleaned as far as the dissection will admit of. The large 9th nerve will be found crossing the external and internal carotids in a curved direction immediately below the digastric muscle, and the superior laryngeal branch of the pneumo-gastric crosses behind them a little lower down.]

The **Common Carotid Artery** (Fig. 149, 14) has the same relations on both sides of the neck from the sterno-clavicular articulation upwards, though its origin is different on the two sides. On the right side it commences at the sterno-clavicular articulation by the bifurcation of the innominata into common carotid and subclavian arteries, but on the left side it begins at the arch of the aorta. Its direction in the neck is upwards and a little outwards, and would be sufficiently indicated by a line from the end of the clavicle to the external auditory meatus; and it ordinarily divides at the level of the upper border of the thyroid cartilage into external and internal carotids, which lie at first side by side, the external being nearer the median line.

It has *in front* of it the sternal origin and anterior border of the sterno-mastoid, and the lower parts of the sterno-hyoid and thyroid muscles; it is crossed by the omo-hyoid muscle and superior thyroid vein, and quite superficially by the platysma; close upon the vessel throughout are the descendens noni nerve and the sheath, and crossing the sheath at the lower part is the middle thyroid vein, and at a variable point the sterno-mastoid artery. *Behind* the artery are the sheath, the sympathetic cord, and its middle cervical ganglion: the inferior thyroid artery, and the longus colli and rectus capitis anticus major muscles. To its *outer* side are the pneumogastric nerve and the internal jugular vein; and to the *inner* side the trachea, larynx, and thyroid body: the pharynx, œsophagus, and recurrent laryngeal nerve.

The point of bifurcation of the common carotid may vary a little, but is more frequently above than below the thyroid cartilage.

Surgery.—*To tie the common carotid artery* (Fig. 139, *e*). This may be accomplished either above or below the omo-hyoid, but above the muscle is the better situation. An incision three inches in length along the anterior border of the sterno-mastoid, beginning at the level of the hyoid bone, will allow that muscle to be turned outwards sufficiently to bring the omo-hyoid into view. In the angle between the two muscles the carotid is to be found, and this part of the operation will be much facilitated by slightly reflecting the omo-hyoid towards the median line (Fig. 137).

The descendens noni nerve may be seen on the sheath of the vessels, and is to be avoided, and the sheath is to be carefully opened on its inner side so as to avoid possible injury to the internal jugular vein. The needle is to be passed from the outer side, care being taken not to include the pneumo-gastric nerve or the sympathetic. On the dead body the vein is frequently empty, and is liable to be injured unless the sheath be opened well to its inner side.

The operation below the omo-hyoid might be performed through a similar incision along the border of the lower part of the sterno-mastoid, but would be facilitated by dividing the sternal origin of the muscle. The sterno-hyoid and thyroid muscles would require to be turned inwards, or even divided, in order to reach the vessel.

The **Internal Carotid Artery** (Fig. 150, 2) ascends to the base of the skull, lying close to the pharynx and upon the prevertebral muscles and sympathetic cord. It is first to the outer side of the external carotid and at the same level, and is crossed by the 9th nerve and the digastric and stylo-hyoid muscles; but it crosses the superior laryngeal branch of the pneumo-gastric. Its course is then deeper than the external carotid, from which it is separated by the stylo-glossus and stylo-pharyngeus muscles and glosso-pharyngeal nerve, and it finally enters the carotid canal in the petrous bone.

The internal carotid will be more fully traced out in the dissection of the pharynx.

The **External Carotid Artery** (Fig. 149) is derived from the Common Carotid opposite the upper border of the thyroid cartilage, and, lying to the inner side of the internal carotid, ascends to between the ear and the jaw, where it gives off its terminal branches. It is crossed by two muscles and a nerve (the digastric and stylo-hyoid muscles and the 9th nerve), and is also separated by two muscles and a nerve (the stylo-glossus and stylo-pharyngeus muscles and glosso-pharyngeal nerve) from the internal carotid at the upper part, where that vessel lies deeper than it. It is crossed by the lingual and facial veins, and enters the substance of the parotid gland behind the angle of the jaw, where it lies beneath the facial nerve and the commencement of the external jugular vein.

Its *branches* may be divided into three sets, each containing three branches.

Anterior set: 1, Superior Thyroid; 2, Lingual; 3, Facial.

Posterior set: 1, Occipital; 2, Posterior Auricular; 3, Sterno-Mastoid.*

Ascending set: 1, Temporal; 2, Internal Maxillary; 3, Ascending Pharyngeal.

The anterior and posterior sets can now be partly dissected, the ascending will be given in other dissections.

The **Superior Thyroid Artery** (Fig. 149, 12) runs forward beneath the depressor muscles of the hyoid bone, and then downward to the thyroid body, which it supplies, anastomosing with the inferior thyroid artery from the subclavian, and with the thyroid vessels of the opposite side. It gives off a *hyoid branch* to the parts about the hyoid bone; the *superior laryngeal branch* which pierces the thyro-hyoid membrane with the corresponding nerve; and the *crico-thyroid branch* which anastomoses with its fellow of the opposite side across the crico-thyroid membrane. Also a *superficial descending branch* which supplies the depressor muscles of the hyoid bone, and from which the *sterno-mastoid artery* is frequently derived.

The crico-thyroid branch is considered to be one of the dangers of laryngotomy, but the fact of its being parallel to the knife when entered transversely, as it ordinarily is in the operation, renders its division unlikely, whilst its small size would cause such an accident to be of slight importance.

The **Lingual Artery** (Fig. 149, 8).—Only a very small portion of the lingual artery is now visible, running transversely upon the middle constrictor of the pharynx

* The sterno-mastoid is perhaps more frequently derived from one of the branches of the external carotid than from the trunk itself, but is inserted here to complete the three branches of the posterior set.

mmediately above the hyoid bone, and disappearing beneath the edge of the hyoglossus muscle. The remainder of the vessel will be given in the dissection of the submaxillary region.

The **Facial Artery** (Fig. 149, 4) passes upwards and forwards beneath the digastric and stylo-hyoid muscles and through the substance of the submaxillary gland to the jaw, which it crosses immediately in front of the masseter muscle; it will be followed out in the dissection of the face. Its *inferior palatine* and *tonsillitic* branches may be seen disappearing between the stylo-glossus and stylo-pharyngeus muscles, and will be afterwards dissected; its *submaxillary* branches (two or three) enter the gland; and the *submental* branch, often of large size, runs forward over the mylo-hyoid muscle to the chin, where it supplies the surrounding tissues.

The facial frequently arises in common with the lingual artery, in which case the vessel passes beneath the hypoglossal nerve.

The **Occipital Artery** (Fig. 149, 1) is only seen in its first part. It runs backwards along the inferior border of the digastric, and may always be recognised by the fact that the 9th nerve (hypoglossal) hooks round it. The occipital artery crosses the following important structures, the hypoglossal nerve, internal carotid artery, pneumogastric nerve, internal jugular vein, spinal-accessory nerve, and sympathetic trunk. It gives a small *posterior meningeal branch* to enter the jugular foramen.

The **Posterior Auricular Artery** runs along the upper border of the digastric, and can only be seen by turning that muscle a little down. At the mastoid process it gives the *stylo-mastoid branch* to the stylo-mastoid foramen, and then supplies the pinna.

The **Sterno-Mastoid Artery** is a small branch of uncertain origin, coming either from the external carotid artery near its commencement, or from the occipital or superior thyroid arteries, or both the latter. It is of small size and enters the under surface of the sternomastoid muscle, crossing the carotid sheath.

Fig. 150.

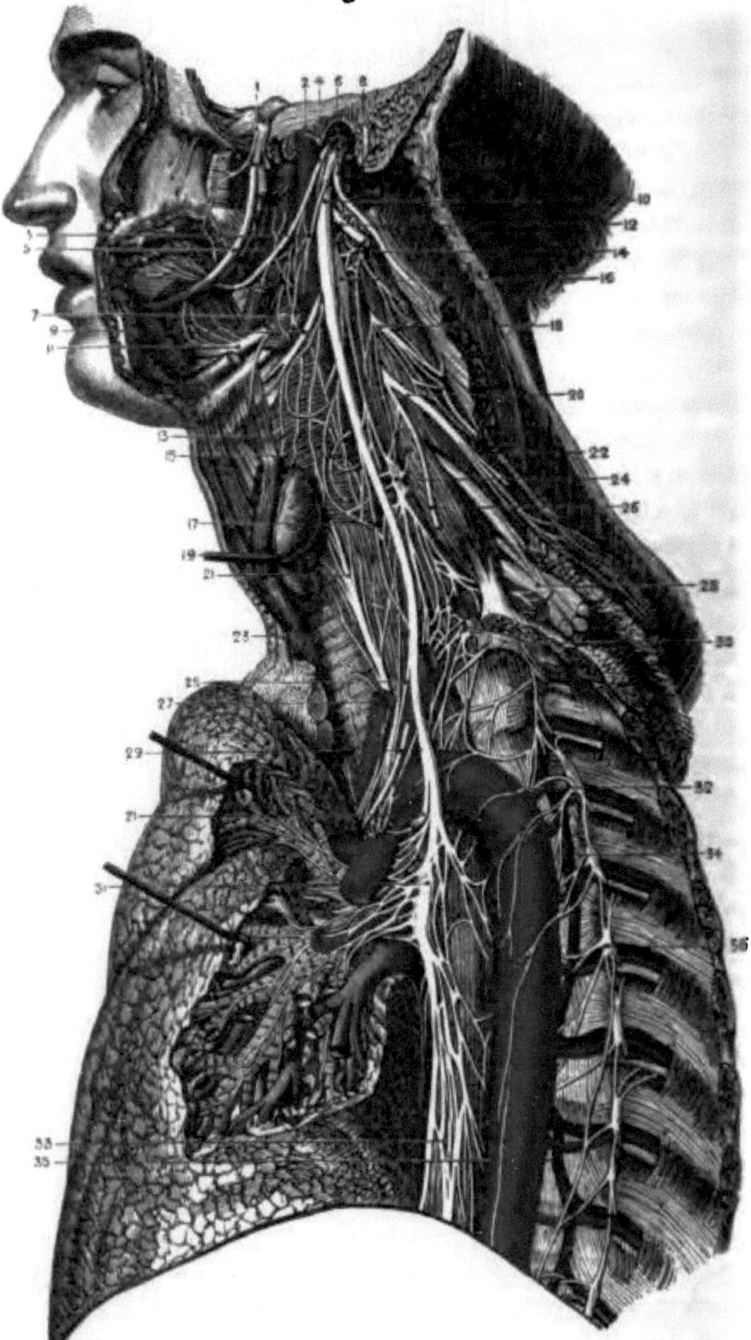

150.—Distribution of the eighth pair of nerves on the left side (from Hirschfeld and Leveillé).

Veins corresponding to the branches of the external carotid artery take nearly the same courses as those vessels, and terminate as follows: The internal maxillary and temporal veins unite in the parotid to form the External Jugular Vein, which afterwards receives the posterior auricular vein. All the other veins open into the Internal Jugular Vein directly, or into an Anterior Jugular if one exists, but the facial vein frequently joins the external jugular.

The **Internal Jugular Vein** (Fig. 149, 5) is deeply placed to the outer side of the internal carotid artery immediately below the base of the skull, being crossed by the stylo-glossus and stylo-pharyngeus muscles and the glosso-pharyngeal nerve. Having been crossed by the digastric and stylo-hyoid and the occipital artery, it is covered by the sterno-mastoid for the rest of its course. It lies to the outer side of the internal and common carotid arteries, having the pneumo-gastric

---

1. Gasserian ganglion of 5th nerve.
2. Internal carotid artery.
3. Pharyngeal branch of pneumogastric.
4. Glosso-pharyngeal nerve.
5. Lingual nerve (5th).
6. Spinal-accessory nerve.
7. Middle constrictor of pharynx.
8. Internal jugular vein (cut).
9. Superior laryngeal nerve.
10. Ganglion of trunk of pneumo-gastric nerve.
11. Hypoglossal nerve on hyoglossus.
12. Ditto communicating with eighth and first cervical nerve.
13. External laryngeal nerve.
14. Second cervical nerve looping with first.
15. Pharyngeal plexus on inferior constrictor.
16. Superior cervical ganglion of sympathetic.
17. Superior cardiac nerve of pneumo-gastric.
18 Third cervical nerve.
19. Thyroid body.
20. Fourth cervical nerve.
21, 21. Left recurrent laryngeal nerve.
22. Spinal-accessory communicating with cervical nerves.
23. Trachea.
24. Middle cervical ganglion of sympathetic.
25. Middle cardiac nerve of pneumo-gastric.
26. Phrenic nerve (cut).
27. Left carotid artery.
28. Brachial plexus.
29. Phrenic nerve (cut).
30. Inferior cervical ganglion of sympathetic.
31. Pulmonary plexus of pneumo-gastric.
32. Thoracic aorta.
33. Œsophageal plexus.
34. Vena azygos superior.
35. Vena azygos minor.
36. Gangliated cord of sympathetic.

nerve between it and them, and is enclosed in the carotid sheath of cervical fascia. After receiving the superior and middle thyroid veins, the jugular vein unites with the subclavian vein to form the *vena innominata.*

At the posterior aspect of the junction of the two veins on the left side of the body will be found the point of entrance of the *Thoracic Duct.* This tube should be carefully preserved, and will be subsequently more fully dissected. A corresponding but much smaller duct (right lymphatic duct) may be found on the right side of the body.

The **Pneumo-Gastric Nerve** (Fig. 150) (8th pair) is only seen in its cervical portion at present. It is enclosed in the carotid sheath, lying between the jugular vein and the artery, and enters the superior aperture of the thorax, passing, on the right side, between the subclavian artery and the innominate vein, and on the left side between the common carotid and subclavian arteries and behind the innominate vein.

Its *Superior Laryngeal* branch (9) curving forwards beneath the internal carotid artery, appears opposite the hyoid bone, and pierces the thyro-hyoid membrane to supply the larynx. Before entering the larynx, the nerve gives a small *external laryngeal* branch, which runs obliquely downwards under the sterno-thyroid muscle to supply the crico-thyroideus, one of the intrinsic muscles of the larynx.

The *Recurrent Laryngeal Nerve* (21) (inferior) is seen running along the side of the trachea and disappearing beneath the lower border of the inferior constrictor of the pharynx. It takes a different course on the two sides of the body; on the right arising from the pneumo-gastric in the neck, and winding round the subclavian artery; on the left arising in the thorax, and turning round the arch of the aorta.

The *Cardiac Nerves* (17) are one or two slender branches of the pneumo-gastric in the lower part of the neck, running into the thorax to join the cardiac plexuses.

The **Hypoglossal Nerve** (Fig. 149, 6) (9th pair) appears

below the posterior belly of the digastric muscle, hooking round the occipital artery and then curving forward superficially to the great vessels. It then passes beneath the

Fig. 151.

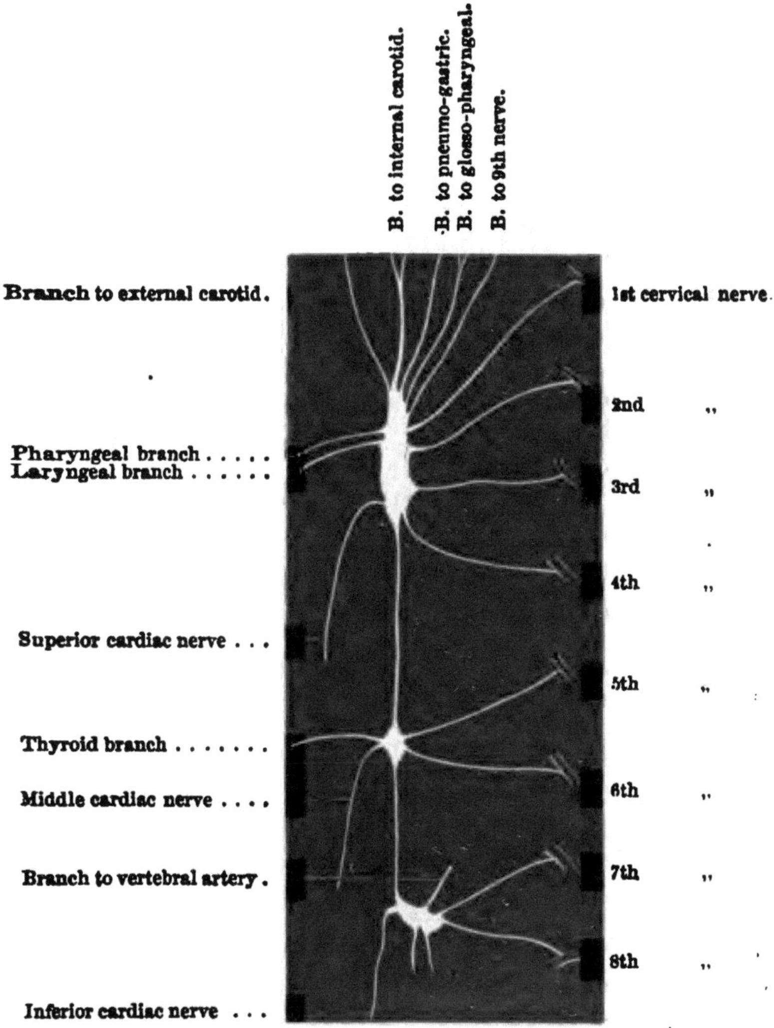

tendon of the digastric close to the hyoid bone, where it is seen in the angle between the two bellies of the muscle

Fig. 151.—Diagram of the superior, middle, and inferior cervical ganglia of the sympathetic (drawn by J. T. Gray).

lying upon the hyo-glossus muscle, and finally disappears beneath the posterior border of the mylo-hyoid muscle.

Its *descending branch* (descendens noni) arises opposite the occipital artery, and runs downward and forward either upon or within the carotid sheath, to supply the sterno-hyoid, sterno-thyroid, and omo-hyoid muscles, and to form one or more loops with the communicans noni branch from the cervical plexus.

The *nerve to the thyro-hyoid muscle* is a very delicate branch derived from the 9th nerve just before it passes beneath the digastric.

The **Sympathetic** (Fig. 150 & 151) in the neck lies behind the carotid sheath upon the prevertebral muscles. It is a greyish nerve which has three cervical ganglia developed upon it, of which only the upper two can now be seen, the lowest being on the neck of the first rib.

The *superior cervical ganglion* (16) is a fusiform enlargement of the nerve behind the internal carotid artery, and lying upon the rectus capitis anticus major. It gives branches of *communication* to the 8th and 9th nerves and also to the four upper cervical nerves: and branches of *distribution* to the internal carotid artery; to the branches of the external carotid artery (nervi molles); to the pharynx, joining the pharyngeal plexus; to the superior laryngeal nerve (occasionally); and the *superior cardiac nerve* to the cardiac plexus.

The *middle cervical* or *thyroid ganglion* (24), is opposite the 5th cervical vertebra and usually over the inferior thyroid artery. From this ganglion branches of *communication* go to the fifth and sixth cervical nerves; thyroid branches are *distributed* upon the inferior thyroid artery: and the *middle cardiac nerve* passes into the thorax, communicating with the recurrent laryngeal nerve.

The **Spinal-Accessory Nerve** (Fig. 149, 3) (8th pair) appears below the digastric and close to the transverse process of the atlas. It pierces the deep surface of the sterno-mastoid muscle, giving branches to it, and then emerging, crosses the posterior triangle to the trapezius.

The **Cervical Plexus** (Fig. 149) is formed by the

ıterior branches of the four upper cervical nerves, but ıe loop from the 1st nerve to the 2nd is of very small .ze and cannot be well seen in this stage of the dissection. 'he 2nd, 3rd, and 4th nerves appear between the rectus apitis anticus major and the middle scalenus. Each of .hese nerves communicates with the one above and below t, and thus the superficial and deep cervical plexuses are 'ormed. The branches of the superficial cervical plexus have been already dissected in the posterior triangle, and can now be traced to their source,—the great auricular, small occipital, and superficial cervical branches to the 2nd and 3rd nerves, and the descending branches to the 3rd and 4th nerves.

The branches of the deep cervical plexus are—

1. *Communicating* branches with the 8th and 9th cranial nerves and the sympathetic, which will afterwards be dissected.

2. *Muscular* branches to the rectus capitis anticus major, rectus minor and rectus lateralis.

3. The *Communicans noni* (7) usually consisting of two branches from the 2nd and 3rd nerves which join the descendens noni branch.

4. The *Phrenic nerve* (11) arising from the 4th nerve, and having in addition a branch of communication with the 5th. It lies upon the scalenus anticus, getting to its inner border at the lower part of the neck, and will subsequently be traced to the diaphragm.

5. *Muscular branches* are derived from the deep plexus, which go to the sterno-mastoid, levator anguli scapulæ and trapezius muscles, and also to the scalenus medius.

Anterior Triangle of the Neck.

After the dissection of the side of the neck is completed, by replacing the sterno-mastoid muscle in its proper position, the boundaries and contents of the anterior triangle will be fully seen.

The **Anterior Triangle** (Fig. 137) of the neck is bounded *in front* by the median line of the neck; and *behind* by the sterno-mastoid muscle. Its *base* is above, and is formed by the lower jaw and a line from the angle of the jaw to the mastoid process; the *apex* is below, at the top of the sternum.

The skin and superficial fascia, the platysma, the deep fascia, and the superficial nerves have been turned aside to expose the contents of the triangle, which are as follows. The digastric and stylo-hyoid muscles are seen to cut off the submaxillary triangle, in which is lodged the submaxillary gland, resting upon the mylo-hyoid muscle between the jaw and the hyoid bone. Below the hyoid bone will be seen the sterno-hyoid, sterno-thyroid and thyro-hyoid muscles; and the anterior belly of the omo-hyoid which runs obliquely across the triangle, dividing it into an upper and lower part.

The common carotid artery comes into view between the sterno-mastoid and omo-hyoid muscles, with the descendens noni nerve superficial to it, and the internal jugular vein to its outer side and concealing the pneumo-gastric nerve. The external and internal carotids are seen at, or near, the upper border of the thyroid cartilage, and are crossed superficially by the hypoglossal nerve; both cross the superior laryngeal nerve, which is seen passing to the larynx between the middle and inferior constrictors of the pharynx to the inner side of the carotid vessels.

The superior thyroid, lingual, and facial arteries are seen in part in the anterior portion of the space; and the occipital artery is seen turning backwards below the digastric, and in relation with the hypoglossal nerve.

THE FACE.

The face is the only region in which it will be convenient to make different dissections on the two sides, viz., the muscles and vessels on one, and the nerves on the other; and the dissectors are therefore advised to adopt this method of proceeding.

Before commencing the dissection, the student should observe the external anatomy of the eye and its apıdages, so that they may be seen in as natural a condin as possible.

On the margin of the orbit is the *supercilium* or eyeow, a ridge of thickened skin covered with hairs. The elids or *palpebræ* are two thin folds composed of cartige, muscle, and fascia, covered by the skin externally and ıed by the conjunctiva continued from the surface of ıe eyeball. The points of junction of the two lids are ılled the inner and outer *canthi*, and along the free order of the lids will be seen the *cilia* or eyelashes, ·hich in health curve upwards from the upper, and down·ards from the lower lid.

Before opening the lids they should be drawn forcibly ›utward towards the temple, in order to make tense and prominent the little *tendo oculi* at the inner canthus.

Fig. 152.

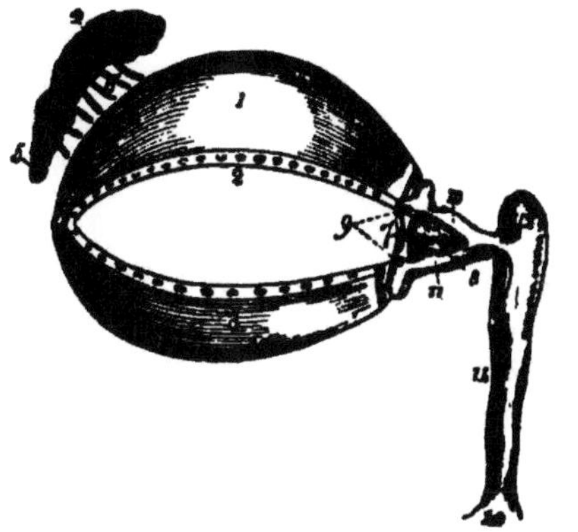

Surgery.—The tendo oculi is important as the guide to the lachrymal sac when it is necessary to introduce a knife into it in cases of lachrymal abscess, and the student should practise this little operation at once

---

Fig. 152.—Appendages of the eye (from Wilson).

1. Superior tarsal cartilage.
2. Lower border of the cartilage on which are seen the openings of the Meibomian glands.
3. Inferior tarsal cartilage; along the upper border of this cartilage the openings of the Meibomian glands are likewise seen.
4. Lachrymal gland; its superior or orbital portion.
5. Inferior or palpebral portion.
6. Lachrymal ducts.
7. Plica semilunaris.
8. Caruncula lachrymalis.
9. Puncta lachrymalia of the lachrymal canals.
10. Superior lachrymal canal.
11. Inferior lachrymal canal.
12. Lachrymal sac.
14. Dilatation of the nasal duct, where it opens into the inferior meatus of the nose.
15. Nasal duct.

as follows:—Drawing the eyelids outward to render the tendon tense, a narrow knife held with the back towards the nose is entered vertically immediately below it, and passed downwards and a little backwards behind the lower margin of the orbit; the handle of the knife being then brought against the forehead, and the point pushed downward and a little outward, it will be felt to be in the nasal duct, and a probe can be passed down into the nose without difficulty.

The lids being separated, close to the inner canthus will be seen a small space called the *lacus lachrymalis*, and on the free margin of each lid close to this will be seen a little eminence (*papilla lachrymalis*) perforated by a minute hole, the *punctum lachrymale* (9). A small probe introduced vertically through the punctum will pass horizontally along the *canaliculus* into the *lachrymal sac* (12), situated in the groove in the lachrymal bone and beneath the tendo oculi. The canaliculi may be opened up upon the probe or with a fine-pointed pair of scissors, when a good-sized probe may be readily introduced into the sac and be pushed down into the nasal duct and nose. If the upper lid is everted over a probe, the conjunctiva lining it, and the *Meibomian glands* beneath it running in parallel rows to the free margin, will be seen.

Close to the inner canthus is a reddish body, the *caruncula lachrymalis* (8), composed of mucous follicles, and external to it is a minute semilunar fold of conjunctiva, called the *plica semilunaris* (7), the representative of the *membrana nictitans* of birds, crocodiles, and batrachia. The conjunctiva is the mucous lining of the lids, and is continued on to the eyeball, where it is loosely attached to the sclerotic but inseparably to the cornea. If the conjunctiva be carefully divided over the sclerotic, the attachments of the muscles of the eyeball may be seen.

The student should recognise with his finger the prominences of the nasal, malar, and jaw bones, and also the margin of the orbit; and it will be well also to examine the state of tension of the eyeball by pressing upon it ently with the forefingers, since this varies very much ccording to the freshness of the subject. Opportunity hould be taken also to examine the external ear and, if ossible, to see the membrana tympani with the speculum.

Fig. 153.

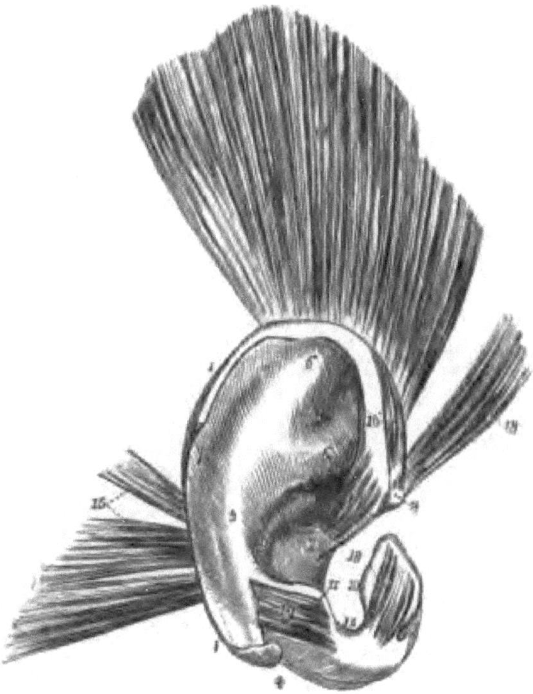

The **Pinna** (Fig. 153) is composed of cartilage covered with skin and is divided into the following parts:—The

---

Fig. 153.—The pinna and its muscles (from Wilson, after Arnold).

1, 1. Helix.
2. Crus helicis.
3. Spina helicis.
4. Processus caudatus helicis.
5. Antihelix.
6, 6. Crura antihelicis.
7. Fossa scaphoidea or fossa of the helix.
8. Fossa of the antihelix.
9. Concha.
10. Tragus.
11. Antitragus.
12. Incisura intertragica.
13. Attrahens aurem.
14. Attollens aurem.
15. Retrahens aurem.
16. Major helicis muscle.
17. Minor helicis muscle.
18. Tragicus.
19. Antitragicus.

outer rim is the *helix* (1) and that next to it the *antihelix* (5), the groove between the two being the *fossa of the helix* (7). The deep cavity leading to the *meatus externus* is the *concha* (9), and the little nodule in front of it, the *tragus* (10), the corresponding nodule, behind being the *anti-tragus* (11). The lowest part of the ear is the *lobule*, which consists of fat.

Fig. 154.

The extrinsic muscles of the ear (Attrahens, Attollens, and Retrahens) have been seen in the dissection of the scalp. Various small muscles have been described as passing from one point to another of the cartilage of the pinna, but do not require special description. They can be sufficiently seen in Figs. 153 and 154 and will not repay the trouble of dissection.

[The cheeks are now to be distended with cotton wool

Fig. 154.—The pinna and its muscles, as seen from behind (from Wilson, after Arnold).

1, 1. Border of the helix.
2. Spine of the helix.
3. Convexity corresponding with the fossa scaphoidea.
4, 4. Convexity of the concha; the fissure between the numbers corresponds with the crus helicis.
5. Ponticulus conchæ.
6, 6. Cartilage of the meatus.
7. Aperture of the meatus.
8. Attrahens aurem.
9. Attollens aurem.
10. Retrahens aurem.
11. Transversus auriculæ.
12. Obliquus auriculæ.

and the lips carefully sewn together. A little piece of wool may also be introduced beneath the eyelids with advantage. An incision is to be made from the temple in front of the ear along the jaw to the chin (the scalp having been already dissected), and the skin is to be carefully reflected towards the median line. A cut around the orbit and mouth will avoid interference with the eyelids and lips. All muscular fibres and vessels are to be carefully preserved, but the small nerves are to be cut away, together with all the loose fat. The parotid gland in front of the ear is to be preserved, and its duct traced across the face below the malar bone.]

MUSCLES OF THE FACE OR MUSCLES OF EXPRESSION.

The **Orbicularis Palpebrarum** (Fig. 156, 4) is a broad muscle surrounding the orbit and attached to its inner angle, being continuous above with the fibres of the occipito-frontalis (*v.* p. 319). It *arises* from the internal angular process of the frontal bone, from the tendo oculi passing between the nasal process of the superior maxilla and the inner extremities of the tarsal cartilages, and from the nasal process itself. Its fibres surround the orbit and are attached to the skin; those passing on the lids are called the palpebral fibres (or *palpebralis* muscle), and are attached externally to the malar bone and external tarsal ligament.

The orbicularis palpebrarum is the muscle which closes the eyelids, and is therefore the direct antagonist of the levator palpebræ.

The **Corrugator Supercilii** will be seen by removing the portion of the orbicularis covering the brow. It *arises* from the inner part of the superciliary ridge, and is *inserted* into the orbicularis about the centre of the eyebrow.

The **Pyramidalis Nasi** (Fig. 155, 1) is a small slip of muscle lying along the nasal bone, continuous with the occipito-frontalis above, and lost on the nasal cartilage, where it joins the tendon of the compressor naris.

The **Levator Labii Superioris Alæque Nasi** (Fig. 156, 8) *arises* from the upper part of the nasal process of the superior maxilla close to the margin of the orbit; it widens as it descends, and divides into two slips, which are *inserted* into the alar cartilage of the nose and into the upper lip.

The **Compressor Naris** (Fig. 155, 3) is to be traced beneath the preceding muscle to its *origin* from the canine fossa of the superior maxilla. The muscle is triangular in shape, and passes to be *inserted* on the bridge of the nose, where it unites with the opposite muscle and the pyramidalis nasi.

Fig. 155.

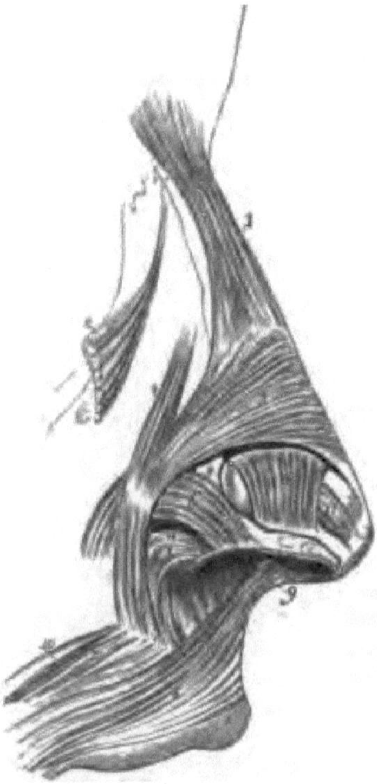

The **Depressor Alæ Nasi** (Fig. 155, 7) is a small muscle which can only be properly seen by everting the lip (at a later stage) and removing the mucous membrane. It *arises* from the myrtiform fossa over the incisor teeth of the upper jaw, and is *inserted* into the posterior part of the columna and the ala of the nose.

The **Orbicularis Oris** (Fig. 156, 7) is the sphincter muscle surrounding the mouth. It is joined by all the

Fig. 155.—Muscles of the nose (from Wilson, after Arnold).

1. Pyramidalis nasi.
2. Upper part of the levator labii superioris alæque nasi turned aside.
3. Compressor naris.
4. Musculus anomalus.
5. Levator proprius alæ nasi anterior.
6. Levator proprius alæ nasi posterior.
7. Part of the depressor alæ nasi.
8. Upper segment of the orbicularis oris.
9. Naso-labialis.
10. Fasciculi accessorii of the orbicularis.
11. A small muscle termed by Arnold compressor narium minor; it extends between the end of the nose and the alar cartilage. Its existence is doubtful.

other muscles which are inserted into the lips, and specially by the buccinator on each side, the fibres of which interlace with it. It is *attached* to each side of the septum of the nose by a small slip (naso-labialis) (Fig. 155, 9), and to the superior maxilla close to the canine tooth on each side, by small slips which have been named the fasciculi accessorii (10). Professor Ellis has described similar attachments to the lower jaw close to the canine teeth.

Fig. 156.

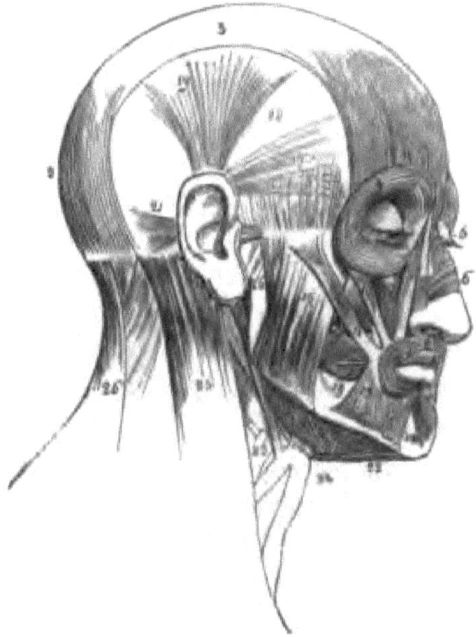

Fig. 156.—Muscles of the head and face (from Wilson).

1. Frontal portion of the occipito-frontalis.
2. Its occipital portion.
3. Its aponeurosis.
4. Orbicularis palpebrarum, which conceals the corrugator supercilii and tensor tarsi.
5. Pyramidalis nasi.
6. Compressor naris.
7. Orbicularis oris.
8. Levator labii superioris alæque nasi; the adjoining fasciculus between figures 8 and 9 is the labial portion of the muscle.
9. Levator labii superioris proprius; the lower part of the levator anguli oris is seen between the muscles 10 and 11.
10. Zygomaticus minor.
11. Zygomaticus major.
12. Depressor labii inferioris.
13. Depressor anguli oris.
14. Levator menti.
15. Superficial portion of the masseter.
16. Part of its deep portion.
17. Attrahens aurem.
18. Buccinator.
19. Attollens aurem.
20. Temporal fascia covering the temporal muscle.
21. Retrahens aurem.
22. Anterior belly of the digastricus.
23. Stylo-hyoid pierced by posterior belly of the digastric.
24. Mylo-hyoideus.
25. Sterno-mastoid.
26. Trapezius.

The **Levator Labii Superioris Proprius** (Fig. 156, 9) *arises* from the superior maxilla and margin of the malar bone above the infra-orbital foramen, and is *inserted* into the upper lip externally to the slip from the common elevator.

The **Levator Anguli Oris** (Fig. 156) *arises* from the canine fossa of the superior maxilla immediately below the infra-orbital foramen, and passes downwards and outwards beneath the preceding muscle to the angle of the mouth, joining the fibres of the orbicularis.

The **Zygomaticus** (Fig. 156) is either a single muscle, or has two slips forming the zygomaticus major and minor.

The Zygomaticus major (11) *arises* from the malar bone close to its junction with the zygoma, and passes obliquely to the angle of the mouth.

The Zygomaticus minor (10), much smaller, *arises* from the malar bone in front of the preceding, and is *inserted* between it and the levator labii superioris, with which it unites.

On the side of the jaw will be seen the insertion of the **Platysma Myoides**, continuous more or less with the orbicularis oris at the angle of the mouth and with the muscles below the lip. Occasionally there is a strong slip passing transversely from the fascia over the masseter to the angle of the mouth, and called the **Risorius Muscle of Santorini.**

The **Depressor Anguli Oris** (Fig. 156, 13) (triangularis menti) is a triangular muscle *arising* from the outer surface of the lower jaw, below and a little external to the mental foramen, and *inserted* into the angle of the mouth.

The **Depressor Labii Inferioris** (Fig. 156, 12) (quadratus menti) *arises* from the outer surface of the lower jaw between the symphisis and the mental foramen, and ascends obliquely to meet its fellow in the median line at the lower lip, thus leaving a triangular interval between them.

The **Levator Menti** (Fig. 156, 14) is a triangular muscle placed internally to the depressor of the lower lip, *arising* from the incisive fossa of the lower jaw and *inserted* into the skin of the chin. It is best dissected by everting the lower lip and removing the mucous membrane.

All the muscles of the face, muscles of expression, are *supplied* by the facial nerve (portio dura of 7th). In health the muscles of the two sides of the face antagonise one another, and an equilibrium is unconsciously maintained, but should the nerve of one side be divided, or become paralysed from disease of the temporal bone, the balance of power will be destroyed, and the face will be at once drawn *to* the healthy side by the preponderating muscles. The paralysed side will remain immoveable, the eye will be open, and the patient will be unable to whistle, etc.

The **Parotid Gland** is the largest of the salivary glands, and is placed between the ear and the lower jaw, its superficial part overlapping the masseter muscle and being called the *socia parotidis.* It reaches as high as the zygoma and as far back as the mastoid process, and below, it is separated from the submaxillary gland by the stylo-maxillary ligament, its deep surface resting against the styloid process and the muscles attached to it. The duct of Stenson (ductus Stenonis) arises from the anterior part below the socia parotidis, and passes transversely across the masseter, to open into the mouth through the buccinator muscle opposite the second molar tooth of the upper jaw. Immediately below the duct is a large branch of the facial nerve, and above it the transverse facial artery. The gland is traversed by the external carotid artery, which gives off its two terminal branches (temporal and internal maxillary) in its substance; by the external jugular vein; and by the facial and auriculo-temporal nerves, from both of which it receives branches. The structure of the parotid is that of a simple lobulated gland like the other salivary glands.

It receives *parotid branches* from the temporal artery and nervous filaments from the auriculo-temporal nerve.

[The remains of the platysma are to be removed to

Fig. 157.

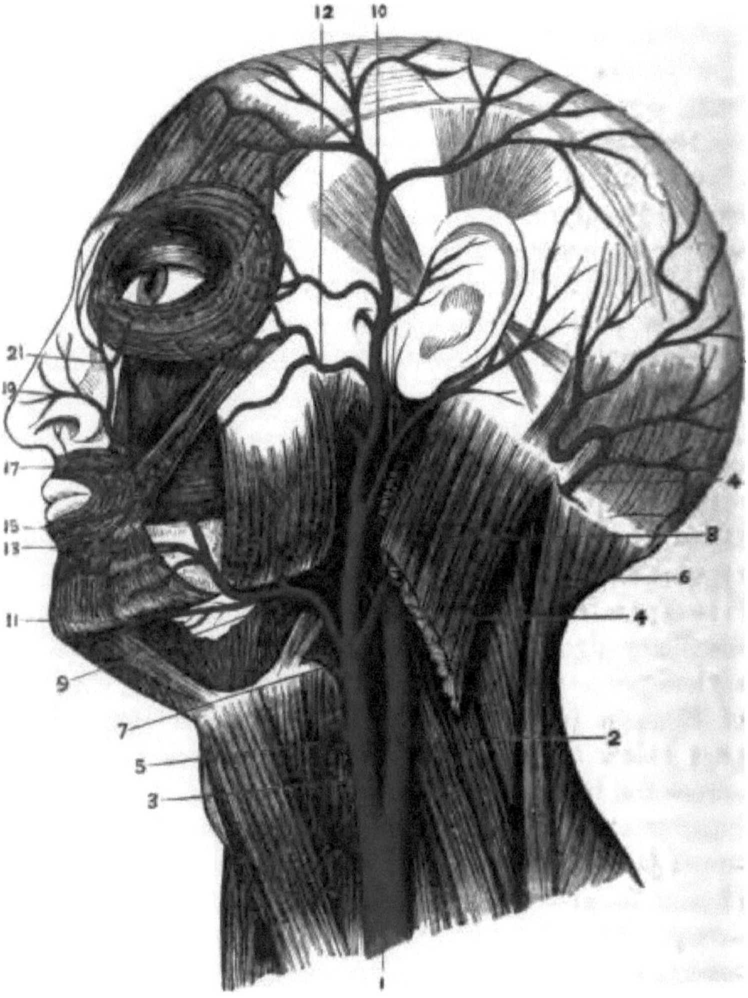

Fig. 157.—The arteries of the face and head (drawn by G. E. L. Pearse).

1. Common carotid.
2. Internal carotid.
3. External carotid.
4, 4. Occipital artery.
5. Superior thyroid artery.
6. Trapezius.
7. Lingual artery.
8. Sterno-mastoid.
9. Facial artery.
10. Temporal artery, dividing into anterior and posterior branches.

show the facial artery and vein in front of the masseter, and the artery is to be followed out to the inner angle of the eye with as little injury to the muscles as possible.]

The **Facial Artery** (Fig. 157) [external carotid], the cervical portion of which has been already dissected (p. 365), appears on the face immediately in front of the masseter muscle, where it lies on the lower jaw and is crossed by the platysma. It then passes upwards and inwards beneath the zygomatici muscles to the side of the nose, lying upon the buccinator, levator anguli oris, and levator labii superioris muscles in its course. At the side of the nose it is crossed by the outer portion of the levator labii superioris alæque nasi, and reaches the angle of the orbit to anastomose with the nasal branch of the ophthalmic artery. This is therefore one of the points of communication between the external and internal carotid vessels.

Branches.—As the artery crosses the jaw it gives—

(1.) *Masseteric* and *buccal* branches to the corresponding muscles.

(2.) The *Inferior labial* (13) branch which runs midway between the chin and the mouth, passing transversely beneath the depressor anguli oris.

(3.) The *Inferior coronary* (15) branch taking the same course as the preceding but nearer the margin of the lip. Its size depends upon that of the inferior labial, with which it anastomoses, and also with the vessel of the opposite side.

(4.) The *Superior coronary* (17) branch often arising in common with the inferior coronary. It pierces the orbicularis oris, and runs beneath the mucous membrane of the lip to anastomose with the vessel of the opposite side. It gives a little branch to the septum of the nose (*artery of the septum*).

---

11. Submental branch.
12. Transverse facial artery.
13. Inferior labial branch.
15. Inferior coronary branch.
17. Superior coronary branch.
19. Lateral nasal branch.
21. Angular branch.

(5.) The *Lateralis nasi* (19) branch passing beneath the levator labii superioris alæque nasi to be distributed to the ala.

(6.) The *Angular artery* (21) which is the terminal branch and anastomoses with the ophthalmic.

Surgery.—The facial artery may be compressed or tied immediately in front of the masseter, where its pulsation can be readily felt through the thin platysma. The position of the coronary arteries immediately beneath the mucous membrane should be borne in mind in operating for hare-lip or cancer, as it is necessary to thrust the hare-lip pin deeply into the substance of the lip, in order to control the hæmorrhage effectually.

The **Facial Vein** commences at the inner angle of the orbit in the angular vein, formed by the frontal and supra-orbital veins. It then runs obliquely across the face to the edge of the masseter, where it lies close to the outer side of the artery, and then passes into the neck to join the internal jugular vein. It is crossed by the zygomatici muscles but lies upon Stenson's duct. Its branches correspond to those of the artery.

The **Transverse Facial Artery** (Fig. 157, 12) arises from the temporal artery in the parotid gland, and appears on the face above the parotid duct, to run transversely inwards towards the nose and anastomose with the facial artery.

In a well-injected body the terminations of the infra-orbital and inferior dental arteries will be found emerging at the infra-orbital and mental foramina, with the 2nd and 3rd divisions of the 5th nerve. The infra-orbital foramen will be found between the levator labii superioris and levator anguli oris, and the mental foramen beneath the depressor anguli oris. The frontal and supra-orbital arteries should also be traced if the scalp has not already fully dissected (*v.* p. 321).

ıe skin being removed from the opposite side of the ..n the same way as above directed, a dissection is to be made of the nerves alone, all other tissues being sacrificed for that purpose. To find the facial nerve, the best way is to cut transversely into the parotid gland about half an inch below the zygoma, and to dissect carefully to some depth to reach a large branch of the nerve, which passes through its substance. One branch having been found, it is to be followed carefully backwards and forwards, when it will lead to other branches; and the parotid gland is to be removed piecemeal to show the ramifications of the nerve, which are then to be traced as far as possible towards the median line.]

The **Facial Nerve** (portio dura of the 7th pair) (Fig. 158, 17) leaves the skull at the stylo-mastoid foramen, and gives off immediately the *posterior auricular* branch (13) to the back of the ear, and the *digastric and stylo-hyoid* (21) branches to the posterior belly of the digastric and to the stylo-hyoid muscles, to which they should now be traced. Passing deeply in the parotid gland, the nerve divides into two large trunks, the *temporo-facial* (22) and the *cervico-facial* (19) divisions, from which numerous branches arise, causing by their interlacement the appearance termed the *pes anserinus* or goose foot.

From each division of the facial nerve three sets of branches are derived, whose direction is indicated by their names.

Temporo-facial division	Temporal (4). Malar (12). Infra-orbital (16).	Cervico-facial division	Buccal (24). Supra-maxillary (30). Infra-maxillary (34).

These branches all ramify upon the face and supply the facial muscles, communicating with the several divisions of the 5th nerve excepting the infra-maxillary, which run below the jaw to the neck and communicate with the superficial cervical nerve (31). The branches are to be traced to all the "muscles of expression" and also to the buccinator, one of the muscles of mastication.

The *three* divisions of the **5th Nerve** appear upon the face, and each of them will be found *three* times in front of the ear.

The **First** or **Ophthalmic Division** (Fig. 158) appears (1) as the *supra-trochlear nerve* (6) at the inner angle of the orbit; (2) at the supra-orbital notch as the *supra-orbital nerve* (8), which has been traced to the forehead;

Fig. 158.

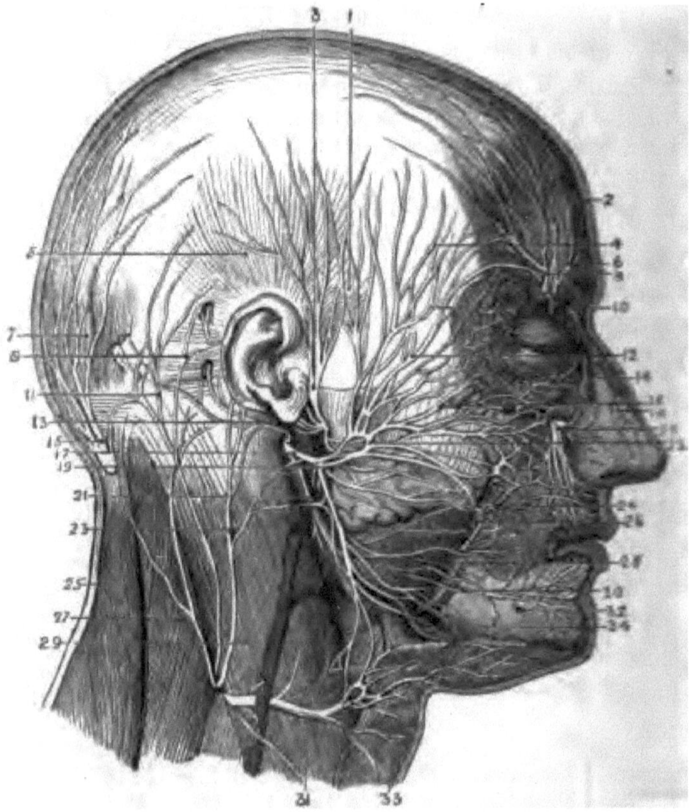

Fig. 158.—Nerves of the face and scalp (from Hirschfeld and Leveillé).

1. Attrahens aurem muscle.
2. Anterior belly of occipito-frontalis.
3. Auriculo-temporal nerve.
4. Temporal branches of facial nerve (7th).
5. Attollens aurem muscle.
6. Supra-trochlear nerve (5th).
7. Posterior belly of occipito-frontalis.
8. Supra-orbital nerve (5th).
9. Retrahens aurem muscle.
10. Temporal branch of orbital nerve (5th).
11. Small occipital nerve.
12. Malar branches of facial nerve.
13. Posterior auricular nerve (7th).
14. Malar branch of orbital nerve (5th) (subcutaneus malæ).
15. Great occipital nerve.
16. Infra-orbital branches of facial nerve (7th).
17. Facial nerve (7th).
18. Nasal nerve (5th).
19. Cervico-facial division of 7th.
20. Infra-orbital nerve (5th).

and (3) as the *nasal nerve* (18) which will now be found emerging from between the nasal bone and lateral cartilage on the side of the nose. (Small terminal twigs of the infra-trochlear and lachrymal nerves may also be sometimes found with care.)

The **Second or Superior Maxillary Division** (Fig. 158) appears (1) as the *infra-orbital nerve* (20) at the infra-orbital foramen between the levator labii superioris and the levator anguli oris, and is to be traced to the upper lip which it supplies with numerous large branches; (2) as the *subcutaneus malæ* (14) branch appearing through the malar bone after piercing the outer wall of the orbit; and (3) as the *temporal branch of the orbital nerve* (10) piercing the temporal fascia immediately above the zygoma and already dissected (*v.* p. 321).

The **Third or Inferior Maxillary Division** (Fig. 158) appears (1) as the *mental nerve* (32) at the mental foramen, and is to be traced to the chin and lower lip; (2) as the *buccinator nerve* (26) upon the surface of the buccinator muscle where it will be found lying beneath, but communicating with, the facial nerve; and (3) as the *auriculo-temporal nerve* (3), piercing the parotid gland and running upwards over the zygoma immediately in front of the ear.

These are all purely sensory nerves with the exception (perhaps) of the buccinator nerve, which is supposed to be of a mixed character. They all join freely with the facial nerve.

[The anatomy of the eyelids and of the nasal cartilages

---

21. Branches to digastric and stylo-hyoid (7th).
22. Temporo-facial division of 7th.
23. Great auricular nerve.
24. Buccal branches of facial nerve.
25. Trapezius.
26. Buccinator nerve (5th).
27. Splenius capitis.
28. Masseter.
29. Sterno-mastoideus.
30. Supra-maxillary branches of facial nerve (7th).
31. Superficial cervical nerve.
32. Mental nerve (5th).
33. Platysma.
34. Infra-maxillary branches of facial nerve (7th).

can be examined on either side of the face by removing the superjacent tissues.]

The **Eyelids** (Fig. 152).—On removing the thin palpebral fibres of the orbicularis from the lids, a piece of thin fibrous tissue will be seen extending from the margin of the orbit to the lid, which is called the *palpebral ligament*. It is continuous with the periosteum of the skull at the margin of the orbit, and is attached to the edge of the tarsal cartilage in each lid. The *tarsal cartilage* is larger in the upper than the lower lid, being semilunar in shape in the former, and nearly straight in the latter. Each is attached externally by a little *tarsal ligament* to the margin of the orbit, whilst internally the tendo oculi divides to be attached to both cartilages, and thus holds them in position. The fibres of the orbicularis must be cut away from the tendo oculi to show it thoroughly, and it will be seen to be attached to the nasal process of the superior maxilla immediately in front of the groove for the lachrymal sac.

Beneath the palpebral ligament in the upper lid will be found the expansion of the tendon of the levator palpebræ superioris attached to the tarsal cartilage. Both lids are lined with conjunctiva, on removing which the Meibomian glands may be seen to groove the cartilages.

The structures composing the upper eyelid are—

1. Skin.
2. Orbicularis palpebrarum.
3. Palpebral ligament.
4. Tendon of levator palpebræ.
5. Tarsal cartilage.
6. Meibomian glands.
7. Conjunctiva.

The structures in the lower lid are the same, *minus* the levator palpebræ.

This will be the best opportunity to dissect the **Tensor Tarsi Muscle** of Horner, by dividing the outer tarsal ligaments and detaching the eyelids so as to turn them over the nose and expose their internal surface. By removing the conjunctiva over the tendo oculi the two little slips of muscular fibre will be seen above and below it. The muscle *arises* from the ridge on the lachrymal bone, and is *inserted* into the tarsal cartilages and canaliculi.

Cartilages of the Nose (Figs. 159 and 160).—Only the lateral cartilages and cartilages of the aperture on each side can now be seen, the cartilage of the septum being

Fig. 159.

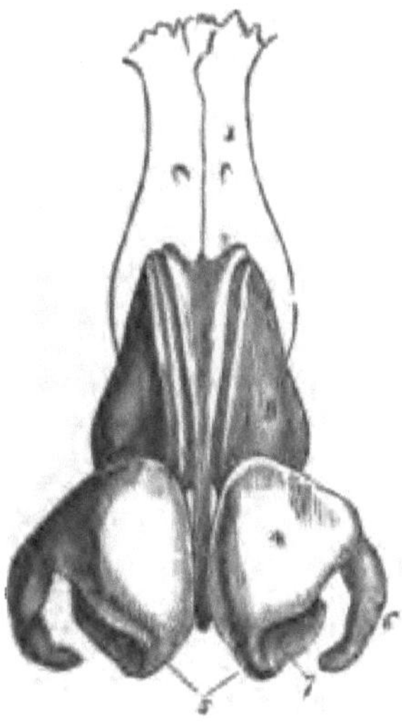

Fig. 160.

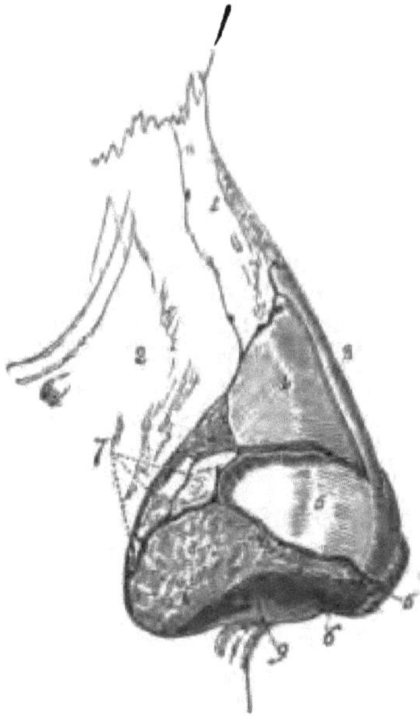

Fig. 159.—The fibro-cartilages of the nose (from Wilson).

1. One of the nasal bones.
2. Cartilage of the septum.
3. Lateral cartilage.
4. Alar cartilage.
5. Central portions of the alar cartilages which constitute the columna.
6. Cartilagines minores or sesamoid cartilages.
7. The nostril.

Fig. 160.—The fibro-cartilages and bones of the nose viewed from the side (from Wilson, after Arnold).

1. Nasal bone.
2. Nasal process of the superior maxillary bone.
3. Cartilage of the septum.
4. Lateral cartilage.
5, 5. Alar cartilage.
6. Inner portion of the alar cartilage.
7. Sesamoid cartilages.
8. Areolar tissue of the ala nasi.
9. Aperture of the nostril.

dissected with the nose. The *lateral cartilage* is triangular, its posterior border being continuous with the nasal bone and the nasal process of the maxillary bone, and its anterior border being separated from that of its fellow of the opposite side by the edge of the septum.

The *cartilage of the aperture* (alar) is semicircular in shape and has no attachment to bone, being embedded in the dense cellular tissue of the ala nasi, in which there are frequently two or three little additional nodules, *sesamoid cartilages*. The cartilages of the two sides adjoin at the tip of the nose and are attached to the lateral cartilages by fibrous tissue.

The Pterygo-Maxillary Region.

[The head being laid on one side, the facial nerve and parotid duct are to be divided and turned forward, and the remains, if any, of the parotid gland are to be cleaned out from between the ear and the jaw, the auriculo-temporal nerve being carefully preserved. The surface of the masseter muscle is then to be cleaned.]

The **Masseter** (Fig. 156, 15) is the most superficial muscle of mastication and is divisible into a superficial and a deep portion, a small part of the latter being seen behind and above the fibres of the former. The superficial part *arises* from the anterior two-thirds of the lower border of the zygomatic arch formed by the malar and temporal bones, and slightly from the superior maxilla; and passes downwards and backwards to be *inserted* into the angle of the jaw and the lower half of the ramus. The deep part *arises* from the posterior third of the lower border of the zygoma and also from the inner surface of the bone; and passes downwards and a little forwards to be *inserted* into the outer side of the upper half of the ramus of the jaw, extending to the root of the coronoid process.

[The temporal fascia is to be detached from the upper border of the zygoma, and a cut is to be made with the aw through the malar bone in front of the masseter, and behind it through the zygoma as near its root as possible. The piece of bone with the masseter attached can then be turned down, when the deep fibres of the muscle will be seen attached to the lower jaw, and the *masseteric* branch of artery and nerve will be found behind the coronoid process, and must necessarily be divided. The temporal muscle is to be cleaned and the fat beneath the masseter removed, so as to expose the buccinator muscle thoroughly, together with its branch of nerve and artery which appear in front of the coronoid process and are to be preserved.]

The Temporal Muscle (Fig. 164, 2) *arises* from the whole of the temporal fossa and from the under surface of the temporal fascia, which has been seen to be attached to the whole length of the temporal ridge. Its fibres converge to a tendon, which is *inserted* into the apex of the outer surface of the coronoid process, and into the inner surface of the coronoid process and ramus of the lower jaw as far as the last molar tooth. It is also connected with the pterygo-maxillary ligament.

[With the saw and bone-forceps the coronoid process and the anterior part of the ramus of the jaw are to be carefully cut off, and turned up. The neck of the jaw is then to be sawn through transversely just below the condyle, and the ramus divided transversely at the level of the molar teeth. In making this last cut the saw should only go through half the thickness of the jaw, after which the bone-forceps can be applied at the upper border so as to split the piece of bone, and allow of its removal in two pieces without injury to the nerve.

On removing the portion of ramus thus separated the periosteum will probably be left behind, and on this being dissected away the inferior dental nerve and artery will be seen lying on the internal lateral ligament before entering the dental foramen, and in front of these the lingual (gustatory) nerve. Crossing both nerves transversely at the upper part of the space now exposed is the external pterygoid muscle, with the internal maxillary artery lying either upon or beneath it. If superficial to the muscle it had better be cleaned, divided, and held aside with a hook.]

The External Pterygoid Muscle (Fig. 160, 1) *arises* by two heads, one from the outer surface of the external pterygoid plate, and the tuberosity of the palate and maxillary bones; and the other from the under surface of the great wing of the sphenoid below the pterygoid ridge. Its fibres run transversely to be *inserted* into the hollow on the front of the neck of the jaw, and into the interarticular fibro-cartilage of the joint.

Fig. 160.

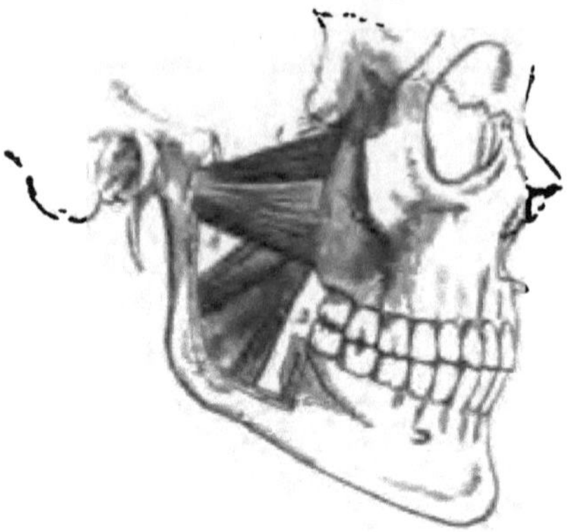

Fig. 161.

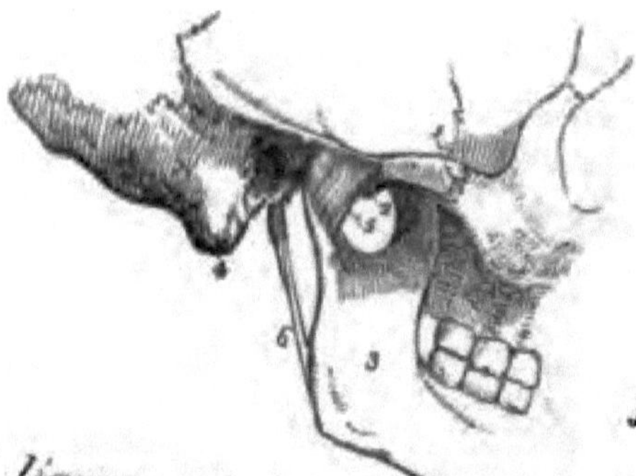

Tempero-Maxillary Articulation. — The *external lateral ligament* (Fig. 161, 5) is a short thick band of fibres passing obliquely downwards and backwards from the root of the zygoma to the neck of the jaw. It overlies the *capsular ligament*, which is attached to the

Fig. 160.—The two pterygoid muscles (from Wilson). The zygomatic arch and the greater part of the ramus of the lower jaw have been removed in order to bring these muscles into view.

1. The sphenoid origin of the external pterygoid.
2. Its pterygoid origin.
3. The internal pterygoid.

Fig. 161.—Internal view of the articulation of the lower jaw (from Wilson).

1. Zygomatic arch.
2. Tubercle of the zygoma.
3. Ramus of the lower jaw.
4. Mastoid portion of the temporal bone.
5. External lateral ligament.
6. Stylo-maxillary ligament.

ins of the articular surfaces. By dividing these ents the interarticular cartilage and the synovial branes above and below it will be brought into view.

Fig. 162.

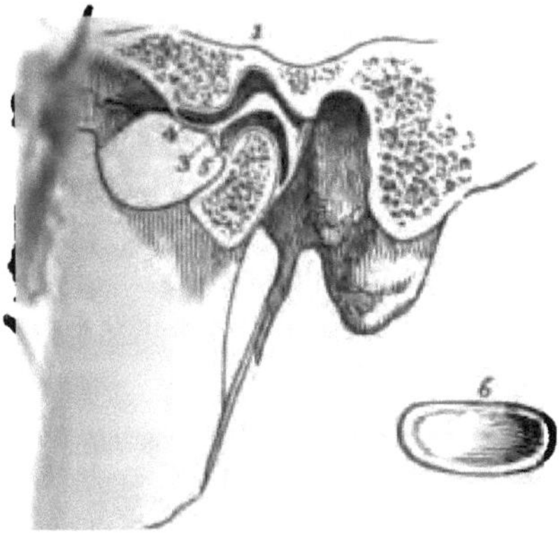

Fig. 163.

The *interarticular fibro-cartilage* (Fig. 162, 3) is seen to fit into the glenoid cavity, and its upper surface is accordingly concavo-convex from before backwards, and the lower surface the converse of this. The cartilage together with the condyle being detached from the temporal bone and turned forward with the external pterygoid muscle, will be found to be oval in outline and thicker at the margins than in the centre.

Fig. 162.—A section of the temporo-maxillary articulation, to show the position of the interarticular fibro-cartilage, and the manner of its adaptation to the articulating surfaces (from Wilson).

1. Glenoid fossa.
2. Eminentia articularis.
3. Interarticular fibro-cartilage.
4. Superior synovial cavity.
5. Inferior synovial cavity.
6. An interarticular fibro-cartilage, removed from the joint, in order to show its oval and concave form; it is seen from below.

Fig. 161.—Internal view of the articulation of the lower jaw (from Wilson).

1. Section through the petrous portion of the temporal bone and spinous process of the sphenoid.
2. Internal surface of the ramus and body of the lower jaw.
3. Capsular ligament.
4. Internal lateral ligament.
5. Aperture through which the mylo-hyoid nerve passes.
6. Stylo-maxillary ligament, a process of the deep cervical fascia.

The *internal lateral ligament* (Fig. 163, 4) (now fully exposed) is a membranous band, narrow above, where it is attached to the spine of the sphenoid bone and is crossed by the internal maxillary artery, and widening below to be attached to the projecting margin of the dental foramen. It is covered by the inferior dental nerve and artery at the lower part, and pierced by the mylo-hyoid branches of artery and nerve near the bone.

The **Internal Pterygoid Muscle** (Fig. 160, 3) is placed beneath the internal lateral ligament. It *arises* from the internal surface of the external pterygoid plate and from the tuberosity of the palate bone, and is *inserted* into the inner surface of the angle of the lower jaw, its fibres running parallel to the superficial portion of the masseter. In many subjects a portion of the internal pterygoid arises from the outer surface of the tuberosity of the superior maxilla, anterior to the fibres of the pterygoideus externus (Fig. 164).

The **Buccinator** (Fig. 164, 14) forms the greater part of the cheek, and has been partly seen in the dissection of the face. It *arises* from the alveolar border of the superior maxilla opposite the molar teeth; from a corresponding portion of the inferior maxilla; and between the jaws from the anterior border of the pterygo-maxillary ligament. The muscle is *inserted* into the angle of the mouth, joining the orbicularis oris, and its middle fibres decussate with one another as they pass to their insertion. The buccinator is perforated by the duct of the parotid gland opposite the second molar tooth of the upper jaw.

The **Pterygo-Maxillary Ligament** (Fig. 165) is a white fibrous band extending between the internal pterygoid plate and the lower jaw, close to the last molar tooth. By its anterior border it gives attachment to the buccinator, and by its posterior border to the superior constrictor of the pharynx; it thus establishes the connection between the cavities of the mouth and the pharynx, the mucous membrane covering its inner surface.

The temporal, masseter, external and internal pterygoids, and buccinator are classed together as the Muscles of Mastication. The temporal, masseter, and internal pterygoid muscles close the jaws, but the masseter and internal pterygoid draw the lower jaw forward, whilst the temporal draws it back, thus producing the antero-posterior movement. The lateral movements of the jaw are produced by the alternate action of the two external pterygoid muscles, which, if acting together, protrude the chin.

The principal action of the buccinator is to keep the food between the teeth during mastication, but it also expels air from the mouth (and hence the name "trumpeter"), as is well seen in using the blowpipe.

The muscles of mastication are all *supplied* by the motor portion of the third division of the 5th nerve, but the buccinator is also, and principally, supplied by the facial nerve (7th); as is shown by the fact that in paralysis of that nerve the buccinator fails in its action, and food collects between the teeth and the cheek.

The **Internal Maxillary Artery** (Fig. 165, 13) is seen in this dissection in its first and second portions, the third being in the spheno-maxillary fossa. One of the terminal branches of the external carotid artery, it arises in the parotid gland immediately behind the neck of the jaw.

The **First Part** of the internal maxillary artery, with its accompanying veins, lies between the neck of the jaw and the internal lateral ligament. It gives off a descending branch, the *Inferior dental*, and an ascending branch, the *Middle meningeal*.

a. The *Inferior dental artery* descends on the internal lateral ligament to the inferior dental foramen, by which it enters the lower jaw; it then runs in a canal in the substance of that bone supplying the teeth, and appears on the face at the mental foramen with the third division of the 5th nerve. Immediately before entering the dental foramen it gives off the *mylo-hyoid branch*, which pierces the internal lateral ligament with a nerve of the same name, and runs in a groove on the internal surface of the lower jaw to the cutaneous surface of the mylo-hyoid and to the digastric muscle, joining the submental artery.

b. The *Middle meningeal artery* (Fig. 165, 11) ascends beneath the external pterygoid muscle, which must be turned forwards with the condyle, as above directed, to see the whole of its course. It passes between the fibres of the auriculo-temporal nerve to the foramen spinosum, by which it enters the skull. Before doing so it gives off the *meningea parva* to enter the foramen ovale, and a *tympanic branch* to enter the Glasserian fissure.

The **Second Part** of the internal maxillary artery varies in position in different subjects. It is always intermuscular, but it may lie between the temporal and external pterygoid or between the external and internal pterygoid muscles. If it lies superficially to the external pterygoid muscle it passes between its two heads to reach the pterygo-maxillary fossa. Its branches are all *Muscular*, *i.e.*, to the muscles of mastication. There are two *deep temporal* branches (anterior and posterior) running in the substance of the temporal muscle; a *masseteric* branch of small size passing through the sigmoid notch to the under surface of the muscle (now necessarily divided); two *pterygoid* branches, uncertain in origin and course; and a *buccinator* branch which generally pierces the external pterygoid muscle to accompany the nerve to the surface of the buccinator.

The *Posterior dental artery*, one of the branches of the third part of the internal maxillary artery, can be seen entering one of the foramina on the posterior surface of the superior maxilla, in which it ramifies, supplying that bone and the molar teeth.

The **Internal Maxillary Vein** receives branches corresponding to those of the artery, which it accompanies to the parotid gland to join the temporal vein and form the external jugular.

The **Inferior Maxillary Nerve** (Figs. 164 and 165) (3rd division of 5th) as soon as it emerges from the foramen ovale, divides into a motor and a sensory portion, the latter, however, having a few motor fibres mingled with it.

The **Motor Portion** (Fig. 164) is of small size, is placed externally to the sensory portion, and is distributed to the

Fig. 164.

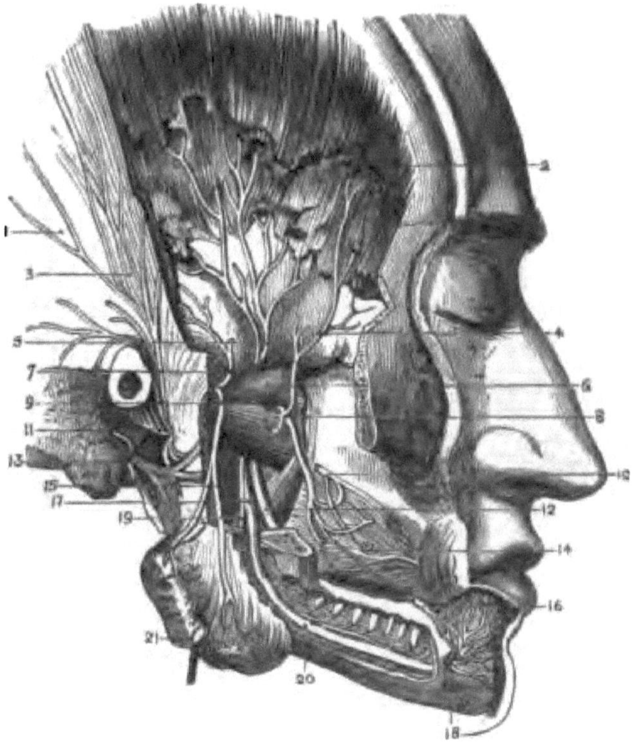

Fig. 164.—Pterygo-maxillary region and fifth nerve (from Hirschfeld and Leveillé).

1. Temporal fascia.
2. Temporal muscle.
3. Temporal branches of auriculo-temporal nerve.
4. Deep temporal branch of buccinator nerve.
5. Deep temporal nerve.
6. Pterygoideus externus.
7. Deep temporal branch of masseteric nerve (inconstant).
8. Buccinator nerve.
9. Masseteric nerve.
10. Buccal branch of 7th.
11. Auriculo-temporal nerve.
12. Lingual nerve.
13. Facial nerve (7th) at stylo-mastoid foramen.
14. Buccinator muscle.
15. Pterygoideus internus.
16. Supramaxillary branch of 7th.
17. Inferior dental nerve.
18. Its mental branches.
19. Its mylo-hyoid branch.
20. Inferior dental canal opened.
21. Masseter (turned down).

muscles of mastication, running with the corresponding branches of the internal maxillary artery. Its branches are, *anterior* and *posterior temporal* (5), *masseteric* (9), *pterygoid*, and *buccal* (8), and are difficult to find, except the last, which is of good size and may be traced to the surface of the buccinator muscle, where it communicates with the facial nerve.

The **Sensory Portion** (Fig. 165) divides close to the base of the skull into three trunks, the *Lingual Nerve* (or gustatory) the most anterior; the *Inferior Dental Nerve* in the middle; and the *Auriculo-Temporal Nerve* most posterior and smallest.

[The auriculo-temporal nerve is to be carefully dissected out from the tough fibrous tissue of the articulation, and traced through the parotid to the pinna and temple.]

The **Lingual or Gustatory Nerve*** (Fig. 165, 16) lies at first between the two pterygoid muscles, and then passes forward between the internal surface of the ramus of the jaw and the internal pterygoid muscle; it will be afterwards traced to the side and tip of the tongue. Beneath the external pterygoid the nerve is joined at an acute angle by the small *chorda tympani nerve* (7), a branch of the 7th nerve which, leaving the facial nerve in the temporal bone, passes through the tympanum and emerges at the Glasserian fissure, to which it should now be traced.

The **Inferior Dental Nerve** (Fig. 165, 18) has a branch of communication with the lingual trunk, and then descends upon the pterygoideus internus, and afterwards upon the internal lateral ligament, to the dental foramen. After entering the bone, it supplies all the teeth of the lower jaw, either directly or by its *incisive* branch (22), and emerges at the mental foramen with its accompanying artery. Immediately before entering the dental foramen

* Since the glosso-pharyngeal nerve has been conclusively shown to be the special nerve of taste, it is desirable that the term "gustatory" as applied to the lingual branch of the fifth should be abandoned.

it gives off the *mylo-hyoid* branch (17), which pierces the internal lateral ligament with the accompanying artery, and, after running in a groove on the inner surface of the

Fig. 165.

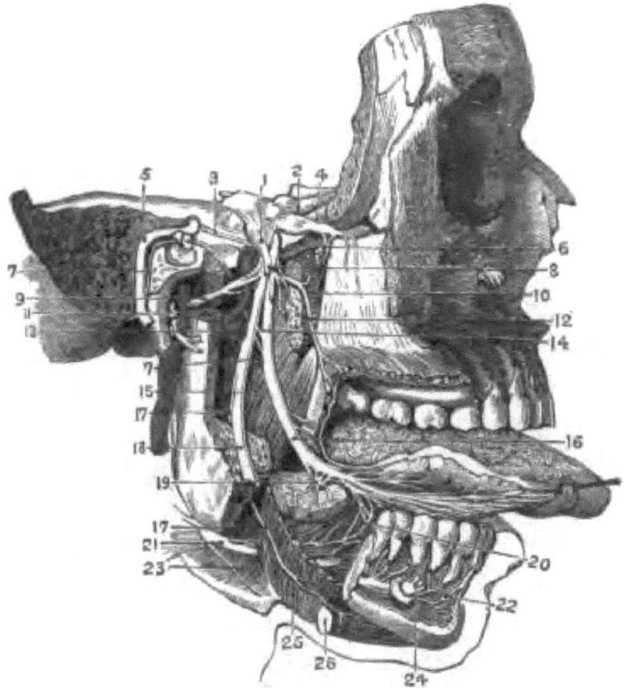

Fig. 165.—Third or inferior maxillary division of the fifth nerve (from Hirschfeld and Leveillé).

1. Third division of fifth nerve.
2. First division of fifth nerve.
3. Superficial petrosal nerve.
4. Second division of fifth.
5. Facial nerve (7th).
6. Meckel's ganglion.
7, 7. Chorda tympani.
8. Muscular branches (divided).
9. Auriculo-temporal nerve.
10. Pterygoideus externus.
11. Middle meningeal artery.
12. Buccinator nerve.
13, 13. Internal maxillary artery.
14. Communicating branch of lingual and inferior dental nerves.
15. Pterygoideus internus.
16. Lingual nerve.
17, 17. Mylo-hyoid nerve.
18. Inferior dental nerve.
19. Submaxillary ganglion on the deep portion of submaxillary gland.
20. Communication between fifth and ninth nerves.
21. Hypoglossal nerve (9th).
22. Incisive branches of inferior dental nerve.
23. Hyo-glossus muscle.
24. Mental branch of inferior dental nerve.
25. Mylo-hyoid muscle.
26. Anterior belly of digastric.

bone, is distributed on the cutaneous surface of the mylo-hyoid muscle and also supplies the anterior belly of the digastric.

The **Auriculo-Temporal Nerve** (Fig. 165, 9) passes horizontally backwards behind the temporo-maxillary articulation, and often embraces the middle meningeal artery. It then turns upwards in front of the ear through the parotid gland, and is distributed on the temple (*v.* p. 321). It gives branches to the ear (auricular), to the parotid (of which it is the excitory nerve according to Bernard), and to the joint; and communicates with the facial nerve and with the sympathetic.

THE SUBMAXILLARY REGION.

The Submaxillary region has been partially dissected with the anterior triangle of the neck. It is seen to be bounded superficially by the two bellies of the digastric muscle and by the lower jaw, thus being triangular in shape, and to contain the submaxillary gland and a portion of the facial artery.

[The facial artery is to be divided at the jaw and together with the submaxillary gland turned back, when the mylo-hyoid muscle forming the floor of the submaxillary triangle will be exposed, and on its surface will be seen the mylo-hyoid branch of artery and nerve derived from the inferior dental trunks, the artery anastomosing with the submental branch of the facial. By cutting through the digastric muscle at its insertion, detaching it and the stylo-hyoid from the hyoid bone, and turning them back, the mylo-hyoid muscle will be fully exposed, and should be put on the stretch by means of a hook attached to the larynx.]

The **Mylo-hyoid Muscle** (Fig. 165, 25) *arises* from the mylo-hyoid ridge on the inner surface of the lower jaw, and its fibres pass downwards and forwards to be *inserted* into the body of the hyoid bone, and to meet the muscle of the opposite side in a median tendon between the hyoid bone and the lower jaw, thus forming the floor of the mouth.

The mylo-hyoid is an elevator of the hyoid bone, and is *supplied* (with the anterior belly of the digastric) by the mylo-hyoid branch of the inferior dental nerve.

[The mylo-hyoid is to be detached from the lower jaw and turned forward, when three structures will be seen lying in the following order from above downwards on the hyo-glossus muscle, viz. the lingual nerve [5th], the deep portion of the submaxillary gland with Wharton's duct, and the hypoglossal nerve [9th]. In addition, immediately below the 9th nerve there is very frequently a large vein corresponding to the lingual artery. In front of the hyo-glossus will be seen the genio-hyoid stretching between the chin and the hyoid bone, and a few fibres of the genio-hyo-glossus, together with a branch (ranine) of the lingual artery. The jaw being divided close to the genio-hyoid can be turned up, and the tongue being drawn out of the mouth and all the parts put on the stretch with hooks, the dissection can be proceeded with by cleaning the submaxillary gland and its duct.]

The **Submaxillary Gland** (Fig. 165, 19) consists of two portions, one, the larger, placed superficially on the mylo-hyoid in the submaxillary triangle, and the other winding round the posterior margin of the mylo-hyoid to rest on the hyo-glossus. From this deep portion the duct (Wharton's) arises, and lies at first between the lingual and hypoglossal nerves, but at the anterior part of the hyo-glossus will be found ascending beneath the lingual nerve to lie at a higher level than it. If followed beneath the mucous membrane of the mouth, the duct will be found to cross again over the lingual nerve at the side of the tongue, and to open into the mouth close to the frænum linguæ.

The **Sublingual Gland** (Fig. 165) is seen beneath the mucous membrane by the side of the tongue, and lies on the genio-hyo-glossus close to Wharton's duct, into which several of its ducts (ductus Riviniani) open, the others opening into the mouth in a crescent near the frænum.

The **Lingual Nerve** (Fig. 166, 1) (Gustatory of 5th) appears at the anterior border of the pterygoideus internus, where it lies against the pterygo-maxillary ligament, and afterwards runs between the jaw and the mucous membrane of the mouth to reach the hyo-glossus, upon which it is placed above Wharton's duct and the 9th nerve. Upon the hyo-glossus it crosses the duct and lies below it, but if followed to the side of the tongue will be found to pass again beneath the duct, and to be distributed to the side and tip of the tongue, supplying the fungiform and filiform papillæ. Branches of communication pass

Fig. 166.

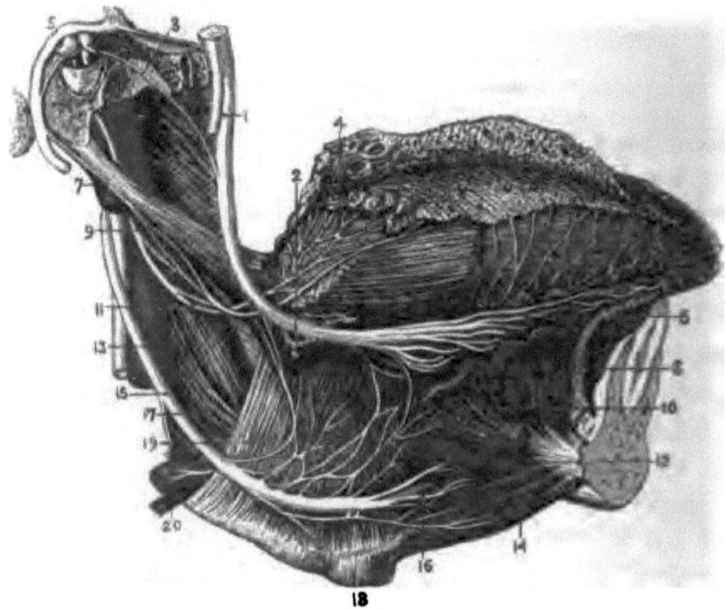

Fig. 166.—Nerves of the tongue (from Hirschfeld and Leveillé).

1. Lingual nerve (5th).
2. Submaxillary ganglion.
3. Chorda tympani nerve.
4. Distribution of glosso-pharyngeal nerve to circumvallate papillæ.
5. Facial nerve (7th).
6. Distribution of lingual nerve (5th).
7. Communication between facial and glosso-pharyngeal nerves.
8. Ranine artery.
9. Stylo-glossus.
10. Genio-hyo-glossus.
11. Glosso-pharyngeal nerve.
12. Sublingual artery.
13. Stylo-pharyngeus.
14. Genio-hyoideus.
15. Hypoglossal nerve (9th).
16. Distribution of ninth nerve.
17. Middle constrictor of pharynx.
18. Hyoid bone.
19. Hyo-glossus.
20. Lingual artery.

across the hyo-glossus to the 9th nerve, and a little below the level of the trunk, opposite the deep portion of the submaxillary gland, may be found the following ganglion.

The *submaxillary ganglion* (2) is of small size and lies below the lingual nerve. It has three roots, the sensory derived from the lingual; the motor from the 7th by means of the *chorda tympani*, which is prolonged on the lingual nerve to the ganglion; and a sympathetic root from the plexus on the facial artery. The branches of distribution are given to the submaxillary gland. The chorda tympani is afterwards prolonged to the inferior lingualis muscle of the tongue.

The **Hypoglossal Nerve** (Fig. 166, 15) (9th)* has been seen crossing the upper part of the anterior triangle after hooking round the occipital artery, to disappear beneath the posterior border of the mylo-hyoid. It is now seen to lie on the hyo-glossus which separates it from the lingual artery, and after communicating with the lingual nerve it passes on to the genio-hyo-glossus muscle and the under surface of the tongue. It gives branches to the stylo-glossus, hyo-glossus, genio-hyoid, and genio-hyo-glossus muscles, and also to the intrinsic muscles of the tongue itself.

The **Hyo-Glossus** (Fig. 167, 6) is a square muscle which *arises* from the body of the hyoid bone close to the median line, from the upper margin of the whole length of its great cornu, and from the lesser cornu. It is *inserted* into the side of the tongue, where its fibres decussate with those of the stylo-glossus. It is sometimes described as consisting of three separate portions—basio-glossus, kerato-glossus, and chondro-glossus.

The hyo-glossus draws the sides of the tongue downward. It is *supplied* by the ninth nerve.

The **Stylo-Glossus** (Fig. 167, 9) *arises* from near the tip of the styloid process and from the stylo-maxillary ligament, and passing across the internal carotid artery and jugular vein reaches the side of the tongue, into

* Lingual nerve of Harrison.

which it is *inserted*, decussating with the hyo-glossus and joining the fibres of the lingualis.

Fig. 167.

The **stylo-glossus** draws the tongue upwards and backwards, and therefore back into the mouth when it has been protruded, thus assisting the anterior fibres of the genio-hyo-glossus. It is *supplied* by the ninth nerve.

The **Stylo-Pharyngeus** (Fig. 167, 11) *arises* from the root of the styloid process, and passes downward to disappear between the upper and middle constrictors of the pharynx opposite the hyoid bone. Its *insertion* will be seen in the dissection of the pharynx.*

* It will assist the student in remembering the origins of the

Fig. 167.—Styloid muscles and muscles of the tongue (from Wilson).

1. Temporal bone of the left side.
2, 2. The right side of the lower jaw divided at its symphysis; the left side having been removed.
3. Tongue.
4. Genio-hyoideus.
5. Genio-hyo-glossus.
6. Hyo-glossus; its basio-glossus portion.
7. Its kerato-glossus portion.
8. Anterior fibres of the lingualis issuing from between the hyo-glossus and genio-hyo-glossus.
9. Stylo-glossus with part of the stylo-maxillary ligament.
10. Stylo-hyoideus.
11. Stylo-pharyngeus.
12. Os hyoides.
13. Thyro-hyoid membrane.
14. Thyroid cartilage.
15. Thyro-hyoid muscle arising from the oblique line of the thyro-cartilage.
16. Cricoid cartilage.
17. Crico-thyroid membrane.
18. Trachea.
19. Commencement of the œsophagus.

[The hyo-glossus is to be divided near the hyoid bone to expose the lingual artery, the branches of which are to be cleaned. The tongue being put on the stretch, all remains of mucous membrane covering the genio-hyoid and genio-hyo-glossus muscles are to be removed and the muscles cleaned.]

The **Lingual Artery** (Fig. 168) has been seen for a short distance in the anterior triangle, viz. from its origin from the external carotid until crossed by the hyo-glossus.

Fig. 168.

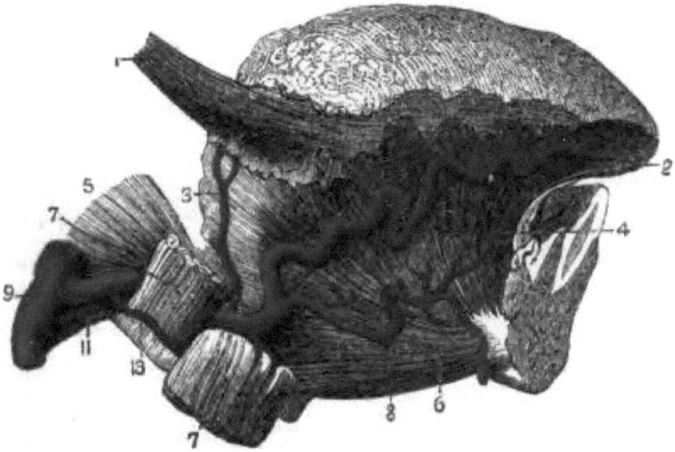

It is now seen to lie against the origin of the middle constrictor of the pharynx immediately above the hyoid

muscles from the styloid process if he notices that the origins are inversely as the insertions, *i.e.* that the muscle which arises highest is inserted lowest down. Thus the order of the origins from above downwards is stylo-pharyngeus, stylo-hyoid, and stylo-glossus.

Fig. 168.—Lingual artery and branches (from Hirschfeld and Leveillé).

1. Stylo-glossus.
2. Ranine artery.
3. Dorsalis linguæ artery.
4. Genio-hyo-glossus muscle.
5. Middle constrictor.
6. Genio-hyoid muscle.

7, 7. Hyo-glossus (cut).
8. Sublingual artery.
9. External carotid artery.
11. Lingual artery.
13. Hyoid branch.

bone, and then to pass on to the genio-hyo-glossus, by which it is conducted to the tip of the tongue.

Its *branches* are four in number :—

1. The *Hyoid* (13) branch runs along the upper border of the hyoid bone.

2. The *Dorsalis linguæ* (3) ascends beneath the hyo-glossus to the dorsum of the tongue.

3. The *Sublingual* (8) generally of good size, comes off at the anterior margin of the hyo-glossus and is directed to the chin across the genio-hyo-glossus muscle, which muscle it supplies together with the sublingual gland.

4. The *Ranine* (2) is the termination of the lingual artery, and after leaving the genio-hyo-glossus enters the under surface of the tongue, to be continued in its substance to the tip, where it anastomoses with the vessel of the opposite side.

The **Lingual Vein** has branches corresponding to those of the artery, and passes beneath the hyo-glossus to open into the internal jugular vein. Not unfrequently, however, a large vein lies superficially to the hyo-glossus in addition to a smaller vein accompanying the artery.

The **Genio-hyoid** muscle (Fig. 168, 6) *arises* from the lower of the two genial (or mental) tubercles close to the symphysis menti, and passes downwards immediately beneath the mylo-hyoid and close to the median line, to be *inserted* into the body of the hyoid bone.

The **Genio-hyo-glossus** muscle (Fig. 168, 4) *arises* from the upper of the genial tubercles, and is triangular or fan-shaped, being *inserted* close to the median line all along the under surface of the tongue, and to the body of the hyoid bone above the genio-hyoid.

The genio-hyoid and genio-hyo-glossus are elevators of the hyoid bone or depressors of the jaw, as they alternately take their fixed point above or below. The genio-hyo-glossus has an important action on the tongue, the posterior and middle fibres drawing up the hyoid-bone and thrusting the tongue out of the mouth, and the anterior fibres drawing the tongue into the mouth again, being assisted by the stylo-glossus.

Both the genio-hyoid and the genio-hyo-glossus are *supplied* by the ninth nerve.

The **Stylo-maxillary Ligament** (Fig. 163, 6) will be seen attached to the angle of the jaw, which is turned up. It is a process of deep cervical fascia attached to the apex of the styloid process (where it gives origin to fibres of the stylo-glossus), and inserted into the inner surface of the angle of the jaw, where it is much widened. It intervenes between the parotid and submaxillary glands.

The **Stylo-hyoid Ligament** is a narrow band passing from the styloid process to the lesser cornu of the hyoid bone, where it gives origin to fibres of the middle constrictor. It is frequently ossified to a considerable extent.

The **Inferior Palatine** and **Tonsillitic** branches of the *facial artery* can be seen ascending, the former between the stylo-glossus and stylo-pharyngeus muscles, and the latter either with it or between the stylo-glossus and pterygoideus internus, to the outside of the pharynx, where they lie upon the superior constrictor. They will be more fully seen in the dissection of the pharynx.

The **Glosso-pharyngeal Nerve** (Fig 166, 11) (8th) is seen emerging from between the jugular vein and internal carotid artery, and winding round the stylo-pharyngeus muscle; if the stylo-glossus be now divided, the nerve may be followed beneath that muscle to the base of the tongue, where it supplies the *papillæ cicumvallatæ*. It gives branches to the stylo-pharyngeus and to the pharynx in this part of its course, and will be again seen in another dissection.

The Thorax.

[The arms having been previously removed, the dissectors of the thorax should detach the remains of the pectoral, serrati, and abdominal muscles from the outer surfaces of the ribs and their cartilages, and dissect the external intercostal muscles.]

The **External Intercostal Muscles** (Fig. 83, 4), eleven

in number, are placed between the ribs, being attached to the lower and upper borders of the adjacent bones, the fibres running obliquely downwards and forwards. The external intercostals extend from the tubercles of the ribs to the line of junction with their cartilages, beyond which point a thin fascia is prolonged to the sternum. Between the floating ribs the fibres cease at the extremities of the bones.

The **Internal Intercostal Muscles** (Fig. 83, 5) take a direction opposite to that of the external muscles, and can now be seen partially near the sternum, where the external muscles are wanting, but are fully seen subsequently from within the thorax.

[The knife is to be carried through the cartilages of all the true ribs as near the bones as possible, and then along the intercostal space between the 5th and 6th ribs to the lower part of the sternum. The sternum being sawn across at this point can then be turned up, the pleuræ being divided near the median line so as to preserve the anterior mediastinum, the internal mammary vessels being necessarily cut across with the bone.]

The **Triangularis Sterni Muscle** (Fig. 169, 3) lies at the back of the sternum, but is ordinarily damaged by the incision made for injecting the subject, and the muscle is therefore best seen on a fresh sternum from a post-mortem examination. It *arises* from the inner surface of the ensiform cartilage and lower part of the sternum, and from the cartilages of the last three or four true ribs; and is *inserted* into the cartilages of the 2nd, 3rd, 4th, and 5th ribs. The triangularis sterni acts as a muscle of expiration by depressing the ribs, and is *supplied* by the intercostal nerves.

[The sternum, which is still attached by the muscles of the front of the neck, is to be divided in the centre, and the two halves are to be turned aside so as to preserve the internal mammary arteries for after examination.]

On opening the thorax as directed the lungs will be exposed by the necessary division of the pleuræ, and will be seen to occupy the sides of the chest, having probably, unless attached by old adhesions, fallen back so as to leave the pericardium exposed. The great vessels of the heart will be seen emerging from the upper part of the pericardium, and if the lungs are fully collapsed the roots of the lungs will be visible on each side. The pericardium will in all probability have been opened at the upper part to allow of the injection of the body, in which case a portion of the heart may be seen within it, and it may be advantageously closed with a stitch.

The **Pleuræ** (Fig. 169) are two shut sacs of serous membrane lining the thorax and covering the lung, but have been opened in removing the sternum. One of the lungs is to be drawn forward, when its pleura can be readily traced over the inner surface of the sternum and ribs (costal or parietal layer) on to the back of the lung; thence over the lung (pulmonary layer) to the front of the root of the lung, by which it is directed to the pericardium and carried forward to the sternum a little to one side of the median line. Frequently it will be impossible to trace the pleura in this manner, owing to the adhesions which have formed between the pleura costalis and pleura pulmonalis as the result of inflammation. When the lung is drawn forward, below its root will be found a fold of pleura connecting the lower lobe with the side of the pericardium, which is called the *ligamentum latum pulmonis.* The pleura will be seen to be reflected on the upper surface of the diaphragm below, and to be prolonged in a conical form above the first rib, where it may be felt in relation with the subclavian artery. The diaphragm rises higher on the right side than on the left, owing to the presence of the liver, and the heart has a direction to the left; hence the right bag of the pleura is the wider, whilst the left is the longer of the two, and the lungs correspond in shape.

The **Mediastina** (Fig. 169) are the spaces between the pleuræ, and are three in number, anterior, middle, and posterior.

The *Anterior Mediastinum* which is now exposed is

the small space bounded by the sternum in front, the pericardium behind, and the pleuræ at the sides, and is placed a little to the left of the median line. It contains (1) the triangularis sterni muscle, (2) the origins of the sterno-hyoid and thyroid muscles, (3) the remains of the thymus gland (a ductless gland largely developed in the fœtus); (4) the *left* internal mammary vessels; some cellular tissue and a few lymphatics.

Fig. 169.

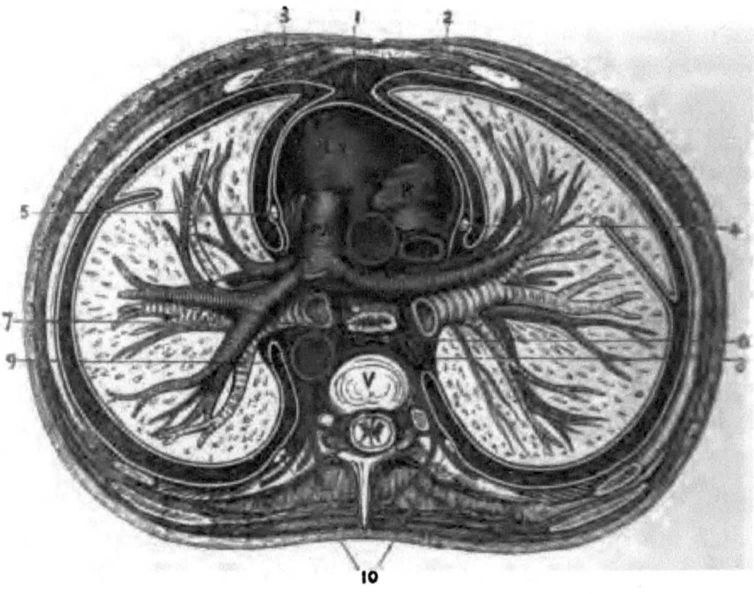

Fig. 169.—Diagram of a transverse section of the thorax (altered from Wilson and Carter, by J. T. Gray).

1. Anterior mediastinum.
2. Internal mammary vessels.
3. Triangularis sterni muscle.
4. Right phrenic nerve between pleura and pericardium.
5. Left phrenic nerve between pleura and pericardium.
6. Thoracic duct in posterior mediastinum.
7. Œsophagus with left vagus in front and right vagus behind.
8. Vena azygos major.
9. Thoracic aorta giving off intercostal arteries.
10. Gangliated cord of sympathetic.

R.V. Right ventricle. R.A. Right auricle of heart in middle mediastinum.
P. A. Pulmonary artery.
A. Aorta.
C. Vena cava superior.
V. Dorsal vertebra.

The *Middle Mediastinum* is the space between the two lungs, and contains therefore (1) the heart and large vessels, (2) the pericardium, (3) the roots of the lungs, and (4) the phrenic nerves.

The *Posterior Mediastinum* is behind the pericardium, and will be afterwards dissected. It contains (1) the thoracic aorta, (2) the trachea, (3) the œsophagus with the two pneumogastric nerves, (4) the vena azygos major, (5) the thoracic duct, (6) the greater splanchnic nerves at the lower part of the space, with some cellular tissue and lymphatics.

[To examine the middle mediastinum the ribs should be divided about three inches from their cartilages and the pieces turned back to give more room, but care should be taken to divide the first and second ribs in front of the attachments of the scaleni muscles. The lungs are to be drawn away from the heart, and the layer of pleura reflected on the pericardium is then to be carefully removed (the phrenic nerves lying between the two being preserved), and the roots of the lungs cleaned without injuring the pericardium.]

The **Pericardium** (Fig. 169) is a fibro-serous membrane enclosing the heart, and is of a conical shape, the base being at the diaphragm and the apex lost on the great vessels. The fibrous layer is pierced by and gives an investment to all the vessels connected with the heart, except the inferior vena cava, which pierces the central tendon of the diaphragm within the line at which the fibres of the pericardium are incorporated with those of the diaphragm. The serous lining consists of a parietal and a visceral layer, the former being inseparably united with the fibrous layer, and the latter covering the heart and being reflected on all the vessels for a short distance, but binding the aorta and the pulmonary artery together in a single tube of membrane.

The **Phrenic Nerves** (Fig. 149, 11) will be seen in the whole of their course and should be carefully preserved. Arising from the 4th and 5th cervical nerves they lie on the scaleni antici muscles in the neck, and then to the inner sides of those muscles upon the subclavian arteries. The two nerves enter the thorax on each side, and passing between the subclavian vein and artery, cross the internal mammary arteries from without inwards. The left phrenic nerve is then superficial to the left pneumo-gastric nerve and crosses the arch of the aorta internally to it; and both nerves pass in front of the roots of the lungs and between the pleuræ and pericardium to the diaphragm, which they pierce and supply. Accompanying each nerve may occasionally be seen the small *comes nervi phrenici* artery from the internal mammary.

[The pericardium is to be dissected from the roots of the lungs and entirely removed from the great vessels, except a small piece which should be left to mark the point at which it was attached to the aorta.]

The **Roots of the Lungs** (Figs. 170 and 177) are each formed by a pulmonary artery, two pulmonary veins, and a bronchus, bound together by cellular tissue; together with the small bronchial vessels, the pulmonary plexus of the pneumo-gastric nerve, and some lymphatics. The order of the vessels from before backwards on both sides of the body is the same, viz., veins, artery, and bronchus; from above downwards on the right side the order is the reverse, viz., bronchus, artery, veins, but on the left side it is artery, bronchus, veins. The reason of this difference is that the left bronchus passes beneath the arch of the aorta, and therefore becomes lower than the pulmonary artery on that side. The right pulmonary vessels pass beneath the arch of the aorta.*

The **Lungs** (Fig. 170) will vary very much in condition in different bodies. If perfectly healthy their surfaces will be smooth and they will be somewhat contracted, but crepitant throughout. Every variety of disease may be

* It may be noticed that the order of structures from before backwards in the root of the lung corresponds to the arrangement in the hilum of the kidney, viz., vein, artery, ureter.

found, the most common being tubercle and abscess in the upper lobes with adhesion to the wall of the thorax, or pneumonic condensation through more or less of the organ. The lungs of subjects dying in large cities not unfrequently present black pigment in the lung tissue and

Fig. 170.

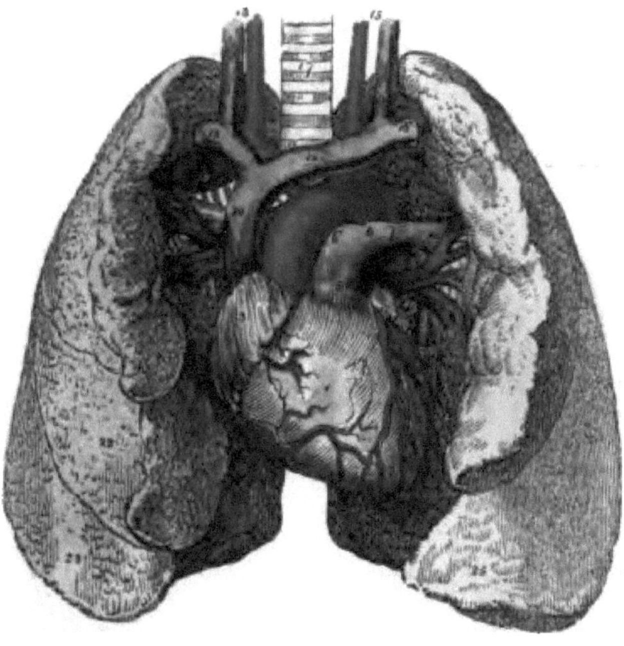

Fig. 170.—The heart and lungs (from Wilson).

1. Right ventricle.
2. Left ventricle.
3. Right auricle.
4. Left auricle.
5. Pulmonary artery.
6. Right pulmonary artery.
7. Left pulmonary artery.
8. Remains of the ductus arteriosus.
9. Arch of the aorta.
10. Superior vena cava.
11. Arteria innominata, and in front of it the right vena innominata.
12. Right subclavian vein, and, behind it, its corresponding artery.
13. Right common carotid artery and jugular vein.
14. Left vena innominata.
15. Left carotid artery and vein.
16. Left subclavian vein and artery.
17. Trachea.
18. Right bronchus.
19. Left bronchus.
20, 20. Pulmonary veins.
21. Superior lobe of the right lung.
22. Middle lobe.
23. Inferior lobe.
24 Superior lobe of the left lung.
25. Inferior lobe.

the lymphatic glands. Each lung has a thin anterior margin and a thicker posterior one, which lies against the spinal column and is longer than the anterior border. The anterior margin of the left lung is notched so as to leave the apex of the heart uncovered. The outer surface of the lung is convex and the inner concave; about the middle of this latter surface and near the posterior border is a slit, the hilum, at which the root of the lung enters, whilst in front of it is a hollow corresponding to the heart and larger in the left lung than the right. The base of the lung is moulded to the diaphragm and is therefore lower behind than in front; whilst the apex is

Fig. 171.

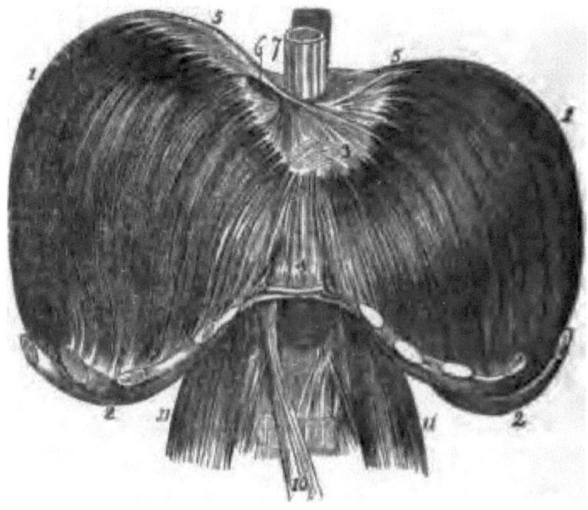

Fig. 171.—Upper or thoracic surface of the diaphragm (from Wilson).

1, 1. The lateral segments of the thoracic portion; arising from 2, 2. The cartilages of the ribs; and inserted into 3. The central tendon. The right segment is seen to be higher than the left.

4. The fasciculus which arises from the ensiform cartilage.

5, 5. Lateral leaflets of the central tendon.

6. Opening for the inferior vena cava.

7. The œsophagus.

8. The thoracic aorta.

9. The abdominal aorta.

10. The tendon of the right crus of the diaphragm; that of the left is seen immediately above on the lumbar vertebræ.

11, 11. The psoas muscles.

rounded and reaches into the neck above the first rib extending to a point an inch and a half above the clavicle. The right lung is the larger (owing to the direction of the heart to the left), but is shorter than the left, owing to the position of the liver. The left lung is divided into two lobes (upper and lower) by a fissure running obliquely across the external surface from near the apex to the anterior border, but the right has three lobes, owing to the existence of a second fissure running horizontally forward from the middle of the first fissure to the anterior margin of the lung.

The upper or thoracic surface of the **Diaphragm** (Fig. 171) is convex, the muscle being in the position of expiration. The right side rises higher into the thorax than the left, reaching to the fourth rib in the former and to the fifth rib in the latter case. The cordiform tendon, to which the pericardium is attached, will be seen in the centre, and the muscular fibres arching around it, leaving a shallow space in front between them and the ribs which is occupied by the edges of the lungs when fully expanded, and a much deeper and wider space behind in which the posterior parts of the lung are found. The vena cava inferior will be seen piercing the tendon to enter the right auricle; the other structures which perforate the diaphragm will be found in the posterior mediastinum.

The **Heart** (Fig. 172) is placed obliquely between the lungs, the base being to the right and the apex to the left side; and resting on the diaphragm which is now convex (the position of expiration) it is nearly horizontal. The base corresponds to the interval between the 6th and 9th dorsal vertebræ, and the apex beats between the 5th and 6th ribs, the larger portion of the organ being to the left of the median line. For purposes of auscultation the usual directions given are, the upper border of the third costal cartilage for the base of the heart, and a spot two inches below the nipple and one to the sternal side for the apex.

The anterior surface of the heart is convex and is formed almost entirely by the right ventricle and auriculaappendage, but the irregular border of the left auricular appendage appears to the left of the pulmonary artery. The posterior surface of the heart is flattened, and is formed by part of the right auricle and the left auricle and ventricle, which last forms the apex of the heart. Each surface of the heart is grooved vertically and horizontally,

Fig. 172.

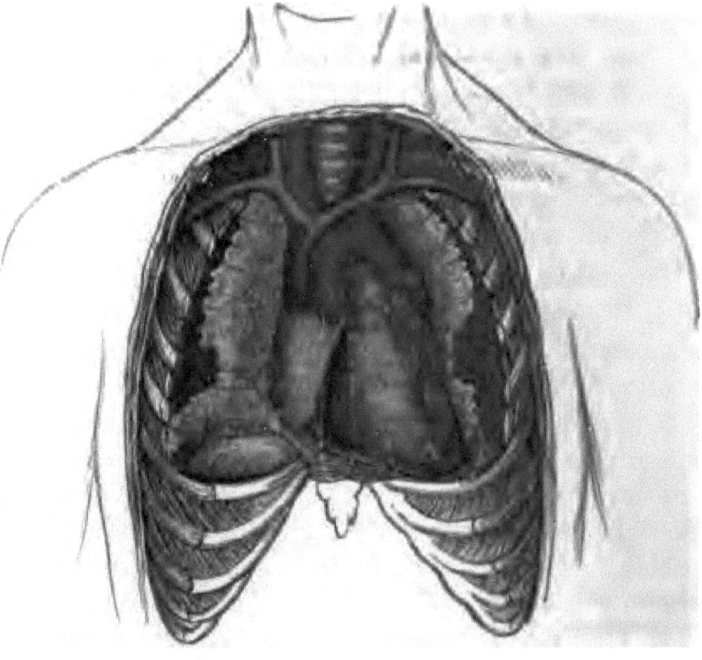

marking the divisions between the auricles and ventricles. The anterior ventricular groove is near the left border of the heart, whilst the posterior ventricular groove is to the right, and they thus indicate the oblique position of the septum. In these grooves will be seen the coronary vessels.

The **Coronary Arteries** (Fig. 173) (right and left) are

Fig. 172.—Diagram of the relations of the heart and great vessels to the wall of the thorax. The collapsed lungs are drawn aside slightly, to expose the parts more clearly. (Drawn by J. T. Gray.)

A. The space over which mitral murmurs are heard.
B. The space over which aortic murmurs are heard.

the first branches of the aorta and supply the substance of the heart.

Fig. 173.

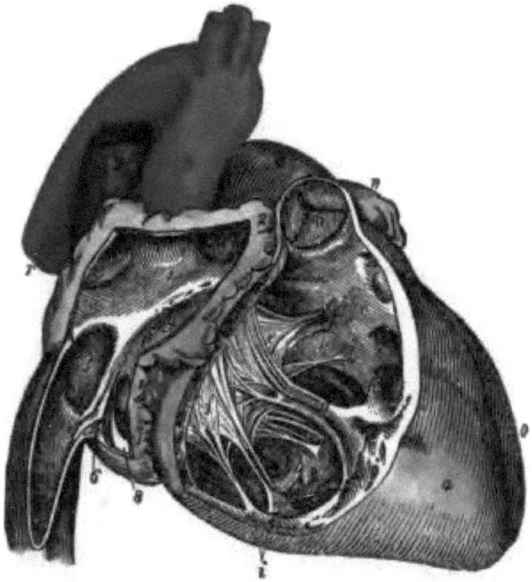

Fig. 173.—Right side of the heart laid open (from Wilson).

1. Cavity of right auricle.
2. Appendix auriculæ, in its cavity are seen the musculi pectinati.
3. Superior vena cava, opening into the upper part of the right auricle.
4. Inferior vena cava.
5. Fossa ovalis; the prominent ridge surrounding it is the annulus ovalis.
6. Eustachian valve.
7. Opening of the coronary sinus.
8. Coronary valve.
9. Entrance of the auriculo-ventricular opening. Between the figures 1 and 2, two or three foramina Thebesii are seen.

a. Right ventricle.
b. Cavity of right ventricle.
c. Conus arteriosus or infundibulum.
d. Pulmonary artery.
e, f. Tricuspid valve; *e* is placed on the left curtain, *f* on the anterior curtain.
g. One of the musculi papillares, to the apex of which the anterior and right curtains are connected by chordæ tendineæ.
h. Columnæ carneæ.
i. Two musculi papillares of the right curtain.
k. Attachment by chordæ tendineæ of the left limb of the anterior curtain.
l, l. Chordæ tendineæ.
m. Semilunar valves of the pulmonary artery.
n. Apex of left appendix auriculæ.
o. Left ventricle.
p. Ascending aorta.
q. Its transverse portion, with the three arterial trunks which arise from the arch.
r. Descending aorta.

The *left coronary artery* runs from right to left in the auriculo-ventricular groove, and gives a large branch to the front of the heart, which appears to the left of the pulmonary artery and runs down the anterior ventricular groove to the apex.

The *right coronary artery* runs from left to right in the auriculo-ventricular groove, and gives a large branch to the back of the heart, which runs along the septum to the apex. Both the vertical and horizontal branches of the two arteries anastomose, and supply branches freely to the substance of the heart.

The **Coronary Veins** do not correspond precisely to the arteries. The *anterior coronary vein* runs up the anterior ventricular groove with the artery, but leaves it to pass along the horizontal groove at the back of the heart to the right auricle. It receives the *posterior coronary vein* from the posterior ventricular groove, and is then called the *coronary sinus*, which opens into the right auricle.

The *superficial cardiac plexus* of the sympathetic is situated immediately below the arch of the aorta, and gives branches to the front and back of the heart, which are seldom seen. To it may be traced the left superficial cardiac nerve of the sympathetic, a cardiac branch from the left pneumo-gastric, and branches from the deep cardiac plexus.

The **Great Vessels of the Heart** (Fig. 170) have the following positions. The vena cava superior is to the right, and the pulmonary artery to the left, whilst between them the arch of the aorta is seen. The inferior vena cava can be seen piercing the diaphragm at the back of the heart, by drawing the organ upwards.

The **Cavities** of the heart are to be opened *in situ* and in the order in which the blood enters them.

[The right auricle is to be opened by one incision from the superior to the inferior vena cava, and another into the auricular appendage.]

Auricle (Fig. 173, 1).—The main cavity of the

auricle is smooth internally, but in the appendix are the *musculi pectinati*, or muscular bands "resembling a comb." The *endocardium* or lining membrane of the heart is seen to be continuous with the lining membrane of the veins, and will be traced subsequently into the arteries.

The large openings into the right auricle are (1) the *superior vena cava* (3), which enters at the upper and anterior part; (2) the *inferior vena cava* (4), which enters at the lower and back part; and (3) the *coronary sinus* (7), which enters close above (4) the *auriculo-ventricular* opening (9).

The *foramina Thebesii* are numerous small openings which are found in the wall of the auricle, and return blood from the muscular tissue of the heart.

The *tubercle of Lower* is a projection which is occasionally found in the wall of the auricle, between the superior and inferior venæ cavæ.

The *coronary valve* (8) is a thin fold at the orifice of the coronary sinus, which serves to prevent regurgitation into it.

The *Eustachian valve* (6) is a fold placed to the left of the vena cava inferior and immediately above the opening of the coronary sinus, which served in the fœtus to direct the current of blood from the inferior vena cava through the foramen ovale.

The *fossa ovalis* and *annulus ovalis* (5) are remains of fœtal structure, found on the inner wall of the right auricle in the position of the *foramen ovale* or communication between the two auricles in the fœtus.

The *annulus ovalis* is a muscular ring which is generally well marked, and which surrounds the shallow *fossa ovalis*, formed by a thin membrane thrown across the foramen ovale. This membrane is produced from the anterior and posterior margins of the foramen immediately after birth, and the two portions gradually obliterate the opening by overlapping one another; not unfrequently however a small oblique opening will be found at their point of junction, through which a probe can be introduced.

[To open the right ventricle, its flaccid wall should be grasped with the left hand and the scalpel made to transfix it about halfway down, and well to the right side of the septum; the knife being then carried towards the apex, the cavity of the ventricle will be opened. The left forefinger is to be passed up into the pulmonary artery, and will serve to guide the incision, which is to be prolonged into that vessel if possible *between* two of the semilunar valves. All clots being removed, the whole of the ventricle will be displayed.]

The **Right Ventricle** (Fig. 173, *a*) consists of a main cavity, the walls of which are irregular, owing to the projections of the muscular substance of the heart; and of a smooth funnel-shaped portion (*infundibulum* or *conus arteriosus*) leading upwards and to the left into the pulmonary artery. The projections on the wall of the ventricle are the *columnæ carneæ* (*h*) (fleshy columns) of which three varieties are described: one in which the column merely stands out in relief, being attached to the wall of the ventricle in its whole length; a second in which the column is attached at both ends but is free in the middle, so that a probe may be passed between it and the wall; and a third variety called the *musculi papillares* (*g*). These last project into the cavity of the ventricle and give attachment by their extremities to the *chordæ tendineæ*, or fibrous cords attached to the flaps of the auriculo-ventricular valve.

The right auriculo-ventricular valve (*e*) consists of three portions, and is hence called *tricuspid*. The flaps are formed by a reduplication of the *endocardium* or lining membrane of the heart, between the layers of which are some tendinous and muscular fibres, the former being continuous with the chordæ tendineæ. The entire valve is attached above to a fibrous ring (zona tendinosa), which bounds the auriculo-ventricular opening, and is divided below into three portions, anterior, posterior, and internal.*

* The tricuspid valve is very irregular in its divisions, sometimes consisting of only two flaps like the mitral valve, and at others being divided into four or even six small portions.

The anterior division of the valve is the largest and is in contact with the anterior wall, whilst the posterior flap lies against the posterior wall, which is formed by the posterior part of the septum ventriculorum. The internal or left flap, which is the smallest of the three, shuts off the infundibulum from the general cavity of the ventricle.

The auricular surface of the tricuspid valve is extremely smooth, for the purpose of facilitating the flow of blood into the ventricles; whilst the surface which corresponds to the walls of the ventricle is remarkably rough, from the prominences formed by the chordæ tendineæ (Power).

The tricuspid valve acts during contraction of the ventricle (systole) and prevents the regurgitation of blood into the auricle; though even in health there is said to be a slight reflux, which has been termed the "safety-valve action" (King).

The internal division of the tricuspid valve serves to prevent the blood from flowing into the pulmonary artery until the ventricle is fully distended and able to contract forcibly on its contents.

The *pulmonary artery* is attached to a fibrous ring which intervenes between it and the muscular substance of the heart, but the lining membrane of the artery is continuous with that of the ventricle.

The *semilunar valves* (*m*) of the pulmonary artery, two anterior and one posterior, are three reduplications of the lining membrane, strengthened by fibrous tissue which is collected principally at the attached border of each valve, the thin portion near the free border being called the *lunula*. The attached border is convex and is fixed to the wall of the artery; the free border is subdivided into two slightly concave portions by a little fibrous body called the *corpus Arantii*.

The semilunar valves act during dilatation of the ventricle (diastole), and prevent the regurgitation of the blood from the pulmonary artery.

The blood (which is venous or dark-coloured) is carried by the pulmonary artery to its bifurcation, and then by the right and left pulmonary arteries to the lungs, where it is aërated; and is brought back to the heart by the four pulmonary veins as arterial or red blood. The pulmonary veins open into the left auricle.

Fig. 174.

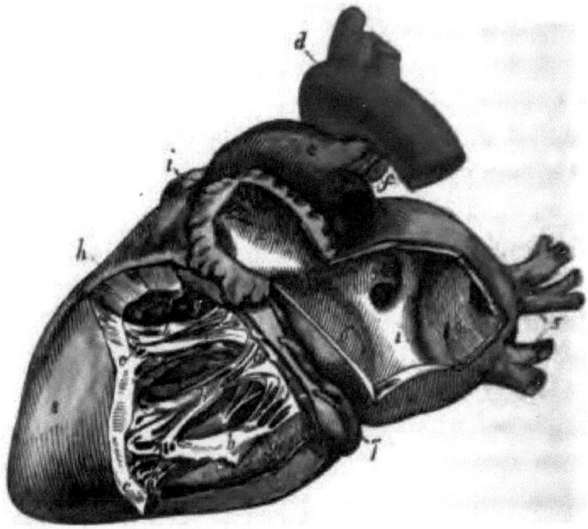

[The left auricle is to be opened by one incision on its posterior aspect placed vertically midway between the pulmonary veins, and another into the auricular appendage. The heart must be drawn well over to the right side to expose the cavity properly.]

Fig. 174.—Left side of the heart laid open from behind (from Wilson).

1. Cavity of the left auricle.
2. Cavity of the appendix auriculæ, near the apex of which are seen musculi pectinati.
3. Opening of the two right pulmonary veins.
4. The sinus, into which the left pulmonary veins sometimes open.
5. Left pulmonary veins.
6. Auriculo-ventricular opening.
7. Coronary vein, lying in the auriculo-ventricular groove.
8. Left ventricle.
9, 9. Cavity of the left ventricle; the figures rest on the septum ventriculorum.

a. Mitral valve; its flaps are connected by chordæ tendinæ to *b*, *b*, Musculi papillares.
c, *c*. Fixed columnæ carneæ, forming part of the internal surface of the ventricle.
d. Arch of the aorta.
e. Pulmonary artery.
f. Obliterated ductus arteriosus.
g. Left pulmonary artery.
h. Right ventricle.
i. Apex of the appendix of right auricle.

The **Left Auricle** (Fig. 174, 1) closely resembles the right auricle, but is altogether on a smaller scale, and its border is thinner and more "crenate" than on the opposite side. The openings are those of the four pulmonary veins, two on each side, and the left auriculo-ventricular opening.

The *musculi pectinati* of the auricular appendix are like those of the right side but smaller, and on the septum of the auricles will be seen the *annulus ovalis* and *fossa ovalis* corresponding to those on the right side, but the ring of muscular fibre is not so well developed.

[To open the left ventricle, the left forefinger should be introduced through the auriculo-ventricular opening, and the knife thrust through the wall of the ventricle near the apex to meet it. The knife is then guided upwards between the flaps of the mitral valve, and an incision made through the front of the ventricle into the auriculo-ventricular opening. The finger is next to be passed from below into the aorta followed by the knife, which is to be carried through the front wall of the ventricle close to the septum, thus isolating the right flap of the mitral valve. The incision is to be prolonged into the aorta between two of the semilunar valves, and it will be found to be necessary to divide the pulmonary artery which lies in front of the aorta, but care should be taken to do so above the pulmonary semilunar valves.]

The **Left Ventricle** (Fig. 174, 8) resembles the right, but its wall is thicker and its cavity reaches to the apex of the heart. The *columnæ carneæ*, *musculi papillares*, and *chordæ tendineæ* resemble those of the right side, but are more fully developed.

The left auriculo-ventricular valve consists of two portions and is hence called bicuspid, or (from the resemblance to a mitre) mitral. The flaps of the valve are composed of the lining membrane of the heart, strengthened by tendinous fibres derived from the chordæ tendineæ like those on the right side, and are attached to the fibrous ring bounding the auriculo-ventricular opening. The right flap of the valve is anterior to the left flap, and is placed between the cavity of the ventricle and the orifice of the aorta, against which it lies during diastole of the ventricle. It is perfectly smooth on both surfaces so as not to impede the current of blood into the aorta, thus differing from the left flap, of which the surface next to the wall of the heart is rough (Power). Between the anterior flap of the mitral valve and the aortic valves is a small space with aponeurotic walls—the *intervalvular space* of Sibson—which receives the aortic valves when distended in "diastole."

The mitral valve acts during contraction of the ventricle (systole) and prevents the regurgitation of blood into the auricle. The anterior division of the mitral valve prevents the blood from flowing into the aorta until the ventricle is fully distended and able to contract forcibly on its contents, and the pressure of the blood in the intervalvular space on the anterior flap of the mitral valve keeps the latter closed up to the end of the "systole" or contraction of the ventricle.

The closure of the mitral valve accompanies the "first sound" of the heart, which is best heard at the apex. A *bruit* caused by disease of the valve would be heard in the same situation (Fig. 172, A), and if systolic depends upon regurgitation through the valve, if diastolic upon roughness of the surface over which the blood passes from the auricle.

A fibrous ring intervenes between the muscular tissue of the heart and the aorta, the lining membrane of which is continuous with the endocardium.

The *Aortic semilunar valves* resemble those of the pulmonary artery but are more fully developed, and the *corpora Arantii* are better seen than on the right side. The aortic valves occupy a position the converse of those of the pulmonary artery, viz., one in front and two behind; and above each of the three valves there is a dilatation of the aorta, called the *aortic sinus* or *sinus of Valsalva.* At the bottom of the anterior and left posterior sinuses will be seen the orifices of the coronary arteries, the first branches of the aorta.

The aortic valves act during dilatation of the ventricle (diastole) and prevent regurgitation into the ventricle, their closure accompanying the "second sound" of the heart, which is best heard over the base of the heart and along the sternum (Fig. 172, B). If from disease of the valves their closure is not perfect and regurgitation occurs, a "diastolic aortic bruit" is produced: if the surfaces of the valves should be so roughened as to offer an obstruction to the flow of blood during contraction of the ventricle, a "systolic aortic bruit" will be heard. A horizontal section above the valves (Fig. 175) shows their relation to one another. The left ventricle occupies the posterior aspect of the heart, and the mitral orifice and valve are therefore behind. In front of this is the aortic orifice, separated only by a fibrous septum. To the left of the aorta, and a little in front, is the pulmonary artery,

Fig. 175.

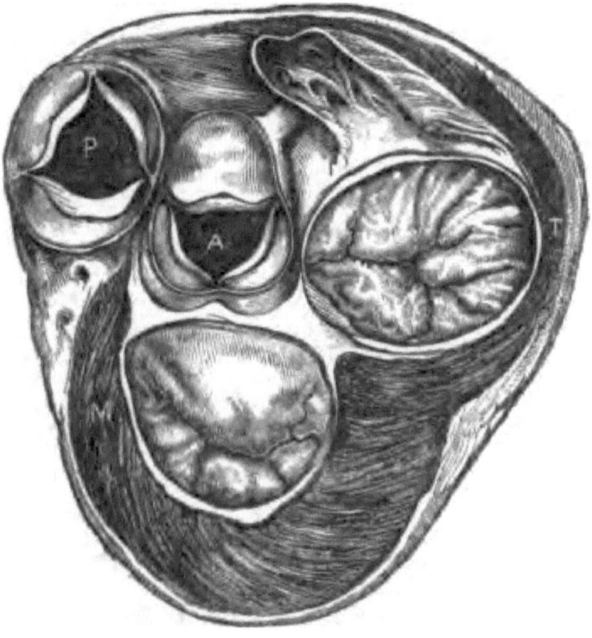

Fig. 175.—A section of the heart at the level of the valves, seen from above (from Sibson's Medical Anatomy).

P. Pulmonary artery. M. Mitral valve.
A. Aorta. T. Tricuspid valve.

with its valves, and to the right is the tricuspid or right auriculo-ventricular valve.

The **Fœtal Circulation** (Fig. 176).—This will be the best opportunity for contrasting the adult circulation with that of the fœtus. In the fœtus the blood is brought from the placenta by the umbilical vein, which enters the body at the umbilicus. It then passes along the longitudinal fissure of the liver, and at the transverse fissure divides into two branches, one of which joins the portal vein, and the other, which is the *ductus venosus*, joins the inferior vena cava. In the inferior vena cava the placental blood is joined by that returned from the lower extremities, and afterwards, through the hepatic veins, by that circulated through the liver; and is then poured into

Fig. 176.

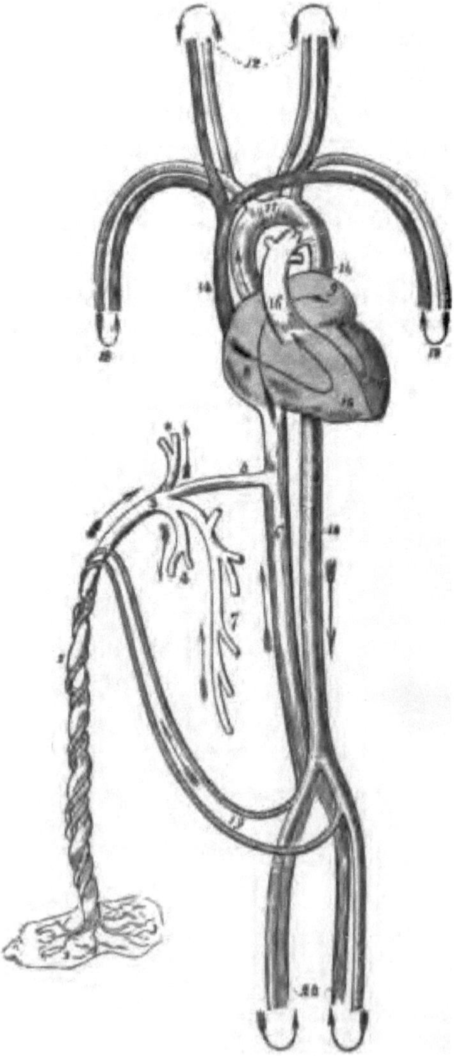

Fig. 176.—Diagram of the Fœtal circulation (from Wilson).

1. Umbilical vein proceeding from the placenta (2).
3. Umbilical vein, dividing into branches; two (4, 4), to be distributed to the liver; and one (5), the ductus venosus, which enters the inferior vena cava (6).

the right auricle. The Eustachian valve, of large size in the fœtus, directs the current across the auricle to the foramen ovale, which is patent, and the blood enters the left auricle. From the left auricle the blood necessarily passes into the left ventricle, and is thence propelled through the aorta to the head and upper extremities. From these it is returned by the superior vena cava, which enters the upper part of the right auricle, and the blood descends at once into the right ventricle, thus taking a course at right angles to the former one. From the right ventricle the blood is propelled into the pulmonary artery, and a small portion reaches the lungs through the right and left pulmonary arteries, but by far the larger portion passes through the *ductus arteriosus* (a short tube connected with the pulmonary artery close to the bifurcation), and enters the descending portion of the arch of the aorta. Through the aorta the blood reaches the iliac arteries, and a small portion passes by the external iliacs to the lower extremities, but the rest passes by the internal iliacs to the hypogastric arteries, which run to the umbilicus and then wind round the umbilical vein to the placenta.

The **Pulmonary Artery** (Fig. 177, 13) has already been seen to arise from the right ventricle and to lie in front of the aorta. It then passes to the left side of the aorta, where it bifurcates into right and left pulmonary arteries which go to their respective lungs; the right being the longer of the two and necessarily passing beneath the arch of the aorta, and the left crossing the descending aorta. The position of each of the arteries in the root of

---

7. Portal vein, communicating with the right hepatic branch.
8. Right auricle.
9. Left auricle.
10. Left ventricle.
11. The arch of the aorta.

The arrows, 12 and 13, represent the return of the blood from the head and upper extremities through the jugular and subclavian veins.

14. Superior vena cava.
15. Right ventricle.
16. Pulmonary artery.
17. Ductus arteriosus.
18, 18. Descending aorta.
19. Hypogastric arteries.
20. External iliacs.

the lung has been already seen, the right being below and the left above its bronchus.

Fig. 177.

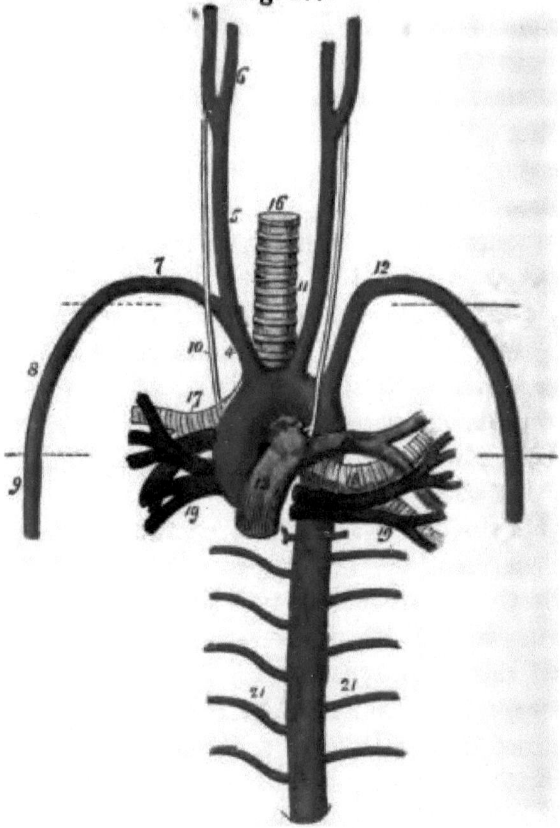

Fig. 177.—Diagram of the large vessels of the heart and lungs (from Wilson).

1. Ascending aorta.
2. Transverse portion of the arch.
3. Thoracic or descending aorta.
4. Arteria innominata.
5. Right common carotid.
6. External and internal carotids.
7. Right subclavian artery.
8. Axillary artery.
9. Brachial artery.
10. Right pneumogastric nerve.
11. Left common carotid.
12. Left subclavian artery.
13. Pulmonary artery.
14. Left pulmonary artery.
15. Right pulmonary artery.
16. Trachea.
17. Right bronchus.
18. Left bronchus.
19, 19. Pulmonary veins.
20. Bronchial arteries.
21, 21. Intercostal arteries; the branches from the front of the aorta above and below the number 3 are pericardiac and œsophageal.

Connecting the pulmonary artery with the descending portion of the arch of the aorta is a fibrous cord, which is the obliterated *ductus arteriosus.*

The **Pulmonary Veins** (Fig. 177, 19) are four in number, two to each lung. Their positions in the roots of the lungs have been seen and they have been traced to the left auricle of the heart. The right veins are the longer and pass beneath the arch of the aorta, the left veins crossing the descending aorta.

The **Arch of the Aorta** (Fig. 177).—The aorta has been seen to arise from the left ventricle, and it at first takes a course upwards and to the right side, and then backwards and to the left, thus forming an arch. For convenience of description the arch is divided into three portions—ascending, horizontal or transverse, and descending.

1. The *ascending portion* (1), beginning opposite the third costal cartilage of the left side, reaches as high as the upper border of the second costal cartilage of the right side close to the sternum. It is almost entirely enclosed within the pericardium, and is crossed at first by the pulmonary artery. *In front* of the pericardium are the sternum and triangularis sterni muscle; *behind* the ascending aorta is the root of the right lung; to its *right* side is the vena cava; and to its *left* side is the bifurcation of the pulmonary artery. Its branches are the two coronary arteries, which have been already traced.

2. The *horizontal* or *transverse portion* has a direction backwards and to the left side, reaching from the second costal cartilage of the right side to the left side of the fourth dorsal vertebra. *In front* of this portion are the sternum and triangularis sterni muscle and, from left to right, the left pneumogastric and left phrenic nerves and the superficial cardiac branch of the sympathetic; *behind* are the trachea, the œsophagus and thoracic duct, the right pneumo-gastric and the left recurrent laryngeal nerves. *Above* is the left innominate vein; and *below* are the left bronchus and right pulmonary vessels, the left recurrent laryngeal nerve, and the obliterated ductus

arteriosus. The branches of this portion are the innominate, the left carotid, and left subclavian arteries.

3. The *descending portion* extends to the lower border of the fifth dorsal vertebra where the thoracic aorta begins, but the division between the two vessels is arbitrary. It is invested almost entirely by the left pleura, which binds it to the side of the vertebræ.

The **Vena Cava Superior** (Fig. 170, 10) is formed by the junction of the right and left brachio-cephalic or innominate veins on the right of the arch of the aorta. The vena cava receives the vena azygos major opposite the right bronchus, and then, crossing the root of the right lung, pierces the pericardium to enter the upper part of the right auricle.

Great Vessels of the Root of the Neck.

The three great branches of the second part of the arch of the aorta are the Innominate, the Left Carotid, and the Left Subclavian arteries.

The **Innominate Artery** (Fig. 177, 4) passes upwards and to the right side, and at the sterno-clavicular articulation divides into the right common carotid and right subclavian arteries. It has *in front* of it the upper piece of the sternum, with the remains of the thymus gland and the origins of the sterno-hyoid and sterno-thyroid muscles ; and is crossed nearly at right angles by the left brachio-cephalic vein, and obliquely by the right inferior thyroid vein. *Behind* it at first is the trachea, but afterwards the prolongation of the right pleura into the neck. To the *right* side are the right pneumo-gastric nerve, right innominate vein, and right phrenic nerve; to the *left* side, the origin of the left common carotid and afterwards the trachea.

The innominate artery ordinarily gives off no branch, but occasionally a small branch (*thyroidea ima*, middle thyroid artery of Harrison) arises from it or from the aorta close to it, and runs up the front of the trachea to the thyroid body.

The **Left Common Carotid Artery** (Fig. 177, 11).—The left common carotid runs upwards and to the left side at a level anterior to that of the left subclavian artery, and from the left sterno-clavicular articulation its relations correspond to those of the right carotid artery (p. 361).

The *thoracic portion* has *in front* of it the upper piece of the sternum, with the remains of the thymus gland and the origins of the sterno-hyoid and thyroid muscles; and is crossed by the left innominate vein. It lies *against* the trachea at first, then upon the œsophagus and thoracic duct, and lastly on the longus colli muscle; having the innominate artery, the trachea, and the left recurrent laryngeal nerve to its *right* side, and the left pneumo-gastric nerve with its cardial branches and the left subclavian artery to its *left* side.

The **Subclavian Arteries** (Fig. 177) differ on the two sides, the right beginning at the sterno-clavicular articulation, and the left at the arch of the aorta, having therefore a course in the thorax. Both arteries may be divided into three parts, of which the second and third correspond on the two sides of the body.

The *first portion on the left side* extends from the arch of the aorta to the inner border of the scalenus anticus, and may be conveniently subdivided into a thoracic and a cervical part. The *thoracic* part is at first nearly vertical in its direction and lies to the left side of, but in a plane posterior to that of, the left carotid artery. It is *crossed* transversely by the left innominate vein and obliquely by the left pneumo-gastric and cardiac nerves; it lies upon the œsophagus and thoracic duct, and afterwards on the pleura for the rest of its course, being also closely invested with pleura on the left side. The *cervical* portion is curved or nearly horizontal, and has *in front* of it the sterno-mastoid, sterno-hyoid, and sterno-thyroid muscles and the inner end of the clavicle, being crossed by the internal jugular and vertebral veins, and by the phrenic nerve close to the scalenus. It *lies against* the apex of the pleura, which intervenes between it and the first rib and also closely invests the artery below.

The *first portion on the right side* extends from the bifurcation of the innominate artery at the sterno-clavicular articulation to the inner border of the scalenus anticus. Its course is nearly horizontal, and it has *in front* of it the inner end of the clavicle with the sterno-mastoid, sterno-hyoid, and sterno-thyroid muscles, being crossed by the pneumo-gastric, cardiac, and phrenic nerves, and the internal jugular and vertebral veins. It lies *against* the recurrent laryngeal nerve and the apex of the pleura, which intervenes between it and the first rib and also invests its lower border. The innominate vein is in front of, but quite below the level of, this part of the artery.

The Branches of the first portion of the subclavian artery are (1) Vertebral, (2) Internal Mammary, and (3) Thyroid Axis, and their distribution is the same on both sides of the body.

1. The **Vertebral Artery** is seen now in only a small part of its course. It ascends between the scalenus anticus and longus colli muscles (being crossed by the inferior thyroid artery and, on the left side, by the thoracic duct); and enters the foramen in the transverse process of the 6th cervical vertebra (usually), being accompanied by a branch from the inferior cervical ganglion of the sympathetic. The artery passes through the transverse processes of all the upper cervical vertebræ, giving off *muscular* and *spinal* branches in its course, and is seen in the suboccipital region to wind inwards upon the atlas and enter the foramen magnum to supply the brain.

The *Vertebral vein* has no course in the skull, but commences in small branches about the atlas. It takes the same course as the artery, receiving corresponding branches and also the *ascending cervical* and *deep cervical* veins, and after crossing the subclavian artery opens into the commencement of the innominate vein.

2. The **Internal Mammary Artery** (Fig. 169, 2) arises from the lower surface of the subclavian artery, and at once descends into the thorax, being crossed superficially

and obliquely by the phrenic nerve. The left artery has been seen to pass through the anterior mediastinum, but the right is excluded by the pleura, which binds it to the costal cartilages. Both arteries enter the fibres of the triangularis sterni muscle and divide opposite the seventh costal cartilage into two terminal branches—superior epigastric and musculo-phrenic.

a. The *superior epigastric branch* enters the fibres of the rectus abdominis muscle and anastomoses with the epigastric branch of the external iliac, thus establishing a communication which becomes of great importance in any case of obstruction of the aorta or iliac arteries.

b. The *musculo-phrenic branch* supplies the diaphragm and runs outwards to anastomose with the intercostal and lumbar arteries, as well as with the phrenic branches of the abdominal aorta.

The other branches of the internal mammary are (*c*) *comes nervi phrenici*, a small branch accompanying the phrenic nerve and seldom seen; (*d*) *mediastinal* and (*e*) *pericardiac* branches to those parts, from which small branches pass to form the "sub-pleural mediastinal plexus" (Turner); (*f*) *anterior intercostals* to the intercostal spaces, anastomosing with other intercostal branches; and (*g*) *perforating* branches to the pectoral muscles and to the mamma.

The *venæ comites* of the internal mammary artery unite to open into the corresponding innominate or brachio-cephalic vein.

3. The **Thyroid Axis** (Fig. 178, 25) is a short thick trunk arising close to the scalenus and from the upper surface of the subclavian, which divides immediately into three branches, (*a*) inferior thyroid, (*b*) transversalis colli, and (*c*) transversalis humeri.

a. The *Inferior thyroid artery* runs upward and inward, across the vertebral artery and behind the carotid sheath and sympathetic trunk, to the thyroid body, where it anastomoses with its fellow of the opposite side and with both the superior thyroid arteries. Besides *œsophageal* and *tracheal* branches, it usually gives off a branch close to its origin, the *ascending cervical*, which ascends upon the vertebræ between the scalenus anticus and rectus capitis anticus major, supplying the prevertebral muscles and anastomosing with branches of the vertebral artery.

The *inferior thyroid veins* pass down the front of the trachea, and frequently have a transverse communicating branch. They open into the innominate veins, the right crossing obliquely over the innominate artery.

b. The *transversalis colli artery* runs transversely outward in front of the scalenus anticus and phrenic nerve, and has been seen in the posterior triangle of the neck to divide into superficial cervical and posterior scapular branches.

This artery is frequently of small size or altogether wanting, the posterior scapular arising from the third part of the subclavian.

c. The *transversalis humeri artery* runs outward in front of the scalenus anticus and phrenic nerve immediately behind the clavicle, and has been seen in the posterior triangle of the neck to become the suprascapular artery. It is often known as the suprascapular artery in its whole course.

The **Second Part of the Subclavian Artery** (Fig. 178, 27) is placed behind the scalenus anticus, and has the same relations on both sides of the body. It has *in front* of it the platysma and cervical fascia, with the clavicular origin of the sterno-mastoid and the scalenus anticus, and rests *against* the scalenus medius and the first dorsal nerve. *Above* it are the lower cervical nerves, and *below* is the pleura with a small portion of the inner border of the first rib.* The *subclavian vein* is quite below the level of the artery at this point, and separated by the scalenus anticus. The only branch of the second part of the subclavian artery is the superior intercostal artery, which is

* The phrenic nerve is commonly given as one of the anterior relations of this part of the artery, but it reaches the inner border of the scalenus above the vessel and is in relation with the first part of the subclavian.

to be traced into the thorax by removing the pleura from the upper intercostal spaces.

The *superior intercostal artery* descends into the thorax in front of the necks of the ribs, giving branches to the first and second intercostal spaces and anastomosing with the upper intercostal artery from the aorta. The branches to the intercostal spaces divide into anterior and posterior branches, and are distributed like the aortic intercostals (*q. v.*). The *deep cervical branch* arises from the superior intercostal close to its origin, and passes backwards between the first rib and the transverse process of the seventh cervical vertebra to be distributed to the muscles of the back.

The *superior intercostal vein* opens into the innominate vein.

The **Third Part of the Subclavian Artery** has been dissected in the posterior triangle of the neck (p. 340).

The **Subclavian Vein** (Fig. 180) is the continuation of the axillary vein, and has been seen to lie below the level of its artery in the third part of its course (p. 340). It then passes in front of the scalenus anticus, which muscle separates it from the second portion of the subclavian artery; and lastly lies in front of and a little below the first part of the artery, with the phrenic nerve intervening on both sides of the body, and on the right side the pneumo-gastric nerve also. Each subclavian vein joins the internal jugular vein of the same side to form the innominate vein, and at the point of junction of these two veins the thoracic duct opens on the left side, and the right lymphatic duct on the right side of the body.

Branches.—The external and anterior jugular veins open into the subclavian vein outside the scalenus anticus.

The **Right Innominate Vein** (Fig. 180) commences at the inner end of the clavicle by the junction of the subclavian and internal jugular veins, and then descends on the outer side of the innominate artery to join the opposite vein in the vena cava. It is closely invested by the right pleura, and has the phrenic nerve on its outer side.

The **Left Innominate Vein** (Fig. 180) commencing at a

Fig. 178.

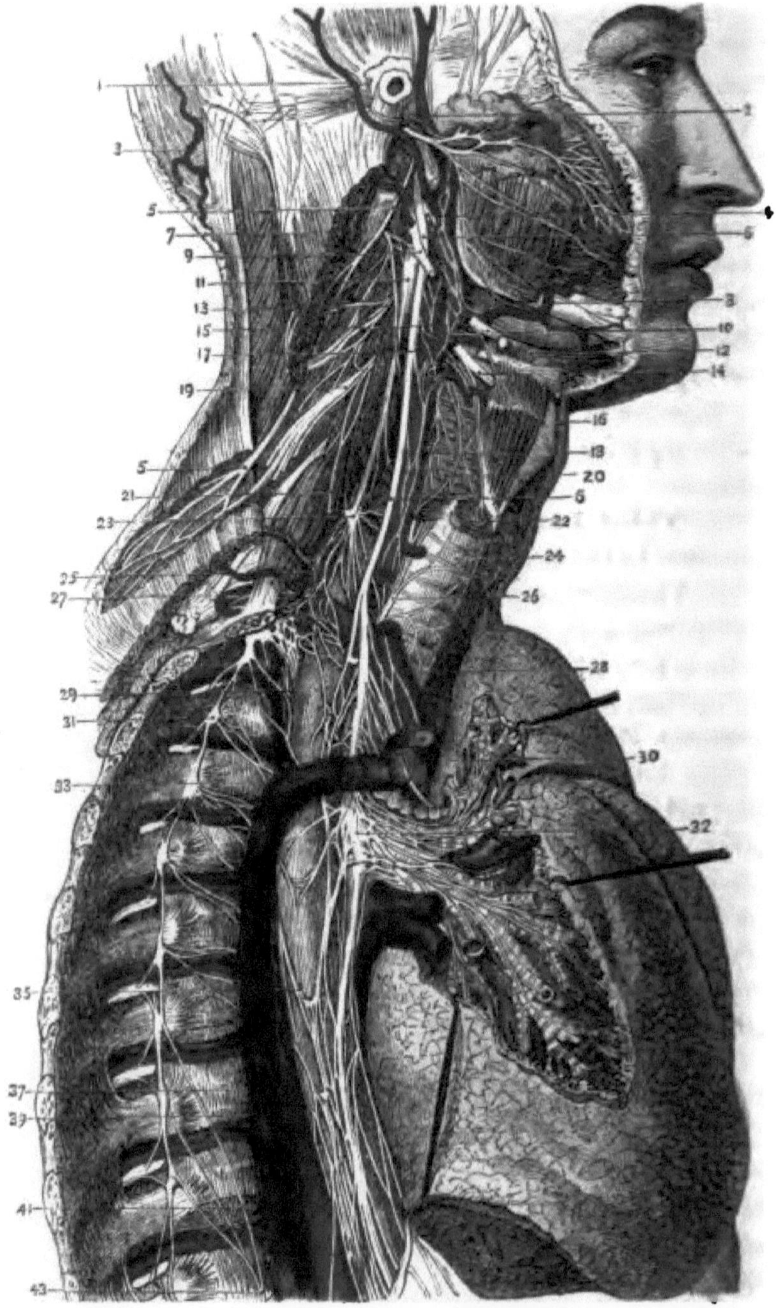

Fig. 178.—Distribution of the eighth nerve of the right side (from Hirschfeld and Leveillé).

corresponding point to and in the same way as the right vein, runs obliquely downwards and to the right side, lying above the level of the arch of the aorta and crossing its large branches. It unites with the right innominate vein on the right of the sternum, between the cartilages of the first and second ribs, to form the vena cava superior (p. 430).

Branches.—Each innominate vein receives the vertebral vein close to its origin, and afterwards the internal mammary, the inferior thyroid, and the superior intercostal branches. The left vein receives in addition small thymic and pericardiac branches.

[The ascending portion of the arch of the aorta and the venæ cavæ are to be divided, and the remains of the heart removed with the pulmonary vessels, which are to be cut close to the lungs. The arch of the aorta is to be held to one side by hooks, and the bifurcation of the trachea with the deep cardiac plexus dissected out.]

The **Deep Cardiac Plexus** (Fig. 178) is situated on each side of the trachea close to its bifurcation. The *rignt half* of the plexus receives all the cardiac nerves of that side, viz., three cardiac nerves from the three cervical

---

1. Posterior auricular artery.
2. Temporal artery.
3. Occipital artery.
4. Glosso-pharyngeal nerve.
5, 5. Spinal-accessory nerve.
6, 6. Pneumo-gastric nerve.
7. Sterno-mastoid (cut).
8. Facial artery.
9. Hypoglossal nerve with communication from 2nd cervical nerve.
10. Lower end of ditto.
11. Superior cervical ganglion of sympathetic.
12. Digastricus.
13. Third cervical nerve.
14. Superior laryngeal nerve.
15. Internal carotid.
16. Thyro-hyoideus.
17. External carotid.
18. Common carotid.
19. Fourth cervical nerve.
20. Inferior constrictor of pharynx.
21. Phrenic nerve on scalenus anticus.
22. Crico-thyroideus.
23. Middle cervical ganglion.
24. Trachea.
25. Thyroid axis.
26. Recurrent laryngeal nerve.
27. Subclavian artery.
28. Innominate artery.
29. Œsophagus.
30. Vena cava superior (cut).
31. Gangliated cord of sympathetic.
32. Posterior pulmonary plexus.
33. Phrenic nerve (cut).
35. Œsophageal plexus.
37. Vena azygos major.
39. Thoracic duct.
41. Thoracic aorta.
43. Great splanchnic nerve.

ganglia of the sympathetic, the three cardiac branches of the pneumo-gastric, and the cardiac branch of its recurrent laryngeal nerve. The branches of this half of the plexus are distributed to the right side of the heart and the right lung, and many of them have been necessarily destroyed. The *left half* (Fig. 179) of the plexus receives the same nerves as the right, with the exception of the superior cardiac nerve of the sympathetic and the inferior cardiac branch of the pneumo-gastric, which have been already traced to the superficial cardiac plexus. The branches of this half of the plexus are distributed to the left side of the heart and the left lung, and also communicate with the superficial cardiac plexus.

The **Trachea** (Fig. 178, 24) is now sufficiently exposed for the examination of its relations, but its structure will be given with that of the lungs. The trachea extends from the lower border of the larynx, about the level of the fifth cervical vertebra, to the level of the fifth dorsal vertebra where it bifurcates into the bronchi. The trachea occupies the middle line, lying in front of the œsophagus and vertebral column, and has the following structures in front of it in the neck: the sterno-hyoid and thyroid muscles with the deep cervical fascia; the isthmus of the thyroid body with the inferior thyroid veins; and the *arteria thyroidea ima* from the innominate, if it exists. In the thorax it lies in the posterior mediastinum, and has in front the sternum with the remains of the thymus gland; the arch of the aorta and the nerves crossing it; the innominate and left carotid arteries (for a very short distance); and the left innominate vein.

The **Right Bronchus** (Fig. 177, 17) is larger than the left, and has been seen to be posterior and superior to the pulmonary vessels in the root of the lung. It is about one inch long, and takes a more horizontal course than the left. The vena azygos major hooks round the right bronchus to open into the vena cava.

The **Left Bronchus** (Fig. 177, 18) is nearly twice as long as the right and takes an oblique course beneath the arch of the aorta, crossing the œsophagus and the descending aorta, and lying posterior to, but visible between, the pulmonary artery and veins (p. 412).

The **Thyroid Body** (Fig. 179, 19) has been seen incidentally in the dissection of the neck, but is now fully exposed for examination. It consists of two symmetrical conical *lobes*, placed on each side of the upper part of the trachea, and united opposite the second and third rings of the trachea by the *isthmus*. The lobes are subject to great variations in size, and when much hypertrophied constitute Bronchocele or Goître.

The thyroid body is covered by the sterno-hyoid and thyroid muscles, and occasionally a few muscular fibres pass from the hyoid bone to the isthmus, constituting the *levator glandulæ thyroideæ* of Soemmering. It is abundantly supplied with blood by the superior thyroid and inferior thyroid arteries of each side, and occasionally by an additional branch from the innominate.

The arteries freely anastomose in the substance of the body, and return their blood by three veins on each side, viz., the superior and middle thyroid which join the internal jugular vein, and the inferior thyroid which has been traced down the front of the trachea to the innominate vein.

The thyroid body is composed of numerous closed vesicles containing a yellow fluid, but its function is not understood.

[The right lung is to be drawn forward and the pleura divided where it is reflected from the lung to the wall of the thorax, and the parts in the posterior mediastinum are to be cleaned.

The muscular œsophagus will be at once seen, and the right pneumo-gastric nerve is to be traced to it and to the back of the right bronchus. On displacing the œsophagus the side of the thoracic aorta will come into view, but it will be better seen in the dissection of the left side.

The vena azygos major will be seen to the right of the aorta, and between the two will be found the slender and collapsed thoracic-duct. The intercostal vessels will be seen crossing the back of the space, and near the diaphragm will be found the splanchnic nerves from the sympathetic cord, which is itself outside the mediastinum.]

Fig. 179.

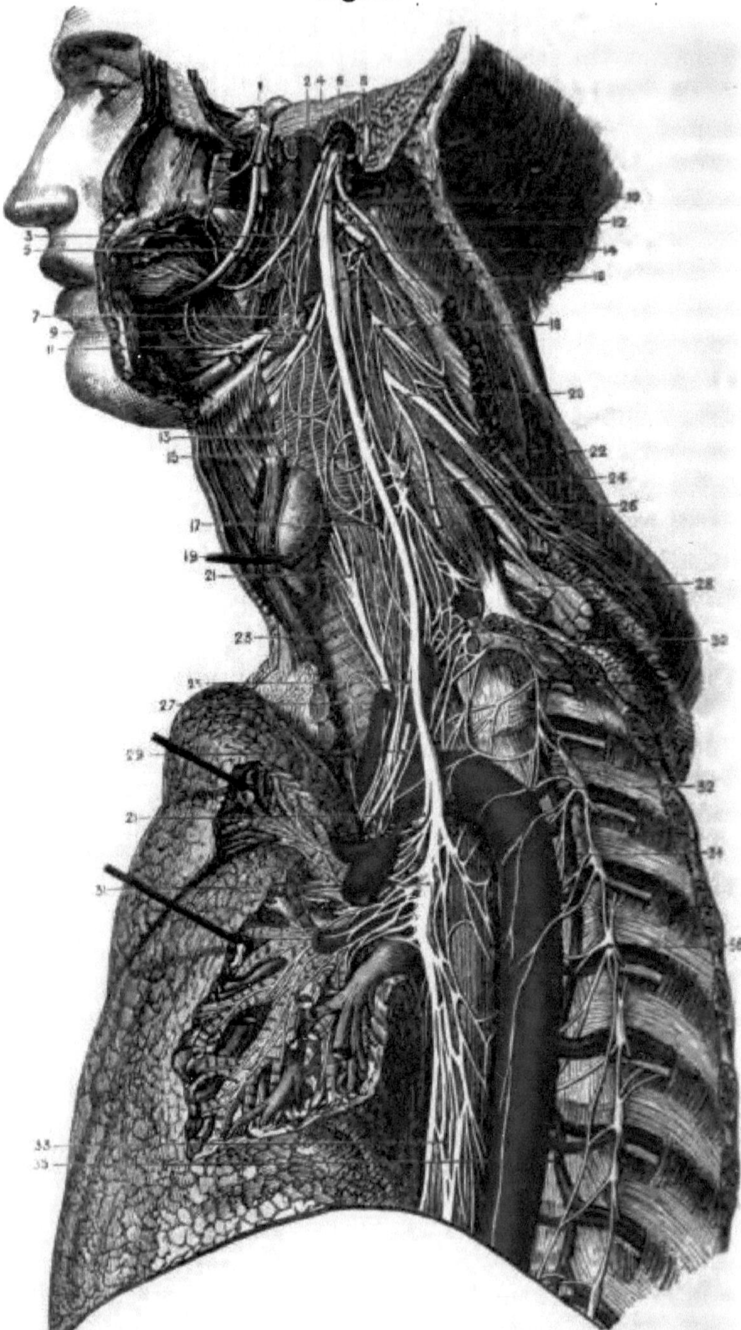

Fig. 179.—Distribution of the eighth pair of nerves on the left side (from Hirschfeld and Leveillé).

The **Posterior Mediastinum** (Fig. 169) is the interpleural space behind the pericardium, bounded by the vertebræ behind, the pericardium in front, and the reflection of the pleura on each side. It contains the trachea, œsophagus, and the two pneumo-gastric nerves; the thoracic aorta, vena azygos major and thoracic duct; and, at the lower part, the great splanchnic nerves.

The **Œsophagus** (Fig. 178, 29) is a muscular tube continuous with the pharynx. It begins opposite the 5th cervical vertebra, and can now be seen to lie to the left side of the median line in the anterior triangle. It then passes through the superior aperture of the thorax, being in relation with the left common carotid artery, and reaches the front of the spine, passing behind the arch of the aorta and being crossed by the left bronchus. It is now seen to pass in front of the thoracic aorta, which it crosses very obliquely to reach the left side and pass through the œsophageal opening in the diaphragm.

---

1. Gasserian ganglion of 5th nerve.
2. Internal carotid artery.
3. Pharyngeal branch of pneumo-gastric.
4. Glosso-pharyngeal nerve.
5. Lingual nerve (5th).
6. Spinal-accessory nerve.
7. Middle constrictor of pharynx.
8. Internal jugular vein (cut).
9. Superior laryngeal nerve.
10. Ganglion of trunk of pneumo-gastric nerve.
11. Hypoglossal nerve on hyoglossus.
12. Ditto communicating with eighth and first cervical nerve.
13. External laryngeal nerve.
14. Second cervical nerve looping with first.
15. Pharyngeal plexus on inferior constrictor.
16. Superior cervical ganglion of sympathetic.
17. Superior cardiac nerve of pneumo-gastric.
18. Third cervical nerve.
19. Thyroid body.
20. Fourth cervical nerve.
21, 21. Left recurrent laryngeal nerve.
22. Spinal-accessory communicating with cervical nerves.
23. Trachea.
24. Middle cervical ganglion of sympathetic.
25. Middle cardiac nerve of pneumo-gastric.
26. Phrenic nerve (cut).
27. Left carotid artery.
28. Brachial plexus.
29. Phrenic nerve (cut).
30. Inferior cervical ganglion of sympathetic.
31. Pulmonary plexus of pneumo-gastric.
32. Thoracic aorta.
33. Œsophageal plexus.
34. Vena azygos superior.
35. Vena azygos minor.
36. Gangliated cord of sympathetic.

The œsophagus consists of two layers of *muscular* fibres, the outer longitudinal and the inner circular, which are of the striped variety in the upper part but of the unstriped variety in the lower part of the tube. Within the muscular coats are a delicate *fibrous* coat and a simple *mucous membrane* with tesselated epithelium. The long meshes of nerve upon the œsophagus are derived from the two pneumo-gastrics, and form the *plexus gulæ*.

Fig. 180.

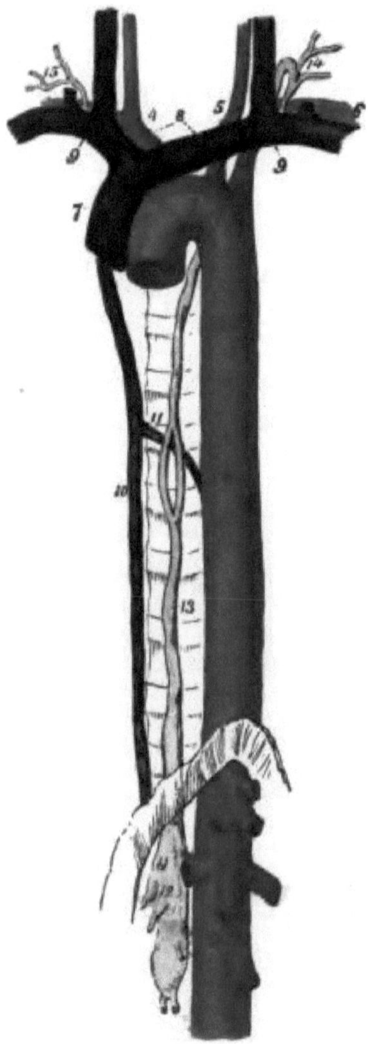

The **Thoracic Duct** (Fig. 180) is a delicate tube about eighteen inches long, embedded in loose tissue between the aorta and the vena azygos major. By cutting away the remains of the diaphragm carefully it may be traced from the *receptaculum chyli* (12) opposite the second lumbar vertebra, and will be found to pass through the aortic opening to the right side of the aorta. It continues to the right of the aorta as high as the fourth dorsal vertebra, and then crosses obliquely to the left of the spine behind the arch of the aorta, and runs along the left

Fig. 180.—The course and termination of the thoracic duct (from Wilson).

side of the œsophagus through the superior aperture of the thorax. In the neck the duct reaches as high as the sixth cervical vertebra, and then, crossing the left vertebral artery and thyroid axis, curves downwards in front of the scalenus anticus and phrenic nerve to enter the left subclavian vein at its junction with the left internal jugular. The duct occasionally passes higher in the neck and opens into the jugular at the point where some irregular vein joins it. The duct is dilated near its termination, and is often double both there and in the thorax. It has valves at various points, and by opening the subclavian vein a pair may be found at its entrance into the vein, which prevent the reflux of blood into the duct.

Some intercostal *lymphatic glands* may be found between the ribs close to the vertebræ, which open into the duct. Œsophageal and bronchial glands may also be seen in connection with the tubes of the same name.

The **Vena Azygos Major** (Fig. 178, 37) commences in one of the lumbar veins and is to be traced through the aortic orifice of the diaphragm. It lies to the right of the thoracic duct upon the vertebræ, crossing the right intercostal arteries and receiving all the right intercostal veins except the first and second. About the level of the sixth dorsal vertebra it receives the *vena azygos minor* passing from the left side behind the aorta, and lastly

---

1. Arch of the aorta.
2. Thoracic aorta.
3. Abdominal aorta.
4. Arteria innominata, dividing into right carotid and right subclavian.
5. Left carotid.
6. Left subclavian.
7. Superior vena cava.
8. The two venæ innominatæ.
9. Junction of the internal jugular and subclavian vein at each side.
10. Vena azygos major.
11. Vena azygos minor.
12. Receptaculum chyli; several lymphatic trunks are seen opening into it.
13. Thoracic duct, the course of the duct behind the arch of the aorta and left subclavian artery is shown by a dotted line.
14. The duct making its turn at the root of the neck and terminating in the posterior aspect of the junction of the internal jugular and subclavian vein.
15. Termination of the trunk of the ductus lymphaticus dexter.

arches forward over the right bronchus to open into the superior vena cava outside the pericardium.

The **Right Pneumo-gastric Nerve** (Fig. 178, 6) has been seen to enter the thorax between the subclavian artery and right innominate vein (p. 368). It is now seen to run backward to the right side of the trachea, along which it passes to the bifurcation to form the *posterior pulmonary plexus* at the back of the right bronchus. The nerve then supplies the œsophagus, forming with the nerve of the opposite side a plexus of long meshes which has been called the *plexus gulæ.* Lastly the right nerve reaches the back of the stomach.

Cardiac branches from the trunk of the pneumo-gastric nerve and from its recurrent laryngeal branch arise in the thorax, and the cervical cardiac branches may also be traced out, and will be afterwards seen to join the deep cardiac plexus.

[The left lung is now to be drawn forward and the pleura removed in the same manner as on the right side. The œsophagus with branches from the left pneumo-gastric will be seen near the diaphragm, and upon displacing it the thoracic aorta will be brought into view with the left splanchnic nerves and vena azygos minor.]

The **Left Pneumo-gastric Nerve** (Fig. 179, 31) enters the thorax between the left carotid and subclavian arteries, and passes behind the left innominate vein. It then crosses the arch of the aorta, around which it gives its *recurrent branch* (21), and can now be traced to the back of the left bronchus, where it breaks up into numerous branches to the left lung and, after giving branches to the œsophagus which unite with those of the opposite side in the *plexus gulæ*, terminates on the anterior surface of the stomach. From the left recurrent branch cardiac nerves pass to the deep cardiac plexus.

The **Thoracic Aorta** (Fig. 179, 32) is the continuation of the arch of the aorta and extends from the lower border of the fifth dorsal vertebra to the twelfth dorsal vertebra, opposite which it passes through the aortic opening in the diaphragm to become the abdominal aorta. In its course it lies at first to the left but afterwards in front of the bodies of the vertebræ, crossing the vena azygos minor. It is crossed by the root of the left lung and, very obliquely, by the œsophagus, which overlies it near the diaphragm. The artery has the thoracic duct and the vena azygos major to its right side, and is closely invested by the pleura on the left side.

Branches.—From the front of the aorta *pericardiac*, *bronchial*, *œsophageal*, and *mediastinal* branches are given off which can now be seen. From the back part of the aorta the right and left *intercostal* arteries arise, which will be afterwards traced.

a. The *pericardiac* branches are irregular.

b. The *bronchial* arteries, one or two to each lung, run on the posterior surface of each bronchus and supply blood to the tissues of the lungs. A *bronchial vein* accompanies each artery; the right opening into the vena azygos major, and the left into the left superior intercostal vein.

c. The *œsophageal* arteries are four or five small branches to the gullet.

d. The *mediastinal* are small twigs to the cellular tissue and glands of the posterior mediastinum. They anastomose with the pericardiac and œsophageal arteries, and form part of the sub-pleural mediastinal plexus of Turner.

[The ascending portion of the arch of the aorta and the venæ cavæ are to be divided, and the trachea cut just above the bifurcation. The lungs are then to be removed from the chest and kept for subsequent examination. The intercostal vessels, the azygos veins, and the gangliated cord of the sympathetic with its branches are now to be dissected out by removing the pleuræ, and opportunity may be taken to follow out the thoracic duct if this was not done satisfactorily before.]

The **Aortic Intercostal Arteries** (Fig. 179) are nine or ten in number on each side, and arise from the back part of the aorta. They supply the lower intercostal spaces, anastomosing with the superior intercostal artery above, and the arteries of the right side are necessarily longer than those of the left, owing to the position of the aorta to the left side of the median line. The upper arteries necessarily ascend to reach their proper intercostal spaces, but the lower ones run transversely, passing beneath the œsophagus, thoracic duct, vena azygos major, and gangliated cord of the sympathetic on the right side; and beneath the vena azygos minor and the gangliated cord of the sympathetic on the left side.

Each artery gives off a *posterior branch* close to the vertebræ, which passes backwards between the transverse processes to the muscles of the back, giving off a small spinal branch through the intervertebral foramen to the spinal cord and body of each vertebra (Fig. 169).

The intercostal arteries lie against the external intercostal muscles at first and are only covered by pleura. A vein and nerve are in relation with each artery, the vein being highest and the nerve lowest in most of the spaces, but the artery being below the nerve at first in the three or four upper spaces. The arteries then disappear beneath the internal intercostals, by the removal of one or two of which the vessels and nerves can be traced out. The artery soon reaches the lower border of the adjacent rib, along the groove in which it runs, being thus protected from injury in the operation of paracentesis thoracis. Lateral and anterior cutaneous branches are given off by both arteries and nerves, which have been already seen.

The **Intercostal Veins** (Fig. 181) open into the azygos vein on each side. The vena azygos major of the right side has been already seen.

The **Vena Azygos Minor** (Fig. 181, 18) commencing in the left lumbar veins, pierces the left crus of the diaphragm and receives the lower intercostal veins of the left side; it has been seen to pass behind the aorta and open into the vena azygos major. The upper intercostal veins of the left side either open into the *superior intercostal vein* or form a separate vein (vena azygos minor superior) which, communicating with the superior intercostal vein above, either terminates below in the vena azygos minor, or crosses the spine separately to open into the vena azygos major.

The **Intercostal Nerves** (Fig. 179) are twelve in number and, with the exception of the first, accompany the intercostal arteries and are distributed to the front and sides of the chest. The first nerve gives only a small branch to the first intercostal space, and then passes through the superior aperture of the thorax to join the brachial plexus.

The **Gangliated Cord of the Sympathetic** (Fig. 178, 31) is placed over the heads of the ribs on each side of the thorax just outside the posterior mediastinum and beneath the pleura, being continuous with the cervical portion of the sympathetic, the inferior cervical ganglion of which should now be dissected on the neck of the first rib. The thoracic ganglia are generally said to correspond to the ribs in number, but there is seldom a distinct ganglion for the first rib, it being united with the inferior cervical ganglion; and occasionally two of the lower ganglia are united.

The *inferior cervical ganglion* has branches of communication with the two cervical nerves, and gives branches upon the vertebral artery and an *inferior cardiac nerve* to the deep cardiac plexus (Fig. 179).

The *thoracic ganglia* may be divided into two sets, upper and lower. From the upper six ganglia communicating branches are given to the six upper intercostal nerves, and to the pulmonary and aortic plexuses. From the six lower ganglia branches of communication are given to the six lower intercostal nerves, and the three splanchnic nerves arise.

Splanchnic Nerves (Fig. 178, 43).—The *great splanchnic nerve* is derived from four ganglia (7th, 8th, 9th, 10th) by separate fibres. The nerve runs inwards and thus enters the lower part of the mediastinum, and after piercing the crus of the diaphragm joins the solar plexus in the abdomen.

The *lesser splanchnic nerve* is derived from the 10th and

Fig. 181.

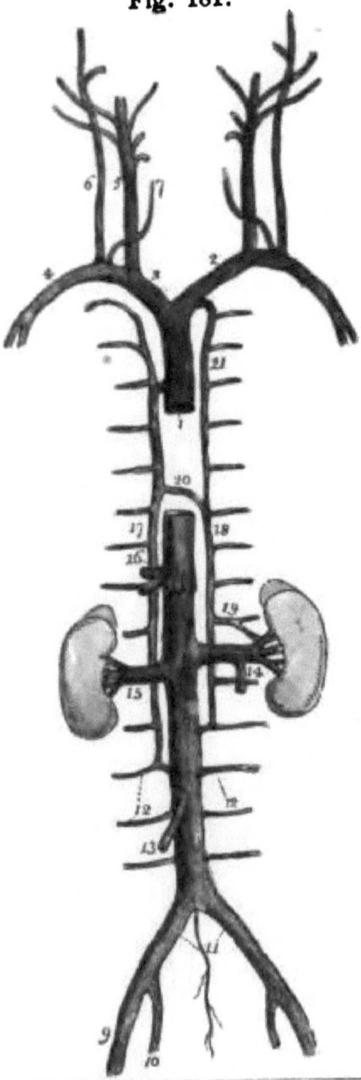

11th ganglia, and also pierces the crus of the diaphragm to join the solar or renal plexus.

The *least splanchnic nerve* is derived from the 12th ganglion, and goes to the renal plexus. It is seldom found, in which case the lesser nerve is connected with the ganglion.

The **Internal Intercostal Muscles** (Fig. 179) can be seen beneath the pleura without any further dissection. Beginning at the sternum the muscles reach as far as the angles of the ribs, at which points the intercostal vessels and nerves lying against the external intercostals are visible. The fibres of the internal intercostals take a direction contrary to that of the external intercostal muscles, *i.e.*, they run forwards and upwards.

The relation of the parts

Fig. 181.—Veins of the trunk and neck (from Wilson).

1. Superior vena cava.
2. Left vena innominata.
3. Right vena innominata.
4. Right subclavian vein.
5. Internal jugular vein.
6. External jugular.
7. Anterior jugular.
8. Inferior vena cava.
9. External iliac vein.
10. Internal iliac vein.
11. Common iliac veins.
12, 12. Lumbar veins.
13. Right spermatic vein.
14. Left spermatic vein.
15. Right renal vein.
16. Trunk of the hepatic veins.
17. Vena azygos major.
18. Vena azygos minor.
19. A branch of communication with the left renal vein.
20. Termination of the lesser in the greater vena azygos.
21. Left superior intercostal vein.

passing through the superior aperture of the thorax can be now fully understood, and will be found in the following table and the accompanying diagram taken from nature (Fig. 182).

Fig. 182.

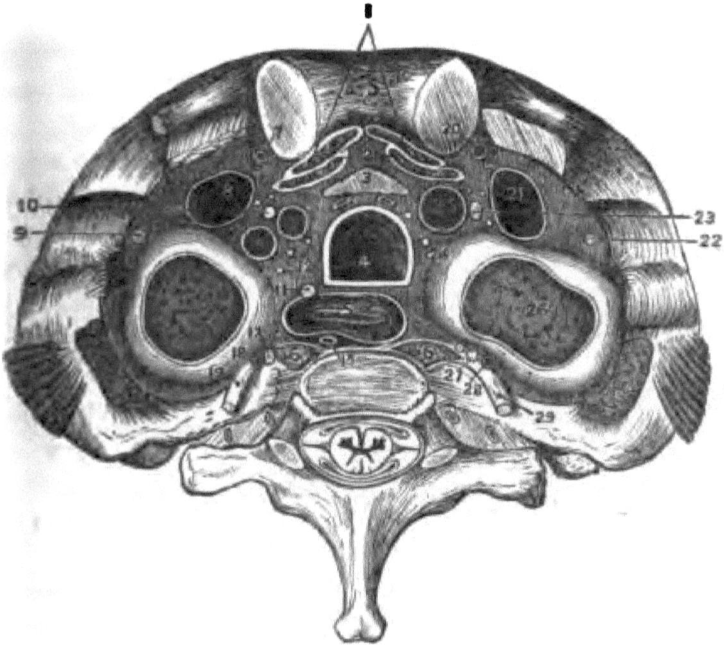

MEDIAN LINE.

1. Sterno-hyoid muscles.
2. Sterno-thyroid muscles.
3. Remains of thymus gland.
4. Trachea.
5. Œsophagus.
6. Longi colli muscles.

LEFT SIDE.	RIGHT SIDE.
7. Internal mammary artery.	20. Internal mammary artery.
8. Innominate vein.	21. Innominate vein.
9. Phrenic nerve.	22. Phrenic nerve.
10. Pneumo-gastric nerve.	23. Pneumo-gastric nerve.
11. Recurrent laryngeal nerve.	
12. Cardiac nerves.	24. Cardiac nerves.
13. Left carotid artery. } 14. Left subclavian artery. }	25. Innominate artery.
15. Thoracic duct.	
16. Apex of lung and pleura.	26. Apex of lung and pleura.
17. Sympathetic.	27. Sympathetic.
18. Superior intercostal artery.	28. Superior intercostal artery.
19. First dorsal nerve.	29. First dorsal nerve.

[The lungs which have been removed and laid aside are now to be dissected, and the structure of the trachea and lungs is to be examined.]

The **Trachea** is about four inches and a half in length, and is convex in front but flattened posteriorly, being composed of a series of cartilages, the extremities of which are connected behind by fibrous and muscular tissue. There are from sixteen to twenty cartilages, each measuring about two lines in depth but decreasing in depth from above downwards. The last cartilage is peculiar, in being cut obliquely on each side so as to be adapted to the commencement of the bronchi. The cartilages are connected together by fibrous tissue, and the first is similarly connected to the cricoid cartilage.

On dissecting away the fibrous tissue at the back of the trachea together with numerous mucous glands, involuntary muscular fibres will be seen connecting the extremities of the cartilaginous rings, constituting what has been termed the *trachealis muscle.* Within this again is an elastic layer closely connected with the mucous membrane.

On looking into the lower end of the trachea a slight septum will be seen between the two bronchi but placed to the left of the median line, thus favouring the passage of foreign bodies into the right bronchus, the orifice of which is seen to be larger than that of the left.

[The trachea and bronchi are to be laid open from behind with scissors, and the divisions of the bronchi should be followed for a short distance into the substance of the lungs.]

The Lungs.—The mucous membrane of the air tubes is of a pinkish colour and has ciliated epithelium. Numerous mucous glands are embedded in the submucous areolar tissue.

On tracing the bronchi they will be found to divide (for the most part dichotomously) again and again, the cartilaginous rings becoming merely plates, and at length disappearing, and the air tubes thus becoming eventually membranous. The minute *bronchial tubes* terminate in *intercellular passages*, in which the mucous membrane is covered with squamous epithelium. Opening out of the intercellular passages are the *air-cells* or alveoli, the septa between which are formed by reduplications of the lining membrane.

The air-cells collected around the extremity of each minute bronchial tube form a *lobule*, and these aggregated together form the substance of the lung, but the air-cells of one lobule have no connection with those of another.

Fig. 183.

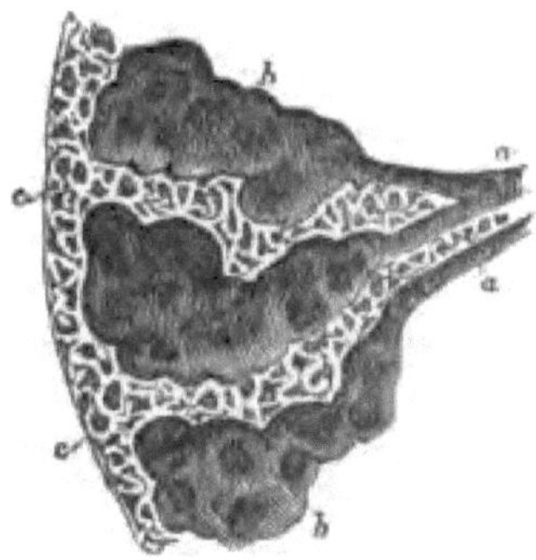

The pulmonary artery subdivides like the bronchus, giving a branch to each lobule, which ends in a plexus of capillaries distributed beneath the mucous membrane of the air-cells and their septa, and also on the walls of the intercellular passages; the capillaries of each lobule being distinct.

The pulmonary veins convey the arterialized blood from the lobules, and correspond to the branches of the arteries. They have no valves.

The bronchial arteries may be traced upon the bronchial tubes for some distance. They supply the substance of the lung, and their blood is returned partly by the

---

Fig. 183.—A diagram showing the dilatation of the ultimate bronchial tubes into intercellular passages, and the enlargement of the latter near the surface of the lung (from Wilson).

a, a. Bronchial tubes.
b, b. Intercellular passages, on the walls of which the air-cells are seen opening.
c, c. Air-cells near the surface of the lung.

bronchial veins and partly by the pulmonary veins (Waters).

Prevertebral Region.

[The carotid arteries with the jugular veins, and the pneumo-gastric and sympathetic nerves are to be divided at the level of the top of the sternum, and the trachea with the œsophagus is to be severed a little lower down. The neck is then to be bent forcibly backward so as to make the cut surface of the skull rest upon the table, and the œsophagus and trachea with the vessels and nerves being drawn forcibly upward, the cellular tissue between the pharynx and the front of the vertebral column is to be cautiously dissected through until the under surface of the base of the skull is exposed. The saw is now to be applied close behind the mastoid process and an oblique cut made, which is to be carried through the whole thickness of the temporal bone into the jugular foramen, and prolonged through the remaining portion of the parietal bone to the cut which was made in removing the brain. A similar cut having been made on the opposite side, a broad chisel is to be applied to the basilar process of the occipital bone where it is exposed behind the pharynx, and it is to be divided. The chisel being again applied on each side of the middle line will unite this cut with those made by the saw, and the preparation will then be divided into two parts; the anterior part of the skull with the pharynx and deep vessels and nerves is to be wrapped up for subsequent examination, and the muscles attached to the vertebral column with the posterior part of the skull are now to be examined.]

The Scaleni muscles have been seen already in part but can now be fully dissected.

The **Scalenus Anticus** (Fig. 184, 2) *arises* from the tubercle on the inner border and upper surface of the first rib (scalene tubercle), and ascends to be *inserted* into the anterior tubercles of the transverse processes of the 3rd, 4th, 5th, and 6th cervical vertebræ. The phrenic nerve will probably still be found on the anterior surface of the muscle, and behind it the brachial nerves emerge and the subclavian artery passes.

The **Scalenus Medius** (Fig. 184, 7) lies behind the brachial nerves, *arising* from the rough marking upon the upper surface of the first rib behind the groove for the subclavian artery. It ascends to be *inserted* into the posterior tubercles of the transverse processes of the lower six cervical vertebræ.

Fig. 184.

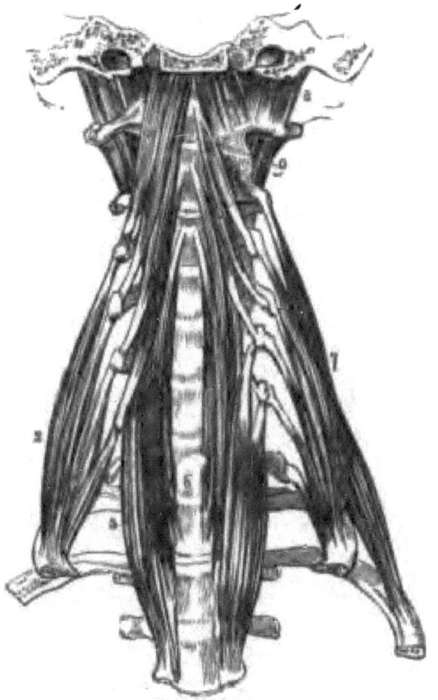

The **Scalenus Posticus** (Fig. 184), which is the smallest of the three muscles, *arises* from a rough mark on the outer surface of the second rib, posterior to the attachment of the serratus magnus; and is *inserted* into the posterior tubercles of the transverse processes of the lowest three cervical vertebræ.

The **Rectus Capitis Anticus Major** (Fig. 184, 1) *arises* from the anterior tubercles of the transverse processes of the 3rd, 4th, 5th, and 6th cervical vertebræ (thus corresponding to the insertion of the scalenus anticus), and is *inserted* into the under surface of the basilar process of the occipital bone close to the median line. The insertions

---

Fig. 184.—Prævertebral muscles of the neck (from Wilson).

1. Rectus capitis anticus major.
2. Scalenus anticus.
3. Lower oblique part of the longus colli of the right side.
4. Rectus capitis anticus minor.
5. Upper oblique portion of the longus colli.
6. Vertical portion of longus colli.
7. Scalenus medius; behind which is seen the scalenus posticus.
8. Rectus lateralis, left side.
9. One of the intertransversales.

of this and the following muscles are very generally damaged by the division of the base of the skull.

The **Rectus Capitis Anticus Minor** (Fig. 184, 4) is beneath the preceding muscle, which must be turned aside to show it. It *arises* from the front of the lateral mass of the atlas and partly from its transverse process, and ascends obliquely inward to be *inserted* into the under surface of the basilar process of the occipital bone, posterior to and further from the median line than the rectus major.

The **Rectus Capitis Lateralis** (Fig. 184, 8) is now exposed although not a prevertebral muscle. It *arises* from the upper surface of the transverse process of the atlas, and is *inserted* into the under surface of the jugular process of the occipital bone.

The **Longus Colli** (Fig. 184) lies on the front of the cervical vertebræ, and is most conveniently divided into three portions, two oblique and one vertical.

The *inferior oblique portion* (3) *arises* from the bodies of the 1st and 2nd dorsal vertebræ, and passes obliquely upward to be *inserted* into the transverse processes of the 5th and 6th cervical vertebræ.

The *superior oblique portion* (5) *arises* from the anterior tubercles of the transverse processes of the 3rd, 4th, and 5th cervical vertebræ, and passes upward and outward to be *inserted* into the anterior tubercle of the atlas.

The *vertical portion* (6) *arises* from the bodies of the three lower cervical and three upper dorsal vertebræ, and is *inserted* into the bodies of the 2nd, 3rd, and 4th cervical vertebræ.

All the prevertebral muscles draw forward the upper part of the vertebral column or bow the head, when acting symmetrically; or when the muscle of one side acts alone it draws the spine to that side. The scaleni muscles, when the vertebræ are fixed, act upon the ribs and raise them, thus being extraordinary muscles of inspiration.

The small *Intertransverse muscles* may be seen between the transverse processes. The anterior ones pass between the anterior tubercles of the transverse processes, and the posterior ones have been already seen in the dissection of the back. Between them the anterior divisions of the cervical nerves appear.

The **Vertebral Artery** will be seen between the scalenus anticus and the longus colli muscles, and may be more conveniently traced through the foramina in the transverse processes than at an earlier period (*v.* p. 432).

Dissection of the Pharynx.

Before dissecting the anterior half of the skull with the pharynx, the dissector should examine the fauces and upper part of the pharynx from the mouth. The soft palate with the uvula in the median line will be readily recognized, and passing from the soft palate on each side will be seen the two pillars of the fauces with the tonsil between them. The anterior pillar extends from the soft palate to the tongue, being vertical in direction and formed by the palato-glossus muscle. The posterior pillar passes obliquely backwards and is lost in the pharynx, being formed by the palato-pharyngeus muscle. The tonsil is generally much shrunken in a subject which has arrived at this stage of dissection.

[The pharynx and upper part of the œsophagus are to be carefully distended with cotton-wool or tow, and the preparation being placed with the face downwards, is to be secured over a small block with hooks, one set of which should draw the œsophagus down and keep the pharynx tense. The vessels and nerves at the back of the pharynx are to be examined before the muscular bag itself is dissected.]

The vessels and nerves now to be examined have all been seen in part in previous dissections, and then from either the front or the side. They are now all seen from behind, and this must be borne in mind thoroughly, or will lead to misconception of the description. The section of the base of the skull is seldom precisely similar on the two sides and it will generally be found advisable therefore to trace the parts first brought into view on one side, and the carotid artery, etc., on the other, as in the illustration (Fig. 185).

The **Sympathetic Nerve** (Fig. 185) with its superior and middle cervical ganglia is at once exposed, and some of its branches may be very conveniently traced.

Fig. 185.

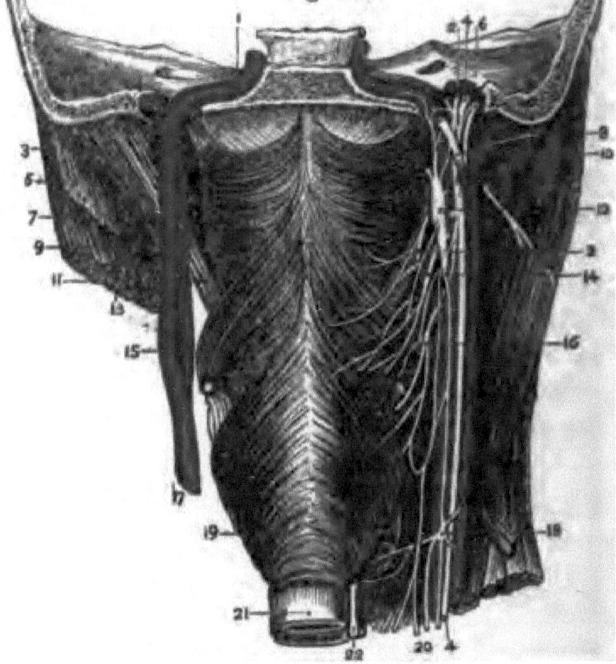

Fig. 185.—Dissection of the pharynx with the carotid vessels and the eighth, ninth, and sympathetic nerves (drawn by J. T. Gray).

1. Fibrous bag of pharynx.
2, 2. Glosso-pharyngeal nerve.
3. Posterior belly of digastric.
4, 4. Pneumo-gastric nerve.
5. Splenius capitis.
6. Spinal-accessory nerve.
7. Superior constrictor of pharynx.
8. Internal jugular vein.
9. Ascending pharyngeal artery.
10. Hypoglossal nerve.
11. Stylo-pharyngeus.
12. Superior ganglion of sympathetic.
13. Sterno-mastoid.
14. Pharyngeal branch of pneumo-gastric.
15. Middle constrictor of pharynx.
16. Superior laryngeal nerve.
17. Common carotid artery.
18. Middle ganglion of sympathetic.
19. Inferior constrictor of pharynx.
20. Cardiac nerves.
21. Œsophagus.
22. Recurrent laryngeal nerve.

The *Superior cervical ganglion* (12) is fusiform and nearly an inch in length. It lies behind the internal carotid artery, and has small branches of communication with the following cranial nerves—the glosso-pharyngeal, the pneumo-gastric, and the hypoglossal. The branches of communication with the cervical nerves have been already seen (p. 370). The branches of distribution are, (1) the *nervi molles* distributed upon the external carotid artery and its branches; (2) the *pharyngeal branch* which can now be traced to the pharynx, where it enters into the formation of the pharyngeal plexus; (3) the *laryngeal branch* to the superior laryngeal nerve; (4) the *superior cardiac nerve* which has been already seen.

The *Middle cervical ganglion* (18) is of small size, and gives off (1) thyroid branches upon the inferior thyroid artery and (2) the middle cardiac nerve.

The **Ninth** or **Hypoglossal Nerve** (Fig. 185, 10) is necessarily cut off at the anterior condyloid foramen in making the dissection, and should therefore be traced from below, where it will be found in relation with the occipital artery. The nerve is at first posterior to the internal carotid artery and jugular vein, and then passes between them, and also between the pneumo-gastric and spinal-accessory nerves, with the former of which it has a communication, as well as with the superior cervical ganglion of the sympathetic. A small branch connected with the ninth nerve at one end and loose at the other is the communicating branch from the first and second cervical nerves (Fig. 186, 19).

The **Jugular Vein** (Fig. 185, 8) commences outside the skull by the junction of the lateral sinus with the inferior petrosal sinus. Its course in the neck has been already seen, and it should now be divided close to the skull and removed.

The **Spinal-accessory Nerve** (Fig. 185, 6) emerges from the foramen jugulare, where it is closely connected with the pneumo-gastric nerve, and may be traced to the deep surface of the sterno-mastoid muscle.

The **Pneumo-gastric Nerve** (Fig. 185, 4) leaves the

foramen jugulare in the same sheath as the spinal-accessory nerve, with which it has communicating branches.

Fig. 186.

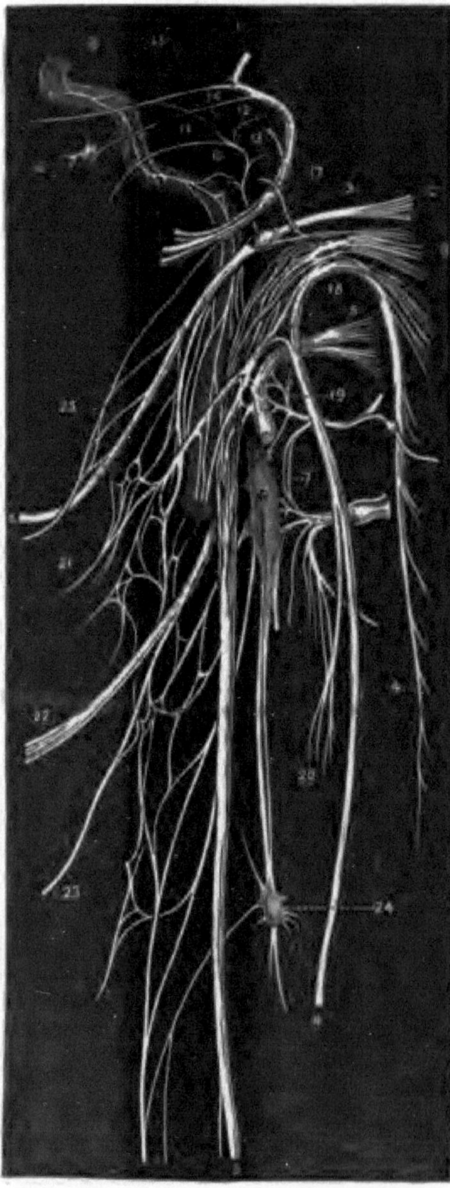

Two ganglia are found upon the pneumo-gastric nerve, viz., the superior or ganglion of the root, and the inferior or ganglion of the trunk.

The *ganglion of the root* is very small and is placed in the jugular foramen. It has minute branches of communication with the glosso-pharyngeal, spinal-accessory, sympathetic, and seventh nerves. The communication with the latter is through a minute *auricular* branch which enters a hole near the root of the styloid process and passes through the temporal bone to the pinna (Fig. 186, 17).

The *ganglion of the trunk* is nearly an inch long and of a pink colour, and has branches of communication with the hypoglossal, the sympathetic, and the loop of the first and second cervical nerves. It gives off the following branches:—

(1). The *pharyngeal branch of the pneumo-gastric* (14), which receives a communicating branch from the spinal-accessory nerve and then passes in front of (or sometimes behind) the internal carotid artery to the pharynx, where it assists in forming the pharyngeal plexus upon the middle constrictor muscle (Fig. 186, 21).

(2). The *superior laryngeal nerve* (16), which takes an oblique course behind the internal carotid to the larynx, where it has been already seen to give off the *external laryngeal* branch and then to pierce the thyro-hyoid membrane (Fig. 186, 22).

The **Glosso-pharyngeal Nerve** (Fig. 185, 2) lies in a little special notch in the lower border of the petrous bone as it leaves the jugular foramen, being thus isolated from the remainder of the eighth nerve. It then passes forward over the internal carotid artery and reaches the stylo-pharyngeus muscle, at the lower border of which it has already been seen in the submaxillary region and has been traced to the tongue (p. 407).

Fig. 186.—Diagram of the eighth, ninth and sympathetic nerves (from Hirschfeld and Leveillé).

1. Facial nerve.
2. Glosso-pharyngeal nerve with its petrous ganglion.
3. Pneumo-gastric nerve.
4. Spinal-accessory nerve.
5. Hypoglossal nerve.
6. Superior cervical ganglion of sympathetic.
7. Loop between 1st and 2nd cervical nerves.
8. Carotid branch of sympathetic.
9. Tympanic nerve (Jacobson).
10. Its branch to carotid plexus.
11. Its branch to Eustachian tube.
12. Its branch to fenestra ovalis.
13. Its branch to fenestra rotunda.
14. Its union with small superficial petrosal nerve.
15. Its union with large superficial petrosal nerve.
16. Otic ganglion.
17. Auricular nerve of pneumogastric.
18. Junction of pneumo-gastric with spinal-accessory.
19. Junction of 9th nerve and 1st cervical nerve.
20. Junction of mastoid branch of spinal-accessory and 2nd cervical nerve.
21. Pharyngeal plexus.
22. Superior laryngeal nerve.
23. External laryngeal nerve.
24. Middle cervical ganglion of sympathetic.
25. Junction of digastric nerve (7th) with glosso-pharyngeal.

It presents two small ganglia, one at the upper part of the foramen, the *superior* or *jugular ganglion* which is of very small size, and the other at the lower part of the foramen which is larger and is called the *inferior* or *petrous ganglion* or *ganglion of Andersch* (Fig. 186, 2).

The upper ganglion involves only some of the fibres of the nerve, but they all pass through the lower ganglion.

The lower ganglion has branches of communication with the pneumo-gastric nerve, one going to its superior ganglion and another to the auricular nerve ; also with the superior ganglion of the sympathetic ; and with the facial nerve by a branch which pierces the posterior belly of the digastric (Fig. 186, 25).

The glosso-pharyngeal nerve gives off the following branches in its course to the tongue :—

(1). *Carotid branches* which join the sympathetic plexus on that vessel and communicate with the pharyngeal branch of the pneumo-gastric.

(2). *Muscular branches* to the stylo-pharyngeus.

(3). *Pharyngeal branches* which assist in forming the pharyngeal plexus.

(4). *Tonsillitic branches* to the tonsils and the soft palate.

[By cautiously cutting away the temporal bone with the bone-forceps the dissector may, in a favourable subject, see some of the branches of Jacobson's nerve.]

The *Tympanic branch of the glosso-pharyngeal nerve* (Jacobson's nerve) (Fig. 186, 9) arises from the petrous ganglion, and enters an aperture in the ridge of bone between the carotid foramen and the jugular fossa. It pierces the floor of the tympanum and grooves the promontory on its inner wall, giving branches to the fenestra ovalis, the fenestra rotunda, and the lining membrane of the tympanum and Eustachian tubes.

The communicating branches of Jacobson's nerve are three in number; one joining the carotid plexus in the carotid canal (10); the second joining the great superficial petrosal nerve (15); and the third running through the temporal bone to end in the otic ganglion as the small superficial petrosal nerve of Arnold (14) (*See* also Fig. 190).

Opportunity may be taken at this point to examine the ossicles of the tympanum (p. 483).

The **Pharyngeal Plexus** (Figs. 185 and 186) is to be found upon the middle and inferior constrictors of the pharynx, both of which it supplies. The branches forming the plexus are derived from the nerves which have been examined, viz. the glosso-pharyngeal, the pneumo-gastric (pharyngeal and superior laryngeal branches), and the sympathetic.

[In all probability the carotid canal in the temporal bone will have been opened on one side in making the section of the skull, but if not, this may now be done with the bone-forceps.]

The **Internal Carotid Artery** (Fig. 185) has been already seen from the front in the dissection of the neck, and is now seen from behind. It ascends by the side of the pharynx, being separated from it, however, by the ascending pharyngeal branch of the external carotid artery. *Behind* the artery is the trunk of the sympathetic, and crossing its posterior aspect is the superior laryngeal nerve of the pneumo-gastric. *In front* of it is the styloid process with the stylo-glossus and stylo-pharyngeus muscles, and the glosso-pharyngeal nerve; and to the *outer side* the pneumo-gastric nerve with the jugular vein. The danger of the close proximity of the carotid to the pharynx has been exaggerated, since (as will be afterwards seen) any ordinary incision may be made in the tonsils or back of the pharynx without any risk of injuring the vessel, which lies quite to the side.

The carotid takes a tortuous course in the temporal bone and cranium, making two sigmoid turns, one, the longer, in the petrous bone, and the other by the side of the sella turcica. In this part of its course it is more or less surrounded by a plexus of nerves derived principally from the sympathetic.

The **Ascending Pharyngeal Artery** (Fig. 185, 9) is one of the ascending branches of the external carotid artery. It arises near the bifurcation of the common carotid, and ascends by the side of the pharynx, and to the inner side of the internal carotid artery, to the base of the skull. It gives branches to the prevertebral muscles, anastomosing with the ascending cervical artery, and divides into meningeal and pharyngeal branches.

The *meningeal branches* are very small, and enter the skull by the foramen lacerum medium and the foramen jugulare, to supply the dura mater.

The *pharyngeal branches* supply the pharynx and turn over the upper border of the superior constrictor to supply the palate; they anastomose with the inferior palatine branch of the facial artery.

The ascending pharyngeal vein opens into the internal jugular.

[The constrictor muscles of the pharynx are to be cleaned in the direction of their fibres, beginning at the lower border of the inferior constrictor. In order to see the origin of the superior constrictor, it will be necessary to remove the internal pterygoid on one side. The pharyngeal plexus must necessarily be destroyed in the course of the dissection, but the superior and inferior laryngeal and the glosso-pharyngeal nerves are to be preserved.]

The **Inferior Constrictor** (Fig. 187, 9) is the most superficial of the three muscles of the pharynx, the upper oblique border overlapping the middle constrictor, and the lower straight border being continuous with the œsophagus. It *arises* from the side of the cricoid cartilage in front of the articular facet, and from the ala of the thyroid cartilage behind the oblique line. All the fibres are *inserted* into the median raphé. The recurrent laryngeal nerve passes beneath the lower border of the inferior constrictor, and the superior laryngeal nerve and artery intervene between it and the middle constrictor.

The **Middle Constrictor** (Fig. 187, 10) is a fan-shaped muscle which with its fellow of the opposite side forms a trapezium in the median line. It *arises* from the upper surface of the great cornu of the hyoid bone, from the lesser cornu, and from the stylo-hyoid ligament; and its fibres ascend and descend obliquely *to be inserted* into the median raphé of the pharynx.

The middle constrictor is overlapped by the inferior constrictor, and itself covers the superior constrictor in part. It is separated from the inferior constrictor by the superior laryngeal nerve, and from the superior constrictor by the stylo-pharyngeus muscle and glosso-pharyngeal nerve.

Fig. 187.

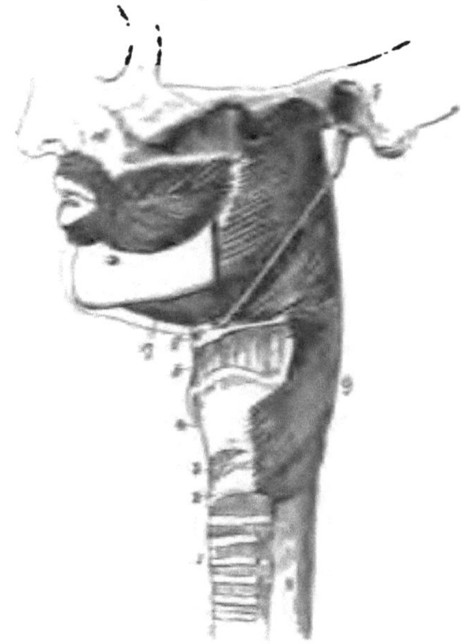

The **Superior Constrictor** (Fig. 187, 11) *arises* from the lower third of the internal pterygoid plate and from the hamular process of the sphenoid bone; from the pterygo-maxillary ligament opposite the attachment of the buccinator; from the inner surface of the lower jaw above the posterior extremity of the mylo-hyoid ridge, and slightly from the side of the tongue. The fibres curve backward, leaving an interval between the muscle and the base

Fig. 187.—Side view of the muscles of the pharynx (from Wilson).

1. Trachea.
2. Cricoid cartilage.
3. Crico-thyroid membrane.
4. Thyroid cartilage.
5. Thyro-hyoid membrane.
6. Os hyoides.
7. Stylo-hyoid ligament.
8. Œsophagus.
9. Inferior constrictor.
10. Middle constrictor.
11. Superior constrictor.
12. Stylo-pharyngeus, passing down between the superior and middle constrictor.
13. Fibrous bag of the pharynx seen above the constrictor.
14. Pterygo-maxillary ligament.
15. Buccinator.
16. Orbicularis oris.
17. Mylo-hyoideus.

of the skull in which the fibrous bag of the pharynx is visible, and are *inserted* into the median raphé, being overlapped at the lower part by the middle constrictor. Distinct tendinous fibres may occasionally be traced to the tubercle on the under surface of the basilar process to which the fibrous bag of the pharynx is attached. The superior constrictor is separated from the middle constrictor by the stylo-pharyngeus muscle and glosso-pharyngeal nerve. Above its upper curved border the ascending pharyngeal artery sends a branch to the palate together with a branch of the inferior palatine artery, and the levator palati muscle and the Eustachian tube cross obliquely inward beneath the fibrous bag.

The three constrictors are *supplied* by the pharyngeal plexus of nerves; the superior has also branches from the glosso-pharyngeal, and the inferior from the external laryngeal nerve.

The **Stylo-pharyngeus Muscle** (Fig. 187, 12) has already been seen at its origin, and can now be followed to its insertion by dividing some of the fibres of the middle constrictor. It *arises* from the root of the styloid process of the temporal bone, and passes between the superior and middle constrictors to be *inserted* into the bag of the pharynx and into the posterior border of the thyroid cartilage. It is *supplied* by the glosso-pharyngeal nerve.

The *Fibrous bag of the Pharynx* is very thin at the lower part where it lies between the muscular fibres and the mucous membrane, but above the border of the superior constrictor it is much stronger, and is expanded from side to side, covering in the Eustachian tubes and the levatores palati muscles. At the base of the skull it is attached to the basilar process of the occipital bone, to the Eustachian tube, and to the under surface of the petrous portion of the temporal bone; and is carried forwards to the root of the pterygoid process and the internal pterygoid plate of the sphenoid bone, becoming continuous with the pterygo-maxillary ligament.

[The pharynx is to be opened from behind by an incision in the median line from the basilar process to the commencement of the œsophagus. The fibrous bag is to be detached from the occipital bone on each side and held widely open with hooks, and the whole of the cotton wool is to be removed from the interior of the pharynx.]

The **Interior of the Pharynx** (Fig. 188) presents seven openings in the following order from above downwards; (1 and 2) the two posterior nares separated by the vomer; (3 and 4) the two Eustachian tubes; (5) the *isthmus faucium* or opening of the mouth; (6) the superior aperture of the larynx guarded by the epiglottis; and (7) the opening into the œsophagus.

Fig. 188.

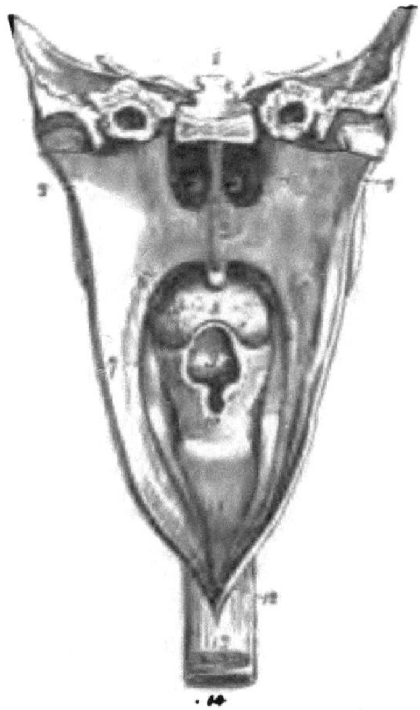

The *soft palate* or *velum pendulum palati* (5) intervenes between the nose and the mouth, and consists of a fibrous membrane which is attached to the palate bones and is strengthened by expansions from the several muscles of the palate; it is covered by mucous membrane, which is ciliated on the upper but not on the lower surface. The centre of its free border is prolonged into the uvula, and on each side will be seen the two folds called the *pillars of the fauces*, formed by mucous membrane reflected upon the palato-glossi and palato-pharyngei muscles.

The mucous membrane of the pharynx is continuous

Fig. 188.—Pharynx laid open from behind (from Wilson).

1. Section of base of skull.
2, 2. Walls of pharynx drawn aside.
3, 3. Posterior nares separated by the vomer.
4. Right Eustachian tube.
5. Soft palate.
6, 6. Posterior pillars of fauces.
7. Anterior pillar of fauces.
8. Base of tongue.
9. Epiglottis.
10. Aperture of larynx.
11. Back of larynx.
12. Opening of œsophagus.
13. Œsophagus.
14. Trachea.

with both that of the nose and of the mouth, and its epithelium varies in different situations, being squamous throughout the canal below the level of the palate, but columnar and ciliated above that point.

This is the best opportunity for studying the **Process of Deglutition.** The food having been duly masticated and insalivated, the mouth is closed in order that the lower jaw may afford a fixed point from which the muscles of the tongue may act. The bolus of food is then carried back to the fauces by the movement of the tongue, and is there grasped by the pillars of the fauces and prevented from returning. The pharynx is now raised by the stylo-pharyngei muscles and is at the same time widened to receive the food, which is prevented from ascending into the nares by the raising and tension of the soft palate, due to the action of the levator and circumflexus palati of both sides. The constrictors, by their successive contraction from above downwards, next force the food towards the œsophagus. At the same moment the larynx is raised by the action of the elevators of the hyoid bone, and the aperture of the larynx is compressed against the epiglottis and the base of the tongue; and by this means the food is prevented from entering the windpipe. An additional security is provided by the epiglottis, which when healthy is folded down over the aperture of the larynx by the passage of the food; but that the epiglottis is not essential is shown by cases in which the cartilage has been destroyed by ulceration, the process of deglutition still remaining perfect. Lastly the contraction of the œsophagus carries the food down to the stomach.

The Palate.

[The soft palate is to be stretched by inserting a hook into the uvula, and the mucous membrane is to be removed from the upper surface of the palate and the neighbouring bones, so as to expose the muscles above the palate and the Eustachian tube. The levator palati will be found passing obliquely inwards, the azygos uvulæ in the median line of the palate, and the tensor palati can be best seen by dividing the levator and detaching the fibres of the superior constrictor from the hamular process, when the muscle will be found upon the internal pterygoid plate.]

The **Levator Palati Mollis** (Fig. 189, 5) arises from the under surface of the apex of the petrous portion of the temporal bone and from the inner surface of the Eustachian tube. The muscle passes inward above the border of the superior constrictor muscle, and is *inserted* into the soft

Fig. 189.

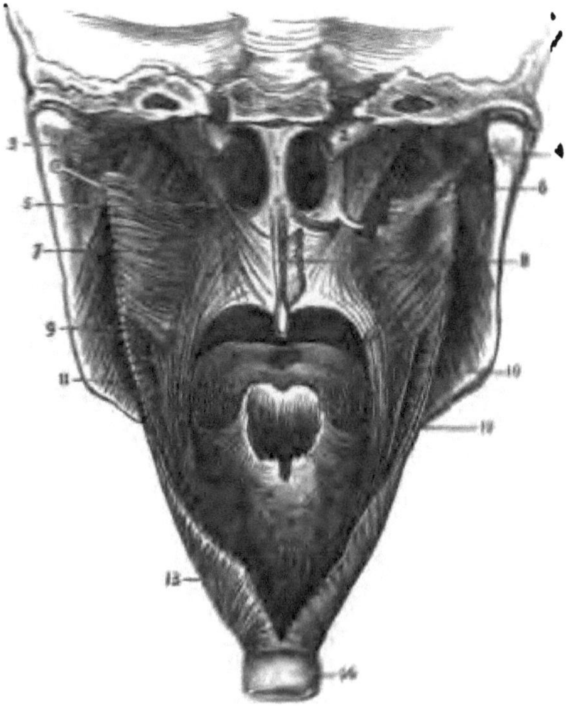

Fig. 189.—**Muscles of the palate.**

1. Septum narium.
2. Eustachian tube.
3. Pterygoideus externus.
4. Pterygoideus internus.
5. Levator palati mollis.
6. Circumflexus palati.
7. Superior constrictor of pharynx.
8. Azygos uvulæ.

palate by a broad expansion which meets that of its fellow-muscle in the median line.

The **Circumflexus** or **Tensor Palati** (Fig. 189, 6) *arises* from the scaphoid fossa at the root of the internal pterygoid plate and slightly from the under surface of the wing of the sphenoid; also from the outer surface of the Eustachian tube. The tendon descends vertically to wind round the hamular process of the sphenoid bone, where it is lubricated by a minute bursa, and then takes a horizontal direction to the soft palate. It is *inserted* into the palate by an expansion beneath the levator, and also into the transverse ridge on the under surface of the palate bone.

The **Azygos Uvulæ** (Fig. 189, 8) consists of two small muscular slips placed parallel to and on each side of the median line. It *arises* from the posterior nasal spine of the palate bone, and is *inserted* into the uvula.

The levator palati raises the soft palate, and the circumflexus makes it tense on each occasion of swallowing. The azygos uvulæ can have but a slight and unimportant action upon the uvula. The levator palati and azygos uvulæ are *supplied* by palatine branches from the spheno-palatine (Meckel's) ganglion; the tensor palati receives a branch from the otic ganglion.

Surgery.—The muscles above the palate, and especially the levator palati, are of interest surgically in relation to the operation of staphyloraphy, or that for closure of a congenital fissure of the palate. Sir W. Fergusson has clearly shown that the two segments of a fissured soft palate are drawn asunder by the levatores palati on every occasion of swallowing, and he therefore proposed and carried out the division of these muscles, by means of a knife consisting of a lancet-shaped blade set at right angles to the handle, which is passed through the fissure.

Opportunity is to be taken, before the pillars of the fauces are dissected, to observe the important surgical fact that a bistoury, if made to transfix the tonsil from before backwards, will pass internally to the carotid artery, unless the point of the instrument is directed purposely to one side, in which case it would be possible to injure this important vessel.

The **Eustachian Tube** (Fig. 189, 2) is the communication between the pharynx and the tympanum or middle ear. The osseous portion of the canal is in the temporal bone, but the cartilaginous portion is now seen to be nearly an inch in length and to terminate in a broad trumpet-shaped extremity. The cartilage of which the tube is formed is triangular in shape and is doubled upon itself, the deficiency at the lower part being completed by fibrous tissue. In the recent condition the thick mucous membrane converts the opening into a mere vertical slit, which is generally closed, but during the process of deglutition is opened by the action of the circumflexus palati muscle, and thus the equilibrium of the air in the tympanum is maintained.

[The palate is to be drawn up so as to put the pillars of the fauces on the stretch as much as possible, and the mucous membrane is to be removed to expose the palato-glossus and palato-pharyngeus muscles.]

The **Palato-glossus** is placed in front of the tonsil and is very small and indistinct. The muscle *arises* from the middle line of the soft palate in common with its fellow, and descends to the side of the tongue, where it is *inserted*, joining the fibres of the stylo-glossus and hyo-glossus muscles.

The **Palato-pharyngeus** (Fig. 189, 9) is larger than the palato-glossus, and is placed behind the tonsil. It *arises* in the palate by two slips which are separated by the levator palati muscle, and the fibres of which meet those of the opposite muscle in the median line. The muscle passes obliquely downwards to the pharynx to be *inserted* into the posterior border of the thyroid cartilage with the stylo-pharyngeus, and to be lost in the wall of the pharynx itself.

The palato-glossus muscle is the constrictor of the fauces and grasps the bolus of food when it has passed out of the mouth, thus preventing its return. Both it and the palato-pharyngeus are *supplied* by branches from Meckel's ganglion.

The **Amygdala** or **Tonsil** is placed between the palato-glossus and palato-pharyngeus muscles and is usually much shrunken after death. It consists of a number of mucous follicles collected together, the orifices of which may be seen on the internal surface. The outside of the tonsil is in close relation with the superior constrictor of the pharynx and with the ascending pharyngeal artery. It is to be noticed that it is anatomically impossible for any enlargement of the tonsil to obstruct the Eustachian tube and thus produce deafness.

[The tongue and the larynx are to be detached by dividing all the structures between them and the palate, and are to be carefully preserved for subsequent examination. By inverting the skull the hard palate will then be brought into view.]

The **Hard Palate** (Fig. 191, 7) is continuous with the soft palate, but its mucous membrane is much more dense, being inseparably united in great part with the periosteum of the maxillary and palate bones. The mucous membrane presents a median ridge indicative of the congenital division of the parts, and is thrown into more or less transverse folds near the anterior part where it is prolonged on to the gums. Numerous mucous glands lie immediately beneath the mucous membrane and open upon its surface.

The **Gums** (Fig. 190) are composed of dense fibrous tissue inseparably united with the periosteum of the alveolus, and covered by the mucous membrane of the mouth which is prolonged into the sockets of the teeth, where it becomes continuous with the peridental membrane covering them.

The **Teeth** (Fig. 190) of the upper jaw are 16 in number, viz., 4 incisors, 2 canines, 4 bicuspids, and 6 molars, the most posterior molars being the *dentes sapientiæ* or wisdom teeth. Each tooth consists of a crown, a neck, and a fang, and upon extracting a molar tooth it will be found to have three fangs, two being on the outer and one on the inner side.

The teeth of the lower jaw correspond in number to those of the upper jaw, but have been necessarily inter-

Fig. 190.

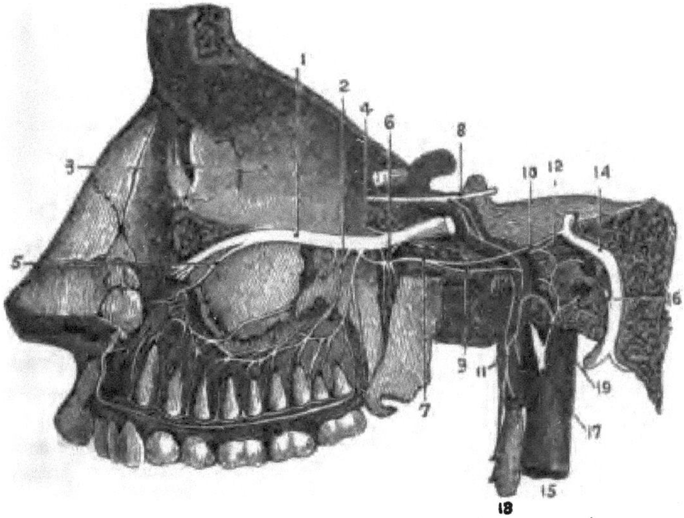

fered with in the progress of the dissection. The lower molar teeth differ from those of the upper jaw in presenting only two fangs.

The **Lips** (Fig. 190) are formed externally by skin, and internally by mucous membrane, which is reflected on to them from the gums; and between the two are the fibres of the orbicularis oris with some cellular tissue and mucous glands and the coronary arteries. The fold of

Fig. 190.—Dissection of the superior maxillary nerve and Meckel's ganglion (from Hirschfeld and Leveillé).

1. Superior maxillary nerve.
2. Posterior dental nerves.
3. Inner wall of orbit.
4. Orbital branch (cut).
5. Anterior dental nerve.
6. Meckel's ganglion.
7. Vidian nerve.
8. Sixth nerve.
9. Carotid branch of Vidian.
10. Superficial petrosal nerve.
11. Carotid plexus of sympathetic.
12. Lesser petrosal nerve.
13. Superficial cervical ganglion.
14. Facial nerve.
15. Internal jugular vein.
16. Chorda tympani.
17. Glosso-pharyngeal nerve.
19. Jacobson's nerve.

mucous membrane connecting each lip with the alveolus in the median line is called the *frænum labii*, and is more prominent in the upper than in the lower lip.

The Superior Maxillary Nerve.

[The skull being placed with the base downwards, a cut with the chisel is to be carried in a straight line from the sphenoidal fissure to the foramen ovale. The side of the skull is then to be sawn through at right angles to the first incision, meeting it at the foramen ovale, and the piece of bone is to be removed. The remains of the malar bone are next to be removed by sawing into the spheno-maxillary fissure externally to the infra-orbital canal; and with the bone-forceps any remaining bone is then to be taken away, so as to expose the whole extent of the superior maxillary nerve.]

The **Superior Maxillary Nerve** (Fig. 190, 1) (second division of the 5th) leaves the cranium at the foramen rotundum, and, having crossed the spheno-maxillary fossa, enters the infra-orbital canal and appears on the face at the infra-orbital foramen. It gives off the following branches:—

1. *Orbital branch* (4) which enters the orbit by the spheno-maxillary fissure and divides into two branches (malar and temporal), which have been seen in the dissection of the face and scalp.

2. *Spheno-palatine branches* (6) which descend to Meckel's ganglion placed opposite the spheno-palatine foramen, and will be afterwards dissected.

3. *Posterior dental branches* (2) which supply the gums, and enter the canals on the posterior aspect of the upper jaw to supply the molar and bicuspid teeth and to communicate with the anterior dental nerve.

4. *Anterior dental branch* (5) which arises from the nerve in the infra-orbital canal and can only be seen by laying the canal open. It descends in a special canal in the wall of the antrum to the incisor and canine teeth, and has a communication with the posterior dental nerve.

The branches of the dental nerves can only be traced to the teeth by removing the outer plate of the alveolus, but the expenditure of time and trouble necessary for this will not be repaid by the results.

The *facial branches* of the superior maxillary nerve have been seen in the dissection of the face (p. 387).

The **Infra-orbital Artery** accompanies the superior maxillary nerve. It is the terminal branch of the internal maxillary artery and gives a branch to the orbit, and an *anterior dental* branch which accompanies the anterior dental nerve to the teeth.

The *infra-orbital vein* communicates with the facial vein and ends in the internal maxillary vein.

THE CAVITY OF THE NOSE.

[In making a section of the nasal cavities it will be found to be almost impossible to preserve the septum and the turbinate bones of both sides uninjured, and the better plan therefore is to preserve the septum carefully at the expense of the turbinate bones of one side, and afterwards to remove it so as to obtain a good view of the turbinate bones and meatuses of the nose on the opposite side. The saw being placed on one side of the septum and parallel to it, is to be carried through the cribriform plate of the ethmoid bone and the palatine processes of the maxillary and palate bones, and the soft palate having been detached, the skull will be divided into two portions, one of which can be used for the examination of the nose, and the other for the dissection of Meckel's ganglion, etc.]

The **Septum Narium** (Fig. 192, *i*) is a vertical plate, which upon removal of the mucous membrane will be found to be partly bony and partly cartilaginous. The bones entering into its formation are the following:—the crest of the nasal bone, the nasal spine of the frontal bone, the large perpendicular plate of the ethmoid bone, the vomer, a part of the rostrum of the sphenoid bone, and the crests of the maxillary and palate bones upon which the vomer rests below. The *cartilage of the septum* is triangular in shape and fits into the interval between the vertical plate of the ethmoid and the vomer, but it not unfrequently extends backwards for a considerable distance between those bones when they are not so fully developed as usual.

The whole septum may be bent considerably to one side as the result of congenital malformation or of violence, and the cartilage may be perforated either congenitally or from disease. The vomer presents an oblique groove for the naso-palatine nerve, which will be afterwards seen upon the opposite side of the bone.

[The septum is to be removed piecemeal with the bone-forceps, so as to leave the mucous membrane on the opposite side untouched. The naso-palatine nerve and artery, and some branches of the olfactory nerve at the upper part of the membrane, may be seen if putrefaction is not too far advanced. The mucous membrane is then to be detached below and turned up, when the nasal cavity will be exposed.]

The **Nasal Fossa** (Fig. 191) is a cavity bounded *above* by a *roof* which slopes anteriorly and posteriorly but is horizontal in the middle, being formed anteriorly by the nasal bone and the nasal spine of the frontal bone; in the middle by the horizontal cribriform plate of the ethmoid bone; and posteriorly by the under surfaces of the body of the sphenoid bone and of the sphenoidal process of the palate bone. The *floor* is slightly concave from side to side and is formed by the palatine processes of the maxillary and palate bones. The *inner* wall is formed by the septum which has been examined. The *outer* wall is divided into three *meatuses* by the projection from it of the three turbinate bones, of which the two upper are portions of the ethmoid bone, but the lowest is a separate bone articulated upon the orifice of the antrum.

The *Superior Meatus* (*b*) is between the superior and middle turbinate bones, and is the smallest of the three, extending for not more than one-third of the length of the outer wall. The posterior ethmoidal cells and sphenoidal sinus open into this meatus.

The *Middle Meatus* (*g*) is between the middle and inferior turbinate bones and extends for the whole length of the outer wall, curving upwards anteriorly. The anterior ethmoidal cells, the frontal sinus (through the infundibulum of the ethmoid bone), and the antrum of Highmore open into the meatus.

The *Inferior Meatus* (*k*) is between the inferior turbinate bone and the floor of the nasal fossa. It extends the whole length of the fossa, and at the anterior part has the nasal duct opening into it.

Fig. 191.

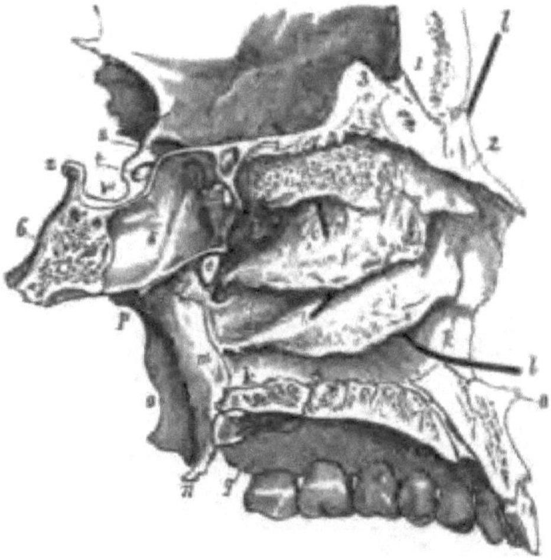

It should be noticed that the orifice of the Eustachian tube is close to the posterior extremity of the inferior turbinate bone, and that in using the Eustachian catheter, the point of the instrument has therefore to be raised before it can enter the tube.

Fig. 191.—The meatuses of the nose on the left side (from Wilson).

1. Frontal bone.
2. Nasal bone.
3. Crista galli of the ethmoid.
4. Cribriform plate of the ethmoid.
5. Part of the sphenoidal sinus.
6. Basilar portion of the sphenoid bone.
7, 7. Palatine process of the superior maxillary bone.
8. Nasal spine.
9. Palatine process of the palate bone.

a. Superior turbinate bone.
b. Superior meatus.
c. A probe passed into the posterior ethmoidal cells.
d. Opening of the sphenoidal sinus into the superior meatus.
e. Spheno-palatine foramen.
f. Middle turbinate bone.
g, *g*. Middle meatus.
h. A probe passed into the infundibulum leading from the frontal sinus and anterior ethmoidal cells; the triangular aperture immediately above the letter is the opening of the antrum.
i. Inferior turbinate bone.
k, *k*. Inferior meatus.
l, *l*. A probe passed up the nasal duct.
m. Internal pterygoid plate.
n. Its hamular process.
o. External pterygoid plate.
p. Root of the pterygoid processes.
q. Posterior palatine foramina.

The *Mucous Membrane* lining the nasal fossæ is termed the *pituitary* or *Schneiderian membrane*, and is thickest at the lower part of the cavity, where it is also ciliated and furnished with numerous mucous glands.

The upper part of the mucous membrane both on the

Fig. 192.

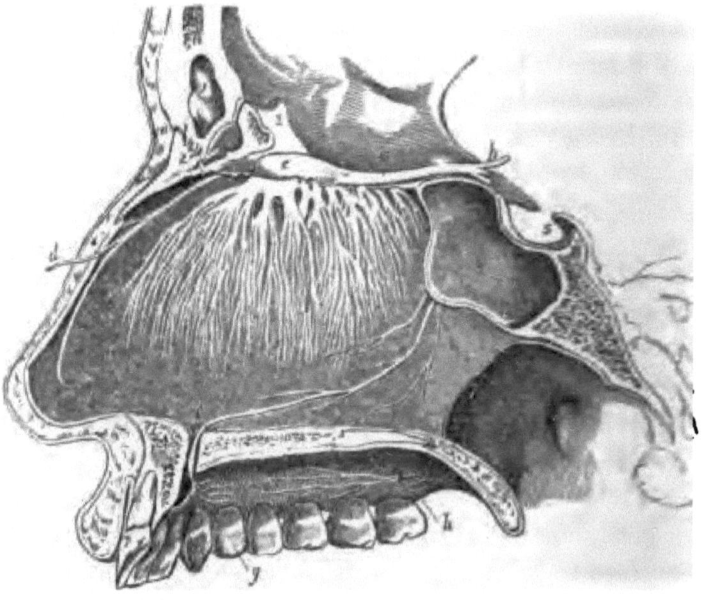

Fig. 192.—Left olfactory nerve, with its distribution on the septum narium (from Wilson).

1. Frontal sinus.
2. Nasal bone.
3. Crista galli of ethmoid bone.
4. Sphenoidal sinus of left side.
5. Sella turcica.
6. Basilar process of sphenoid and occipital bone.
7. Posterior opening of the right naris.
8. Opening of the Eustachian tube in the upper part of the pharynx.
9. Soft palate divided through its middle.
10. Cut surface of the hard palate.

a. Olfactory nerve.
b. Its three roots of origin.
c. The olfactory bulb.
d. Nasal nerve (ophthalmic div. of 5th).
e. Naso-palatine nerve (from the spheno-palatine ganglion).
f. The anterior palatine foramen.
g. Branches of the naso-palatine nerve to the palate.
h. Anterior and posterior palatine nerves.
i. Septum narium.

outer and inner walls of the fossa is peculiar, and has been named the Olfactory Region, from the fact that the branches of the olfactory or first nerve are distributed to it. It is impossible to distinguish this region except immediately after death, but it comprises the upper and a portion of the middle turbinate bones, and the corresponding portion of the septum. The mucous membrane of the olfactory region is of a brown colour in the recent state, and its epithelium is of the tesselated variety.

The **Olfactory Nerve** (Fig. 192, *a*) is the special nerve of smell, and its branches, which are derived from the *olfactory bulb*, reach the nasal cavity by piercing the cribriform plate of the ethmoid bone. The branches are distributed to the olfactory region, *i.e.*, upon the superior and middle turbinate bones and the upper third of the septum, but are very difficult to follow, owing to their being destitute of the white substance of Schwann.

The nerves of common sensation to the nose are derived from Meckel's ganglion and will be afterwards dissected.

[With a strong pair of scissors the turbinate bones may be readily cut so as to expose the several openings into the three meatuses.]

The *nasal branch of the fifth nerve* (Fig. 192, *d*) may be found in a groove on the under surface of the nasal bone in its passage from the orbit to the face, and gives a branch to the septum (p. 349).

Dissection of Meckel's Ganglion, etc.

Before beginning this dissection the student should ascertain the position of the spheno-palatine foramen (Fig. 191, *e*), opposite to which the spheno-palatine ganglion is placed, and through which the internal maxillary artery enters the nose. This will be found on either half of the skull immediately behind the superior meatus of the nose.

[The mucous membrane lining the vertical plate of the palate bone is to be removed, and the thin plate of bone chiselled through immediately below the spheno-palatine foramen, when the palatine nerves and arteries will be exposed enclosed in a tube of membrane. An attempt may be made to open up the Vidian canal in the root of the pterygoid process, in order to expose the Vidian nerve and artery, but the proceeding is very difficult and seldom satisfactory in its results.]

The **Spheno-palatine** or **Meckel's Ganglion** (Fig. 193, 1) is a minute red body situated opposite the spheno-palatine foramen, which is connected with the superior maxillary nerve by one or two branches (p. 472); with the facial nerve by the Vidian nerve; and with the sympathetic upon the internal carotid artery by a branch which joins the Vidian nerve (Fig. 190, 9).

The **Branches** (Fig. 193) of the ganglion are ascending, descending, internal, and posterior.

1. The *ascending branches* are two or three of very small size, which pass to the periosteum of the orbit through the spheno-maxillary fissure.

2. The *descending branches* go to the palate, and are three in number :—

The *anterior* or *great palatine nerve* (14) descends through the posterior palatine canal to the hard palate, where it forms a junction with the naso-palatine nerve passing through the anterior palatine canal (Fig. 192, *g*). Whilst in the canal the nerve gives off *inferior nasal branches* to the inferior turbinate bone.

The *middle* or *external palatine nerve* (16) is very small, and descends to the soft palate and tonsil.

The *posterior* or *small palatine nerve* (17) descends through a small canal behind the great nerve and is distributed to the soft palate and uvula, supplying the levator palati and the azygos uvulæ muscles.

3. The *internal branches* pass through the spheno-palatine foramen to the nose, and are the superior nasal and the naso-palatine nerves.

The *superior nasal branches* (4) supply the mucous membrane of the upper and middle turbinate bones.

The *naso-palatine nerve* (8) crosses the nasal fossa to

the septum, along which it passes through a groove in the vomer to the anterior palatine canal. In the canal the left nerve lies in front of the naso-palatine arteries and the right behind them, and both descend to the front of the hard palate to unite with the great palatine nerve. (Fig. 192, *e*).

Fig. 193.

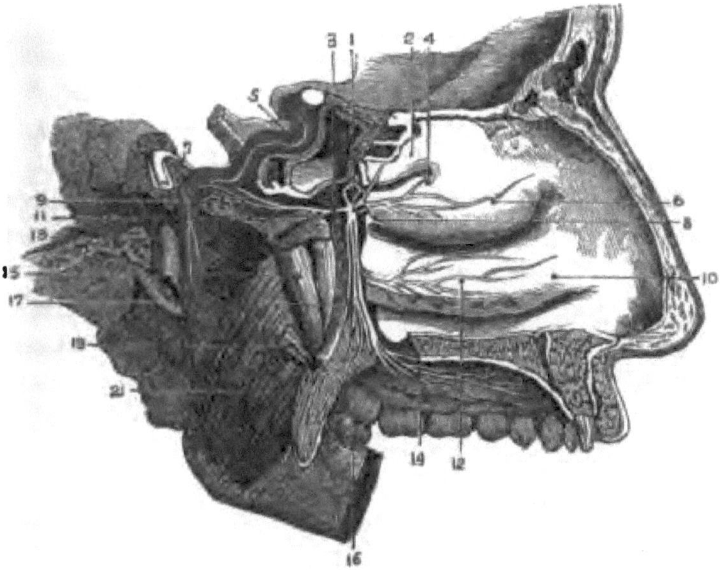

4. The *posterior branches* are the Vidian and the pterygo-palatine nerves.

The *Vidian nerve* (3) runs through the Vidian canal in

Fig. 193.—Meckel's ganglion and its branches (from Hirschfeld and Leveillé).

1. Spheno-palatine ganglion (Meckel).
2. Superior turbinate bone.
3. Vidian nerve.
4. Nasal branches.
5. Carotid artery in temporal bone.
6. Middle turbinate bone.
7. Superficial petrosal nerve joining the 7th.
8. Naso-palatine nerve (cut).
9. Carotid branch of Vidian.
10. Inferior turbinate bone.
11. Pharyngeal nerve.
12. Inferior nasal branches.
13. Pterygoideus externus.
14. Anterior or great palatine nerve.
15. Tensor palati.
16. Middle or external palatine nerve.
17. Posterior palatine nerve.
19. Levator palati (cut).
21. Pterygoideus internus.

the pterygoid process and divides into superficial petrosal and carotid branches.

The *superficial petrosal branch* (7) pierces the cartilage which fills up the foramen lacerum medium basis cranii, and runs in a groove on the surface of the petrous portion of the temporal bone to the hiatus Fallopii, through which it passes to join the facial nerve.

The *carotid branch* (9) joins the sympathetic plexus upon the internal carotid artery.

The *pterygo-palatine* or *pharyngeal nerve* (11) is very small, and passes through the pterygo-palatine canal to the mucous membrane of the upper part of the pharynx.

The **Internal Maxillary Artery** gives off the following branches in the spheno-maxillary fossa: the descending palatine, Vidian, pterygo-palatine, spheno-palatine, and infra-orbital, the last of which has been already examined.

1. The *posterior* or *descending palatine artery* accompanies the great palatine nerve through the posterior palatine canal. It gives branches to the soft palate and uvula, which descend through the smaller palatine canals, and then runs forward on the hard palate, supplying it and anastomosing with the nasal or spheno-palatine artery at the anterior palatine canal.

2. The *Vidian artery* accompanies the nerve through the Vidian canal, and supplies the upper part of the pharynx and the Eustachian tube.

3. The *pterygo-palatine artery* is very small and runs backwards through the pterygo-palatine canal.

4. The *nasal* or *spheno-palatine artery* enters the nose by the spheno-palatine foramen and gives branches to the mucous membrane of the upper part of the cavity, which anastomose with the anterior and posterior ethmoidal branches of the ophthalmic artery. The *naso-palatine* branch runs down the septum of the nose with the naso-palatine nerve and anastomoses with the artery of the septum from the facial trunk and also with the descending palatine artery through the anterior palatine canal.

The Otic Ganglion.

[If the part is not too much decomposed, the otic ganglion may be exposed, from within, on the side upon which the internal pterygoid muscle has not been cut away. The levator palati muscle and the Eustachian tube are to be carefully removed, when upon the inner surface of the pterygoid muscle will be found the nerve to that muscle, which if traced upwards will lead to the ganglion.]

The **Otic Ganglion** (Fig. 194, 7) (Arnold's) is a minute body lying on the inner surface of the third division of the fifth nerve and especially connected with the branch to the internal pterygoid muscle. Like all the cranial ganglia, the otic has motor, sensory, and sympathetic roots; the motor root from the fifth nerve; the sensory roots from the auriculo-temporal nerve and from Jacobson's nerve, by means of a small superficial petrosal nerve of Arnold; and the sympathetic root from that upon the middle meningeal artery.

The otic ganglion gives branches to two muscles, the tensor tympani and the circumflexus palati.

The Tympanum.

[If the tympanum has not been damaged in the previous dissections, an attempt may be made to trace the facial nerve through the temporal bone, thus: the base of the skull being turned upward, the saw is to be carried through a line drawn from the stylo-mastoid foramen to the foramen ovale. By this cut the cavity of the tympanum will be divided, and it will be possible to examine its walls and ossicula, together with the seventh nerve and the chorda tympani.]

The **Tympanum** or **Middle Ear** (Fig. 194) is an irregularly six-sided cavity, which may be conveniently said to resemble a room with a passage (meatus auditorius externus) leading to it.

The *outer* wall of the chamber is formed by the membrana tympani, which is placed obliquely at the end of the meatus; the *inner* wall corresponds to the outer wall of the *vestibule; in front*, is the opening of the Eustachian tube with the special tube for the tensor tympani muscle formed by the processus cochleariformis; *behind*, is the opening into the mastoid cells. The *roof* of the chamber is a thin portion of bone separating it from the cavity of the cranium; the *floor* is a thicker portion corresponding to the jugular fossa.

Fig. 194.

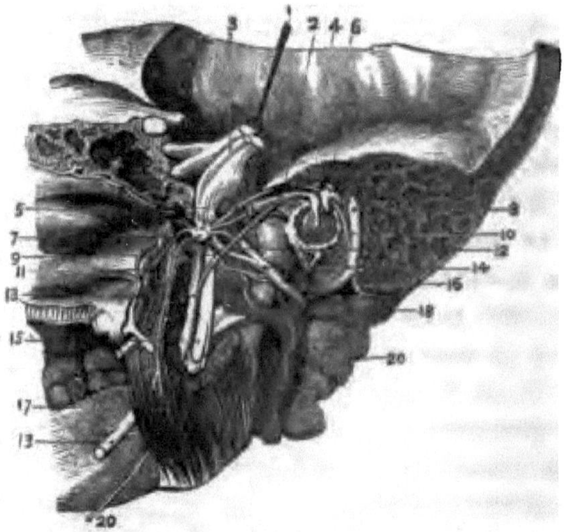

The inner wall is the most important, and presents the following points for examination (Fig. 190): (1) the *fenestra ovalis* in which the base of the stapes articulates; below this (2) the *fenestra rotunda*, an opening into the cochlea; anterior to these a slight eminence grooved by nerves, (3) the *promontory* with Jacobson's nerve; and, posterior to the fenestra ovalis and fenestra rotunda and close to the opening of the mastoid cells, (4) the *pyramid*, a

---

Fig. 194.—Internal view of Otic ganglion and Tympanum (from Hirschfeld and Leveillé).

1. Sensory portion of 5th nerve with Gasserian ganglion.
2. Tensor tympani muscle.
3. Motor portion of 5th passing beneath the ganglion.
4. Malleus.
5. Small superficial petrosal nerve of Arnold.
6. Incus.
7. Otic ganglion.
8. Facial nerve.
9. Chorda tympani.
10. Membrana tympani.
11. Tensor palati muscle.
12. Middle memingeal artery.
13, 13. Lingual nerve (5th).
14. Auriculo-temporal nerve.
15. Inferior dental nerve (5th).
16. Pterygoideus externus.
17. Pterygoideus internus.
18. Internal maxillary artery.
20, 20. Mylo-hyoid nerve.

onical projection of bone pierced at the top by a small hole through which the *stapedius* muscle works.

The *Aqueduct of Fallopius*, or canal for the facial nerve, forms a slight projection curving behind the pyramid and above the fenestra ovalis, and may be opened with bone-forceps in the part which has not been already divided with the saw. It extends from the bottom of the meatus auditorius internus to the stylo-mastoid foramen. In the upper part of the canal is the *intumescentia ganglioformis*, an enlargement of the facial nerve at the point where it is joined by the great petrosal nerve through the hiatus Fallopii. The facial nerve in this part of its course gives off a minute branch to the stapedius muscle and the chorda tympani.

The *Chorda Tympani* (Fig. 194, 9) enters the tympanum close to the pyramid, and passes forward, between the handle of the malleus and the long process of the incus, to an opening close to the Glasserian fissure by which it leaves the temporal bone to join the gustatory nerve.

The **Ossicula Auditûs** (Fig. 194) are the malleus, incus, and stapes.

The *Malleus* (hammer) (4) consists of a head, neck, handle (manubrium), and two processes (processus gracilis and processus brevis). The head articulates with the incus; the manubrium is inserted between the mucous and fibrous layers of the membrana tympani; the processus gracilis is inserted into the Glasserian fissure; the processus brevis gives attachment to the tensor tympani muscle.

The *Incus* (anvil) (6) consists of a body and two processes. The body articulates with the head of the malleus; the short process is attached to the margin of the orifice of the mastoid cells; the long process is nearly parallel to the handle of the malleus, and has at its extremity a small nodule of bone, the *os orbiculare*, which in the fœtus is separate, but becomes united in adult life and articulates with the stapes.

The *Stapes* (stirrup) is articulated by its head with the long process of the incus and at right angles to it. The base (at which the two crura unite) is attached to the fenestra ovalis; the neck gives attachment to the small stapedius muscle.

The **Muscles** of the Tympanum are three in number, viz., the tensor tympani, laxator tympani, and stapedius, but some anatomists add a fourth, the laxator tympani minor.

The *Tensor Tympani* (Fig. 194, 2) *arises* from the under surface of the apex of the petrous portion of the temporal bone and from the Eustachian tube, and runs backwards in a distinct canal formed by the processus cochleariformis to be *inserted* into the root of the handle and the processus brevis of the malleus. It is *supplied* by a branch from the otic ganglion.

The *Laxator Tympani arises* from the under surface of the spine of the sphenoid bone and from the Eustachian tube, and entering the tympanum through the Glasserian fissure is *inserted* into the neck of the malleus. It is *supplied* by a branch from the chorda tympani.

The *Stapedius arises* from the interior of the pyramid, and emerges from its apex to be *inserted* into the neck of the stapes. It is *supplied* by a branch of the facial nerve.

The *Laxator Tympani Minor arises* from the upper margin of the meatus externus, and is *inserted* into the handle and processus brevis of the malleus. It is regarded as a ligament by many anatomists.

THE TONGUE.

[The tongue and larynx which were laid aside are to be examined without separating them. The branches of nerves and the several muscles which were necessarily divided in removing the tongue are to be identified before the examination of the organ itself is proceeded with.]

The **Tongue** (Fig. 195) is connected with the os hyoides by muscular fibres and by a membrane (hyo-glossal) which is deeply placed between the muscles. On its under surface and sides will be found the attachments of the extrinsic muscles of the tongue, viz., the hyo-glossus, stylo-glossus, palato-glossus, and superior constrictor (glosso-pharyngeus); and close to the median line will be seen the large genio-hyo-glossus of each side.

All the anterior part of the dorsum of the tongue is covered by papillæ, but behind a V-shaped row of large (circumvallate) papillæ the surface only presents the orifices of numerous mucous glands and follicles.

The **Papillæ** (Fig. 195) of the tongue are of three kinds, the Circumvallate, Fungiform, and Filiform or Conical.

Fig. 195.

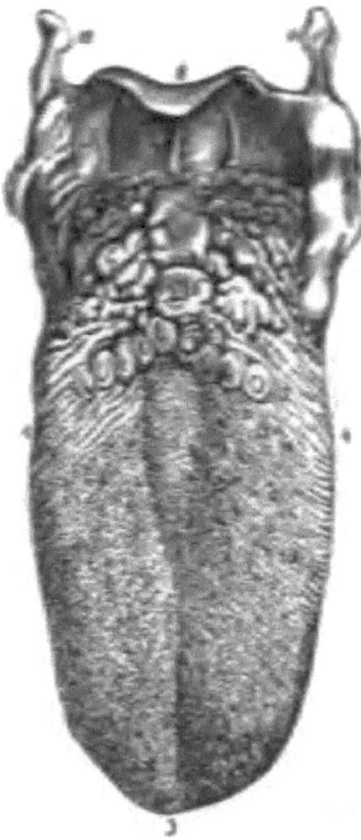

The *papillæ circumvallatæ* (5) are from eight to fifteen in number, and are arranged in two oblique rows which meet at the *foramen cæcum*, a deep mucous follicle. Each papilla is surrounded by a fossa, and is covered on its free surface with secondary papillæ. The circumvallate papillæ are the special organs of taste, and are supplied by the glosso-pharyngeal nerve.

The *papillæ fungiformes* are scattered over the tongue, but particularly at the tip and sides. They are globular in form and have slender attachments to the surface of the tongue; they are of a dark red colour and are covered with secondary papillæ.

Fig. 195.—The tongue with its papillæ (from Wilson).

1. The raphé, which sometimes bifurcates on the dorsum of the tongue, as in the figure.
2, 2. Lobes of the tongue; the rounded eminences on this part, and near its tip, are the papillæ fungiformes. The smaller papillæ, among which the former are dispersed, are the papillæ conicæ and filiformes.
3. Tip of the tongue.
4, 4. Its sides.
5, 5. The V-shaped row of papillæ circumvallatæ.
6. Foramen cæcum.
7. Mucous glands of the root of the tongue.
8, 8. Epiglottis.
9, 9. Fræna epiglottidis.
10, 10. Greater cornua of the os hyoides.

The *papillæ conicæ* and *filiformes* exist all over the tongue, but are most extensively developed at the tip. The conical papillæ are covered with minute secondary papillæ, and the filiform are prolonged into thread-like processes which in the carnivora are developed into spines.

The orifices of mucous follicles may be found among the papillæ on the dorsum of the tongue, and behind the *papillæ circumvallatæ* are numerous compound racemose lingual glands which lubricate the posterior portion of the organ.

Beneath the mucous membrane is a fibrous layer or *corion*, which gives attachment to many of the muscular fibres. In the centre of the tongue and dividing it into two symmetrical halves is a fibrous septum, the existence of which is marked by a raphé in the median line.

The **Intrinsic Muscles** are the Lingualis Superior, the Lingualis Transversus, and the Lingualis Inferior.

The *Lingualis Superior* extends along the surface of the tongue immediately beneath the mucous membrane, being connected with the septum internally.

The *Lingualis Transversus* constitutes the chief bulk of the tongue and is placed beneath the lingualis superior. Its fibres are attached to the median septum and pass outwards to the mucous membrane, giving passage to the ascending fibres of the lingualis inferior.

The *Lingualis Inferior* is to be seen on the under surface of the tongue by removing the hyo-glossus. It is larger than the lingualis superior and extends along the under surface of the tongue, giving ascending fibres which pass between those of the transverse muscle, and being closely connected with the stylo-glossus muscle.

Glands of the Frænum (Blandin).—By removing the mucous membrane on the under surface of the tongue near the tip a couple of oval glands may be seen. They are similar in structure to the sublingual glands although distinct from them, and are sometimes united in front, forming a single mass of an arched form (Deville).

The **Nerves** (Fig. 166) of the tongue have been already examined but may now be followed out minutely. They are the *Hypoglossal nerve* (9th) to the muscular structure of the tongue; the *Lingual nerve* (5th) to the fungiform and filiform papillæ on the dorsum and apex of the tongue; the *Glosso-pharyngeal nerve* (8th) to the circumvallate papillæ at the base of the tongue and to the surface behind them.

THE LARYNX.

The **Larynx** or organ of voice is closely connected with the hyoid bone and tongue. On tracing the mucous membrane from the dorsum of the tongue it will be found to form three little *fræna* between the base of that organ and the prominent cartilage of the epiglottis; these are the *glosso-epiglottidean folds* (Fig. 195, 9). After being reflected over the epiglottis, the mucous membrane forms the two *arytæno-epiglottidean folds*, which bound the superior aperture of the larynx; and is then carried to the interior of the larynx and over its posterior surface into the œsophagus.

The **Superior Aperture of the Larynx** (Fig. 188) is bounded in front by the epiglottis, behind by the arytænoid cartilages, and on each side by the arytæno-epiglottidean folds. On looking through it the Glottis or Rima Glottidis will be seen, *i.e.*, the space between the two inferior or true vocal cords, which varies from a mere chink to a triangular aperture with the apex forwards, according to the movements of the arytænoid cartilages. Above the true vocal cords will be seen the less distinct false cords, and through the glottis may be seen the interior of the trachea. This view of the parts deserves especial notice as it is exactly that which is seen reversed when reflected in the laryngoscope.

[The larynx is to be fastened upon a table by driving a nail or pin through the tongue and putting the trachea on the stretch with hooks, the anterior surface being exposed; the extrinsic muscles of the larynx, the sterno-hyoid, sterno-thyroid, thyro-hyoid, and inferior constrictor of the pharynx, are then to be carefully removed without injuring the crico-thyroid muscle or the external laryngeal nerve.]

Intrinsic Muscles of the Larynx.

The **Crico-thyroid Muscle** (Fig. 178, 22) is placed on each side of the crico-thyroid membrane and is somewhat triangular in shape, being narrow below and expanded above. It *arises* from an oval surface on the side of the cricoid cartilage, and ascends to be *inserted* into the posterior part of the lower border and into the inferior cornu of the thyroid cartilage. It is *supplied* by the external laryngeal branch of the superior laryngeal nerve.

[The preparation is to be turned over and the mucous membrane is to be dissected off the back of the larynx, the remnant of the œsophagus being removed. The crico-arytænoidei postici muscles on the back of the cricoid cartilage are readily dissected; but in order to clean the cross-fibres of the arytænoideus it will be necessary to over-distend the larynx so as to stretch the muscle; and this may be most readily done by the finger of an assistant.]

The **Arytænoideus** (Fig. 196, 6) is an example of a single muscle in the median line, its fibres crossing one another and being equally distributed on each side. It occupies the concave posterior surfaces of the two arytænoid cartilages, the fibres passing from one to the other, and being divisible into superficial and deep layers. The superficial fibres cross obliquely from the apex of one cartilage to the base of the other and *vice versâ* so as to form an X, whilst the deeper fibres pass transversely beneath them. The arytænoideus is *supplied* by both the superior and the inferior laryngeal nerves.

The **Crico-Arytænoideus Posticus** (Fig. 196, 7) *arises* from the posterior surface of the cricoid cartilage on one side of the median ridge, and its fibres pass obliquely upward and outward to be *inserted* on the posterior aspect of the prominent outer angle of the base of the arytænoid cartilage. It is *supplied* by the inferior laryngeal nerve.

[To see the crico-arytænoideus lateralis and the thyro-arytænoideus muscles, the superior and inferior cornua of one half of the thyroid cartilage should be cut through and the ala drawn forward, any tissue being detached from its inner surface to within a quarter of an

Fig. 196.

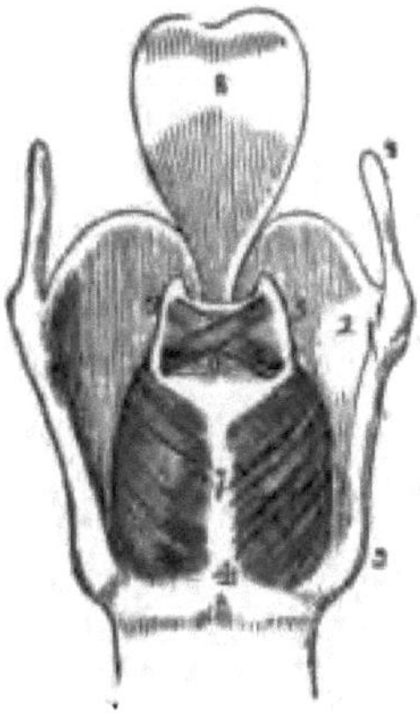

Fig. 197.

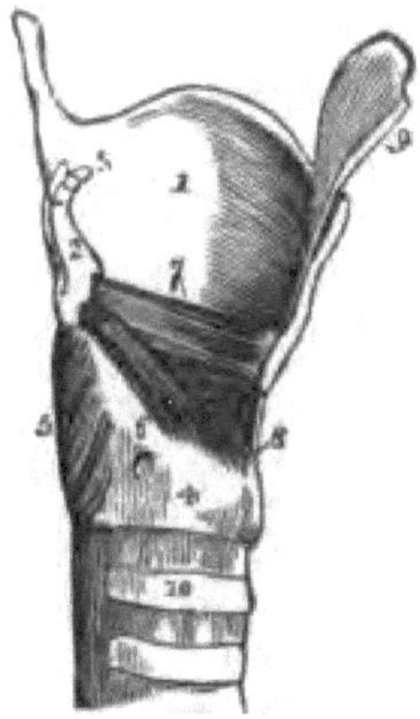

inch of the median line. A vertical cut is then to be made through it there with strong scissors or the bone-forceps, and the piece is to be removed. The larynx being distended as before, the muscles can be readily dissected.]

The **Crico-Arytænoideus Lateralis** (Fig. 197, 6) *arises* from the upper border of the side of the cricoid cartilage, and its fibres pass obliquely upward and backward to be *inserted* into the anterior aspect of the prominent

---

Fig. 196.—Posterior view of the larynx (from Wilson).

1. Thyroid cartilage, right ala.
2. Ascending cornu.
3. Descending cornu.
4. Cricoid cartilage.
5, 5. Arytænoid cartilages.
6. Arytænoideus muscle, consisting of oblique and transverse fasciculi.
7. Crico-arytænoidei postici muscles.
8. Epiglottis.

Fig. 197.—Side view of the larynx, one ala of the thyroid cartilage removed (from Wilson).

1. Left ala of thyroid cartilage.
2. Right arytænoid cartilage.
3. Corniculum laryngis.
4. Cricoid cartilage.
5. Crico-arytænoideus posticus muscle.
6. Crico-arytænoideus lateralis.
7. Thyro-arytænoideus.
8. Crico-thyroid membrane.
9. One half the epiglottis.
10. Upper part of trachea.

outer angle of the base of the arytænoid cartilage. This muscle is always more or less closely connected with the thyro-arytænoideus, which is above. It is *supplied* by the inferior laryngeal nerve.

The **Thyro-Arytænoideus** (Fig. 197, 7) has a more horizontal direction than the preceding muscle. It *arises* from the receding angle of the thyroid cartilage immediately external to the inferior or true vocal cord, and reaches down to the crico-thyroid membrane, to which some of its fibres are attached. It passes backwards, the upper fibres being more oblique than the lower ones, to be *inserted* into the anterior surface of the arytænoid cartilage external to the attachments of the vocal cords. It is *supplied* by the inferior laryngeal nerve.

[If a second larynx can be procured, a very satisfactory way of seeing the thyro-arytænoid muscles is to cut the thyroid cartilage horizontally at the level of the true vocal cords, when the muscles can be readily displayed immediately outside them.]

Above the preceding muscles a few muscular fibres may be traced to the epiglottis from both the thyroid and the arytænoid cartilages. These are the *thyro-epiglottideus* and the *arytæno-epiglottideus*, or the two sets of fibres have been considered as one muscle under the name of *thyro-arytæno-epiglottideus*. The upper muscle is contained in the arytæno-epiglottidean fold of mucous membrane; the lower one spreads over the mucous pouch of the larynx and was termed *compressor sacculi laryngis* by Mr. Hilton.

Actions of the Laryngeal Muscles.

The cartilages are moved upon one another by the several muscles, producing an effect upon either the tension of the vocal cords, or the size of the opening between them—the glottis.

The *Crico-thyroidei* draw the thyroid cartilage downward on the cricoid cartilage, and thus tighten the vocal cords.

The *Thyro-arytænoidei* being parallel to, and attached

to the same points as, the vocal cords, must necessarily by their contraction approximate the points of attachment, and so relax the cords.

The *Crico-arytænoidei postici* swing the arytænoid cartilages outwards and thus dilate the glottis, as can be easily shown by making traction upon the two muscles simultaneously with the points of a pair of forceps.

The *Crico-arytænoidei laterales* are antagonists to the postici and bring the arytænoid cartilages back to their old positions, and afterwards approximate them, thus narrowing the glottis.

The *Arytænoideus* draws the arytænoid cartilages together and thus (according to Mr. Bishop) narrows the glottis; but it may be doubted whether the superficial fibres which are inserted farthest out would not swing the cartilages upon their axes and thus widen the glottis.

[A side view of the interior of the larynx is now to be obtained by cutting away the crico-arytænoideus lateralis with the vocal cords and mucous membrane of the same side, but the cricoid cartilage is to be left entire.]

On a Side-view (Fig. 198) of the interior of the larynx will be seen the horizontal projections of the mucous membrane formed by the superior and inferior vocal cords beneath it. Between the two cords is the cavity of the *ventricle* of the larynx, and from this a probe can be carried into the *laryngeal pouch* or *sacculus laryngis*, which projects upward between the upper cord and the thyroid cartilage. The mucous membrane is continued from the ventricle into the sac, in which it is provided with numerous mucous glands.

[The mucous membrane is to be dissected from the vocal cords and their attachments are to be clearly shown.]

The **Vocal Cords** or **Thyro-arytænoid Ligaments** (Fig. 198) are four in number, two on each side of the larynx, and are composed of yellow or elastic fibrous tissue.

The *superior ligament* or *false cord* (*h*) has an arched

direction, and is much more slender than the inferior ligament. It is attached in front to the receding angle of the thyroid cartilage, and behind to the projection on the upper part of the front of the arytænoid cartilage.

Fig. 198.

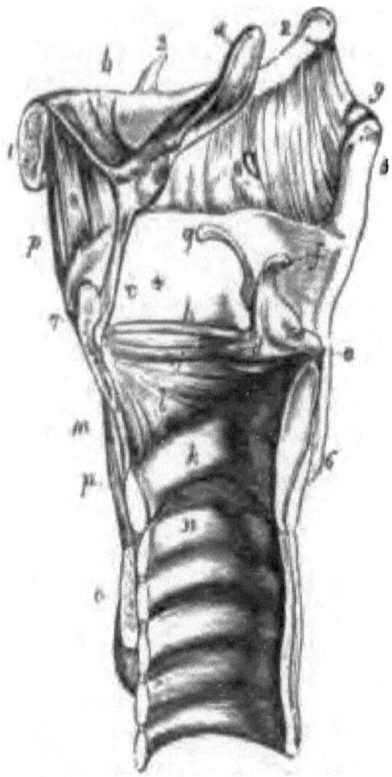

The *inferior ligament* or *true cord* (*i*) is horizontal in direction, being attached in front to the receding angle of the thyroid cartilage below the superior ligament, and behind to the prominent anterior angle of the base of the arytænoid cartilage. The interval between the two true vocal cords is the *rima glottidis*.

The **Arteries** of the larynx are derived from the superior and inferior thyroid arteries.

The *superior laryngeal artery*, derived from the superior thyroid, enters the

Fig. 198.—Vertical section of the larynx, showing its ligaments (from Wilson).

1. Body of os hyoides.
2. Its great cornu.
3. Lesser cornu.
4. Ala of the thyroid cartilage.
5. Superior cornu.
6. Inferior cornu.
7. Pomum Adami.

8, 8. Thyro-hyoid membrane.

9. Lateral thyro-hyoid ligament.

a. Epiglottis.
b. Hyo-epiglottidean ligament.
c. Thyro-epiglottidean ligament.
d. Arytænoid cartilage; inner surface.
e. Outer angle of base of arytænoid cartilage.
f. Corniculum laryngis.
g. Cuneiform cartilage.
h. Superior thyro-arytænoid ligament.
i. Inferior thyro-arytænoid ligament, or chorda vocalis; the elliptical space between the two thyro-arytænoid ligaments is the ventricle of the larynx.
k. Cricoid cartilage.
l. Lateral portion of the crico-thyroid membrane.
m. Central portion of the same membrane.
n. Upper ring of trachea.
o. Section of isthmus of thyroid gland.
p, p. Levator glandulæ thyroideæ.

larynx through the thyro-hyoid membrane, and divides into ascending and descending branches which supply the muscles and mucous membrane.

The *inferior laryngeal artery*, derived from the inferior thyroid artery, reaches the back of the cricoid cartilage and anastomoses with the superior artery.

The **Nerves** (Fig. 179) of the larynx are the superior and the inferior (or recurrent) branches of the pneumogastric.

The *superior laryngeal nerve* (9) has been seen to give a branch (*external laryngeal*) to the crico-thyroid muscle, and then to pierce the thyro-hyoid membrane with the superior laryngeal artery. It supplies the mucous membrane of the larynx and back of the tongue, and gives a branch to the arytænoideus muscle, besides a communicating branch to the inferior laryngeal nerve.

The *inferior* or *recurrent nerve* (21) is the motor nerve of the larynx, and has been seen to take a different course on the two sides, winding around the subclavian artery on the right, and around the aorta on the left side, but in the larynx the nerves have similar distributions. Each nerve running up the side of the trachea passes beneath the border of the inferior constrictor of the pharynx and beneath the ala of the thyroid cartilage, to end in branches to the muscles of the larynx. It supplies all the muscles of the larynx except the crico-thyroid, and thus one muscle, the arytænoideus, is supplied from two sources.

[In order to see the ligaments which bind the several parts of the larynx together it will be necessary to remove cautiously all the muscular fibres and mucous membrane.]

The **Thyro-hyoid Ligament** (Fig. 198, 8) connects the upper border of the thyroid cartilage with the hyoid bone. It is composed of yellow elastic tissue and may be divided into a central portion or *membrane*, and rounded lateral portions (9) which pass from the cornu of the hyoid bone to the superior cornu thyroid cartilage, and in which a nodule of cartilage is often developed. The thyro-hyoid membrane is perforated by the superior laryngeal nerve and artery.

The **Crico-thyroid Membrane** or **Ligament** (Fig. 198, *l, m*) is composed of elastic tissue, and passes between the borders of the cricoid and thyroid cartilages in the median line, but at the sides is prolonged to the inferior vocal cords, with which it is incorporated.

The **Crico-tracheal Ligament** is between the lower border of the cricoid cartilage and the upper ring of the trachea, and is a simple membrane like that between the several rings of the trachea.

Capsular Ligaments surround the articular extremities of the cartilages, which are also furnished with synovial membranes. These exist in the articulations between the *cricoid* and the inferior cornua of the *thyroid cartilage;* and between the *cricoid* and the bases of the two *arytænoid cartilages.*

The **Thyro-epiglottidean Ligament** (Fig. 198, *c*) connects the lower border of the epiglottis with the back of the thyroid cartilage opposite the notch in its upper border.

The **Hyo-epiglottidean Ligament** (Fig. 198, *b*) is a band of yellow elastic tissue passing between the apex of the epiglottis and the upper border of the hyoid bone.

Cartilages of the Larynx.

[The large cartilages of the larynx, viz. the thyroid, cricoid, two arytænoid, and the cartilage of the epiglottis, are readily recognised, but two pairs of minute cartilages, the cornicula laryngis and the cuneiform cartilages, are to be looked for in the arytæno-epiglottidean folds of mucous membrane.]

The **Thyroid Cartilage** (Fig. 178, 1) (θυρεὸς εἶδος, like a shield) is the largest cartilage of the larynx, and consists of two halves or *alæ* which meet at an angle in front and form the *pomum Adami.* Each ala is quadrilateral in shape, and has a rounded posterior border into which the stylo-pharyngeus and palato-pharyngeus muscles are inserted, and which is prolonged into a *superior* and an *inferior cornu.* The superior cornu is the longer of the two and is non-articular, but the inferior cornu has been seen to articulate with the cricoid cartilage, and to give attachment to the crico-thyroid muscle.

On the side of the ala is an *oblique ridge*, extending from a tubercle at the root of the superior cornu to another tubercle about the middle of the lower border of the cartilage. The ridge gives attachment to the sterno-thyroid and thyro-hyoid muscles, and the surface behind it to the inferior constrictor of the pharynx. The inner surfaces of the alæ are smooth, and in the receding angle formed by their junction seven structures are attached, viz. the thyro-epiglottidean ligament, the two pairs of thyro-arytænoid ligaments (false and true vocal cords), and the two thyro-arytænoid muscles, just external to the inferior ligaments or true cords (Fig. 198).

The **Cricoid Cartilage** (Fig. 197, 4) (κρίκος εἶδος, like a ring) is a ring of cartilage, shallow in front, (where it has been seen to be connected with the thyroid cartilage by a membrane, and has the crico-thyroideus attached to its surface), but deep behind, where it fills up a part of the space left between the posterior borders of the thyroid. The upper border of the deep portion presents two oval articular surfaces for the arytænoid cartilages, external to which are the origins of the crico-arytænoidei laterales. On each side and near the lower border of the cartilage are two small facets for articulation with the inferior cornua of the thyroid cartilage. The posterior surface is divided in the middle line by a vertical ridge to which some of the fibres of the œsophagus are attached, the concave surface on each side giving origin to the crico-arytænoidei postici muscles.

The **Arytænoid Cartilages** (Fig. 198, *d*) (ἀρυταίνα, a pitcher*) are two in number and are of a pyramidal shape. The base of each cartilage is triangular and articulates

* This derivation has reference to the appearance of both cartilages taken together and covered by mucous membrane. In animals,

with the upper border of the cricoid cartilage; its anterior angle gives attachment to the true vocal cord, and its external angle to the crico-arytænoideus posticus and crico-arytænoideus lateralis.

The apex is curved backward and inward and is surmounted by the corniculum laryngis. The posterior surface of the cartilage is concave and gives attachment to the arytænoideus muscle; the anterior surface presents a small tubercle for the attachment of the false vocal cord, and also gives attachment to the thyro-arytænoideus muscle; the internal surface is smooth and covered with mucous membrane.

The **Cornicula Laryngis** or **Cartilages of Santorini** (Fig. 198, *f*) are two small conical cartilages connected with the apices of the arytænoid cartilages and with the arytæno-epiglottidean folds. They are composed of yellow fibro-cartilage.

The **Cuneiform Cartilages** (Fig. 198, *g*) are two small bodies found in the arytæno-epiglottidean folds of mucous membrane, and are sometimes wanting. They are composed of yellow fibro-cartilage.

The **Epiglottis** (Fig. 198, *a*) is composed of yellow fibro-cartilage, shaped like a cordate leaf, and has been seen to be connected with the thyroid cartilage and the hyoid bone. It is covered with mucous membrane, upon removing which it will be found to be perforated by numerous holes in which mucous glands are lodged.

The large cartilages of the larynx are very apt to undergo calcareous degeneration in advanced life, and the epiglottis is not unfrequently thickened by disease or injured by ulceration.

---

which were the principal subjects of dissection among the ancients, the opening of the larynx, with the arytenoid cartilages, bears a curious resemblance to the mouth of a pitcher with a large spout (Wilson).

LIGAMENTS OF THE VERTEBRÆ.

[The ligaments of the spinal column are to be examined by removing all the remaining muscular fibres.]

The Ligaments of the *Bodies* of the vertebræ are—1, the Anterior Common Ligament; 2, the Posterior Common Ligament; and 3, the Intervertebral Substance or Fibro-cartilage.

The **Anterior Common Ligament** (Fig. 199, 1) is a broad band of fibres extending down the front of the

Fig. 199.

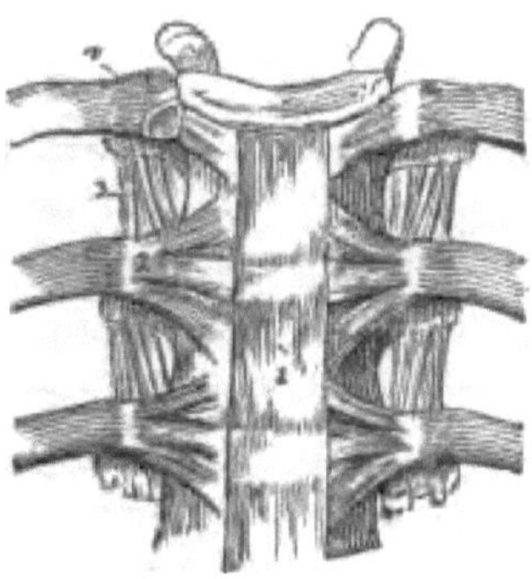

Fig. 200.

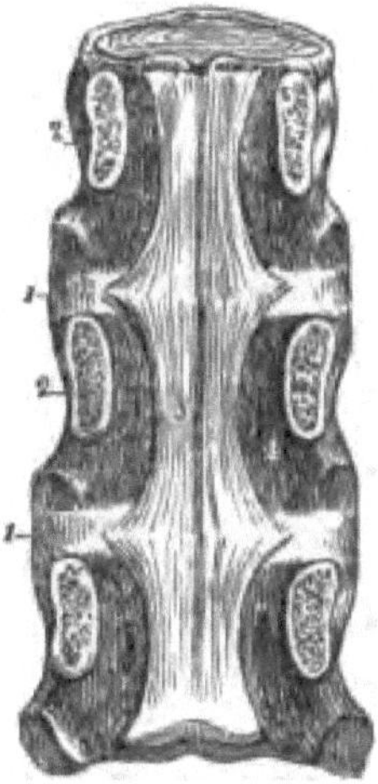

vertebral column from the axis to the sacrum. The fibres are attached principally to the intervertebral substances, and only slightly to the upper and lower

---

Fig. 199.—Anterior ligament of the vertebræ, and ligaments of the ribs (from Wilson).

1. Anterior common ligament.
2. Anterior costo-vertebral or stellate ligament.
3. Anterior costo-transverse ligament.
4. Interarticular ligament connecting the head of the rib to the intervertebral substance, and separating the two synovial membranes of this articulation.

Fig. 200.—Posterior ligament of the vertebræ (from Wilson).

1, 1. Intervertebral substances.
2. Pedicles of vertebræ sawn through.
3. Posterior common ligament.

margins of the vertebræ; the deep fibres passing between the adjacent bones, but the superficial crossing three or more vertebræ. The ligament when divided will be seen to be thicker opposite the bodies of the vertebræ than on the intervertebral substances; and it is broader in the lumbar than in the cervical region.

[To see the posterior common ligament it will be necessary to remove the spinal cord, if this has not been already done, as follows: the saw is to be applied over the pedicles of, say, six dorsal vertebræ, and these being divided on each side, the remains of the spinal cord can be removed, when the posterior ligaments will be seen on the backs of the bodies of the vertebræ; and the ligamenta subflava can be dissected upon the vertebral arches which have been removed.]

The **Posterior Common Ligament** (Fig. 200, 3) is found upon the posterior aspect of the bodies of the vertebræ from the axis to the sacrum. It is remarkable in being broad opposite the intervertebral substances, with which it is closely connected, and narrow opposite the bodies of the vertebræ, over which it stretches to allow of the passage of vessels to the bone. It is broader in the cervical than in the lumbar region, but thickest in the dorsal region, and is composed of superficial and deep fibres like the anterior ligament.

An **Intervertebral Substance** (Fig. 200, 1) is found between the bodies of all the vertebræ from the axis to the sacrum, and is largest in the lumbar region. Each substance is composed of fibrous tissue and fibro-cartilage externally, but has a soft elastic material in the centre. The substances vary according to the curves of the spinal column, being deepest in front in the cervical and lumbar regions and the reverse in the dorsal region.

On dissecting the superficial fibres of an intervertebral substance they will be found to pass obliquely between the adjacent vertebræ, alternate layers taking opposite directions. By cutting through the attachment to one of the vertebræ with a strong knife, the pulpy material in the centre will be shown, and will expand considerably above the level of the fibro-cartilaginous rings around it.

On making a vertical section of two or more vertebræ with their intervertebral substances, the structure of the intervertebral substance will be best seen, and will be found to consist of fibro-cartilaginous plates placed in front of and behind the central pulpy material. The plates of both series are arranged one within another, the superficial ones being slightly bent towards the surface and the deeper ones towards the centre.

The *Articular Processes* are surrounded by **Capsular Ligaments** (Fig. 201, 2) which are looser in the cervical than in the dorsal and lumbar regions; they permit a simple gliding movement (arthrodia) between the processes, which are invested by a simple synovial membrane.

The *Arches* are united by the **Ligamenta Subflava** (Fig. 201, 1). These are to be seen on the internal surfaces of the arches which have been removed, and will be found to consist of a series of yellow elastic ligaments, extending between the arches of the vertebræ on each side of the middle line from the axis to the sacrum. Owing to the overlapping of the vertebræ the ligaments are attached to the posterior surface of the arch below and to the anterior surface of the arch above, extending from the articular process to the root of the spinous process on each side.

The *Spinous Processes* are united by (1) the Interspinous and (2) the Supraspinous ligaments.

The **Interspinous Ligaments** are placed between the spinous processes of the vertebræ, extending from the root to the tip of each, and being stronger in the lumbar than in the dorsal, and in the dorsal than in the cervical region.

The **Supraspinous Ligament** (Fig. 206, 1) is a fibrous cord extending along the tips of the spinous processes from the last cervical vertebra to the sacrum. It is continuous with the *ligamentum nuchæ*, and is thicker in the lumbar than in the dorsal region. It is composed of superficial and deep fibres, which have the same arrange-

Fig. 201.

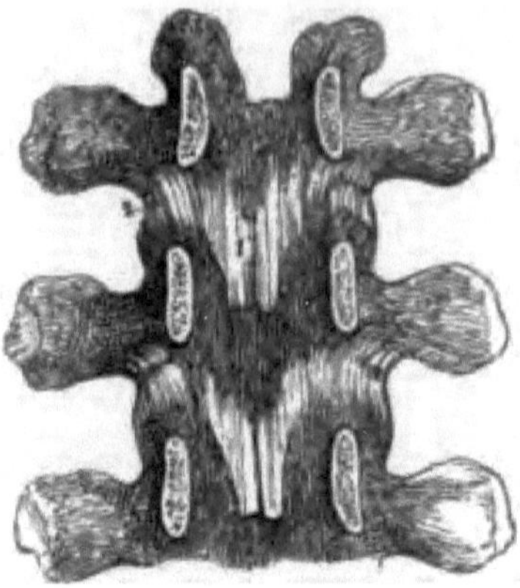

ment as those of the anterior and posterior common ligament.

The *Transverse Processes* are connected by **Intertransverse Ligaments** (Fig. 206) which are frequently absent in the cervical region, but are round and thick in the dorsal, and thin and membranous in the lumbar region.

Fig. 202.

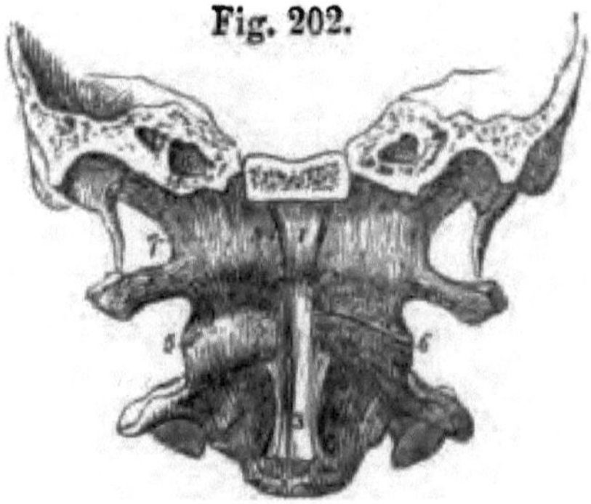

Ligaments of the Atlas, Axis, and Occiput.

[In order to examine the special ligaments of the first two cervical vertebræ and the occipital bone, it will be convenient to separate the whole of the cervical from the

Fig. 201.—Internal view of the arches of three vertebræ (from Wilson).

1. One of the ligamenta subflava. 2. The capsular ligament of one side.

Fig. 202.—Anterior view of the ligaments connecting the atlas, axis, and occipital bone (from Wilson).

1. Anterior round occipito-atloid ligament.
2. Anterior broad occipito-atloid ligament.
3. Commencement of the anterior common ligament.
4. Anterior atlo-axoid ligament, continuous inferiorly with the commencement of the anterior common ligament.
5. One of the atlo-axoid capsular ligaments; that on the opposite side has been removed to show the approximated surfaces of the articular processes (6).
7. One of the occipito-atloid capsular ligaments. The most external of these fibres constitute the lateral occipito-atloid ligament.

dorsal spine, reserving the latter with the ribs for subsequent examination.]

The ligaments of the first two vertebræ and the occipital bone may be conveniently divided into (1) those external to and (2) those internal to the vertebral canal. The external ligaments more or less resemble the ligaments common to the vertebral column, but the internal ligaments have special relation to the movements of nutation and rotation which are peculiar to this region.

1. The *External Ligaments* are the Anterior, Posterior, and Lateral Occipito-Atloid, and the Anterior and Posterior Atlo-Axoid ligaments; besides Capsular ligaments for the articular processes of the vertebræ and the condyles of the occipital bone.

The **Anterior Occipito-Atloid Ligament** (Fig. 202) consists of a superficial and a deep portion. The superficial portion (1) is round, and is attached to the basilar process of the occipital bone above, and to the anterior tubercle of the atlas below. The deep portion (2) is broad and membranous, and is attached to the occipital bone close to the foramen magnum above, and to the upper margin of the anterior arch of the atlas below.

The **Posterior Occipito-Atloid Ligament** (Fig. 203, 3) resembles the ligamenta subflava, but is composed of white fibrous tissue. It is thin and membranous, and is attached above to the posterior margin of the foramen magnum, and below to the posterior arch of the atlas. On each side this ligament is perforated by the vertebral artery and suboccipital nerve, over which some of its fibres arch.

The **Lateral Occipito-Atloid Ligaments** (Fig. 203, 4) are short strong bands passing between the roots of the transverse processes of the atlas and the jugular processes of the occipital bone.

The **Anterior Atlo-Axoid Ligament** (Fig. 202, 4) resembles the anterior occipito-atloid ligament, and consists of a superficial and a deep portion. The superficial portion is round, and is continuous with the anterior common ligament, being attached to the

The **Ascending Pharyngeal Artery** (Fig. 185, 9) is one of the ascending branches of the external carotid artery. It arises near the bifurcation of the common carotid, and ascends by the side of the pharynx, and to the inner side of the internal carotid artery, to the base of the skull. It gives branches to the prevertebral muscles, anastomosing with the ascending cervical artery, and divides into meningeal and pharyngeal branches.

The *meningeal branches* are very small, and enter the skull by the foramen lacerum medium and the foramen jugulare, to supply the dura mater.

The *pharyngeal branches* supply the pharynx and turn over the upper border of the superior constrictor to supply the palate; they anastomose with the inferior palatine branch of the facial artery.

The ascending pharyngeal vein opens into the internal jugular.

[The constrictor muscles of the pharynx are to be cleaned in the direction of their fibres, beginning at the lower border of the inferior constrictor. In order to see the origin of the superior constrictor, it will be necessary to remove the internal pterygoid on one side. The pharyngeal plexus must necessarily be destroyed in the course of the dissection, but the superior and inferior laryngeal and the glosso-pharyngeal nerves are to be preserved.]

The **Inferior Constrictor** (Fig. 187, 9) is the most superficial of the three muscles of the pharynx, the upper oblique border overlapping the middle constrictor, and the lower straight border being continuous with the œsophagus. It *arises* from the side of the cricoid cartilage in front of the articular facet, and from the ala of the thyroid cartilage behind the oblique line. All the fibres are *inserted* into the median raphé. The recurrent laryngeal nerve passes beneath the lower border of the inferior constrictor, and the superior laryngeal nerve and artery intervene between it and the middle constrictor.

The **Middle Constrictor** (Fig. 187, 10) is a fan-shaped muscle which with its fellow of the opposite side forms a

To expose these ligaments it will be necessary to saw ·ough the occipital bone obliquely on each side of the amen magnum. The posterior part of the bone being noved, the arches of the atlas and axis are to be cut rough on each side and removed, and any remains of e spinal cord dissected away, when the occipito-axoid ;ament will be exposed.]

The **Occipito-Axoid Ligament** (Fig. 205, 5) (apparatus gamentosus colli) ; continuous with ıe posterior comnon ligament. It s attached below ;o the back of the ɔody of the axis, and above to the basilar groove of the occipital bone above the foramen magnum, through which it passes.

Fig. 205.

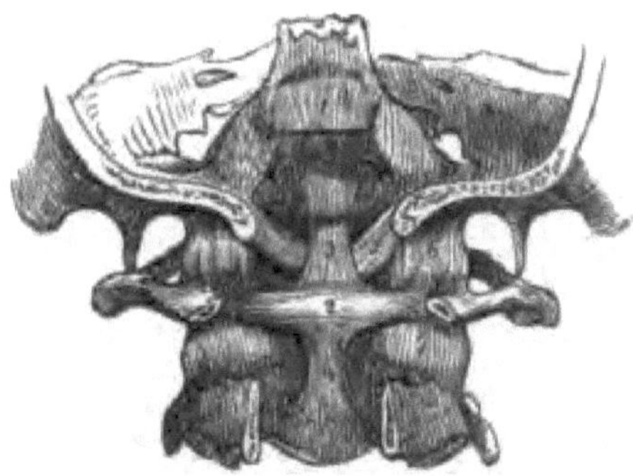

[The occipito-axoid ligament is to be divided and turned up and down, when the cruciform ligament of the atlas and axis will be brought into view, and immediately above the transverse portion will be seen the two oblique odontoid ligaments.]

The **Cruciform Ligament** (Fig. 205) consists of a strong transverse ligament and two small vertical slips.

The **Transverse Ligament of the Atlas** (Fig. 205, 2) is attached to a tubercle on the inner surface of the articular

Fig. 205.—Posterior view of the ligaments connecting the atlas, axis, and occipital bone (from Wilson).

1. Superior part of the occipito-axoid ligament, which has been cut away in order to show the ligaments beneath.
2. Transverse ligament of the atlas.
3, 4. Ascending and descending slips of the transverse ligament, which complete the cruciform arrangement.
5. One of the odontoid ligaments; the fellow ligament is seen on the opposite side.
6. One of the occipito-atloid capsular ligaments.
7. One of the atlo-axoid capsular ligaments.

process of each side, and thus forms a ring with the anterior arch of the atlas in which the odontoid process of the axis is firmly held. The vertical slips of the cruciform ligament pass from the transverse ligament to the basilar process of the occipital bone and the body of the axis respectively, and by detaching the upper one, the odontoid ligaments will be fully seen.

The **Odontoid or Check Ligaments** (Fig. 205, 5) are two strong bands of fibres which pass obliquely from the apex of the odontoid process to the margin of the foramen magnum on each side, and they thus check the rotation of the cranium and atlas upon the odontoid process. Between these is a *central odontoid ligament* (*ligamentum suspensorium dentis*) passing vertically between the margin of the foramen magnum and the top of the odontoid process.

On cutting all the ligaments of the articulation except the transverse ligament of the atlas, the odontoid process will be found to be still firmly held in position, owing to the fact that the ligament surrounds the contracted neck of the odontoid process, which is considerably smaller than the head. There are two *synovial membranes* in connection with the odontoid process, one in front, between it and the posterior surface of the atlas; the other behind, between it and the transverse ligament.

ARTICULATION OF THE RIBS WITH THE VERTEBRÆ.

The ligaments of the ribs may be divided into two sets, (1) those connecting the ribs with the bodies of the vertebræ (costo-vertebral), and (2) those connecting the ribs with the transverse processes (costo-transverse).

1. The costo-vertebral ligaments are Anterior, Capsular, and Interarticular.

The **Anterior Costo-vertebral or Stellate Ligament** (Fig. 199, 2) consists of three short fasciculi which radiate from the anterior surface of the head of the rib. The *superior* fasciculus passes to the vertebra above; the *middle* fasciculus to the intervertebral substance; the *inferior* fasciculus to the vertebra below. The whole ligament is closely connected with the anterior common ligament of the vertebræ. This arrangement only holds good in the case of those ribs which articulate with two vertebræ. In the case of the first, tenth, eleventh, and twelfth ribs, which articulate with a single vertebra each, the stellate ligament is attached principally to that vertebra, but sends a few fibres to the vertebra above.

The **Capsular Ligament** is a thin layer of fibres surrounding the articulation where the anterior ligament is wanting.

The **Interarticular Ligament** (Fig. 199, 4) is seen by removing the stellate ligament, and is a short band passing between the ridge on the head of the rib and the intervertebral substance. It divides the articulation into two parts, each of which has a separate *synovial membrane*, but it is absent from the articulations of the first, tenth, eleventh, and twelfth ribs, which have a single synovial membrane.

Fig. 206.

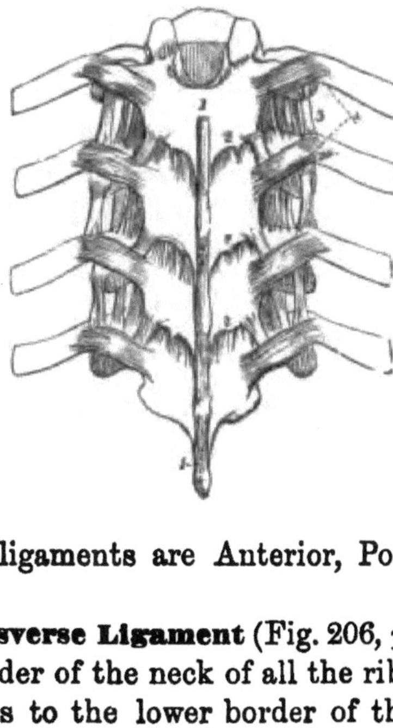

2. The Costo-transverse ligaments are Anterior, Posterior, and Middle.

The **Anterior Costo-transverse Ligament** (Fig. 206, 3) is attached to the upper border of the neck of all the ribs except the first, and ascends to the lower border of the transverse process of the vertebra above. It separates

Fig. 206.—Posterior view of part of the dorsal vertebral column, showing the ligaments connecting the vertebræ with each other and the ribs with the vertebræ (from Wilson).

1, 1. Supraspinous ligament.
2, 2. Ligamenta subflava, connecting the laminæ.
3. Anterior costo-transverse ligament.
4. Posterior costo-transverse ligaments.

the anterior from the posterior division of the intercostal nerves.

The **Posterior Costo-transverse Ligament** (Fig. 206, 4) is a short thick band passing obliquely upward from the apex of the transverse process of the vertebra to the rough non-articular portion of the tubercle of the rib.

[In order to see the middle costo-transverse ligament it will be necessary to make a horizontal section through the transverse process of a vertebra and the neck of the corresponding rib, when the short ligament will be found between the two.]

The **Middle Costo-transverse Ligament** consists of short strong fibres passing between the posterior surface of the neck of the rib and the corresponding transverse process.

A *synovial membrane* exists between the facet upon each of the transverse processes of the ten upper dorsal vertebræ and the articular tubercle of the corresponding rib.

The anterior extremity of each rib has a hollow in it into which the costal cartilage fits, being firmly united by the periosteum.

Chondro-Sternal Articulation. The cartilages of the true ribs fit into depressions on the side of the sternum, and are attached by *anterior* and *posterior* ligaments. A band attaching the seventh rib to the xiphoid cartilage is called the costo-xiphoid ligament.

The fifth, sixth, seventh, eighth, ninth, and tenth cartilages are connected by fibrous bands.

Synovial Membranes.—The first costal cartilage is continuous with the sternum and has no synovial cavity. The second and third cartilages have a double synovial membrane, owing to the existence of an interarticular ligament. The fourth, fifth, sixth, and seventh cartilages have a single synovial membrane at their sternal articulations, and there are in addition three synovial membranes between the sixth, seventh, eighth, and ninth costal cartilages respectively.

PART V.

DISSECTION OF THE BACK AND SPINAL CORD.

IN dissecting the Back it is customary for the dissectors of the arms to take the first two layers of muscles, and the dissectors of the head and neck the remainder of the region, including the spinal cord.

[An incision is to be made from the occiput to the sacrum in the median line, and another along each crest of the ilium at right angles to it. The dissectors should then raise the skin with all the infiltrated fascia, going at once down to the superficial muscles, through which the cutaneous nerves appear. On the left side of the subject the arm is to be drawn forward so as to put the latissimus dorsi on the stretch, and the dissector should begin to clean the lower part of that muscle, and work upward to the trapezius; on the right side the arm should be pulled down at first, and the head drawn over to the opposite side with hooks, to put the upper fibres of the trapezius on the stretch, and the dissector must begin at the upper border of that muscle (already exposed in the posterior triangle) and work downward to the latissimus dorsi. On both sides the arm and scapula will require to be moved from time to time to put the different sets of fibres on the stretch in turn.]

The **Cutaneous Nerves** are derived from the posterior divisions of the spinal nerves and are of small size. The cervical and upper dorsal nerves pierce the trapezius close to the spine, the second and third turning up to the occiput; the lower dorsal and the lumbar nerves pierce the latissimus dorsi along the oblique line at which the muscular fibres commence; while the sacral nerves give small branches through the tendinous expansion near the spine.

The **First Layer of Muscles** (Fig. 207) consists of the Trapezius and Latissimus Dorsi. The trapezius partially overlaps the latissimus, but between them and the base of the scapula is a small triangular interval in which the lower fibres of the rhomboideus major can be seen, and also, below that muscle, the posterior surfaces of two or three ribs and their intercostal muscles.

The **Trapezius** (Fig. 207, 1) of one side is triangular, but the two muscles together resemble a trapezium. It *arises* from the external occipital protuberance and inner third of the superior curved line of the occipital bone; from the ligamentum nuchæ and seventh cervical spinous process; and from the spinous processes of all the dorsal vertebræ. The fibres converge to be *inserted* into the outer third of the posterior border of the clavicle, and into the upper border of the acromion process and spine of the scapula. The fibres thus have different directions, and the *action* of the muscle upon the scapula is said to differ according to the fibres brought into use; the action of the entire muscle is to draw the scapula towards the spine, and thus the two trapezii throw back the shoulders. When the scapulæ are fixed by other muscles, the two trapezii throw the head back, or one muscle acting by itself would draw the head and spine to that side.

The trapezius has been seen to be *supplied* by the spinal accessory nerve and branches of the deep cervical plexus, it also receives branches from the posterior divisions of the nerves which pierce it.

The *Ligamentum Nuchæ* is a band of white fibrous tissue which extends from the prominent spinous process of the seventh cervical vertebra to the external occipital protuberance, being connected by small slips with the spines of the intervening vertebræ. It intervenes between, and gives origin to, the muscles of the two sides of the neck, and in the lower animals being composed of yellow elastic tissue it gives an important support to the head.

The **Latissimus Dorsi** (Fig. 207, 4) *arises* by a tendinous origin from the posterior half of the outer lip of the crest of the ilium and the back of the sacrum; from all the lumbar and the six lower dorsal spines; and from the outer surfaces of the three lower ribs, interdigitating with the external oblique muscle of the abdomen. The broad muscular fibres become collected together as they wind round the inferior angle of the scapula, to which they sometimes have a slight attachment, and then pass forward and upward in front of the teres major to be *inserted* into the bottom of the bicipital groove of the humerus. The fibres of the muscle are twisted so that those which were highest at their origin are lowest at the insertion and *vice versâ*.

The latissimus dorsi, when taking its fixed point below, rotates the humerus inwards and then draws it behind the back, the two muscles, when acting together, crossing the arms. When the humerus is fixed, the muscle would draw up the trunk as in climbing, and may act as an extraordinary muscle of inspiration upon the lower ribs. It is *supplied* by the long subscapular nerve, and by the dorsal nerves which pierce it.

[The trapezius is to be divided by a vertical incision near the spine, care being taken not to divide the subjacent tendinous origins of the rhomboidei in the cervical region. In reflecting the trapezius its tendon will be seen to glide over the smooth triangular surface at the inner end of the spine of the scapula. The posterior surface of the levator anguli scapulæ and the two rhomboidei muscles are now to be cleaned.]

Second Layer of Muscles.

The **Levator Anguli Scapulæ** (Fig. 207, 10) *arises* from the posterior tubercles of the transverse processes of the upper four cervical vertebræ, and is *inserted* into the base of the scapula opposite the supraspinous fossa. The anterior surface of this muscle has been seen in the posterior triangle of the neck, where the levator covers the splenius colli muscle.

The **Rhomboideus Minor** (Fig. 207, 11) is a narrow

Fig. 207.

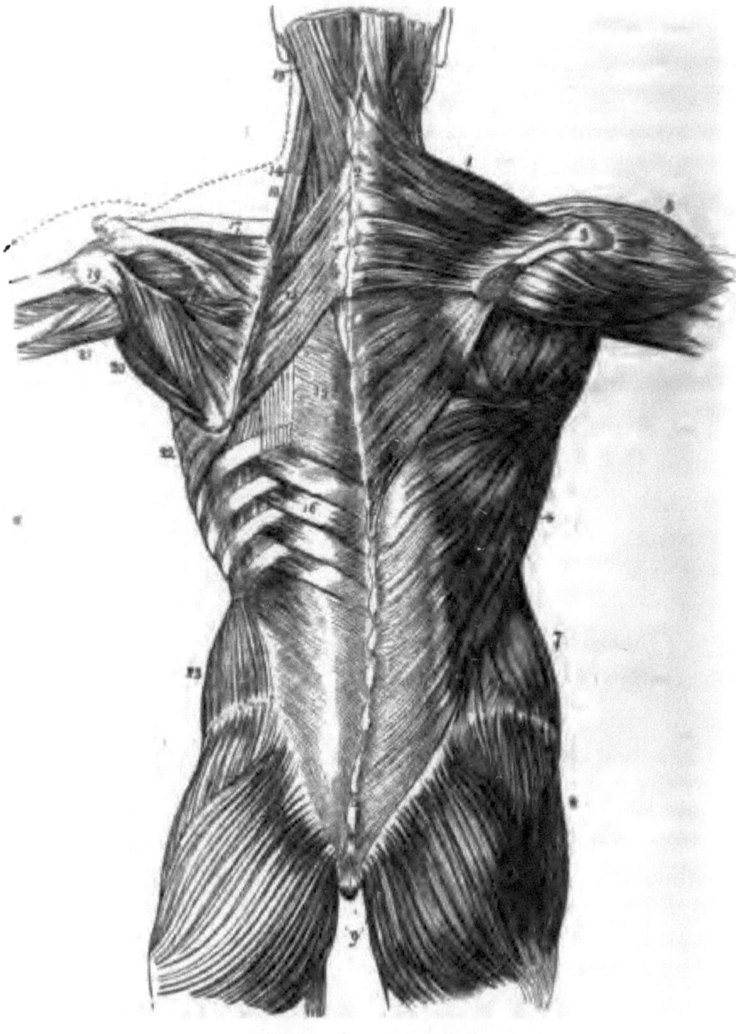

Fig. 207.—First, second, and part of the third layer of muscles of the back; the first layer occupies the right; the second, the left side (from Wilson).

1. Trapezius.
2. Ligamentum nuchæ.
3. Acromion process and spine of the scapula.
4. Latissimus dorsi.
5. Deltoid.
6. Muscles of the dorsum of the right scapula: infraspinatus, teres minor, and teres major.
7. Obliquus externus.
8. Gluteus medius.

muscle *arising* from the ligamentum nuchæ and from the spine of the seventh cervical vertebra. It is *inserted* into the base of the scapula opposite the spine.

The **Rhomboideus Major** (Fig. 207, 12) *arises* from the spinous processes of the upper four dorsal vertebræ, being often united with the minor; and is *inserted* into the base of the scapula opposite the infraspinous fossa, the middle fibres being attached to a fibrous arch, which is connected with the bone at its extremities.

The three muscles of the second layer act upon the scapula, the levator drawing up the superior angle, and the rhomboidei drawing the lower part of the bone towards the spine, and thus tending to rotate the scapula and depress the glenoid cavity. The muscles are *supplied* by branches from the upper part of the brachial plexus.

[The posterior belly of the omo-hyoid is to be traced out to the scapula (p. 36) and the levator, rhomboidei and the upper half of the latissimus dorsi are then to be divided, when the scapula can be drawn away from the ribs and the posterior surface of the serratus magnus muscle with a quantity of loose cellular tissue between it and the ribs will be brought into view. This muscle is to be divided by a vertical incision, and the clavicle being sawn through (if this has not already been done), the arm will be attached only by the vessels and nerves, which may be cut through and the limb removed. Between the rhomboidei and the serratus will be seen an artery, which is the posterior scapular. The dissector of the arm will proceed at once with the dissection of the scapular muscles (p. 32). The tendinous origin of the latissimus is then to be removed to expose the subjacent muscles.]

---

9. Glutei maximi.
10. Levator anguli scapulæ.
11. Rhomboideus minor.
12. Rhomboideus major.
13. Splenius capitis; the muscle internal to, and overlaid by, the splenius, is the complexus.
14. Splenius colli, partially seen; the common origin of the splenius is seen attached to the spinous processes below the origin of the rhomboideus major.
15. Vertebral aponeurosis.
16. Serratus posticus inferior.
17. Supraspinatus.
18. Infraspinatus.
19. Teres minor.
20. Teres major.
21. Long head of triceps.
22. Serratus magnus.
23. Obliquus internus abdominis.

The **Third Layer of Muscles** (Fig. 207) consists of the Serratus posticus, superior and inferior, and the Splenius which is divided into splenius capitis and splenius colli.

The **Serratus Posticus Superior** *arises* from the ligamentum nuchæ and the spinous processes of the seventh cervical and upper two dorsal vertebræ; and passes downwards to be *inserted* into the superior borders of the upper four ribs, externally to their angles.

The **Serratus Posticus Inferior** (Fig. 207, 16) *arises* from the spinous processes of the last two dorsal and upper two lumbar vertebræ; and passes upwards to be *inserted* into the lower borders of the last four ribs, external to their angles.

The serrati postici are antagonistic muscles, the superior raising the ribs, and being therefore a muscle of inspiration, the inferior depressing the ribs, and being therefore a muscle of expiration.

[The serrati muscles are to be divided and turned aside, when the posterior portion of the *fascia lumborum* (p. 213) covering the deeper muscles, and united more or less with the origins of the latissimus dorsi and serratus posticus inferior as well as with the subjacent muscles, will be seen. The name of *vertebral aponeurosis* has been given to the continuation of this structure beneath the serratus posticus superior and over the deep muscles; this is to be defined and afterwards must be removed.]

The **Splenius** (Fig. 207) is single at its *origin*, which is from the lower half of the ligamentum nuchæ and from the spinous processes of the seventh cervical and upper six dorsal vertebræ.

The *splenius capitis* (13) or upper portion is *inserted* across the mastoid process and into the outer third of the superior curved line of the occipital bone, immediately beneath the sterno-mastoid muscle.

The *splenius colli* (14) or lower portion is *inserted* into the posterior tubercles of the transverse processes of the upper four cervical vertebræ.*

* The splenius is inserted into the posterior tubercles of the vertebræ which gave origin to the levator anguli scapulæ.

The splenii draw the head and neck backward and thus assist in maintaining the erect posture.

The splenius capitis was seen in the posterior triangle of the neck, but the splenius colli was hidden by the levator anguli scapulæ. At the upper border of the splenius the nearly vertical fibres of the complexus appear.*

The **Fourth Layer of Muscles** (Fig. 208) consists of the Spinalis dorsi, the Erector spinæ with its subdivisions, and the Complexus.

The **Spinalis Dorsi** (Fig. 208, 4) lies close to the spinous processes, *arising* from the last two dorsal and upper two lumbar spinous processes, and being *inserted* into the spinous processes of the upper six dorsal vertebræ.

The **Erector Spinæ** (Fig. 208, 1) is the great muscle filling up the hollow at the back of the sacrum and lumbar vertebræ. It *arises* from the posterior fifth of the inner lip of the crest of the ilium; from the posterior surface of the sacrum; from the transverse and spinous processes of all the lumbar vertebræ; as well as from the aponeurosis covering the muscle.

Opposite the last rib the erector divides into the sacro-lumbalis and longissimus dorsi muscles, of which the sacro-lumbalis is the farther from the median line.

The **Sacro-Lumbalis** (Fig. 208, 2) is the smaller of the two divisions, and passes upward and outward to be *inserted* into the angles of the six lower ribs.

The **Accessorius** (musculus accessorius ad sacro-lumbalem) is placed internally to the sacro-lumbalis, and consists of muscular slips which pass *from* the lower six *to* the upper six ribs near their angles.

The **Cervicalis Ascendens** (Fig. 208, 5) is placed internally to the accessorius, of which it is the continuation,

* The student is advised not to burden his memory with the *attachments* of the remaining muscles of the back, with the exception of the small muscles of the suboccipital region. He should however dissect the principal ones so as to be able to identify them and to know their general positions.

Fig. 208

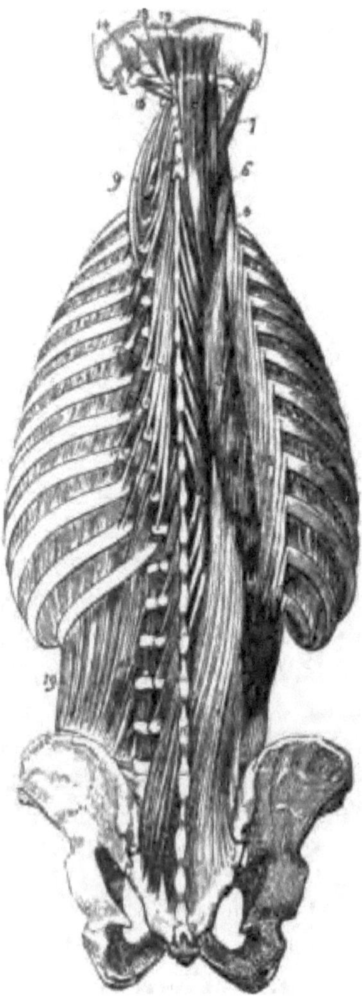

arising from the upper four ribs, and being *inserted* into the posterior tubercles of the transverse processes of the lower four cervical vertebræ, with the exception of the seventh.*

The **Longissimus Dorsi** (Fig. 208, 3), the inner division of the Erector spinæ, is *inserted* into the transverse processes of all the dorsal vertebræ, and into the ten lower ribs externally to their tubercles.

The **Transversalis Cervicis** (Fig. 208, 6) is the continuation of the longissimus, and lies to its inner side, *arising* from the transverse processes of the upper six dorsal vertebræ. It is *inserted* into the posterior tubercles on the transverse processes of the lower six cervical vertebræ.

The **Trachelo-mastoid**

* The muscles forming the outer division of the erector spinæ may be remembered by the first syllable of SACro-lumbalis, thus—Sacro-lumbalis, Accessorius, Cervicalis ascendens.

Fig. 208.—Fourth and fifth, and part of the sixth layer of the muscles of the back (from Wilson).

1. Common origin of the erector spinæ.
2. Sacro-lumbalis.
3. Longissimus dorsi.
4. Spinalis dorsi.
5. Cervicalis ascendens.
6. Transversalis cervicis.
7. Trachelo-mastoideus.

(Fig. 208, 7) is to the inner side of the transversalis, and between it and the complexus. It *arises* with the transversalis from the transverse processes of the upper four dorsal vertebræ and from the articular processes of the lower four cervical vertebræ; and is *inserted* into the posterior part of the apex of the mastoid process beneath the splenius capitis.

The **Complexus** (Fig. 208, 8) is a large muscle placed close to the vertebræ and reaching to the occiput, and is remarkable for a tendinous intersection about its middle, and for being pierced by the great occipital and third nerves. It *arises* from the transverse processes of the upper six dorsal and last cervical vertebræ, and from the articular processes of the 4th, 5th, and 6th cervical vertebræ. Its fibres ascend nearly vertically to be *inserted* into the occipital bone between the curved lines.

The most internal and superficial part of the complexus has been called the **Biventer Cervicis**, from the fact that it consists of two fleshy portions or bellies with an intervening tendon.

The erector spinæ is one of the most important muscles for maintaining the erect position of the trunk, and the complexus in addition draws back the head.

Occipital Artery (Fig. 209, 3).—The second portion of the occipital artery is now exposed in its course between the muscles attached to the occipital bone. It was traced along the lower border of the digastricus in the neck (p. 365) and, leaving that muscle, it now runs beneath the splenius capitis and trachelo-mastoideus, lying above the border of the superior oblique muscle and then getting on to the complexus close to its insertion. It lastly pierces the trapezius to reach the back of the

---

8. Complexus.
9. Transversalis cervicis, showing its origin.
10. Semispinalis dorsi.
11. Semispinalis colli.
12. Rectus posticus minor.
13. Rectus posticus major.
14. Obliquus superior.
15. Obliquus inferior.
16. Multifidus spinæ.
17. Levatores costarum.
18. Intertransversales.
19. Quadratus lumborum.

scalp, being accompanied by the great occipital nerve. which appears through the trapezius (1).

The *Arteria princeps cervicis* is given off from this portion of the occipital artery and divides into a branch to the trapezius and a descending larger one, which passes beneath the complexus to anastomose with the profunda cervicis artery.

[The complexus is to be divided transversely at the point where the great occipital nerve pierces it so as to preserve the nerve uninjured, and the muscle is then to be dissected up from the condensed cellular tissue beneath

Fig. 209.

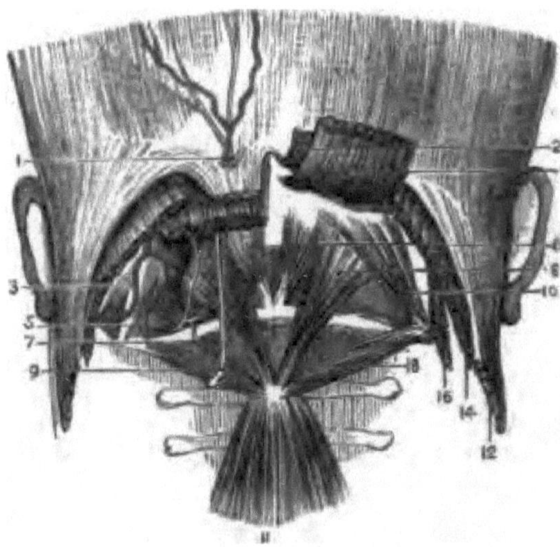

it and turned outward. The small muscles with the vessels and nerves of the suboccipital region are then to be dissected out, together with the semispinalis muscle which fills the vertebral groove in the cervical and dorsal

Fig. 209.—Suboccipital region (drawn by J. T. Gray).

1. Occipital artery and nerve piercing trapezius.
2. Trapezius.
3. Occipital artery.
4. Complexus.
5. Vertebral artery.
6. Rectus capitis posticus minor.
7. Suboccipital nerve.
8. Rectus capitis posticus major.
9. Great occipital nerve.
10. Obliquus superior.
11. Semispinalis colli.
12. Sterno-mastoideus.
14. Splenius capitis.
16. Trachelo-mastoideus.
18. Obliquus inferior.
20. Semispinalis colli.

regions. All vessels and nerves are to be carefully preserved.]

The **Fifth Layer of Muscles** (Fig. 208) consists of the two Recti, the two Obliqui, and the Semispinalis muscles.

The **Rectus Capitis Posticus Major** (Fig. 209, 8) *arises* from the spinous process of the axis, and passes upward and outward to be *inserted* into the outer part of the inferior curved line upon the occipital bone, and into a small portion of the space below it.

The **Rectus Capitis Posticus Minor** (Fig. 209, 6) is placed close to the median line and is slightly overlapped by the preceding muscle. It *arises* from the posterior tubercle of the atlas, and is *inserted* by fleshy fibres into the space below the inferior curved line on the occipital bone, close to the median line and to its fellow muscle of the opposite side.

The **Obliquus Inferior** (Fig. 209, 18) *arises* from the spinous process of the axis, and is *inserted* into the posterior tubercle of the transverse process of the atlas. Around the lower border of this muscle the great occipital nerve makes its appearance.

The **Obliquus Superior** (Fig. 209, 10) is the continuation upwards of the preceding muscle. It *arises* from the transverse process of the atlas, and is *inserted* into the space between the curved lines of the occipital bone, externally to the complexus and overlapping the rectus major. The occipital artery has been seen to run along the upper border of this muscle.

The recti muscles draw the head backward, and the rectus major, owing to its oblique direction, assists slightly in producing rotation. The obliqui produce rotation, the inferior oblique rotating the atlas upon the axis, and the muscle of one side acting with the superior oblique of the other, which latter acts slightly upon the skull.

Vertebral Artery (Fig. 209, 5).—The horizontal portion of the vertebral artery will be found in the triangle formed by the rectus posticus major and the two oblique muscles. It ascends through the foramina in the transverse processes of the vertebræ beginning at the sixth, and having perforated the atlas is seen to wind in the groove behind the superior articular process of that bone and pierce the posterior occipito-atloid ligament to enter the vertebral canal. Lastly it reaches the interior of the skull through the foramen magnum. While in the transverse processes the artery gives small branches to the muscles and to the spinal cord, and in its horizontal portion it occasionally gives off a branch to the back of the skull to supply the muscles.

The *Vertebral vein* does not accompany the artery in this part of its course, but only commences close to the transverse process of the atlas by the junction of one or two minute radicles. The vein accompanies the artery through the transverse processes of the vertebræ, and ends in the innominate vein.

The **Suboccipital Nerve** (Fig. 209, 7) (posterior division of 1st cervical) emerges between the occiput and atlas, and generally beneath the vertebral artery. It is of small size, and supplies branches to the recti and obliqui muscles and one to the complexus.

The **Great Occipital Nerve** (Fig. 209, 9) is the posterior division (internal branch) of the second cervical nerve, and leaves the vertebral canal between the atlas and axis. It winds below the inferior oblique, giving a branch to that muscle, and then turns upward to pierce the complexus and trapezius muscles and supply the skin of the occipital region. Its external branch is of small size.

The *posterior division of the third nerve* gives a branch which takes the same direction as the great occipital nerve and is joined with it.

The **Deep Cervical Artery** lies upon the semispinalis colli muscle, and is exposed when the complexus is reflected. It is the posterior branch of the superior intercostal artery (p. 435) and reaches the back by passing between the transverse process of the last cervical vertebra and the neck of the first rib. It supplies the deep muscles, anastomosing with the muscular branches of the vertebral and with the princeps cervicis of the occipital artery. There is thus an anastomosis established between a branch of the carotid and a branch of the subclavian artery, which would enlarge if the common carotid artery were tied.

The **Semispinalis** (Fig. 208) is divided into the semispinalis colli and semispinalis dorsi.

Semispinalis colli (11) *arises* from the transverse processes of the six upper dorsal vertebræ and the articular processes of the four lower cervical vertebræ, and is *inserted* into the spines of the cervical vertebræ from the 2nd to the 5th inclusive.

Semispinalis dorsi (10) *arises* from the transverse processes of the dorsal vertebræ from the sixth to the tenth inclusive, and is *inserted* into the spinous processes of the last two cervical and upper four dorsal vertebræ.

Spinal Nerves.—The Posterior Divisions of all the spinal nerves (with the exception of the first cervical and fourth and fifth sacral) divide into internal and external branches which supply all the muscles of the back.

The Internal branches of all the cervical nerves (except the first) and of the upper six dorsal nerves become cutaneous near the spinous processes.

The External branches of the lower six dorsal and of all the lumbar nerves, become cutaneous at an oblique line corresponding to the margin of the aponeurosis of the latissimus dorsi.

The External branches of the upper sacral nerves become cutaneous near the median line. The fourth and fifth sacral nerves are very small and join with the coccygeal nerve.

The *Coccygeal nerve* emerges from the lower end of the vertebral canal and is distributed to the coccyx.

The Posterior Divisions of the **Intercostal and Lumbar Arteries** divide like the nerves into internal and external branches, which accompany the nerves to the muscles of the back.

The **Sixth Layer of Muscles** (Fig. 208) consists of the Interspinales, Intertransversales, Multifidus Spinæ, and Levatores Costarum, which will not repay the trouble of dissection.

The position of the Interspinales and Intertransversales is sufficiently indicated by their names.

The Multifidus Spinæ (16) fills up the vertebral groove

Fig. 210.

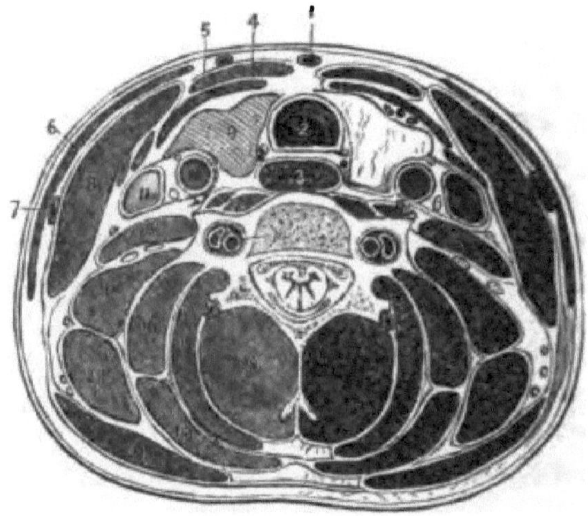

beneath the erector spinæ, *arising* from the articular processes of the cervical and lumbar vertebræ and from

---

Fig. 210.—Section of neck at the fourth cervical vertebra (altered from Béraud).

1. Anterior jugular vein.
2. Trachea.
3. Œsophagus.
4. Sterno-hyoid muscle.
5. Sterno-thyroid muscle.
6. Platysma myoides.
7. External jugular vein.
8. Sterno-mastoid.
9. Thyroid gland.
10. Common carotid artery.
11. Internal jugular vein. The descendens noni nerve is seen in front of these vessels, the pneumo-gastric between them, and the sympathetic behind them.
12. Longus colli, with rectus capitis anticus major to outer side.
13. Scalenus anticus with phrenic nerve.
14. Vertebral vessels in transverse process.
15. Scalenus medius with brachial cords in front.
16. Splenius colli.
17. Complexus.
18. Semi-spinalis colli.
19. Splenius capitis.
20. Levator anguli scapulæ.
21. Trapezius.

the transverse processes of the dorsal vertebræ and sacrum. The muscle is *inserted* into the spinous processes of all the vertebræ except the atlas.

The Levatores Costarum are twelve fan-shaped muscles passing between the dorsal transverse processes and the upper borders of the ribs.

THE SPINAL CORD AND MEMBRANES.

[To open the spinal canal the remains of the muscles of the back should be cleared away as far as possible, when some part of the plexus of *dorsi-spinal veins* may be seen upon the vertebræ. A block then being placed beneath the thorax so as to make the dorsal region prominent, a cut is to be made with the saw on each side of the middle line, so as to divide the laminæ of the vertebræ as far out as possible. Two or three of the arches being now removed with the chisel, the point of a spine chisel or rachet is to be introduced into the canal and the rachet carefully hammered through the arches of the vertebræ for the whole length of the spine, except the upper two cervical vertebræ. The operation being repeated on the opposite side, the arches can be removed with the bone-forceps, and the canal will be thoroughly opened. On the inner surface of the arches will be seen the *ligamenta subflava* which are described with the other vertebral ligaments (p. 500).]

Upon opening the Spinal Canal some loose tissue and fat will be seen, together with the *meningeo-rachidian veins*, which are occasionally full of blood. These extend for the whole length of the spinal cord under the name of posterior longitudinal spinal veins, and communicate both with the veins outside the vertebræ and with the anterior longitudinal spinal veins at the backs of the bodies of the vertebræ.

By removing the fat and veins the dura mater will be exposed.

The **Dura Mater** (Fig. 211, 1) of the spinal cord differs from the dura mater of the brain, in being comparatively smooth on its exterior, and in not forming the periosteum of the vertebræ. It also does not send processes into the spinal cord, nor has it any sinuses formed in it, but has numerous large veins on its outside.

The dura mater, being opened by a longitudinal incision, will be found to be a tube firmly connected with the margin of the foramen magnum, where it is continuous with the dura mater of the skull. The tube is largest in the cervical and lumbar regions, but at the top of the sacrum it becomes impervious, and is continued to the coccyx by a small fibrous process. Sheaths are given to all the spinal nerves, and take corresponding directions, being nearly horizontal in the cervical region and gradually becoming more vertical below. In the sacral canal the sheaths of the nerves lie side by side with the fibrous continuation of the dura mater between them.

Fig. 211.

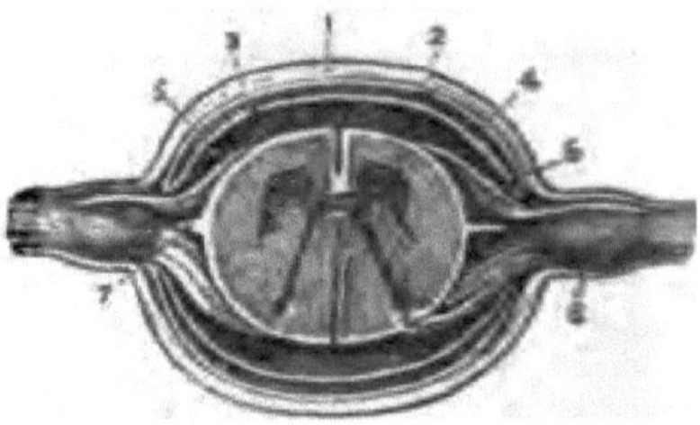

By removing the transverse process of two or three of the dorsal vertebræ, the tubes of dura mater may be traced upon the nerves to the intervertebral foramina, and will be found to be lost upon the nerves themselves.

[By opening one of the tubes of dura mater the position of the ganglion upon the posterior root in the intervertebral foramen will be seen. The cord is next to be removed by carrying the knife along the outside of the dura mater so as to divide the nerves as far out as possible, those which have been thoroughly exposed by the above dissection being divided beyond the ganglion. The cord attaching the dura mater to the coccyx is to be divided, and the sacral nerves cut as long as possible, and lastly the dura mater divided transversely at the level of the axis. The cord when removed is to be laid out on a table with the posterior surface uppermost.]

Fig. 211.—Transverse section of the spinal cord and its membranes (from Hirschfeld and Leveillé).

1. Dura mater.
2. Parietal layer of arachnoid.
3. Visceral layer of arachnoid.
4. Cavity of arachnoid.
5. Subarachnoid space.
6. Anterior root of nerve.
7. Ligamentum denticulatum.
8. Ganglion on posterior root of nerve.

The anterior and the posterior surfaces of the spinal cord when removed from the body, may always be distinguished as follows :—

1. There is a bright shining cord (linea splendens) running down the anterior surface.

2. There is one spinal artery on the anterior, but two arteries on the posterior surface.

3. The posterior roots of the nerves are ganglionic.

The **Arachnoid** (Fig. 211, 2, 3) consists of a parietal and visceral layer like the arachnoid of the brain, with which it is continuous. The parietal layer is seen on the inner surface of the dura mater, to which it is firmly attached; the visceral layer is reflected loosely upon the spinal cord, giving to the several nerves loose sheaths which are reflected upon themselves at the point where the nerves pierce the dura mater.

The **Subarachnoid Space** (Fig. 211, 5) is between the visceral layer of the arachnoid and the spinal cord, and can be shown by lifting up the arachnoid with forceps, or by inserting a blowpipe and forcing air beneath the membrane. This space communicates with the interior of the brain by an opening at the lower extremity of the fourth ventricle. It lodges the *subarachnoid fluid*, which also enters the ventricles, and is hence called *cerebro-spinal fluid.**

[The cord is to be turned over so as to expose its anterior surface, and an incision is to be made through the dura mater to expose the arachnoid and subarachnoid space, as was done behind. The pia mater is then to be examined both on the front and back of the cord.]

The **Pia Mater** encloses the spinal cord, giving prolongations upon the roots of the nerves. It is continuous with

* It has been supposed that a distinct serous membrane existed beneath the true arachnoid, in which the subarachnoid fluid was contained. Mr. Hilton has given prominence to this view lately in his "Lectures on Rest and Pain" (1863), but the question was decided some years since by Dr. Sharpey, who found that microscopically there was no evidence of a second serous membrane.

the pia mater of the brain, but is more fibrous and less vascular than it. It sends a process into the anterior median fissure of the cord, opposite to which is a fibrous band called *linea splendens*, and at the first lumbar vertebra (termination of the spinal cord) ends in the *filum terminale* or *central ligament* of the cord, which lies within the prolongation of the dura mater to be attached with it to the coccyx.

The **Ligamentum Denticulatum** (Fig. 212, 16) is found on each side of the cord between the anterior and posterior roots of the nerves. It is formed by a series of twenty-one or two serrations, connected with the pia mater and with the inner surface of the dura mater midway between the apertures of exit for the nerves. It serves to sling the cord and secure it from shocks.

The *Anterior Spinal Artery* and the *Two Posterior Spinal Arteries* may be traced upon the cord if well-injected. They are branches of the vertebral arteries, the anterior being formed by the junction of a branch from each side. The arteries ramify in the pia mater, anastomosing with the spinal branches which enter the vertebral canal through the intervertebral foramina.

The *Veins* of the cord pass through the intervertebral foramina to join the vertebral, intercostal, and lumbar veins.

The **Spinal Cord** (Figs. 211, 212) extends from the termination of the medulla oblongata at the level of the upper border of the atlas, to the first lumbar vertebra. It is cylindrical in shape, but presents two flattened enlargements, one the *brachial*, which extends from the third cervical to the first dorsal vertebra, and the lower or *lumbar* enlargement near the lower extremity of the cord. From the upper enlargement the nerves to the upper extremity have their origins; from the lower enlargement the lumbar and sacral nerves arise, which, lying side by side before they leave the vertebral canal, form the *cauda equina*.

The cord presents an *anterior median fissure* which

extends for its whole length, and a *posterior median fissure* which is most distinct at the upper and lower parts of the cord.

A *lateral fissure* extends along the line of attachment of the posterior roots of the spinal nerves, thus dividing each half of the cord into an *antero-lateral* and a *posterior column*.

The anterior roots of the nerves emerge from the antero-lateral columns, and these are sometimes considered to mark a further subdivision into *anterior* and *lateral* columns. At the bottom of the median fissures is the *commissure*, which is nearer to the anterior than the posterior surface of the cord.

The **Spinal Nerves** (Fig. 212) are thirty-one in number

Fig. 212.

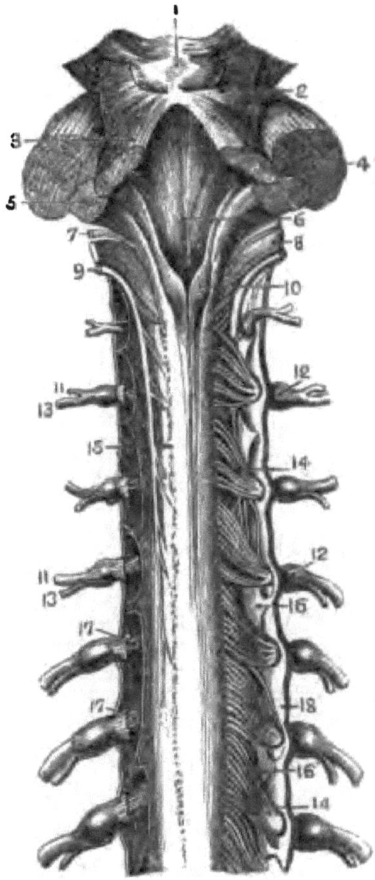

Fig. 212.—Fourth ventricle and upper part of spinal cord and membranes. The posterior roots of the nerve are removed on the left side (from Hirschfeld and Leveillé).

1. Corpora quadrigemina.
2. Fillet of the olivary body.
3. Processus a cerebello ad testes.
4. Processus a cerebello ad pontem.
5. Processus a cerebello ad medullam.
6. Floor of fourth ventricle.
7. Glosso-pharyngeal nerve.
8. Pneumo-gastric nerve.
9. Spinal-accessory nerve.
10. Posterior columns of spinal cord.

11, 11. Anterior divisions of spinal nerves.
12. 12. Ganglia of nerves.
13, 13. Posterior divisions of spinal nerves.
14, 14. Posterior roots of spinal nerves.
15. Line of origin of posterior roots of left side.
16, 16. Ligamentum denticulatum.
17, 17. Anterior root of spinal nerves.
18. Dura mater.

on each side of the cord. Each nerve has been seen to arise by two roots, the posterior (with the exception of the first nerve) being larger than the anterior. The posterior roots have each a ganglion which is generally placed in the intervertebral foramen, at which point the anterior (motor) and posterior (sensory) roots unite to form a *spinal nerve* of mixed endowments.

The first two cervical, with the sacral and coccygeal nerves, are exceptions to the general rule as regards the position of the ganglia. The ganglia of the two cervical nerves lie upon the arches of the atlas and axis, at which points the roots of the nerves unite. The ganglia of the sacral and coccygeal nerves are within the sacral canal.

Each spinal nerve divides into an anterior and a posterior trunk, the anterior divisions being the larger throughout, with the exception of the 1st and 2nd cervical nerves of which the posterior is larger than the anterior division. The majority of the spinal nerves divide just outside the intervertebral foramina, but in the first cervical, the last sacral, and the coccygeal nerves the division takes place within the dura mater; and the upper four sacral nerves divide within the sacral canal, the anterior and posterior trunks emerging at the anterior and posterior sacral foramina.

On a transverse section (Fig. 211) the spinal cord will be found to consist of white nervous matter, in which are seen two grey crescents, with their convexities turned toward one another and connected together by a *grey commissure*, in front of which are some white fibres forming the *white commissure*.

The *anterior cornua* of the grey crescents are larger than the posterior cornua, but do not reach to the periphery of the cord; the slender *posterior cornua* are closely connected with the posterior roots of the spinal nerves in the lateral fissure. A small central canal exists throughout the cord.

It is impossible for the student in his ordinary dissection to investigate the minute anatomy of the spinal cord. ... therefore referred to works on minute an... details.

PART VI.

DISSECTION OF THE BRAIN.

Before dissecting the Membranes or Vessels of the Brain, it will be well to recognise the several parts of the Encephalon, as this will assist the student in following the description.

The upper part of the brain is formed by the Cerebrum or great brain, which is divided into two symmetrical halves by a median *longitudinal fissure.* The under surface or *base* of the brain is much more complicated than the upper surface, and will be subsequently examined more particularly. At the posterior part will be seen the Cerebellum or smaller brain, which consists of two symmetrical halves, like the great brain; these are united by a convex white body, the Pons Varolii, at the lower border of which is the upper part of the spinal cord or the Medulla Oblongata.

These parts will be readily recognised by referring to Fig. 214.

[The brain is to be placed in a plate with the base upward, for the examination of the membranes and the dissection of the vessels.]

The **Membranes of the Brain** are the Dura Mater, the Arachnoid, and the Pia Mater.

The **Dura Mater** has been already seen lining the skull and giving off processes for the support of the brain (p. 327).

The **Arachnoid** is a serous membrane which has been seen to consist of two layers—*parietal*, which lines the dura mater, and *visceral*, which is now to be examined.

The *Visceral layer* is more or less united with the subjacent pia mater, but differs from it, in passing from one convolution to another without dipping into the sulci. It passes into the great longitudinal fissure between the hemispheres of the cerebrum, and may be conveniently traced from the anterior termination of this over the base of the cerebrum (the lobes of which it binds together) to the cerebellum and medulla oblongata, where it becomes continuous with the arachnoid of the spinal cord. It gives sheaths to the cranial nerves as far as the foramina of exit, where they are reflected to join the parietal layer of arachnoid.

Between the arachnoid and the pia mater is the **Subarachnoid Space**, which contains the cerebro-spinal fluid, and which, though existing all over the brain, is only to be seen in the following places :—

The *Anterior Subarachnoid Space* (diamond-shaped space) is immediately in front of the pons Varolii, and is formed by the stretching of the arachnoid from one middle lobe of the cerebrum to the other, as far forward as the optic commissure.

The *Posterior Subarachnoid Space* (fourth ventricle) will be found beneath the cerebellum on lifting up the medulla oblongata. This communicates with the subarachnoid space of the spinal cord, and with the interior of the brain by means of an aperture into the fourth ventricle, which may now be seen by removing the layer of arachnoid.

The **Pia Mater** is a vascular membrane closely investing the brain and passing into the sulci between the convolutions, besides giving processes into the interior of the brain which will be subsequently examined. It becomes more tough and fibrous as it approaches the spinal cord, upon which its vascularity almost entirely disappears.

[The arachnoid is to be removed from the arteries at

the base of the brain, which are to be cleaned and the branches followed. It will now be possible to show the longitudinal fissure between the two anterior lobes of the cerebrum more clearly than before, and also the fissure of Sylvius between the anterior and middle lobes of each side, into which a large artery may be traced.]

The **Arteries of the Brain** (Fig. 213) are derived from four great trunks—the two vertebral and the two internal carotid arteries.

The **Vertebral Artery** (1) has been seen to be a branch of the subclavian artery, and has been traced through the transverse processes of the cervical vertebræ to the atlas. The artery pierces the dura mater between the occiput and the atlas, and enters the skull through the foramen magnum by the side of the medulla oblongata. At the lower border of the pons Varolii the two vertebrals unite in the **Basilar Artery** (6), which passes along the middle line of the pons Varolii to bifurcate at its anterior border into the *posterior cerebral arteries* (8).

The *Branches* of the *Vertebral Artery* are :—

1. A small *Posterior Spinal artery* (3) which runs down the posterior surface of the spinal cord with its fellow of the opposite side.

2. A small *Anterior Spinal artery* (2) which joins its fellow of the opposite side to form a small trunk which runs down the anterior surface of the spinal cord.

3. A small *Posterior Meningeal artery* (4) to the dura mater.

4. The *Posterior Inferior Cerebellar artery* (5) which passes backward to the posterior part of the inferior surface of the cerebellum, and anastomoses with the other cerebellar arteries.

The *Branches* of the *Basilar Artery* are :—

1. The *Anterior Inferior Cerebellar arteries*, a pair of small branches passing to the anterior part of the inferior surface of the cerebellum on each side, and anastomosing with the other cerebellar arteries. These are only two of the following.

2. *Transverse arteries of the Pons*, four or five on each

side, which supply the pons Varolii, and one of which, the *auditory*, accompanies the auditory nerve into the temporal bone.

Fig. 213.

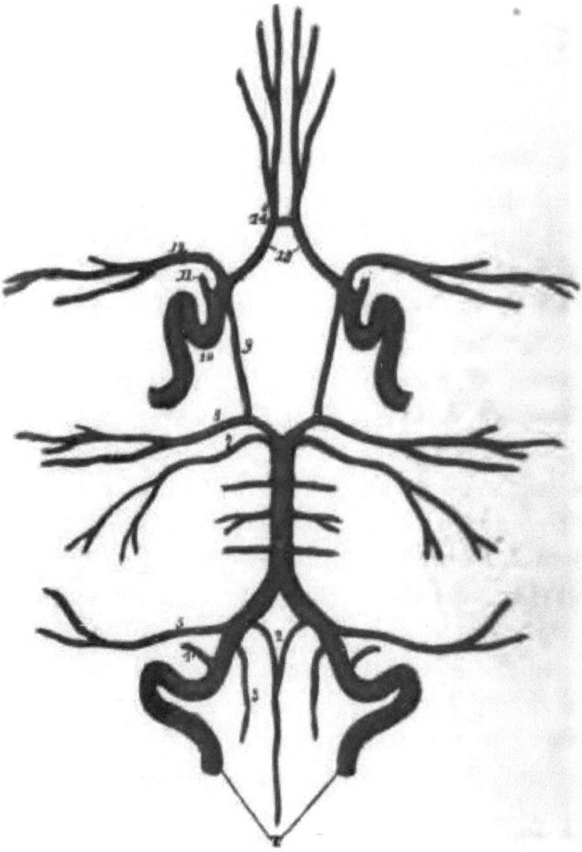

Fig. 213.—Arteries of the Brain and Circle of Willis (from Wilson).

1. Vertebal arteries.
2. Anterior spinal branches uniting to form a single vessel.
3. Posterior spinal artery.
4. Posterior meningeal artery.
5. Inferior cerebellar artery.
6. Basilar artery giving off transverse branches.
7. Superior cerebellar artery.
8. Posterior cerebral artery.
9. Posterior communicating branch of the internal carotid.
10. Internal carotid artery, showing its curves within the skull.
11. Ophthalmic artery divided.
12. Middle cerebral artery.
13. Anterior cerebral arteries, connected by—
14. Anterior communicating artery.

3. The *Superior Cerebellar arteries* (7) which arise close to the bifurcation, having the third pair of nerves between them and the posterior cerebral arteries. Each artery winds round the crus cerebri, parallel to the fourth nerve, to the upper surface of the cerebellum, where it anastomoses with its fellow and with the inferior cerebellar arteries.

4. The *Posterior Cerebral arteries* (8) which are the terminal branches of the basilar trunk. Each artery winds round the crus cerebri, parallel to the superior cerebellar artery but separated from it by the third nerve, and is distributed to the under surface of the posterior lobe of the cerebrum, where it anastomoses with the middle cerebral artery. It gives off small branches to the *locus perforatus posticus* and a *choroid* artery to the *velum interpositum.*

The **Internal Carotid Artery** (Fig. 213, 10) has been traced to the anterior clinoid process of the sphenoid bone, close to which it was divided in removing the brain. The artery reaches the base of the brain close to the outer side of the optic commissure, and at once breaks up into three branches—the anterior and middle cerebral, and the posterior communicating arteries.

1. The *Anterior Cerebral artery* (13) runs forward in the longitudinal fissure and, turning round the corpus callosum, is distributed to the anterior part of the cerebrum. The arteries of the two sides are united at the commencement of the fissure by a short transverse branch, the *anterior communicating artery* (14).

2. The *Middle Cerebral artery* (12) is a large branch which runs in the fissure of Sylvius between the anterior and middle lobes and reaches the surface of the brain, where it anastomoses with the anterior and posterior cerebral arteries. It gives off some small branches near its commencement, which pierce the *locus perforatus anticus.*

3. The *Posterior Communicating artery* (9) is a long slender branch which runs backward to join the posterior cerebral artery at a right angle.

the *genu*, and the central portion continued into the lamina cinerea as the *rostrum*.

Fig. 214.

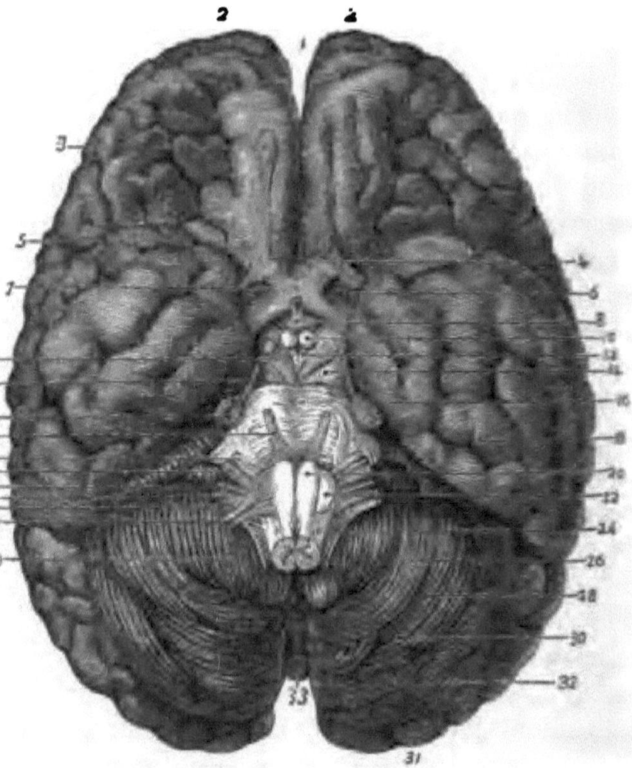

Fig. 214.—The base of the brain (from Hirschfeld and Leveillé).

1. Longitudinal fissure.
2, 2. Anterior lobes of Cerebrum.
3. Olfactory bulb.
4. Lamina cinerea.
5. Fissure of Sylvius.
6. Locus perforatus anticus.
7. Optic commissure.
8. Tuber cinereum and infundibulum.
9. Third nerve.
10. Corpus albicantium.
11. Fourth nerve.
12. Locus perforatus postic
13. Fifth nerve.
14. Crus cereb
15. Sixth ner
16. Pons Varolii.
17. Portio dura of 7th.
18. Middle lobe of cerebrum.
19. Portio mollis of 7th.
20. Anterior pyramid.
21. Glosso-pharyngeal nerve.
22. Olivary body.
23. Pneumo-gastric nerve.
24. Lateral tract.
25. Spinal accessory nerve.
26. Digastric lobe.
27. Hypoglossal nerve.

Cerebellum.

Amygdala.

lender lobe.

lobe.

The **Lamina Cinerea** (Fig. 214, 4) is a thin grey layer, continuous with the under surface of the corpus callosum and prolonged to the margin of the optic commissure. This is often torn in removing the brain.

The **Optic Commissure** or **Chiasma** (Fig. 214, 7) is the point of communication between the two optic nerves. The fibres of each optic tract have the following arrangement:—The outermost fibres pass to the optic nerve of the same side; the middle fibres decussate with those of the opposite side forming a figure of X, and pass to the optic nerve of the opposite side; the most internal fibres are reflected back to the brain along the opposite optic tract behind the cross arrangement, and a few fibres of the optic nerve pass in front of it from one eye to the other.

The **Tuber Cinereum** (Fig. 214, 8) is a grey body placed immediately behind the optic chiasma. Projecting from it is the *infundibulum* (funnel), a tube connected with the pituitary body, which is generally left in the *sella turcica* upon the removal of the brain. The infundibulum communicates with the third ventricle of the brain by its upper end, and in the fœtus opened into the hollow pituitary body, but is closed below in adult life.

The *pituitary body* (if left attached) will be seen to be solid in the adult and to consist of two lobes of a reddish colour. The anterior lobe is the larger and is oblong in shape; the posterior lobe is round.

The relation of the infundibulum and pituitary body to the tuber cinereum will be better understood by referring to the vertical section of the brain (Fig. 223, 20).

The **Corpora Albicantia** or **Mammillaria** (Fig. 214, 10) are two white bodies resembling small peas, which are placed between the two crura cerebri. They are formed by the anterior crura of the *fornix* which, as will be afterwards seen, descend to the base of the brain and there make a sudden twist upon themselves, thus forming the bodies. Upon section grey matter will be found in the interior of each body.

The **Locus Perforatus Posticus** (Fig. 214, 12) (Pons Tarini) is placed in the angle between the two crura cerebri. It is composed of grey matter and is perforated by numerous vessels.

The several structures which have been enumerated in the median line, viz., lamina cinerea, optic chiasma, tuber cinereum, corpora albicantia, and locus perforatus posticus, are all included within the area of the circle of Willis, and also form the floor of the third ventricle of the brain, as will be seen on referring to a vertical section (Fig. 223).

The **Olfactory Peduncle** (First nerve) (Fig. 214, 3) is of a prismatic shape and is very soft, being destitute of neurilemma and only held in its place by a reflection of the arachnoid. The peduncle has three *roots*, two white and one grey. The external root (white) passes across the fissure of Sylvius to the middle lobe and may be traced to the corpus striatum; the internal root (white) goes to the inner side of the anterior lobe; the middle root (grey) passes to the papilla of grey matter in the olfactory sulcus.

The **Olfactory Bulb** (Fig. 214, 3) is formed upon the extremity of the olfactory peduncle, and from it the branches of the olfactory nerve arise which descend to the nose. It is composed of grey matter, is oval in shape, and rests upon the cribriform plate of the ethmoid bone.

The **Optic** or **Second Nerve** (Fig. 214, 7) passes from the optic commissure to the eye-ball, where it is lost in the retina. It is round and firm and is invested by a sheath of dura mater. It has been shown that each optic nerve has fibres derived from the opposite as well as from its own side of the body.

The **Optic Tract** (Fig. 214) is a flattened band connecting the optic commissure with the brain. It can be seen disappearing around the crus cerebri and beneath the middle lobe, and will be afterwards traced to the *corpora* *niculata*, to the optic thalamus, and to the *corpora* *adrigemina*.

The **Locus Perforatus Anticus** (Fig. 214, 6) is situated se behind the olfactory nerve and at the inner end of the fissure of Sylvius. It is composed of grey nervous matter and is perforated by numerous small branches from the middle cerebral artery.

The **Fissure of Sylvius** (Fig. 214, 5) runs outward between the anterior and middle lobes of the cerebrum, and is occupied by the middle cerebral artery. It corresponds to the margin of the lesser wing of the sphenoid bone when the brain is *in situ*, and is divided into two portions on the outer aspect of the brain, which will be subsequently traced. At the bottom of the fissure will be seen a few nearly straight convolutions (*gyri operti*) aggregated together, to which the name Island of Reil or central lobe has been applied.

The **Crus Cerebri** (Fig. 214, 14) (peduncle of the cerebrum) is one of two large white bodies which appear at the anterior border of the pons Varolii and diverge to enter the under surface of the cerebrum. Winding round each crus are the optic tract and the fourth nerve, and between the two crura is the *interpeduncular space* containing the tuber cinereum, the corpora albicantia, and the locus perforatus posticus.

The **Third Nerve** (Fig. 214, 9) (motor oculi) is a good-sized nerve which appears with its fellow between the crura cerebri in front of the pons Varolii. The fibres are connected with the grey substance of the crura (locus niger), and may be traced to the floor of the fourth ventricle.

The **Fourth Nerve** (Fig. 214, 11) (trochlearis vel patheticus) is the smallest of the cranial nerves and winds round the crus cerebri. It will be afterwards traced to the valve of Vieussens.

The **Pons Varolii** or **Annular Protuberance** (Fig. 214, 16) is the large white body immediately behind the crura and in front of the medulla oblongata. It is more or less convex from side to side, but is slightly grooved along the middle line by the basilar artery. On each side is a round thick process of white fibres passing obliquely outwards and backwards to the cerebellum, which is the *crus cerebelli* or *processus a cerebello ad pontem.* Emerging from the side of the pons is the fifth nerve, and overlying it (in this position) is the sixth pair of nerves.

The **Fifth Nerve** (Fig. 214, 13) (trifacial, trigeminal) is the largest of the cranial nerves, and consists of two portions, motor and sensory, of which the motor is the smaller and the anterior. The nerve arises from the side of the pons Varolii, a few fibres of which intervene between the two roots; but its deep origin can be traced to the floor of the fourth ventricle.

The **Sixth Nerve** (Fig. 214, 15) (abducens oculi) arises from the anterior pyramid of the medulla oblongata close to the posterior border of the pons, and slightly from the pons itself. It may be traced deeply to a nucleus in the floor of the fourth ventricle.

The **Medulla Oblongata** (Fig. 214) is the upper part of the spinal cord, and extends from the upper border of the atlas to the lower border of the pons, being about an inch and a quarter in length. The medulla oblongata contains all the elements of the spinal cord, which are continued through it to the great brain, and has in addition a special body, the olive or olivary body. It will be impossible to see more than the superficial anatomy of the medulla at this stage, the rest will be taken with the dissection of the cerebellum.

The medulla is divided into two symmetrical halves by an *Anterior Median Fissure*, which is continuous with the anterior fissure of the spinal cord. On either side of the fissure is the rounded *Anterior Pyramid* (*corpus pyramidale* (20), which is continuous with the anterior column of the cord; the two anterior pyramids decussate across the median fissure at the lower part of the medulla.

The *Lateral Tract* (24) is continuous with the lateral column of the spinal cord. At the lower part of the medulla it is broad and is placed between the anterior pyramid and the restiform body, but at the upper part it is narrowed and pushed aside by the projection of the olivary body between it and the anterior pyramid.

The *Olivary body* (22) is an ovoid projection close to the anterior pyramid and immediately below the pons, but is separated by a groove from the surrounding parts. Some white fibres arching below and over the olivary body have been called the arciform fibres.

The *Restiform body* is only slightly seen at the side of the medulla in this view, but will be afterwards dissected with the fourth ventricle.

The **Seventh Nerve** (Fig. 214) consists of two portions, the *portio dura* or facial nerve, and the *portio mollis* or auditory nerve, the facial being anterior and the auditory posterior; and between the two is a minute nerve which has been named *pars intermedia* and joins the facial nerve.

The *portio dura* (17) is round, and smaller than the portio mollis; it arises from the lateral tract of the medulla oblongata close to the pons and slightly from the pons itself. The deep origin can be traced to the floor of the fourth ventricle.

The *portio mollis* (19) is flattened and very soft owing to the absence of neurilemma. It arises from the floor of the fourth ventricle and from the restiform body, around which it winds to join the facial nerve. The deep origin of the nerve is partly from the auditory nucleus in the floor of the fourth ventricle and partly from the posterior pyramid of the medulla.

The **Eighth Nerve** (Fig. 214) consists of three portions, the Glosso-pharyngeal nerve, the Pneumo-gastric nerve, and the Spinal-accessory nerve, which lie in that order from before backwards.

The *Glosso-pharyngeal nerve* (21) is the smallest of the three nerves, and arises from the lateral tract of the medulla oblongata by three or four fibrillæ.

The *Pneumo-gastric* or *Vagus nerve* (23) arises from the lateral tract below the glosso-pharyngeal by a number of fibrillæ which unite to form a single nerve. This lies parallel to the glosso-pharyngeal nerve and upon (in this view) a small lobe of the cerebellum, which has therefore been called the lobe of the pneumo-gastric or *flocculus*.

The *Spinal-accessory nerve* (25) arises partly from the medulla and partly from the spinal cord. The upper part (accessory to vagus) arises by fine fibrillæ below the origin of the pneumo-gastric for the whole length of the medulla. The spinal portion (Fig. 212, 9) arises by fine fibrillæ from the side of the spinal cord between the ligamentum denticulatum and the posterior roots of the spinal nerves, as low as the sixth cervical nerve.

The deep origins of the eighth pair may be traced to special nuclei in the floor of the fourth ventricle. The spinal fibres are connected with the grey matter of the spinal cord.

The **Ninth** or **Hypoglossal Nerve** (Fig. 214, 27) arises by numerous fibrillæ from the groove between the anterior pyramid and the olivary body. The nerve consists of two bundles which pierce the dura mater separately. The deep fibres may be traced to a special nucleus at the lower part of the fourth ventricle.

[By slicing down the pons Varolii the arrangement of its transverse and longitudinal fibres may be seen. These are in alternate layers, the transverse being continuous with the crura cerebelli and the longitudinal with the fibres of the cord and the crura cerebri. On slicing one of the crura cerebri deeply, a mass of grey matter (locus niger) will be found.]

THE EXTERIOR OF THE BRAIN.

[The brain being placed with the base downwards, the pia mater is to be removed from the upper surface of the cerebrum.]

The **Upper Surface of the Brain** (Figs. 215 and 216) is divided into two symmetrical halves or hemispheres by the *longitudinal fissure*, in which will be found the branches of the anterior cerebral artery. By separating the hemispheres the white corpus callosum will be seen at the bottom of the longitudinal fissure. The surface of the brain is marked by *convolutions*, the intervals between which are called *sulci*, and these have recently acquired sufficient importance to require individual notice, although the arrangement of the convolutions is not precisely the same in any two brains or even on both sides of the same brain.

The *Fissure of Sylvius* (*c*) which was seen at the base of the brain (p. 537) dividing the anterior from the middle lobe can now be traced out, and will be seen to divide into

Fig. 215.

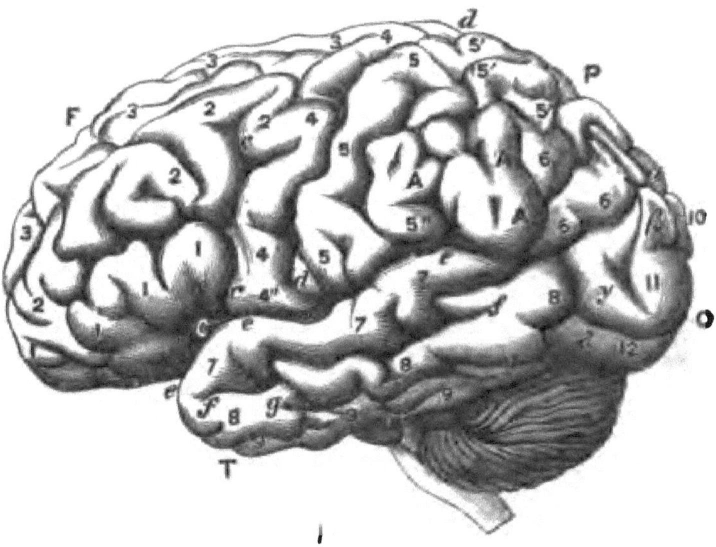

Fig. 215.—Lateral view of the brain of a bush-woman (from Marshall).

C. Central lobe, or Island of Reil.
F. Frontal lobe.
P. Parietal lobe.
O. Occipital lobe.
T. Temporal lobe.
c. Anterior division of fissure of Sylvius (antero-parietal).
d. Fissure of Rolando.
e. Sylvian fissure, posterior division.
f. Parallel fissure.
g. Inferior temporal fissure.
h. Parieto-occipital fissure.
1. Lower frontal convolution.
2. Middle frontal convolution.
3. Upper frontal convolution.
4. Ascending frontal convolution.
5. Ascending parietal convolution.
5'. Lobule of ascending parietal convolution.
4"—5". Supra-marginal convolution.
A. Lobule of supra-marginal convolution.
6. Angular convolution.
7. Upper external temporal convolution.
8. Middle external temporal convolution.
9. Lower external temporal convolution.
10. Upper occipital convolution.
11. Middle occipital convolution.
12. Lower occipital convolution.
α. First or upper external connecting convolution.
β. Second external connecting convolution.
γ. Third external connecting convolution.
δ. Fourth external connecting convolution.

two portions, the posterior or horizontal, which runs between the frontal (F) and temporal (T) lobes, and the anterior or vertical (*c*) which ascends between the frontal convolutions. Posterior to this portion of the fissure of Sylvius is a well-marked groove the *fissure of Rolando* (*d*) which, beginning at the margin of the longitudinal fissure, runs downward and forward nearly to the horizontal sylvian fissure, being bounded in front and behind by two large and well-marked convolutions, and separating the frontal from the parietal lobe.

The *Parallel Fissure* (*f*) is a well-marked fissure parallel to and below the horizontal sylvian fissure, and running between the convolutions of the temporal lobe.

The *Frontal Lobe* (F) contains one well-marked vertical convolution bounding the fissure of Rolando, and called the *ascending frontal convolution* (4), with three more or less horizontal convolutions in front of it, named respectively *lower* (1), *middle* (2), and *upper* (3) *frontal.*

The *Parietal Lobe* (P) contains one well-marked vertical convolution posterior to the fissure of Rolando, and called the *ascending parietal convolution* (5), with three more or less horizontal convolutions behind it, named respectively the parietal lobule, the supra-marginal lobule, and the angular gyrus.

The *Parietal Lobule* (5′) is continuous with the ascending parietal convolution, and lies along the margin of the longitudinal fissure, being connected to the upper part of the occipital lobe by the *first annectent convolution* (*a*).

The *Supra-marginal Lobule* (A) lies below the preceding and above the fissure of Sylvius. It is very variable in shape and size, and is connected posteriorly with the following convolution.

The *Angular Gyrus* (6) is posterior to the supra-marginal lobule and is very irregular in form. Behind it is connected with the occipital lobe by the *second annectent convolution* (*β*).

Between the ascending parietal convolution and the supra-marginal lobule is the commencement of a fissure named by Turner *intra-parietal*, which takes a curved course between the convolutions of the parietal lobe.

The *Temporal Lobe* (T) contains three well-marked horizontal convolutions. The *first* (7) lies between the

Fig. 216.

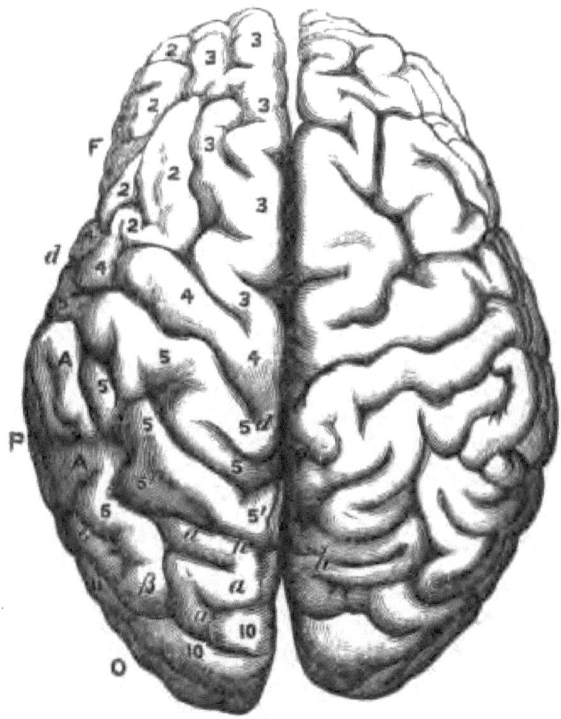

fissure of Sylvius and the parallel fissure, and the *second* (8) and *third* (9) are immediately below, separated from one another by the *inferior temporal fissure*.

The second and third temporal convolutions are con-

Fig. 216.—Upper surface of the brain of a bush-woman (from Marshall).

The references to this figure are the same as to Fig. 215, which should be studied with it.

Note.—In order to facilitate the study of the convolutions of the brain, I have availed myself of Mr. Marshall's permission to copy two drawings from his paper on the brain of a bush-woman ("Philosophical Trans.," 1864), so that the student may have the simplest form for reference when studying the more complex brain of Europeans.

nected with the occipital lobe by the *third* (γ) and *fourth* (δ) *annectent convolutions.*

The *Occipital Lobe* (O) is continuous with the temporal lobe, and contains three badly-defined convolutions placed horizontally, and named first (10), second (11), and third (12).

[A slice of brain substance about an inch thick is to be removed from the right hemisphere, when the centrum ovale minus will be exposed.]

The *Centrum Ovale Minus* is the oval mass of white cerebral matter surrounded by convolutions which forms the hemisphere. The convolutions will be seen to have a cortical layer of grey substance superficial to their white substance, which is continuous with the centre of the hemisphere. Numerous minute points will be seen on the surface of the white matter, which are the *puncta vasculosa* or orifices of minute blood-vessels, from which blood exudes in the recent condition.

The Inner Surface of the Hemisphere.

[By slicing off one hemisphere to the level of the corpus callosum, a side view of the opposite hemisphere will be obtained, but to complete the study of the inner surface of the hemisphere a section of a hardened brain should be referred to.]

The **Inner Surface of the Hemisphere** (Fig. 217) at its anterior part presents two large convolutions, the upper, the *marginal convolution* (1) (convolution of longitudinal fissure), and the lower, the *convolution of the corpus callosum* (2) or *gyrus fornicatus*, separated by the *calloso-marginal fissure* (C M) which reaches the margin of the hemisphere and bounds the marginal convolution posteriorly.

The *Gyrus Fornicatus* (2) is a well-marked convolution which begins at the anterior perforated space in the base of the brain, and winds round the corpus callosum to the posterior part, where it becomes continuous with the per internal temporal convolution or uncinate gyrus.

Posterior to the marginal convolution and continuous below with the *gyrus fornicatus* is the *quadrate lobe* (Q), which is separated from the *occipital lobe* (O) by the *parieto-occipital fissure* (P O), a continuation of the fissure of the same name seen externally.

Fig. 217.

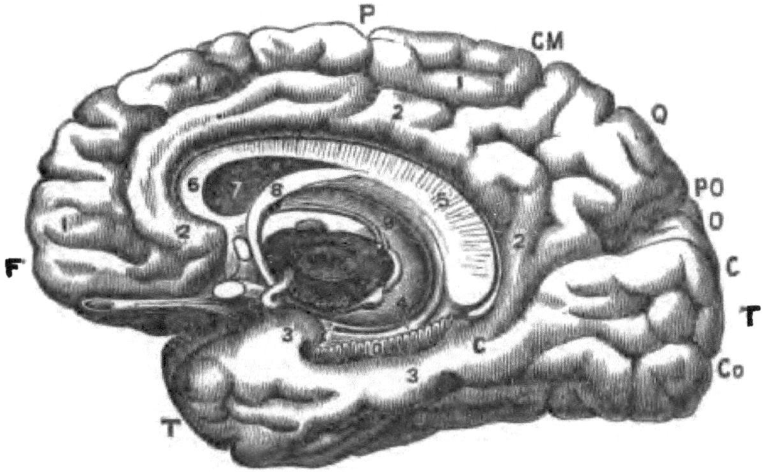

The *Occipital Lobe* (O) is triangular in shape, and is placed between the parieto-occipital and the calcarine fissure (C) which separates it from the internal convolutions of the temporal lobe (T).

The *Calcarine Fissure* (C) of Huxley runs below the occipital lobe to the extremity of the gyrus fornicatus, being joined midway by the parieto-occipital fissure. It is an important fissure, since it corresponds to the pro-

Fig. 217.—The inner and under surfaces of the brain to show the convolutions (from Hirschfeld and Leveillé).

F. Frontal lobe.
P. Parietal lobe.
Q. Quadrate lobe.
O. Occipital lobe.
T. T. Temporal lobe.
C. M. Calloso-marginal fissure.
P. O. Parieto-occipital.
C. C. Calcarine fissure.
Co. Collateral fissure.
1, 1. Marginal convolution.
2, 2. Gyrus fornicatus.
3, 3. Uncinate gyrus.
4. Dentate convolution.
5. United corpus callosum and fornix.
6. Genu of corpus callosum.
7. Cavity of lateral ventricle.
8. Fornix.
9. Thalamus with corpora geniculata.
10. Fascia dentata in dentate fissure.

jection called the *hippocarpus minor* in the posterior cornu of the lateral ventricle.

The *Internal Temporal Convolutions* form the lower part of the inner wall of the hemisphere, and are continuous with those of the temporal lobe externally. The most important is the *Uncinate Gyrus* (3) which is immediately below the calcarine fissure and runs horizontally forward, becoming united with the gyrus fornicatus and passing in front of the fascia dentata. It then makes a sudden bend backward for half-an-inch, this hook-like process from which it derives its name becoming united with the tænia hippocampi.

The *Collateral Fissure* (Co) separates this last convolution from the inferior temporal convolutions, which are very irregular. It causes the prominence in the descending cornu of the lateral ventricle known as the *eminentia collateralis.*

The *Dentate Fissure* (10) is that immediately above the uncinate gyrus which lodges the *fascia dentata*, and corresponds to the *hippocampus major* in the descending cornu of the lateral ventricle. Immediately above this is a small *dentate convolution* (4), united with the tænia hippocampi and continuous below with the uncinate gyrus.

The Interior of the Brain.

The *Centrum Ovale Majus* is seen by slicing the opposite hemisphere to the level of the corpus callosum. It is the great white mass formed by the continuation of the fibres of the corpus callosum into the hemispheres, the grey convolutions surrounding its circumference.

The **Corpus Callosum** (Fig. 218, 4) forms the great commissure of the brain, and consists of fibres passing transversely from one hemisphere to the other. It is nearer the anterior than the posterior extremity of the brain, and in the median line forms the floor of the longitudinal fissure, being slightly arched from before backwards. On a vertical section (Fig. 217, 6) it will be seen to bend abruptly anteriorly, to pass to the base of the brain, thus forming the *genu*, and to be continuous posteriorly with the subjacent fornix by a thickened portion which has been called the *splenium*.

On the surface of the corpus callosum and in the middle line is a longitudinal depression, the *raphé*, which is bounded on each side by the slight projections called the *nerves of Lancisi*. Externally to these are some transverse markings called the *lineæ transversæ*, which are crossed beneath the gyrus fornicatus by the longitudinal fibres forming the *covered band of Reil*.

The Lateral Ventricles.

[To open the cavity in each hemisphere called the lateral ventricle, the corpus callosum is to be carefully cut through about half an inch from the middle line, until the delicate serous membrane lining the ventricle is seen; the handle of the scalpel should then be used to reflect the brain substance and expose the cavity, which bends from the median line in front and towards it behind.]

The **Lateral Ventricle** (Fig. 218) consists of a *central cavity* and three *cornua*, anterior, posterior, and descending. The *anterior* cornu turns outward in the anterior lobe of the cerebrum, and the *posterior* cornu turns inward in the posterior lobe of the cerebrum, thus forming with the central portion a cavity shaped (on the right side) like the italic letter *f*. The ventricle is bounded superiorly by the corpus callosum which forms its *roof*, the *floor* being formed by the following parts from before backwards—1. Corpus striatum; 2. Tænia semicircularis; 3. Thalamus (opticus); 4. Choroid plexus; 5. Corpus fimbriatum; 6. Fornix.

The **Corpus Striatum** (Fig. 218, 3) is a pyriform body with the greater end forward, and constitutes the *superior ganglion of the cerebrum*. It is grey upon the surface, but if cut into, will be found to have white fibres interspersed throughout the deeper portion (and hence the name), becoming grey again at the under surface. The two corpora striata are separated posteriorly by the two thalami.

The **Tænia Semicircularis** (Fig. 218, 6) is a narrow band of white matter, which becomes slightly widened posteriorly, and connects the corpus striatum with the optic thalamus.

The **Thalamus** (opticus) (Fig. 218, 7) is a white body, which will be better seen in a subsequent dissection.

The **Choroid Plexus** (Fig. 218, 8) is a vascular fringe

Fig. 218.

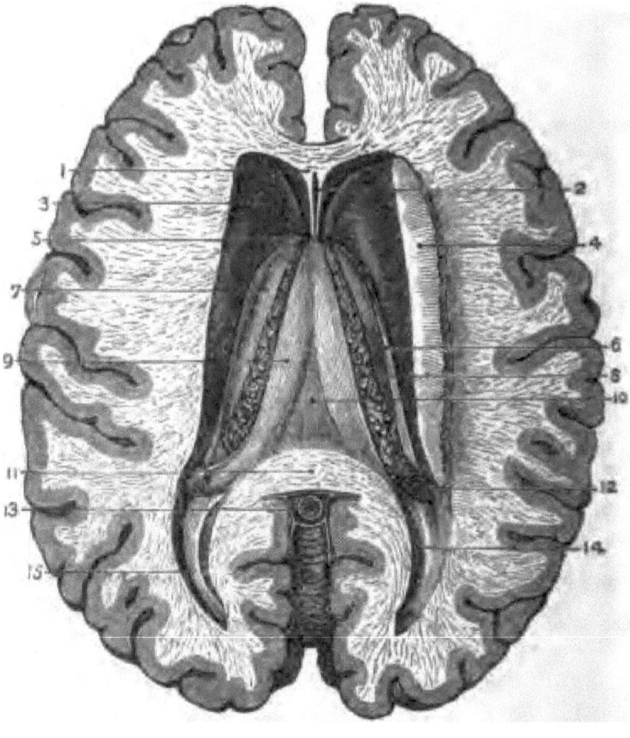

Fig. 218.—Lateral ventricles of the brain (from Hirschfeld and Leveillé).

1. Septum lucidum.
2. Fifth ventricle.
3. Corpus striatum.
4. Corpus callosum, reflected.
5. Foramen of Monro.
6. Tænia semicircularis.
7. Thalamus (opticus).
8. Choroid plexus.
9. Corpus fimbriatum.
10. Fornix.
11. Posterior extremity of corpus callosum.
12. Commencement of descending cornu.
13. Vena Galeni.
14. Hippocampus minor.
15. Posterior cornu of lateral ventricle.

lying upon the surface of the thalamus and continued into the descending cornu of the lateral ventricle. It will subsequently be seen to be the edge of a process of pia mater (velum interpositum) which lies beneath the fornix. By drawing gently upon the choroid plexus of one side it may be shown to be connected with that of the opposite side through the *foramen of Monro* (5), an opening beneath the fornix in the middle line.

The **Corpus Fimbriatum** (Fig. 218, 9) is the thin edge of the fornix, which however does not present any fimbriæ.

[To see the fornix thoroughly, the remnant of the corpus callosum in the middle line should be cut through transversely about its centre, when the posterior part is to be carefully dissected away from the subjacent fornix.]

The **Fornix** (Fig. 218, 10) is a thin white body placed beneath the corpus callosum in the middle line. Anteriorly it is divided into two crura, which pass to the base of the brain, and will be afterwards seen. Posteriorly it is incorporated with the splenium of the corpus callosum, and is continuous with the hippocampus major. The fornix is separated from the corpus callosum in front by the septum lucidum.

The lateral ventricles are separated from one another by the **Septum Lucidum**, a double layer of white cerebral matter grey externally, containing the fifth ventricle, which extends between the anterior part of the corpus callosum and the fornix, being deeper in front than behind (Fig. 223, 6).

The **Fifth Ventricle** (Fig. 218, 2) can be shown by cutting through the septum lucidum with a pair of scissors close to the remains of the corpus callosum, which must be turned forward. It lies between the two layers of the septum lucidum, and is deepest in front. It is lined by a delicate membrane which in the adult forms a shut sac, but in the fœtus communicates with the subjacent third ventricle.

The **Foramen of Monro** (Fig. 218, 5) is the communication between the two lateral and the third ventricles. It is placed beneath the anterior part of the fornix, and the choroid plexus has already been traced to it (Fig. 223, 10).

The **Hippocampus Minor** (Fig. 219, 7) is a projection of variable size from the inner wall of the posterior cornu of the lateral ventricle, corresponding to the *calcarine fissure*. It is white on the surface but grey in the interior.

Fig. 219.

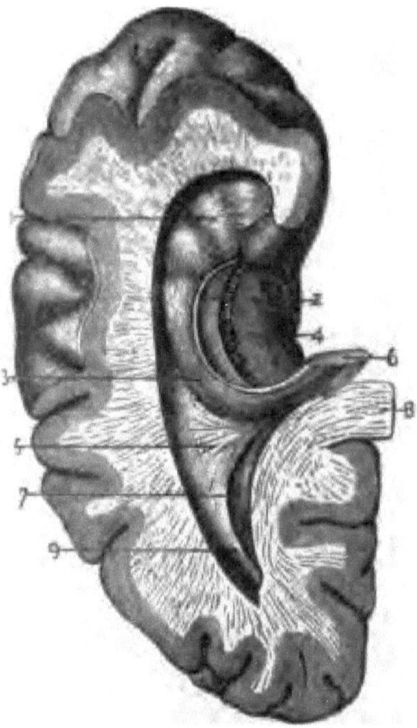

[To see the descending cornu, the side of the brain should be freely cut through opposite the point at which the choroid plexus disappears; this being taken as the guide, the descending cornu can be opened along its side, and the cavity exposed by drawing the parts asunder.]

The **Descending** or **Middle Cornu** (Fig. 219) takes a curved course downward in the middle lobe of the cerebrum and beneath the optic thalamus. Its direction is at first slightly backward, and then outward, downward, forward, and inward,* and it has been said to resemble a bent fore-finger. In it will be found the continuations of some of the structures which have been seen in the main cavity of the ventricle but under different names.

* Artificial Memory, BODFI.

Fig. 219.—Descending cornu of the lateral ventricle of brain (from Hirschfeld and Leveillé).

Pes hippocampi.	6. Fornix (cut).
Tænia hippocampi.	7. Hippocampus minor.
Hippocampus major.	8. Corpus callosum (cut).
Fascia dentata.	9. Posterior cornu.
Pes accessorius.	

The **Hippocampus Major** (Fig. 219, 3) is continuous with the fornix, of which it may be considered to be the posterior peduncle. It is a prominent convex body corresponding to the *dentate fissure*, and winds forward to the extremity of the cornu. Its anterior extremity is enlarged, and is marked with more or less distinct transverse grooves, and to this part the name *Pes Hippocampi* (1) has been given, from its fancied resemblance to the foot of an animal.

At the anterior border of the hippocampus major is a thin band of white cerebral matter continuous with the corpus fimbriatum, but now called the **Tænia Hippocampi** (Fig. 219, 2). By lifting this up with the handle of the scalpel and turning it aside, a serrated free border of grey nervous matter will be seen, which has been called the *Fascia Dentata* (4). This is the grey matter of the convolution which forms the pes hippocampi, as may be seen by making a transverse section of it.

The **Choroid Plexus** of the descending cornu is continuous with the choroid plexus of the lateral ventricle, and can now be seen to be connected with the pia mater through a slit immediately in front of the tænia hippocampi (transverse fissure of Bichât).

The **Pes Accessorius** or **Eminentia Collateralis** (Fig. 219, 5) is a projection of variable size at the commencement of the descending cornu and between the hippocampus major and the hippocampus minor, corresponding to the *collateral fissure*.

[By cutting through the fornix opposite the foramen of Monro, it can be carefully lifted up with the scalpel and turned backward, when some transverse markings on its under surface will be seen (from which it has been called the Lyra), and the velum interpositum will be brought into view.]

The **Velum Interpositum** (Fig. 223, 3) is a triangular process of pia mater carried into the interior of the brain through the great transverse fissure, which is now laid open. The continuity of the pia mater may be traced upon the upper surface of the cerebellum and the under surface of the posterior lobe of the cerebrum, and it may be followed through the descending cornu of the lateral ventricle to the base of the brain, where it appears by the side of the crus cerebri (*v.* page 537). At each side of the velum interpositum are the choroid plexuses of the lateral ventricles, and in the centre are two veins (*Venæ Galeni*) which open into the straight sinus of the dura mater.

The **Great Transverse Fissure** or **Fissure of Bichât** (Fig. 223, 11) which is now opened, is the slit by which the pia mater enters the brain. It is opposite the interval between the cerebrum and the cerebellum, and reaches forward in the brain beneath the fornix and corpus callosum, and above the great ganglia of the brain (corpora striata and optic thalami). The fissure is continued downward on each side in the descending cornu of the lateral ventricle, and reaches the base of the brain at the outer side of each crus cerebri.

[The velum interpositum is to be reflected, when the two small *choroid plexuses of the third ventricle* will be seen on its under surface. The third ventricle will now be seen, and behind it the corpora quadrigemina and the pineal gland, which latter is very liable to be removed with the velum interpositum, unless it is carefully dissected.]

The **Third Ventricle** (Fig. 220) is the space in the middle line between the two optic thalami. Its *roof* is formed by the fornix and velum interpositum, and its *floor* by the structures contained within the circle of Willis at the base of the brain, viz., the lamina cinerea, optic chiasma, tuber cinereum, corpora albicantia, and locus perforatus posticus (from before backwards).

The *anterior* boundary of the ventricle is formed by the *anterior commissure* (5), a white band which may be seen between and in front of the two anterior pillars of the fornix; the *posterior* boundary is the *posterior commissure* (Fig. 223, 14), a slender white band which may be seen immediately in front of, and a little beneath, the pineal gland. Stretching across the ventricle between the two optic thalami is the *middle* or *soft commissure* (7) (frequently torn), which divides the ventricle into two portions to which the names *foramen commune anterius* and *foramen commune posterius* are sometimes given.

The third ventricle communicates with the two lateral ventricles by the foramen of Monro, and with the fourth ventricle by the *iter a tertio ad quartum ventriculum* or aqueduct of Sylvius, which passes beneath the posterior commissure, the pineal gland and the corpora quadrigemina (Fig. 223, 16). In the fœtus the ventricle communicated in addition with the fifth ventricle and with the infundibulum.

The **Thalamus Opticus** (Fig. 220, 6) is now fully exposed, and will be seen to be a large white body placed posteriorly to the corpus striatum and at the side of the third ventricle. It has been seen to form part of the floor of the lateral ventricle by its upper surface, on which is a slight prominence called the anterior tubercle. Along the inner margin is a narrow white band, one of the peduncles of the pineal body, and by its inner surface which bounds the third ventricle, it gives attachment to the middle and posterior commissures of the third ventricle, the posterior piercing its substance.

The thalamus opticus forms the roof of the descending cornu of the lateral ventricle, and by drawing it upward on the side upon which the cornu has been opened, two projections on its under surface may be seen. These are the *Corpora Geniculata* (externum and internum) of which the outer one is the larger. By turning the brain on its side the optic tract may be readily traced to the under surface of the optic thalamus, to which it is attached, and will be found to divide into two parts, which are connected with the corpora geniculata and pass on to the corpora quadrigemina.

The **Pineal Body** or **Gland** (Fig. 220, 10) (conarium) is a pink body of a conical shape, lying between the anterior pair of the corpora quadrigemina and above the posterior commissure of the third ventricle. Its anterior part or base is connected with the margins of the optic thalami by two slender *anterior peduncles* or *habenæ*, and is also connected with the subjacent bodies by slender *inferior*

Fig. 220.

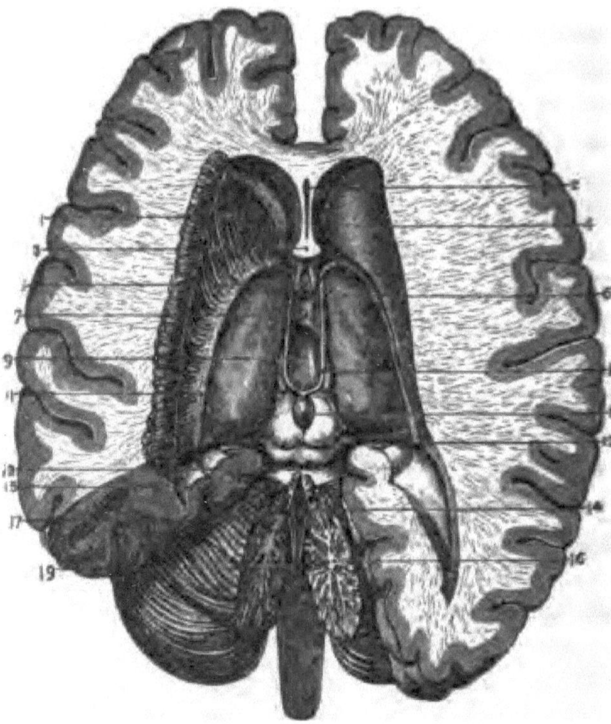

peduncles. The velum interpositum gives a special investment of pia mater to the gland. The pineal body

---

Fig. 220.—Third ventricle of brain (from Hirschfeld and Leveillé).

1. Corpus striatum dissected.
2. Fifth ventricle.
3. Anterior crura of fornix (cut).
4. Corpus striatum.
5. Anterior commissure of third ventricle.
6. Optic thalamus.
7. Middle or soft commissure.
8. Habenæ or peduncles of pineal gland.
9. Third ventricle.
10. Pineal gland.
11. Tænia semicircularis.
12. Corpora quadrigemina.
13. Valve of Vieussens.
14. Fourth ventricle.
15. Anterior extremity of superior vermiform process.
16. Arbor vitæ cerebelli.
17. Anterior extremity of inferior vermiform process (nodulus).
19. Communication of 4th ventricle with subarachnoid space.

contains a cavity in which are some particles of calcareous matter (acervulus).

The **Corpora Quadrigemina** (Fig. 220, 12) are four white prominences placed immediately behind the third ventricle, and named *Nates* and *Testes*, from their fancied resemblance to those parts; but it is to be noted that their position is the reverse of that in man, since the two anterior bodies are the nates and the posterior the testes. Both sets of bodies are connected to the optic thalami by white bands, and the nates are also connected to the pineal gland, which lies upon their upper surface.

The two broad white bands passing from the cerebellum to the testes, are the superior peduncles of the cerebellum (*processus a cerebello ad testes*) (Fig. 221, 3), and between them is a thin layer of white matter, the *Valve of Vieussens* (Fig. 220, 13), to which the fourth pair of nerves may be traced round the superior peduncles of the cerebellum.

The band of white matter passing transversely beneath the corpora quardigemina on each side, and seen immediately in front of the superior peduncles of the cerebellum, is the *Fillet of the Olivary body* (Fig. 221, 2).

[Opportunity may now be taken to trace out the anterior commissure of the third ventricle and the anterior pillar of the fornix, by carefully scraping away the corpus striatum of one side.]

The *Anterior Commissure* is a cylindrical white band which may be traced through the corpus striatum to the roof of the descending cornu of the lateral ventricle.

Tne *Anterior pillar of the Fornix* descends in front of the third ventricle and reaches the base of the brain, where it makes a twist to form the superficial white substance of the corpus albicans of one side, and then ascends to be lost in the grey matter of the optic thalamus (Fig. 217).

THE CEREBELLUM.

The **cerebellum** (Fig. 214, 28) or small brain lies beneath the posterior lobes of the cerebrum, and in the skull is separated from them by the tentorium cerebelli. It is of a darker colour than the cerebrum, and its surface is divided into laminæ instead of convolutions, and these are separated by shallow sulci. The cerebellum is divisible into two lateral halves united by a commissure, and the *horizontal fissure* divides the organ into an upper and a lower part.

The upper surface is flat except in the median line, where there is a slight ridge forming the commissure, and called the *superior vermiform process.* The upper part of each hemisphere is divided into an anterior and a posterior lobe by an indistinct fissure.

The *anterior lobe* is the larger and of a square shape, reaching as far back as the posterior extremity of the vermiform process.

The *posterior lobe* is the small portion behind the level of the vermiform process, and reaches to the horizontal fissure.

The cerebellum is connected to the cerebrum and spinal cord by three peduncles or crura, of which the superior one can now be seen.

The **Superior Peduncle** (Fig. 221, 3) (*processus a cerebello ad testes*) is a broad, flattened white band which is connected below with the inferior vermiform process and passes forwards to the corpora quadrigemina, the two peduncles of opposite sides converging at the posterior border of the testes. The two processes are prolonged beneath the corpora quadrigemina to the optic thalamus, their fibres decussating in their passage.

The **Valve of Vieussens** (Fig. 220, 13) (*Velum medullare anterius*) is the thin layer of white nervous matter stretched between the two superior peduncles of the cerebellum and connected with the anterior extremity of the inferior vermiform process. It is narrow in front but broader behind, where it has a little grey matter connected with it, and is often torn through, in which case the cavity of the fourth ventricle is exposed. The fourth pair of nerves arises from the middle line of the valve close behind the corpora quadrigemina.

[The preparation is to be turned over for the examination of the remaining peduncles and of the under surface of the cerebellum.]

The **Middle Peduncle** (Fig. 221, 4) (*processus a cerebello ad pontem*) has been already seen at the base of the brain. It is the largest of the three peduncles, and the fibres pass across the pons from one hemisphere of the cerebellum to the other, constituting the transverse fibres of the pons, which have been already dissected.

The **Inferior Peduncle** (Fig. 221, 5) (*processus a cerebello ad medullam*) forms part of the restiform body of the medulla, and will be better seen when the fourth ventricle is opened.

Fig. 221.

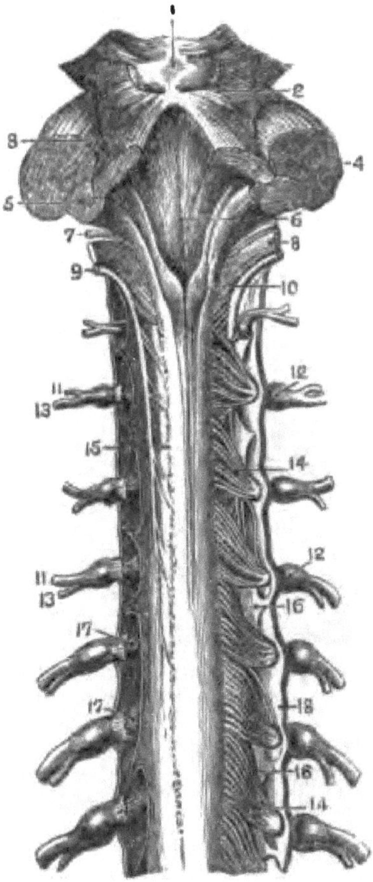

Fig. 221.—Fourth ventricle and upper part of spinal cord and membranes. The posterior roots of the nerves are removed on the left side (from Hirschfeld and Leveillé).

1. Corpora quadrigemina.
2. Fillet of the olivary body.
3. Processus a cerebello ad testes.
4. Processus a cerebello ad pontem.
5. Processus a cerebello ad medullam.
6. Floor of fourth ventricle.
7. Glosso-pharyngeal nerve.
8. Pneumo-gastric nerve.
9. Spinal-accessory nerve.
10. Posterior columns of spinal cord.

11, 11. Anterior divisions of spinal nerves.
12, 12. Ganglia of nerves.
13, 13. Posterior divisions of spinal nerves.
14, 14. Posterior roots of spinal nerves.
15. Line of origin of posterior roots of left side.
16, 16. Ligamentum denticulatum.
17, 17. Anterior root of spinal nerves.
18. Dura mater.

The **Inferior Surface** (Fig. 222) of the cerebellum is divided into two hemispheres by a deep fissure, the *vallecula*, at the bottom of which is the *inferior vermiform process*, which is to be seen by drawing the medulla oblongata well forward.

Fig. 222.

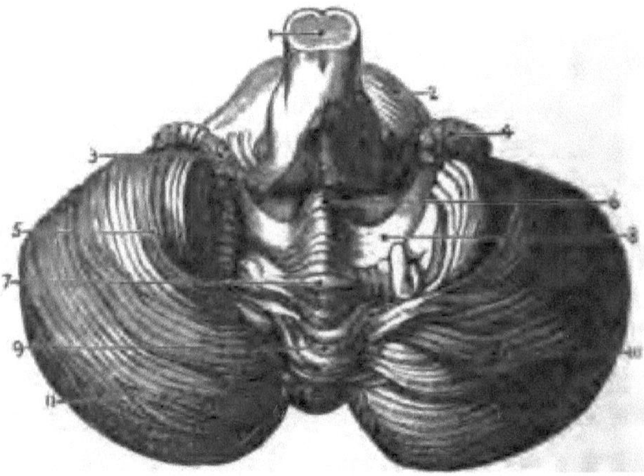

Each hemisphere is divided somewhat arbitrarily into lobes, since the divisions between them are very uncertain. Beginning behind is the *Posterior Lobe* (11); next the *Slender Lobe* (10); and in front of that the *Biventral* or *Digastric Lobe* (5).

The *Amygdala* or *Tonsil* (Fig. 214, 29) is a prominent lobe, close to the vallecula which it partially conceals.

The *Flocculus* (Fig. 221, 4) is a small lobe immediately in front of the biventral lobe, which lies beneath the crus cerebelli, and is hence called the subpeduncular lobe.

The **Inferior Vermiform Process** (Fig. 222) is divided into the following portions. Most anteriorly is the *Nodule* (6), which projects into the fourth ventricle; posterior to

---

Fig. 221.—Under surface of cerebellum, the amygdalæ having been removed (from Hirschfeld and Leveillé).

Medulla oblongata.
Pons Varolii.
Choroid plexus of the 4th ventricle.
s.
lobe of cerebellum.
6. Nodulus of inferior vermiform process.
7. Uvula.
8. Posterior medullary velum
9. Pyramid.
10. Slender lobe.
11. Posterior inferior lobe.

this is the thin ridge called the *Uvula* (7) from the fact of its lying between the two tonsils; behind this is the *Pyramid* (9), and posterior to this again are a few transverse commissural fibres.

By turning aside or cutting away the amygdalæ, a layer of white matter will be brought into view, extending from the flocculus on each side to the tip of the nodule. This is the *Velum Medullare Posterius* (8).

[The brain being again placed with the base downward, an incision is to be made through the valve of Vieussens and the cerebellum, in order to expose fully the cavity of the fourth ventricle.]

The **Fourth Ventricle** (Fig. 221, 6) is situated at the back of the pons Varolii and medulla oblongata, those bodies forming its *floor*. The *roof* is formed by the valve of Vieussens and the inferior vermiform process of the cerebellum; the *sides* by the two superior peduncles of the cerebellum (processus ad testes) above, and below by the restiform body. The cavity of the ventricle is lozenge-shaped, and it communicates above with the third ventricle by the Aqueduct of Sylvius (*iter a tertio ad quartum ventriculum*) which passes beneath the corpora quadrigemina (Fig. 223, 16). Below, the fourth ventricle is closed by a reflection of pia mater, in which there is usually an opening establishing a communication between the ventricles of the brain and the subarachnoid space of the spinal cord. Connected with this process of pia mater is the *Choroid plexus of the fourth ventricle* (Fig. 222, 3) which extends for some distance into the cavity.

In the floor of the ventricle is a median groove, which when traced downward will be found to end in a small hole, the commencement of the central canal of the spinal cord. On each side of this groove is a convex body, the *fasciculus teres*, which is white at the upper part of the ventricle but covered below by grey matter, and is the continuation of the fibres of the lateral tract and restiform body of the medulla.

There are four gangliform projections or nuclei on each

side of the median groove in the floor of the ventricle. The upper one is for the sixth and facial nerves, and immediately below are some white lines (*lineæ transversæ*), which run transversely from the median fissure and are connected with the auditory nerve. The lower nuclei are for the auditory, the eighth, and the ninth nerves.

The lower extremity of the ventricle, which is bounded

Fig. 223.

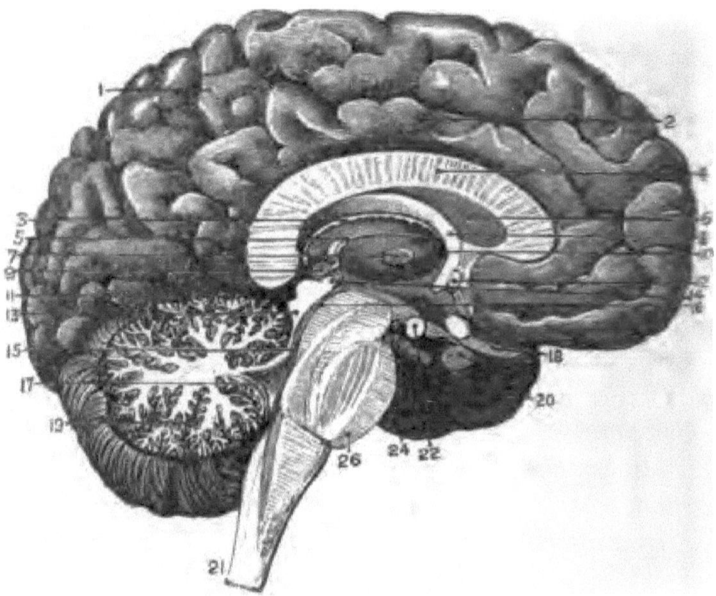

Fig. 223.—Vertical longitudinal section of the brain (from Hirschfeld and Leveillé).

1. Convolution of longitudinal fissure.
2. Gyrus fornicatus.
3. Velum interpositum.
4. Corpus callosum.
5. Peduncle of pineal gland on the margin of optic thalamus.
6. Septum lucidum.
7. Middle commissure of third ventricle.
8. Fornix.
9. Pineal gland.
10. Foramen of Monro.
11. Fissure of Bichât.
12. Anterior commissure.
13. Corpora quadrigemina.
14. Posterior commissure.
15. Valve of Vieussens.
16. Iter a tertio ad quartum ventriculum.
17. Fourth ventricle.
18. Optic nerve.
19. Arbor vitæ cerebelli.
20. Pituitary body and infundibulum.
21. Section of medulla oblongata.
22. Corpus albicantium.
24. Locus perforatus posticus.
26. Section of pons Varolii.

on each side by the enlarged extremities (*processus clavati*) of the Posterior Pyramids of the spinal cord, has been called the *calamus scriptorius*, from its fancied resemblance to a pen, of which the lineæ transversæ form the feathers. By slicing vertically either hemisphere of the cerebellum, the appearance known as the *arbor vitæ* (Fig. 223, 19) will be brought into view. This is due to the peculiar arrangement of white cerebral matter within the grey matter of the external laminæ; and by careful slicing, an irregular grey body (*corpus dentatum*) will be seen in the centre of the white matter of each hemisphere. By making a transverse section of the medulla oblongata a small *corpus dentatum* will also be seen in the olivary body of each side.

When possible, the student, as soon as he has finished the above dissection, should procure another brain, in order that he may make various sections of it, and so more thoroughly understand the relations of the several parts. The most useful section is one made in the median plane (Fig. 223), or a little to one side of it (Fig. 217), by which most of the important parts will be exposed.

PART VII.

DISSECTION OF THE EYE.

IN order to study the general anatomy of the globe of the eye it will be best to procure half-a-dozen bullocks' eyes, since the parts are larger than in the human eye, which also it is difficult to procure in a sufficiently recent condition. It must be borne in mind however that the eye of the bullock differs from that of man in the following particulars: the cornea is oval instead of being nearly circular; the pupil is elongated into a slit instead of being a circular aperture; the choroid coat presents the peculiar coloured appearance known as the *tapetum lucidum*, which is absent in man; and the yellow spot which is present in the human retina is wanting in the eyes of quadrupeds. The following description will be of the human eye, which the student will find no difficulty in following.

[All the fat and the remnants of the muscles of the eye are to be removed with scissors, the optic nerve being carefully preserved.]

The **Human Eyeball** (Fig. 224) is nearly globular in shape, but has a portion of a smaller sphere (the cornea) projecting anteriorly, thus making its antero-posterior greater than the transverse diameter.

The **Sclerotic** (Fig. 224, 11) or external tunic is composed of dense white fibrous tissue, and serves to maintain the shape of the eyeball and to protect the internal parts. It is thicker behind than in front, and is pierced posteriorly by the optic nerve, which enters at a distance of about its own breadth to the inner side of the axis of the eyeball. The opening in the sclerotic for the optic nerve is funnel-shaped, and the fibrils of the nerve pass through a fine fibrous tissue which has been named *lamina cribrosa* from its numerous apertures, one of which in the

Fig. 224.

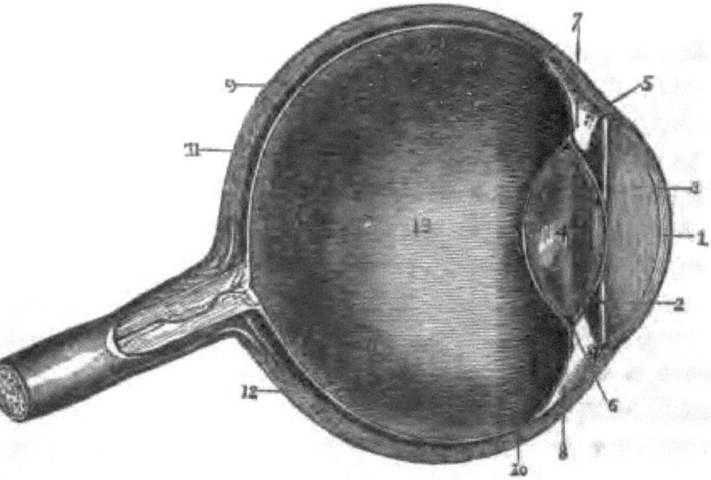

centre for the *arteria centralis retinæ* has been called the *porus opticus.* The sclerotic is pierced at various points by the ciliary vessels and nerves. In front, the sclerotic gives insertion to the recti muscles, from which it receives an expansion (*tunica albuginea*), and in front of this point the conjunctiva is reflected upon its surface but can be readily stripped up as far as the margin of the cornea.

The sclerotic overlaps the margin of the cornea like the case of a watch does the watch-glass, and encroaches

---

Fig. 224.—Longitudinal section of the eye (drawn by H. Power).

1. Cornea.
2. Iris.
3. Anterior chamber communicating with the posterior chamber through the pupil.
4. Lens enclosed in its capsule.
5. Canal of Fontana.
6. Canal of Petit.
7. Ciliary body and process.
8. Ciliary muscle.
9. Retina.
10. Ciliary zone.
11. Sclerotic.
12. Choroid.
13. Vitreous humour.

upon it slightly above and below. The sclerotic and the cornea are inseparably united, the tissue of the two structures being continuous.

By squeezing the eyeball between the finger and thumb the cornea will be rendered white and opaque, but will resume its former appearance when the pressure is relaxed.

The **Cornea** (Fig. 224, 1) is one of the transparent media of the eye; and in man is perfectly circular when seen from within, but appears wider transversely from without, on account of the greater overlapping of the sclerotic above and below.

The curvature of the cornea varies in different individuals, but the anterior and posterior surfaces are always parallel. The cornea is divisible into five layers: 1. anterior epithelium (conjunctiva); 2. anterior elastic lamina; 3. cornea proper; 4. posterior elastic lamina; 5. posterior epithelium (of aqueous humour). The cornea in health is a non-vascular structure, no blood-vessels existing in it, and its nourishment being derived from the surrounding structures. Minute branches of the ciliary nerves have however been traced into and through it.

[Holding the eye lightly with the left hand, the scalpel is to be thrust through the margin of the cornea into the anterior chamber, when the aqueous humour will necessarily escape. With scissors the cornea may then be removed entirely.]

The **Anterior Chamber** (Fig. 224, 3) is the space between the cornea and the iris. It communicates with the posterior chamber through the pupil, and is filled with the aqueous humour.

The **Iris** (Fig. 224, 2) is the highly vascular curtain which separates the anterior from the posterior chamber. It is composed of involuntary muscular fibres, blood-vessels, and pigment, and may be considered to be a prolongation of the choroid coat. In man the pigment of the iris is of various colours, but it is absent altogether in albinos, and occasionally the iris itself is wanting. The

Pupil is a circular opening (in man) in the centre of the iris, which varies in size according to the contraction of the tissue of the iris under the influence of light.

Fig. 225.

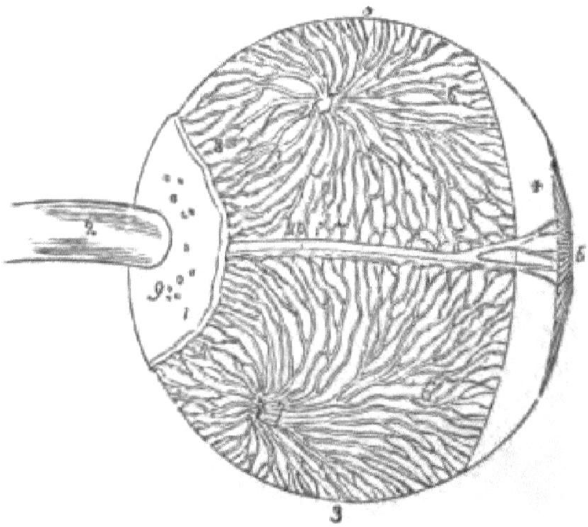

The **Posterior Chamber** is smaller than the anterior, and can be hardly said to exist when the aqueous humour has been let out. It is between the iris and the anterior layer of the capsule of the lens.

[In order to see the choroid coat entire, a fresh eye should be taken, and a puncture having been made through the sclerotic, about its middle, a blowpipe is to be introduced, through which air may be forced between the sclerotic and choroid coats, and they will thus be separated from one another. The sclerotic may then be cautiously divided circularly, and the posterior cup-like

Fig. 225.—The choroid coat and the distribution of the vasa vorticosa (from Wilson, after Arnold).

1. Part of the sclerotic coat.
2. Optic nerve.
3, 3. Choroid coat.
4. Ciliary muscle.
5, Iris.
6, 6. Vasa vorticosa.
7. 7. Trunks of the vasa vorticosa at the point where they have pierced the sclerotica.
8, 8. Posterior ciliary veins which enter the eyeball in company with the posterior ciliary arteries, by piercing the sclerotic at 9.
10. One of the long ciliary nerves, accompanied by a long ciliary vein.

portion of the sclerotic being left as a support to the eye, the anterior part (with the cornea) is to be carefully detached from the choroid. In order to do this it will be necessary to scrape the interior of the sclerotic with the handle of the scalpel, in order to tear through the ciliary muscle which attaches the choroid to the sclerotic coat.]

The **Choroid Coat** (Fig. 225, 3) is a vascular structure containing pigment, expanded over the whole of the posterior portion of the globe of the eye and continuous in front with the iris. It is pierced by the optic nerve, at which point it is closely connected to the sclerotic; but is attached to the inner surface of that coat only by a delicate fibrous tissue called the *membrana fusca*. On the outer surface of the choroid may be seen (in an injected specimen) the loops of blood-vessels to which the name *vasa vorticosa* has been given. Within this is the *tunica Ruyschiana*, a plexus of capillaries; which is again lined by the choroidal epithelium (*membrana pigmenti*). The dark pigment of the choroid is interspersed among the vessels and is washed out when the eye is immersed in water.

Fig. 226.

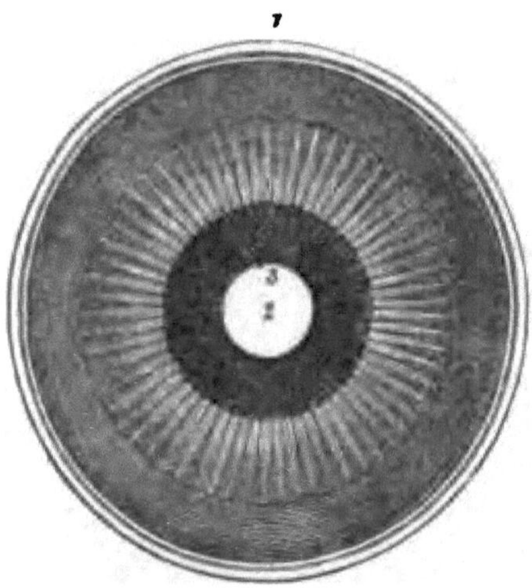

Surrounding the iris is a ring which is the **Ciliary Muscle** (Fig. 224, 8), composed of unstriped fibre and

Fig. 226.—Anterior half of the eye, seen from within (from Wilson).

1. Divided edge of the three coats; sclerotic, choroid (the dark layer), and retina.
2. Pupil.
3. Posterior surface of the iris.
4. Ciliary processes.
5. The anterior border of the retina (ora serrata).

having a grey appearance. The anterior part of the ciliary muscle is connected with the sclerotic coat, and the posterior part is attached to that portion of the choroid which is called the ciliary body, and from which the ciliary processes pass inwards to be connected with the lens. Thus the ciliary muscle is enabled to act upon the lens and to accommodate the focus of the eye to various distances.

[In order to see the ciliary body and processes, a third eye should be divided circularly through the sclerotic and choroid coats without damaging the vitreous humour; the vitreous humour and lens are then to be carefully separated from the anterior part of the choroid, the ciliary processes being detached from the hyaloid membrane, upon which they will leave a circle marked with more or less pigment (circle of Zinn). The ciliary processes and iris are now seen from behind. If the lens and vitreous humour be drawn out of the fundus of an eye and left attached to the ciliary processes, a magnified view of them will be obtained.]

The **Ciliary Body** (Fig. 224, 7) is the thickened portion of the choroid from which the **Ciliary Processes** (Fig. 226, 4) project towards the centre of the eye. These are vascular fringes which resemble a series of plaits in appearance, and which form a circular curtain parallel but posterior to the iris, from which they are separated by the *posterior chamber*. They fit into a corresponding series of plaitings in the hyaloid membrane of the vitreous humour and together constitute the *ciliary zone*.

Between the anterior margin of the sclerotic and the ciliary body is a minute canal, which runs round the entire circumference of the eye and is called the *Canal of Fontana* (Fig. 224, 5).

[The lens will be seen on the posterior half of this section, or may be shown on the first eye by carefully removing the iris.]

The **Lens** (Fig. 224, 4) is situated in front of the vitreous humour, and is contained in a delicate and perfectly transparent *capsule*, which is united behind with the hyaloid membrane of the vitreous humour. The capsule is kept in its place by the *suspensory ligament of the lens*, a transparent membrane which passes from the ciliary processes to the capsule in front of the hyaloid membrane. By tearing through the capsule (as in the operation for extraction of cataract) its existence will be demonstrated and the lens itself allowed to escape.

The lens is bi-convex, but the posterior surface has a greater curve than the anterior. It is perfectly transparent in health, but has a complicated structure consisting of fibres arranged around three axes running in different directions, of which indications are usually visible in the bullock's eye. The exterior of the lens readily breaks down, but the interior or nucleus is very dense.

The **Canal of Petit** (Fig. 224, 6) is a minute space surrounding the lens immediately around the capsule, and is bounded in front by the suspensory ligament and behind by the hyaloid membrane.

The **Vitreous Humour** (Fig. 224, 13) is the transparent body filling all the posterior part of the eyeball. It is a gelatinous substance contained in a transparent membrane called the *hyaloid membrane*, and is traversed by numerous delicate and perfectly transparent septa. The existence of these may be demonstrated by crushing the vitreous humour with the fingers, when the fluid portion will drain away.

On the eye from which the vitreous humour has been removed the retina will fall together, leaving the inner surface of the choroid exposed, and in the bullock's eye the *tapetum lucidum* will be seen. This is the coloured appearance which is peculiar to the lower animals, and is due to the presence of a thick layer of wavy fibrous tissue outside the choroidal epithelium. The object of this is to reflect the rays through the retina a second time, and thus to enable the animal to see with a very small amount of light.

[The retina is best seen by looking through the vitreous humour of an eye from which the iris and lens have been removed.]

The **Retina** (Fig. 224, 9) is the nervous coat of the eye and its essential part. It lines the choroid coat, being connected posteriorly with the optic nerve, and extends in front nearly as far forward as the ciliary body, where it terminates about one-eighth of an inch behind the margin of the cornea in a finely jagged border, the *ora serrata*, being thicker behind than in front. During life the retina is transparent, so that the vascular choroid can be seen through it with the ophthalmoscope, but after death it has a greyish colour and is thrown into folds, owing to the diminution of the tension of the globe.

The entrance of the optic nerve may be seen about one-tenth of an inch to the inner side of the axis of the eye, and radiating from this may be seen the branches of the *arteria centralis retinæ*, which enters the eye at this point, and its accompanying veins. In the axis of the eyeball in the human eye is the *yellow spot of Soemmering*, the most sensitive point, which consists of a small elevation, in the centre of which is a minute depression, the *fovea centralis.*

The retina consists of three layers:—1. The internal or *fibrous* layer is continuous with the optic nerve (the fibrillæ of which here lose their white substance of Schwann), and has also numerous ganglionic cells and nuclei developed in it. 2. The middle or *granular* layer consists of oval bodies collected into two sets between which is granular material. 3. The outer layer or *Jacob's membrane* consists of a peculiar arrangement of particles to which the names *rods* and *bulbs* have been given. These rods are continuous with some fibres which pierce the whole depth of the retina, and are called the fibres of Müller.

The Ear.

The description of the External Ear will be found at page 375, and that of the Middle Ear or Tympanum at page 481.

The Internal Ear or Labyrinth is so difficult of dissection owing to the density of the bone in which it is embedded, that it is impossible for the student to make its dissection with any advantage. He is therefore advised to study its anatomy on prepared specimens, and is referred for its description to works on general and minute anatomy.

INDEX.

Abdomen, 179.
surface of, 202.
Abdominal aorta, 248.
cavity, 224.
contents, table of, 227.
hernia, 213.
regions, 226.
ring, external, 208.
internal, 213.
viscera, 262.
Abducens nerve, 352, 538.
Abductor indicis, 76.
minimi digiti, 71.
pedis, 146.
oculi, 351.
pollicis manûs, 72.
pedis, 146.
Accelerator urinæ muscle, 188.
Accessorius muscle, 149.
ad sacro-lumbalem, 513.
Accessory nerve of the obturator, 260.
Acromial cutaneous nerves, 337.
thoracic artery, 16.
Acromio-clavicular articulation, 76.
Adductor brevis, 100.
longus, 108.
magnus, 112.
opening in, 107.
minimi digiti, 72.
oculi, 351.
pollicis manûs, 73.
pedis, 151.
Air cells of the lung, 451.
Alar ligaments of the knee, 168.
Amygdaloid lobe of cerebellum, 558.
Anastomotic artery of brachial, 28.
of femoral, 107.
Anconeus muscle, 57.
Angular artery, 384.
gyrus, 542.
vein, 384.
Ankle-joint, 172.
Annular ligament of ankle, anterior, 155.
external, 155.
internal, 142.
of wrist, anterior, 69.
posterior, 63.
protuberance, 537, 540.
Annulus ovalis, 419.
Anterior chamber of eye, 564.
elastic layer of cornea, 564.
commissure, 552, 555.
mediastinum, 409.
medullary velum, 556.
triangle of neck, 371.
Anti-helix, 376.
Anti-tragus, 376.
Aorta, 429.
abdominal, 248.
thoracic, 444.
Aortic arch, 429.
opening of diaphragm, 247.
plexus, 261.
sinus, 424.
Aperture of the Eustachian tube, 465.
for the femoral artery, 107.
of the larynx, 465, 487.
of the mouth, 465.
of the nares, 465.
of the œsophagus, 465.
of the thorax, 449.
Aponeurosis, epicranial, 319.
of external oblique, 208.
of the femoral artery, 93.
of internal oblique, 221.
lumbar, 512.
palmar, 66.
of the pharynx, 464.
plantar, 143.
of the soft palate, 465.
temporal, 323.
of the transversalis muscle, 213.
vertebral, 512.
Appendages of the eye, 374.
Appendices epiploicæ, 267.
Appendix auriculæ, 419, 423.
cæci vermiformis, 267.
Aqueduct of Fallopius, 483.
of Sylvius, 533, 559.
Aqueous humour, 564.
Arachnoid membrane of the brain, 528.
of the spine, 523.
Arbor vitæ cerebelli, 561.
uteri, 312.

Arch of aorta, 429.
crural or femoral, 95.
of diaphragm, 247.
palmar, deep, 73.
superficial, 67.
plantar, 153.
of soft palate, 465.
Arciform fibres, 539.
Areola of the mamma, 5.
Arm, dissection of, 5.
Arter: anastomotica brachialis, 28.
magna, 107.
angularis faciei, 384.
aorta, 429.
abdominalis, 248.
thoracica, 444.
articulares inferiores, 126.
superiores, 125.
articularis azygos, 126.
auricularis posterior, 322, 365.
auditorius, 326, 530.
axillaris, 15.
basilaris, 529.
brachialis, 26.
brachio-cephalica, 430.
bronchiales, 445.
buccalis, 396.
bulbi, 193, 201.
calcanea interna, 141.
capsulares, 251.
carotis communis, 361.
sinistra, 431.
externa, 363.
interna, 333, 363, 461, 531.
carpi ulnaris anterior, 50.
posterior, 50, 63.
radialis anterior, 45.
posterior, 63.
centralis retinæ, 351, 569.
cerebelli inferior, 529.
inferior anterior, 529.
superior, 531.
cerebri anterior, 531.
media, 531.
posterior, 531.
cervicalis ascendens, 434.
profunda, 435, 518.
superficialis, 340.
choroidea cerebri, 532.
ciliares, 351.
circumflexa anterior, 18.
externa, 111.
ilii interna, 254.
superficialis, 91, 101, 208.
interna, 111, 121.
posterior, 19, 35.
cœliac axis, 238, 251.
colica dextra, 236.
media, 236.
sinistra, 236.
comes nervi ischiadici, 120.
phrenici, 433.
Arter: communicans cerebri anterior, 53[illegible].
posterior, 531.
palmaris, 68.
plantaris, 153, 15[illegible].
coronaria dextra, 418.
labii inferioris, 3[illegible]3.
superioris, 3[illegible].
sinistra, 418.
ventriculi, 238.
corporis bulbosi, 193, 201.
cavernosi, 195, 202.
cremasterica, 254.
crico-thyroidea, 364.
cystica, 239.
deferentialis, 293.
dentalis anterior, 473.
inferior, 395.
posterior, 396.
diaphragmatica, 251.
digitales manûs, 67.
pedis, 153.
dorsales pollicis, 63.
dorsalis carpi radialis, 63.
ulnaris, 63.
clitoridis, 202.
indicis, 63.
linguæ, 406.
pedis, 160.
penis, 195.
pollicis pedis, *161.*
scapulæ, 35.
epigastrica, 222, 254.
superficialis, 91, 10[illegible], 208.
ethmoidalis, anterior, 351.
posterior, 351.
facialis, 365, 383.
femoralis, 99, 106.
frontalis, 321, 351.
gastrica, 238.
gastro-duodenalis, 239.
gastro-epiploica dextra, 2[illegible]9.
sinistra, 241.
glutea, 119, 294.
hæmorrhoidalis inferior, 186.
media, 293.
superior, 236.
hepatica, 239.
hyoidea lingualis, 406.
thyroideæ, 364.
hypogastrica, 224, 291.
ileo-colica, 236.
iliaca communis, 253.
externa, 254.
interna, 291.
ilio-lumbalis, 294.
incisoria, 395.
infra-orbitalis, 473.
innominata, 430.
intercostales anteriores, 433.
posteriores, 445, 519.
intercostalis superior, 435.

Arter: interossea, 49.
anterior, 50.
posterior, 60.
interosseæ manûs, 74.
pedis, 161.
intestinales, 236.
ischiadica, 120, 294.
labialis inferior, 383.
lachrymalis, 346, 350.
laryngealis inferior, 433, 493.
superior, 364, 492.
lingualis, 364, 404.
lumbales, 252, 519.
magna pollicis manûs, 75.
pedis, 153.
malleolares, 160.
mammaria interna, 432.
masseterica, 383, 396.
maxillaris interna, 364, 395, 480.
mediastinæ, 433, 445.
meningea media, 334, 396.
parva, 334.
meningeæ anteriores, 334.
posteriores, 334, 529.
mesenterica inferior, 236.
superior, 234.
metacarpa radialis, 63.
metatarsea, 161.
musculo-phrenica, 433.
mylo-hyoidea, 395.
nasalis, 351.
lateralis, 384.
septi, 383.
nutritia femoris, 111.
fibulæ, 140.
humeri, 27.
tibiæ, 140.
obturatoria, 280, 293.
occipitalis, 323, 365, 515.
œsophageales, 495.
ophthalmica, 350.
ovariana, 252.
palatina descendens, 480.
inferior, 407.
palmaris superficialis, 67.
profunda, 68.
palpebrales, 351.
pancreaticæ, 241.
pancreatico-duodenales, 234, 240.
perforantes femorales, 111, 128.
mammariæ internæ, 433.
manûs, 74.
pedis, 153.
pericardiacæ, 445.
perinæi superficialis, 187, 197.
peronea, 140.
anterior, 150.
pharyngea ascendens, 462.
phrenicæ inferiores, 251.
superiores, 433.
plantaris externa, 147, 153.
interna, 147.
poplitea, 125.

Arter; princeps cervicis, 516.
pollicis, 74.
profunda cervicis, 435, 518.
femoris, 101, 110.
inferior, 27.
superior, 27, 40.
ulnaris, 68.
pterygoideæ, 396.
pterygo-palatina, 480.
pudica externa, 91, 101, 208.
interna, 120, 193, 201, 294.
pulmonalis, 421, 427.
dextra, 412.
sinistra, 412.
pylorica, 239.
radialis, 43, 61.
indicis, 74.
ranina, 406.
recurrens interossea posterior, 60.
radialis, 45.
tibialis, 159.
ulnaris anterior, 49.
posterior, 49.
renales, 251.
sacra media, 252.
sacro-lateralis, 294.
scapularis posterior, 37, 340.
sciatica, 120, 294.
sigmoidea, 236.
spermatica, 252.
spheno-palatina, 480.
spinales posteriores, 524, 529.
spinalis anterior, 524, 529.
splenica, 239, 241.
sterno-mastoidea, 365.
stylo-mastoidea, 365.
subclavia, 340, 434.
dextra, 432.
sinistra, 431.
sublingualis, 406.
submentalis, 365.
subscapularis, 17, 37.
superficialis cervicis, 340.
volæ, 45.
supra-orbitalis, 321, 346, 350.
renales, 251.
scapularis, 36, 340.
tarsea, 160.
temporales profundæ, 396.
temporalis, 322, 364.
thoracica acromialis, 16.
alaris, 16.
longa, 16.
superior, 16.
thyroidea axis, 433.
ima, 430.
inferior, 433.
superior, 364.
tibialis anterior, 157.
posterior, 138.
tonsillaris, 407.
transversalis colli, 339, 434.

Arter: transversalis faciei, 384.
humeri, 340, 434.
perinæi, 187, 197.
pontis, 529.
tympanica, 396.
ulnaris, 49.
umbilicales, 427.
uterina, 293.
vaginalis, 293.
vertebralis, 432, 455, 517, 529.
vesicalis inferior, 293.
superior, 293.
vidiana, 480.
Articular popliteal arteries, 125, 126.
nerves, 124.
Articulation, acromio-clavicular, 76.
atlo-axoidean, 501.
calcaneo-astragaloid, 174.
cuboid, 175.
of carpal bones, 83.
carpo-metacarpal, 83.
of cervical vertebræ, 500.
chondro-costal, 506.
sternal, 506.
of costal cartilages, 506.
costo-vertebral, 504.
crico-arytænoid, 494.
thyroid, 494.
femoro-tibial or knee, 165.
humero-cubital or elbow, 79.
ilio-femoral or hip, 129.
of lower jaw, 392.
metacarpo-phalangeal, 86.
metatarso-phalangeal, 178.
occipito-atloidean, 500.
phalangeal of fingers, 86.
of pubic symphysis, 315.
radio-carpal or wrist, 83.
ulnar inferior, 82.
superior, 825.
sacro-coccygeal, 315.
iliac, 314.
sacro-vertebral, 314.
scapulo-clavicular, 76.
humeral or shoulder, 78.
sterno-clavicular, 356.
tarsal, 174,
tarso-metatarsal, 176.
temporo-maxillary, 392.
tibio-fibular, 171.
tarsal or ankle, 172.
of vertebræ, 497.
Arcus externus diaphragmatis, 247.
internus diaphragmatis, 247.
Arytæno-epiglottidean folds, 487.
Arytænoid cartilages, 495.
muscle, 488.
Ascending cervical artery, 434.
pharyngeal artery, 462.
Attollens aurem, 319.
oculum, 351.
Attrahens aurem, 319.
Auditory artery, 326, 530.
nerve, 325, 539.
Auricle of the ear, 375.
Auricle of the heart, left, 423.
right, 418.
Auricular artery, posterior, 322, 36[illegible]
nerve, anterior, 400.
great, 322, 335.
posterior, 322, [illegible]
vein, 367.
nerve of vagus, 458.
Auriculo-temporal nerve, 321, 387, [illegible]
ventricular aperture, left, [illegible]
right, [illegible]
Auricularis magnus nerve, 322, 335.
Axilla, 13.
dissection of, 11.
Axillary artery, 15.
glands, 13.
plexus, 19, 342.
vein, 19.
Axis, cœliac, of artery, 238, 251.
thyroid, of artery, 433.
Azygos artery, 126.
ganglion, 300.
veins, 262, 443, 446.
uvulæ muscle, 468.

Back, dissection of, 507.
Bag of pharynx, 464.
Bartholin, glands of, 201.
Base of brain, 532.
of the skull, arteries of, 333.
dissection of, 327.
nerves of, 330.
Basilar artery, 529.
sinus, 333.
Basilic vein, 23.
Beak of the corpus callosum, 534.
Bend of elbow, 30.
Biceps femoris muscle, 127.
flexor cubiti, 24, 31.
Bile-duct, 244.
Biventer cervicis muscle, 515.
Biventral lobe, 558.
Bladder, interior of, 304.
ligaments of, 279, 283.
connections of, 287, 289.
structure of, 303, 510.
Bones of the ear, 483.
muscles of, 484.
Brachial artery, 26.
nerves, 19.
plexus, 19, 342.
veins, 29.
Brachialis anticus, 26.
Brachio-cephalic artery, 430.
vein, left, 435.
right, 435.
Brain, base of, 532.
examination of interior, 546.

Brain, exterior of, 540.
membranes of, 527.
preservation of, 327.
removal of, 323.
vessels of, 529.
Breast, 5, 9.
Broad ligament of uterus, 291.
Bronchial arteries, 445.
glands, 443.
veins, 445.
Bronchus, left, 438.
right, 438.
structure of, 450.
Brunner's glands, 266.
Buccinator nerve, 387, 398.
artery, 396.
muscle, 394.
Bulb of the urethra, 190.
artery of, 193, 201.
nerve of, 189.
Bulbi vaginæ (vestibuli), 200.
Bulbous part of the urethra, 306.
Bulbus olfactorius, 536.
Bursa patellæ, 101.
of ligament of patella, 166.
of gluteus maximus, 115.

Cæcum coli, connections of, 227.
structure of, 267.
Calamus scriptorius, 561.
Calcaneo-astragaloid articulation, 174.
cuboid ligaments, 175.
Calices of the kidney, 277.
Caliciform papillæ, 485.
Canal of Fontana, 567.
of Nuck, 313.
of Petit, 568.
of the tensor tympani, 481.
Capsular arteries, 251.
veins, 251.
ligament of the hip, 130.
of the knee, 167.
of the shoulder, 79.
of the thumb, 84.
Capsule of crystalline lens, 568.
of Glisson, 273.
suprarenal, 256, 278.
Caput coli, 267.
gallinaginis, 305.
Cardiac nerve, inferior, 447.
middle, 370.
of pneumo-gastric, 368, 444.
superior, 370.
plexus, 418, 437.
Carneæ columnæ 420.
Carotid artery, common, 361.
external, 363.
internal, 323, 363, 461, 531.
left common, 431.
plexus, 333.
Carpal artery, radial anterior, 451.
posterior, 63.
Carpal artery, ulnar anterior, 50.
posterior, 50, 63.
Carpo-metacarpal articulation, 83.
Cartilage, arytænoid, 495.
cricoid, 495.
cuneiform, 496.
of the ear, 375.
thyroid, 494.
of the septum of nose, 473.
Cartilages of the eyelids, 388.
of the nose, 389.
of Santorini, 496.
of trachea, 450.
Caruncula lachrymalis, 374.
Carunculæ myrtiformes, 182, 311.
Cava, inferior, 253, 419.
superior, 419, 430.
Cavernous body, 307.
artery of, 195, 202.
plexus, 333, 350.
sinus, 330.
Cavities of the heart, 418.
Cavity of the abdomen, 224.
Central artery of the retina, 351, 569.
point of the perinæum, 183.
Centrum ovale cerebri, 544.
Cephalic vein, 22.
Cerebellar arteries, inferior, 529.
superior, 531.
Cerebellum, form of, 555.
lobes of, 558.
structure of, 561.
Cerebral artery, anterior, 531.
middle, 531.
posterior, 531.
protuberance, form, 537.
structure, 540.
Cerebrum, 532.
Cervical fascia, 337, 355.
ganglion, inferior, 447.
middle, 457.
superior, 457.
nerves, anterior branches, 370.
posterior, 519.
plexus of nerves, 370.
superficial, 335.
Cervicalis, ascendens artery, 434.
muscle, 513.
profunda artery, 435, 518.
superficialis nerve, 335, 355.
Cervico-facial nerve, 385.
Cervix uteri, 312.
vesicæ, 287.
Chamber of the eye, anterior, 564.
posterior, 565.
Chiasma of the optic nerves, 535.
Chondro-costal articulations, 506.
glossus muscle, 403.
sternal articulations, 506.
Chorda tympani nerve, 398, 483.
Chordæ tendineæ, 420.
vocales, 491.
Willisii, 324.
Choroid artery of the brain, 532.

Choroid coat of the eye, 566.
plexuses of the brain, 548, 551.
veins of the brain, 548.
of the eye, 566.
Choroidal epithelium, 566.
Ciliary arteries, 351.
body, 567.
muscle, 566.
processes of the choroid, 567.
nerves of nasal, 349.
of lenticular ganglion, 350.
Circle of Willis, 532.
Circular sinus, 333.
Circulation in the fœtus, 426.
Circumflex artery, anterior, 18.
external, 111.
internal, 111, 121.
posterior, 19, 35.
iliac artery, deep, 254.
superficial, 91, 101, 208.
nerve, 20, 35.
Circumvallate papillæ, 485.
Clavicular cutaneous nerves, 337.
Clitoris, 182.
Coccygeal artery, 120.
muscle, 296.
nerve, 300.
Cœliac axis, 238, 251.
Colic artery, left, 236.
middle, 236.
right, 236.
Colon, course of, 227.
structure of, 267.
Columnæ carneæ, 420.
Columns of the cord, 525.
vagina, 311.
Comes nervi ischiadici artery, 120.
phrenici artery, 433.
Commissure, anterior, 552, 555.
of the cerebellum, 558.
of the cord, 525.
great, 546.
of the optic nerves, 535.
posterior, 552.
soft, 553.
Commissural fibres of the medulla, 538.
Communicating artery of anterior cerebral, 531.
of posterior cerebral, 531.
in the palm, 68.
in the sole, 153, 161.
fibular nerve, 125, 132.
Complexus muscle, 515.
Compressor of the nose, 378.
of the urethra, 192.
of bulb of vagina, 198.
Conarium, 553.
Concha, 376.
Cones of the retina, 569.
Congenital hernia, 217.
Coni vasculosi, 309.
Conical papillæ, 486.
Conjoined tendon, 210, 213.
Conjunctiva, 388, 563.
Conoid ligament, 77.
Conus arteriosus, 420.
Constrictor inferior, 462.
of the fauces, 469.
middle, 462.
superior, 463.
urethræ, 192.
Convolutions of the brain, 540.
of corpus callosum, 5
of longitudinal fiss 544.
Coraco-acromial ligament, 77.
brachialis muscle, 26.
clavicular articulation, 76.
humeral ligament, 79.
Cordiform tendon, 247.
Cords of the abdominal wall, 221.
Cornea, 564.
structure, 564.
Cornicula laryngis, 496.
Cornua of lateral ventricle, 547.
Corona glandis, 220.
Coronary vessels of the heart, 416.
of the lips, 383.
artery of the stomach, 2
ligament of the liver, 27
sinus, 419.
valve, 419.
vein of the stomach, 238.
Corpora albicantia, 535.
Arantii, 421, 424.
cavernosa, 307.
Malpighiana, 277.
mamillaria, 535.
olivaria, 538, 561.
pyramidalia anteriora, 538
posteriora, 561
quadrigemina, 555.
restiformia, 539.
Corpus callosum, 546.
ciliare, 567.
dentatum cerebelli, 561.
medullæ, 561.
fimbriatum, 549.
geniculatum externum, 553.
internum, 553.
Highmori, 308.
luteum, 313.
olivare, 538, 561.
spongiosum urethræ, 190, 30
striatum, 547.
thyroideum, 439.
Corpuscles of Malpighi, 270, 277.
Corrugator supercilii muscle, 377.
Cortical substance of the kidney, 277
Costo-clavicular ligament, 356.
Costo-coracoid membrane, 14.
vertebral ligaments, 504.
Cotyloid ligament, 131.

Covered band of Reil, 547.
Coverings of hernia, femoral, 95.
inguinal, 216.
Cowper's glands, 302.
Cranial aponeurosis, 319.
nerves, 325.
Cremaster muscle, 210.
Cremasteric artery, 254.
fascia, 216.
Cribriform fascia, 92.
Crico-arytænoid articulation, 494.
muscle, lateral, 489.
posterior, 488.
thyroid articulation, 494.
membrane, 494.
muscle, 488.
Cricoid cartilage, 495.
Crucial ligament, anterior, 168.
posterior, 168.
Cruciform ligament, 503.
Crura cerebelli, 537, 540.
cerebri, 537.
of the diaphragm, 247.
of the fornix, 549, 555.
Crureus muscle, 106.
Crural arch, 95.
deep, 95.
canal, 95.
hernia, 92.
nerve, 101, 107, 259.
ring, 94.
sheath, 93.
Crypts of Lieberkühn, 266.
Crystalline lens, 567.
capsule, 567.
Cuneiform cartilages, 496.
Cutaneous nerves of the abdomen, 205, 208.
of the arm, 22.
of the back, 507.
of the buttock, 114.
of the face, 385.
of the foot, back, 154.
sole, 148.
of the forearm, 40.
of the hand, back, 53.
palm, 68.
of the head, 319.
of the leg, back, 132.
front, 154.
of the neck, behind, 507.
fore part, 335.
of the perinæum, 188, 197.
of the scalp, 319.
of the thigh, front, 97.
of the thorax, 11, 12.
Cystic artery, 239.
duct, 273.

Dartos, 186.
Decussation of the pyramids, 536.
Deep cervical artery, 435, 518.
crural arch, 95.
transverse muscle of perinæum, 193, 201.
Deferential artery, 293.
Deglutition, process of, 466.
Deltoid ligament, 173.
muscle, 33.
Dental artery, anterior, 473.
inferior, 395.
posterior, 396.
nerve, anterior, 472.
inferior, 398.
posterior, 472.
Dentate fascia, 551.
ligament, 524.
Descendens noni nerve, 357.
Descent of testicle, 210.
Depressor anguli oris, 380.
labii inferioris, 380.
alæ nasi, 378.
Diaphragm, 245.
arteries of, 251, 433.
openings in, 247.
upper surface of, 415.
Digastric muscle, 359.
nerve, 385.
Digital arteries of plantar, 153.
of radial, 74.
of tibial, anterior, 161.
of ulnar, 67.
nerves of median, 70.
of plantar, 148, 149.
of radial, 53.
of ulnar, 53, 68.
Dissection of the abdomen, 179.
of the abdominal cavity, 224.
of the abdominal wall, 202.
of the arm, 5.
of the axilla, 11.
of the back, 507.
of the base of the skull, 327.
of the bend of the elbow, 30.
of the brain, membranes, and nerves, 527.
of the buttock, 113.
of the cerebellum, 555.
of the cerebrum, 546.
of the external ear, 375.
of the middle ear, 481.
of the eighth nerve, 457.
of the eye, 562.
lids, 388.
of the face, 372.
of femoral hernia, 92.
of the foot, back, 154.
sole, 143.
of the forearm, back of, 52.
front of, 40.
of the fourth ventricle, 559.
of the great vessels of root of neck, 430.

Dissection of the hand, back, 52.
palm, 65.
of the head and neck, 317.
of the heart, 415.
of the inguinal hernia, 213.
region, 206.
of the larynx, 487.
of the lateral ventricles, 547.
of the leg, back, 132.
front, 154.
of the ligaments of atlas, axis, and occiput, 500.
of the ligaments of clavicle and scapula, 76.
of the ligaments of hip joint, 129.
of the ligaments of jaw, 392.
of the ligaments of lower limb, 165.
of the ligaments of pelvis, 313.
of the ligaments of ribs, 504.
of the ligaments of upper limb, 76.
of the ligaments of the vertebræ, 497.
of the lower limb, 87.
of Meckel's ganglion, 477.
of mesenteric vessels, 236.
of the neck, 354.
of the neck, anterior triangle, 371.
posterior triangle, 334.
of the nose, 473.
of the orbit, 344.
of the otic ganglion, 481.
of the palate, 466.
of the parts about the scapula, 32.
of the pelvis, female, 279.
side view, 289.
of the pelvis, male, 279.
side view, 285.
of the male pelvic viscera, 300.
female, 309.
of the perinæum, both sexes, 181.
female, 195.
male, 186.
of the pharynx, 455.
of the popliteal space, 123.
of the prevertebral region, 452.
posterior triangle, 334.
of the pterygoid region, 390.
of the sacral plexus, 297.
of the scalp, 318.
of Scarpa's triangle, 98.
of the soft palate, 466.
of the spermatic cord, 219.
of the spinal cord, 521.
Dissection of the submaxillary re[g]
400.
of the superior max-
nerve, 472.
of the testis, 307.
of the thigh, back, 1[2]
front, 89.
inner [...]
108.
of the third ventricle, [...]
of the thorax, 407.
of the tongue, 484.
of the tympanum, 481.
of the upper limb, 5.
Dorsal artery of the clitoris, 202.
of the foot, 160.
of the penis, 195.
of the tongue, 406.
of the scapula, 35.
nerves, anterior branches, 3[...]
posterior branches, 519.
of the penis, 195, 2[...]
Dorsi-spinal veins, 521.
Douglas, fold of, 221.
Ductus ad nasum, 475.
arteriosus, 427, 429.
communis choledochus, [...] 273.
cysticus, 244, 273.
ejaculatorius, 303.
hepaticus, 273.
lymphaticus, 368.
pancreaticus, 244, 269.
Riviniani, 401.
Stenonis, 381.
thoracicus, 368, 442.
venosus, 426.
Whartonii, 401.
Duodenum, connections, 241.
structure, 265.
Dura mater, 327, 527.
of the cord, 521.
vessels of, 334.

Ear, external, 375.
middle, 481.
Eighth nerve, 326, 457, 539.
Elbow joint, 79.
Eminentia collateralis, 551.
Encephalon, 532.
Endocardium, 420.
Epididymis, 308.
Epigastric artery, 222, 254.
superficial, 91, 101, 208.
region of the abdomen, 226.
Epiglottis, 496.
Epiploon, 228.
Erector clitoridis, 199.
penis, 189.
spinæ, 513.
Ethmoidal arteries, 351.
Eustachian tube, cartilaginous part, 469.

Eustachian valve, 419.
Extensor carpi radialis brevior, 55.
longior, 54.
carpi ulnaris, 56.
digiti minimi, 56.
digitorum brevis, 157.
communis, 55.
longus pedis, 155.
indicis, 59.
ossis metacarpi pollicis, 58.
proprius pollicis, 157.
primi internodii pollicis, 58.
quadriceps femoris, 104.
secundii internodii pollicis, 59.
External cutaneous nerve of arm, 29.
cutaneous nerve of thigh, 97, 259.
saphenous nerve, 132, 154.
vein, 133.
Eye-ball, 562.
brows, 373.
lashes, 373.
lids, 388.
structure, 388.

Face, dissection of, 372.
Facial artery, 365, 383.
nerve, 321, 385.
vein, 384.
Falciform ligament of the liver, 232.
border of saphenous opening, 92.
Fallopian tube, 234, 312.
Falx cerebelli, 328.
cerebri, 327.
Fascia axillary, 11.
brachial, 24.
cervical deep, 337, 355.
costo-coracoid, 14.
cremasteric, 216.
cribriform, 92.
dentata, 551.
of the forearm, 41.
of the groin, 206.
iliac, 257.
intercolumnar, 208.
intermuscular of the humerus, 24.
of the thigh, 97.
lata, 92, 102.
of the leg, 155.
lumborum, 213, 512.
obturator, 279.
palmar, 66.
pelvic, 281.
perinæal, deep, 190, 200.
superficial, 186, 195.
plantar, deep, 143.
popliteal, 123.
of the psoas, 257.
of the quadratus, 257.
recto-vesical, 282.
spermatic, 209.
Fascia Scarpa's, 206.
temporal, 323.
transversalis, 213.
triangular of groin, 209, 215.
Fasciculus teres, 559.
Femoral arch, 95.
artery, 99.
canal, 95.
hernia, 92.
ligament, 92.
ring, 94.
sheath, 93.
vein, 101.
Femoro-tibial articulation, 165.
Fenestra ovalis, 482.
rotunda, 482.
Fibres of Müller, 569.
Fibro-cartilage. *See* Interarticular.
Fibrous coat of eye, 562.
Fifth nerve, 330, 538.
Filiform papillæ, 486.
Fillet of the olivary body, 555.
Fimbriæ of the Fallopian tube, 312.
First nerve, 325, 477, 536.
Fissure, longitudinal, of the cerebrum, 533.
of Sylvius, 537, 541.
transverse, 552.
parallel, 542.
calcarine, 545.
collateral, 546.
dentate, 546.
for vena cava, 272.
Fissures of the cord, 524.
Flexor accessorius muscle, 149.
brevis minimi digiti, 72.
pedis, 151.
carpi radialis, 42.
ulnaris, 43.
digitorum brevis pedis, 145.
longus pedis, 136, 149.
profundus, 47, 71.
sublimis, 46, 70.
pollicis longus, 47, 71.
pedis, 137, 150.
brevis, 72.
pedis, 150.
Flocculus cerebelli, 558.
Fœtal circulation, 426.
Foot, dorsum, 154.
sole, 143.
Foramen of Munro, 549.
ovale, 419, 427.
quadratum, 247.
of Winslow, 230.
Foramina Thebesii, 419.
Forearm, dissection of, front, 40.
back, 52.
cutaneous nerves, 40.
veins, 40.
Fornix, 549.
Fossa, ischio-rectal, 184.

Fossa, nasal, 474.
navicular of the urethra, 306.
of the pudendum, 181.
ovalis, 419.
Fossæ of abdominal wall, 224.
Fourth nerve, 330, 345, 537.
ventricle, 559.
Fovea centralis, 569.
Frænum labii, 472.
linguæ, 401.
præputii, 220.
Frontal artery, 321, 356.
nerve, 345.
Fungiform papillæ, 485.

Galen, veins of, 552.
Gall bladder, 273.
Ganglia, cervical, inferior, 447.
middle, 370, 457.
superior, 370, 457.
lumbar, 262.
sacral, 300.
semilunar, 261.
of spinal nerves, 526.
thoracic, 447.
Ganglion of the vagus, 458.
azygos, 300.
Gasserian, 330.
impar, 300.
jugular, 460.
lenticular, 349.
Meckel's, 478.
ophthalmic, 349.
otic, 481.
petrosal, 460.
spheno-palatine, 478.
submaxillary, 403.
thyroid, 370, 457.
Gastric artery, 238.
plexus, 261.
vein, 244.
Gastro-colic omentum, 228.
duodenal artery, 239.
epiploïc arteries, 239, 241.
hepatic omentum, 228.
Gastrocnemius muscle, 133.
Gemellus inferior muscle, 118.
superior muscle, 118.
Genio-hyo-glossus, 406.
hyoid muscle, 406.
Genital organs, 181.
Genito-crural nerve, 97, 259.
Gimbernat's ligament, 95, 208.
Gland, lachrymal, 347.
parotid, 381.
pineal, 553.
pituitary, 535.
prostate, 303.
sublingual, 401.
submaxillary, 401.
thyroid, 439.
Glands, axillary, 13.
bronchial, 443.
Glands, Bartolini's, 201.
Blandin, 486.
Brunner's, 266.
cervical, 355.
concatenate, 338.
Cowper's, 302.
of frænum, 486.
inguinal, 91, 208.
intercostal, 443.
intestinal, 266.
lingual, 486.
lumbar, 262.
mammary, 5, 9.
mediastinal, 443.
Meibomian, 388.
mesenteric, 236.
odoriferous, 220.
of Pacchioni, 324.
Peyer's, 267.
popliteal, 126.
solitary, 266.
tracheal, 443.
Glans of the clitoris, 182.
of the penis, 220.
Glenoid ligament, 79.
Glisson's capsule, 273.
Globus major epididymis, 308.
minor epididymis, 308.
Glosso-pharyngeal nerve, 326, 407, 4??, 539.
Glottis, 487.
Gluteal artery, 119, 294.
nerve, superior, 121, 299.
nerves, inferior, 122.
Gluteus maximus muscle, 115.
medius muscle, 116.
minimus muscle, 116.
Graafian vesicles, 313.
Gracilis muscle, 108.
Granular layer of retina, 569.
Great omentum, 228.
Grey substance of the cord, 526.
of the corpus striatum, 547.
of the medulla oblongata, 559.
of the third ventricle, 553.
Gubernaculum testis, 210.
Gums, 470.
Gustatory nerve, 398, 401.
Gyrus fornicatus, 544.

Ham, 124.
Hamstrings, 127.
Hand, dissection of, 65.
Hæmorrhoidal artery, inferior, 186.
middle, 293.
superior, 236.
nerve, inferior, 186.
Head, dissection of, 317.
Heart, 415.
cavities of, 418.
great vessels of, 418.

Heart, position, 415.
Helix, 376.
Hepatic artery, 239.
ducts, 273.
plexus, 261.
veins, 274.
Hernia, crural or femoral, 92.
inguinal, 213.
varieties of, 217.
umbilical, 202.
Hesselbach's triangle, 224.
Hip joint, 129.
Hippocampus, major, 551.
minor, 550.
Hollow before elbow, 30.
Humero-cubital articulation, 79.
Hunter's canal, 106.
Hyaloid membrane, 568.
Hydatid of Morgagni, 309.
Hymen, 182.
Hyo-epiglottidean ligament, 494.
glossus muscle, 403.
glossal membrane, 484.
Hypochondriac region of abdomen, 226.
Hypogastric plexus of nerves, 300.
region of the abdomen, 226.
Hypoglossal nerve, 368, 403, 457, 540.

Ileo-cœcal valve, 268.
colic artery, 236.
valve, 268.
Ileum intestine, connections of, 226.
structure of, 267.
Iliac artery, common, 253.
external, 254.
internal, 291.
fascia, 257.
region of the abdomen, 226.
vein, common, 253.
external, 254.
internal, 295.
Iliacus muscle, 109, 258.
Ilio-femoral ligament, 130.
hypogastric nerve, 208, 212.
inguinal nerve, 97, 208, 212.
lumbar artery, 294.
ligament, 314.
Incisor branch of nerve, 398.
Incus, 483.
Indicator muscle, 58.
Infantile hernia, 218.
Inferior cornu of the lateral ventricle, 550.
maxillary nerve, 331, 387, 396.
Infra-orbital artery, 473.
nerve, 387.
vein, 473.
trochlear nerve, 349.
Infra-spinatus muscle, 33.
Infundibulum of the brain, 535.
of the heart, 420.
Inguinal canal, 214.
glands, 91, 208.
Inguinal hernia, 213, 215.
region of the abdomen, 226.
Innominate artery, 430.
veins, 435.
Interarticular cartilage of the jaw, 393.
of the knee, 169.
of the scapula, 76.
sacro-iliac, 315.
sterno-clavicular, 357.
of symphysis pubis, 316.
of the vertebræ, 498.
of the wrist, 86.
Interclavicular ligament, 356.
Intercolumnar fascia, 208.
Intercostal arteries, anterior branches, 433.
posterior branches, 445, 519.
artery, superior, 435.
muscle, external, 407.
internal, 408, 448.
nerves, 447.
cutaneous anterior, 10, 205.
lateral, 11.
veins, 446.
Intercosto-humeral nerve, 12.
Interior of pharynx, 465.
Intermuscular septa of the arm, 24.
of the thigh, 102.
Internal cutaneous nerve of arm, 22, 40.
of thigh, 102.
saphenous vein, 92, 154.
nerve, 102, 154.
Interosseous arteries of the foot, 161.
of the hand, 74.
artery, anterior, 50.
posterior, 60.
ligament of the arm, 82.
of the leg, 171.
muscles of the foot, 164, 165.
of the hand, 75.
nerve, anterior, 51.
posterior, 61.
Interspinal muscles, 520.
Intertransverse muscles, 520.
Intervalvular space, 424.
Intervertebral substance, 498.
Intestine, large, 267.
small, 265.
Intestinal arteries, 236.
canal, divisions, 227.
structure, 262.
Intra-spinal arteries, 524.
veins, 521, 524.
Intumescentia gangliformis, 483.
Iris, 564.

Ischio-rectal fossa, 184.
Island of Reil, 537.
Isthmus faucium, 465.
of the thyroid body, 439.
Iter a tertio ad quartum ventriculum 553.
ad infundibulum, 553.

Jacob's membrane, 569.
Jarjavay's muscle, 201.
Jacobson's nerve, 460.
Jejunum, connections of, 226.
structure, 267.
Joint, ankle, 172.
elbow, 79.
hip, 129.
knee, 165.
lower jaw, 392.
shoulder, 78.
wrist, 83.
Jugular ganglion, 460.
vein, anterior, 357.
external, 335.
internal, 367, 457.

Kidney, 275.
connections of, 256.
structure, 277.
vessels of, 251.
Knee of the corpus callosum, 546.
joint, 165.

Labia pudendi externa, 181.
interna, 182.
Labial glands, 471.
artery, inferior, 383.
Lachrymal artery, 346, 350.
canals, 374.
ducts, 346.
gland, 347.
nerve, 345.
point, 374.
sac, 373.
Lacteals, 236.
Lacunæ of the urethra, 306.
Lamina cinerea, 535.
Laminæ of cerebellum, 556.
of the lens, 568.
Large intestine, connections, 227.
structure and form of, 267.
Laryngeal arteries, 364, 433, 492.
nerve, external, 368, 459.
inferior, 368, 444, 493.
superior, 368, 459, 493.
pouch, 491.
Larynx, 487.
aperture of, 467, 487.
cartilages of, 494.
interior of, 491.
muscles, 488.
nerves, 493.
ventricle, 491.
vessels, 492.
Lateral column of the medulla, 55-
of the cord, 525.
sinus, 329.
ventricles, 547.
Latissimus dorsi, 21, 509.
Laxator tympani, 484.
Leg, dissection of the back, 132.
front, 154.
Lens of the eye, 567.
Lenticular ganglion, 349.
Levator anguli oris, 380.
scapulæ, 509.
ani, 185, 296.
glandulæ thyroideæ, 439.
labii superioris, 380.
alæque nas 377.
menti, 380.
palati, 467.
palpebræ superioris, 347.
Levatores costarum, 521.
Lieberkühn's follicles, 266.
Ligament of the lung, 409.
Ligaments of the bladder, 279, 283.
of the larynx, 491, 493.
of the ovary, 313.
of the uterus, 291, 313.
Ligament, acromio-clavicular, 76.
alar of the knee, 168.
annular, anterior of the ankle, 155.
external of the ankle, 155.
internal of the ankle, 142.
anterior of the wrist, 69.
posterior of the wrist, 63.
anterior, special, of ankle 173.
of elbow joint 81.
of knee joint 165.
of wrist joint 83.
atlo-axoid, anterior, 501.
posterior, 502.
calcaneo-astragaloid, 174.
cuboid, 175.
calcaneo-scaphoid, 175.
capsular of the hip, 130.
of the jaw, 392.
of the knee, 167.
of the shoulder, 79.
of the thumb, 84.
carpal, 83.
carpo-metacarpal, 83.
chondro-sternal, 506.
common, anterior of vertebræ, 497.
common, posterior, 498.
conoid, 77.

[Li]gament, coraco-acromial, 77.
clavicular, 76.
humeral, 79.
costo-clavicular, 356.
transverse, anterior, 505.
middle, 506.
posterior, 506.
vertebral, 504.
cotyloid, 131.
crico-thyroid, 494.
tracheal, 494.
crucial, anterior, 168.
posterior, 168.
cruciform, 503.
deltoid, 173.
dorsal of the carpus, 83.
of Gimbernat, 95, 208.
glenoid, 79.
hyo-epiglottidean, 494.
ilio-femoral, 130.
lumbar, 314.
interarticular, of the clavicle, 357.
of the hip, 131.
of the jaw, 393.
of the knee, 168.
of the pubes, 317.
of the ribs, 505.
of the wrist, 86.
interclavicular, 356.
interosseous of astragalus and os calcis, 174.
interosseous of carpus, 83.
interosseous of metatarsal bones, 176.
interosseous of radius and ulna, 82.
interosseous of the tibia and fibula, 171.
interosseous, inferior of the tibia and fibula, 172.
interspinous, 499.
intertransverse, 500.
intervertebral, 498.
lateral, external of the ankle, 173.
lateral, internal ditto, 173.
lateral, external of the carpus, 83.
lateral, internal ditto, 83.
lateral, external of the elbow, 81.
lateral, internal ditto, 81.
lateral, phalangeal of the hand, 86.
lateral, external of the jaw, 392.
lateral, internal ditto, 394.
Ligament, lateral, external of the knee, 166.
lateral, internal ditto, 167.
lateral, external of the wrist, 83.
lateral, internal ditto, 83.
long plantar, 175.
lumbo-sacral, 314.
metacarpal, 83.
metacarpo-phalangeal, 86.
metatarsal, 176.
mucous, 168.
obturator, 280.
occipito-atloid, anterior, 501.
lateral, 501.
posterior, 501.
occipito-axoid, 503.
odontoid, 504.
orbicular of the radius, 82.
palpebral, 388.
of the patella, 165.
phalangeal, 86.
of Poupart, 95, 208.
posterior of ankle, 173.
of elbow, 81.
of knee, 166.
of wrist, 83.
pterygo-maxillary, 394.
pubic, anterior, 316.
superior, 316.
round of the hip, 131.
round of the radius and ulna, 82.
sacro-coccygeal, anterior, 315.
posterior, 315.
sacro-iliac, anterior, 314.
oblique, 314.
posterior, 314.
sacro-sciatic, large, 314.
small, 315.
sacro-vertebral, 314.
of the scapula, 76.
sterno-clavicular, 356.
stylo-hyoid, 407.
maxillary, 407.
sub-pubic, 316.
supra-spinous, 499.
suspensory of penis, 220.
of liver, 232.
tarso-metatarsal, 176.
thyro-epiglottidean, 494.
hyoid, 493.
arytænoid, 491.
tibio-fibular, superior, 171.
middle, 171.
inferior, 172.
tarsal, 175.
transverse of the atlas, 503.
of the hip, 132.
of the knee, 169.

Ligament, transverse of metacarpus, 75.
of metatarsus, 176.
of scapula, 77.
trapezoid, 77.
triangular of the abdomen, 209. 215.
of the urethra, 190, 200.
of Winslow, 166.
Ligamentum arcuatum, 247.
denticulatum, 524.
latum pulmonis, 409.
longum plantæ, 175.
mucosum, 168.
nuchæ, 508.
patellæ, 165.
subflavum, 499.
teres, 131.
Linea alba, 208, 221.
semilunaris, 221.
Lineæ transversæ, 202, 221.
Lingual artery, 364, 404.
glands, 486.
nerve, 398, 401.
vein, 406.
Lingualis muscles, 486.
Lips, 471.
Lithotomy, parts cut, 185, 189.
Liver, 270.
ligaments, 232, 270.
connections of, 226.
structure, 274.
vessels, 238.
Lobes of the cerebellum, 556, 558.
of the cerebrum, 532.
Lobules of the testis, 308.
of the liver, 272.
Lobulus auris, 376.
caudatus, 273.
quadratus, 273.
Spigelii, 273.
Locus niger, 540.
perforatus anticus, 536.
posticus, 536.
Longissimus dorsi, 514.
Longitudinal fissure of the brain, 533, 540.
fissure of the liver, 271.
sinus, inferior, 328.
superior, 324.
Longus colli muscle, 454.
Lumbar aponeurosis, 512.
arteries, 252, 519.
ganglia, 262.
glands, 262.
nerves, anterior branches, 259.
posterior branches, 519.
plexus, 259, 261.
region of the abdomen, 226.
veins, 253.
Lumbo-sacral ligament, [illegible]
nerve, 361, [illegible].
Lumbricales of the foot, [illegible]
of the hand, 71.
Lungs, connections, 414.
physical characters, [illegible]
roots of, 412.
structure, [illegible]
vessels and nerves, [illegible]
Lymphatic duct, [illegible], 462.
right, [illegible].
Lyra, 551.

Malleolar arteries, 169.
Malleolus internus, parts behind
Malleus, 483.
Malpighian corpuscles of spleen
of kidney
Mamillæ of the kidney, 277.
Mamma, 5, 9.
Mammary artery, internal, 432.
gland, 5, 9.
Masseter muscle, 390.
Masseteric artery, 383, 396.
nerve, 398.
Mastoid cells, 482.
Maxillary artery, internal, 364, 480.
nerve, inferior, 331, [illegible]
superior, 331, 472.
vein, internal, 396.
Meatus auditorius externus, 376.
urinarius, 182, 306.
Meatuses of the nose, 474.
Meckel's ganglion, 478.
Median basilic vein, 22.
cephalic vein, 22.
nerve, 29, 51, 70.
vein, 22, 40.
Mediastina of thorax, 409, 441.
Mediastinal arteries, 433, 445.
Mediastinum testis, 308.
Medulla oblongata, 538.
structure of, 561.
spinalis, 526.
Medullary layer of the retina, 569.
substance of the kidney, 277.
velum, anterior, 555.
posterior, 559.
Meibomian follicles, 388.
Membrana fusca, 566.
pigmenti, 566.
sacciformis, 85.
tympani, 481.
Membrane hyaloid, 568.
Jacob's, 569.
Membranes of the brain, 293, 528.
of the spine, 521, 523.
Membranous urethra, 285, 306.
Meningeal artery, anterior, 334.
middle, 334, 396.

ningeal artery, posterior, 334, 529.
small, 334.
ntal nerve, 387.
senteric artery, inferior, 236.
superior, 234.
glands, 236.
plexus, inferior, 238.
superior, 236.
vein, inferior, 238.
superior, 236.
sentery, 228, 231.
so-cæcum, 231.
colon, left, 232.
right, 231.
transverse, 228.
rectum, 228.
etacarpal artery, 63.
etatarsal artery, 161.
tral valve, 428.
ons Veneris, 203.
otor oculi nerve, 330, 352, 537.
outh, cavity of, 470.
ucous ligament, 168.
ultifidus spinæ muscle, 520.
usculi papillares, 420.
pectinati, 419, 423.
usculo-phrenic artery, 433.
cutaneous nerve, 154, 163.
of the arm, 29, 41.
usculo-spiral nerve, 33, 39.
usc: abductor digiti minimi, 71.
digiti minimi pedis, 146.
indicis, 76.
oculi, 351.
pollicis, 72.
pollicis pedis, 146.
accelerator urinæ, 188.
accessorius pedis, 149.
ad sacro-lumbalem, 513.
adductor brevis, 109.
digiti minimi, 72.
longus, 108.
magnus, 112.
oculi, 351.
pollicis manûs, 72.
pedis, 151.
anconeus, 57.
arytænoideus, 488.
arytæno-epiglottideus, 490.
attollens aurem, 319.
oculum, 351.
attrahens aurem, 319.
azygos uvulæ, 468.
biceps femoris, 127.
flexor cubiti, 24, 31.
biventer cervicis, 515.
brachialis anticus, 26.
buccinator, 394.
cervicalis ascendens, 513.
ciliaris, 566.
circumflexus palati, 468.

Musc: coccygeus, 296.
complexus, 515.
compressor bulbi vaginæ, 198.
naris, 378.
urethræ, 192.
constrictor inferior, 462.
isthmi faucium, 469.
medius, 462.
superior, 463.
urethræ, 192.
vaginæ, 201.
coraco-brachialis, 26.
corrugator supercilii, 377.
cremastericus, 210.
crico-arytænoideus lateralis, 489.
posticus, 488.
thyroideus, 488.
crureus, 106.
deltoides, 33.
depressor anguli oris, 380.
labii inferioris, 380.
alæ nasi, 378.
diaphragma, 245, 415.
digastricus, 359.
erector clitoridis, 199.
penis, 189.
spinæ, 513.
extensor carpi radialis brevior, 55.
longior, 54.
ulnaris, 56.
digiti minimi, 56.
digitorum brevis pedis, 157.
digitorum communis, 55.
digitorum longus pedis, 155.
indicis, 58.
ossis metacarpi pollicis, 58.
pollicis proprius, 157.
primi internodii pollicis, 58.
secundi internodii pollicis, 58.
flexor accessorius, 149.
brevis digiti minimi, 72.
brevis digiti minimi pedis, 151.
carpi radialis, 42.
ulnaris, 43.
digitorum brevis pedis, 145.
profundus, 47, 71.
sublimis, 46, 70.
longus pedis, 136, 149.
pollicis longus, 47, 71.
pedis, 137, 150.
pollicis brevis, 72.

Musc: flexor pollicis pedis, 150.
gastrocnemius, 133.
gemellus inferior, 118.
superior, 118.
genio-hyo-glossus, 406.
hyoideus, 406.
gluteus maximus, 115.
medius, 116.
minimus, 116.
gracilis, 106.
hyo-glossus, 403.
iliacus, 109. 258.
indicator, 58.
infra-spinatus, 33.
intercostales externi, 407.
interni, 408, 448.
interossei manûs dorsales, 75.
palmares, 75.
pedis dorsales, 164.
plantares, 165.
interspinales, 520.
intertransversales, 520.
Jarjavay, 201.
latissimus dorsi, 21, 509.
laxator tympani, 484.
levator anguli oris, 380.
scapulæ, 509.
ani, 185, 296.
glandulæ thyroideæ, 439.
labii superioris, 380.
alæque nasi, 377.
menti, 380.
palati, 467.
palpebræ, 347.
levatores costarum, 521.
linguales, 486.
longissimus dorsi, 514.
longus colli, 454.
lumbricales manûs, 71.
pedis, 150.
masseter, 390.
multifidus spinæ, 520.
mylo-hyoideus, 400.
obliquus abdominis externus, 206, 208.
internus, 209.
capitis inferior, 517.
superior, 517.
oculi inferior, 352.
superior, 347.
obturator externus, 113, 119.
internus, 118, 297.
occipito-frontalis, 319.
omo-hyoideus, 36, 339, 358.
opponens digiti minimi, 72.
pollicis, 72.
orbicularis oris, 378.
palpebrarum, 377.
palato-glossus, 469.
pharyngeus, 469.

Musc: palmaris brevis, 67.
palmaris longus, 42.
pectineus, 108.
pectoralis major, 10.
minor, 14.
peroneus brevis, 162.
longus, 162.
tertius, 156.
plantaris, 134.
platysma myoides, 335, 354, 380.
popliteus, 135.
pronator quadratus, 48.
radii teres, 41.
psoas magnus, 109, 257.
parvus, 258.
pterygoideus externus, 392.
internus, 394.
pyramidalis abdominis, 221.
nasi, 377.
pyriformis, 117, 300.
quadratus femoris, 119.
lumborum, 258.
quadriceps extensor cruris, 104.
rectus abdominis, 221.
capitis anticus major, 453.
minor, 454.
lateralis, 454.
posticus major, 517.
minor, 517.
femoris, 104.
oculi externus, 351.
inferior, 351.
internus, 351.
superior, 347.
retrahens aurem, 319.
rhomboideus major, 510.
minor, 510.
risorius Santorini, 380.
sacro-lumbalis, 513.
sartorius, 103.
scalenus anticus, 452.
medius, 453.
posticus, 453.
semi-spinalis colli, 519.
dorsi, 519.
semi-membranosus, 127.
semi-tendinosus, 127.
serratus magnus, 21.
posticus inferior, 512.
superior, 512.
soleus, 134.
sphincter ani externus, 183.
internus, 183.
vaginæ, 198, 201.
spinalis dorsi, 513.
splenius capitis, 512.
colli, 512.
stapedius, 484.
sterno-cleido-mastoideus, 355.
hyoideus, 358.
thyroideus, 358.
stylo-glossus, 403.
hyoideus, 361.
pharyngeus, 404, 463.

uscle: subanconeus, 80.
subclavius, 14.
subcrureus, 168.
subscapularis, 36.
supinator radii brevis, 57.
longus, 54.
supra-spinatus, 33.
temporalis, 323, 391.
tensor palati, 468.
tarsi, 388.
tympani, 484.
vaginæ femoris, 103.
teres major, 21, 34.
minor, 34.
thyro-arytænoideus, 490.
epiglottideus, 491.
hyoideus, 358.
tibialis anticus, 155.
posticus, 136.
trachelo-mastoideus, 514.
transversalis abdominis, 210.
cervicis, 514.
transversus linguæ, 486.
pedis, 152.
perinæi, 189, 199.
profundus, 193, 201.
trapezius, 506.
triangularis sterni, 408.
triceps extensor cubiti, 38.
vastus, externus, 105.
internus, 105.
zygomaticus major, 380.
minor, 380.
Mylo-hyoid artery, 395.
muscle, 400.
nerve, 398.

Nares, posterior, 465.
Nasal artery, 351.
lateral, 384.
cartilages, 389, 473.
duct, 374, 475.
fossæ, 474.
nerve, 349, 387, 477.
Naso-palatine artery, 480.
nerve, 478.
Neck, anterior triangle of, 371.
posterior, 334.
dissection of, 354.
Nerve of Wrisberg, 20, 22.
Nerve to the latissimus, 20, 37.
levator anguli scapulæ, 343, 371.
obturator internus, 122, 299.
pectineus, 107.
pyriformis, 299.
quadratus femoris, 122.
rhomboid muscle, 343.
serratus magnus, 21, 343.
subclavius, 343.
tensor vaginæ femoris, 121.
Nerve to the teres major, 20, 37.
minor, 35.
thyro-hyoideus, 370.
Nerv: abducens, 352, 538.
accessorius obturatorii, 260.
acromiales cutanei, 337.
articulares poplitei, 124.
articularis poplitei obturatorius, 112.
auditorius, 325, 539.
auricularis magnus, 322, 335.
auricularis pneumogastricus, 458.
auricularis posterior, 322, 385.
auriculo-temporalis, 321, 387, 400.
buccinatorius, 387, 398.
buccales, 385.
cardiacus inferior, 447.
medius, 370.
superior, 370.
cardiaci pneumogastrici, 368, 444.
cervicales descendentes, 337.
nervi facialis, 355, 385.
rami anteriores, 370.
rami posteriores, 519.
cervicalis superficialis, 335, 355.
cervico-facialis, 385.
chorda tympani, 398, 483.
ciliares ganglii ophthalmici, 350.
nasalis, 349.
circumflexus, 20, 35.
claviculares cutanei, 337.
coccygealis, 300.
communicans fibularis, 125, 132.
noni, 371.
corporis bulbosi, 189.
cruralis, 101, 107, 259.
cutanei abdominis anteriores, 205, 208.
laterales, 205.
cutaneus externus brachialis 29, 40.
cutaneus externus lumbalis, 97, 259.
cutaneus externus musculo-spiralis, 37.
cutaneus internus brachialis, major, 22, 40.
cutaneus internus brachialis minor, 20, 22.
cutaneus internus femoris, 102.
cutaneus internus ischiadicus, 122.
cutaneus internus musculo-spiralis, 22.
cutaneus obturatorii, 102.

Nerv: cutaneus medius femoris, 97, 101.
musculo-cutaneus, 41.
palmaris, 52. 73.
plantaris, 124, 149.
radialis, 53.
dorsales manûs, 53.
dentales posteriores, 472.
dentalis, anterior, 472.
inferior, 398.
descendens noni, 357.
digastricus, 385.
digitales mediani, 70.
plantares, 148, 149.
radiales, 53.
ulnares, 53, 68.
dorsales, rami anteriores, 342.
posteriores, 507, 519.
dorsalis clitoridis, 202.
penis, 195, 220.
ulnaris, 52.
facialis, 321, 385, 539.
frontalis, 345.
genito-cruralis, 259.
ramus femoralis, 97.
ramus genitalis, 259.
glosso-pharyngeus, 326, 407, 459, 539.
glutei inferiores, 122.
gluteus superior, 121, 299.
gustatorius, 398, 401.
hæmorrhoidalis inferior, 186.
hypoglossus, 368, 403, 457, 540.
ilio-hypogastricus, 208, 212.
inguinalis, 97, 208, 212.
incisorius, 398.
infra-maxillares nervi facialis, 385.
orbitales nervi facialis, 385.
orbitalis, 337.
infra-trochlearis, 349.
intercostales, 447.
intercosto-cutanei anteriores, 10, 205.
intercosto-cutanei laterales, 11.
intercosto-humeralis, 12.
interosseus anticus, 51.
posticus, 61.
ischiadicus major, 122, 128, 299.
minor, 121, 299.
lachrymalis, 345.
laryngeus externus, 368, 459.
laryngeus inferior, 368, 444, 493.
laryngeus superior, 368, 459, 493.
lingualis, 398, 401.
lumbales, rami anteriores, 259.
rami posteriores, 519.
lumbo-sacralis, 261, 297.
malares nervi facialis, 385.
malaris, 387.
massetericus, 398.
maxillaris inferior, 331, 387, 396.
Nerv: maxillaris superior, 331, 387, 4..
medianus, 29, 51, 70.
mentalis 387.
molles, 457.
motor oculi, 330, 352, 537.
musculo-cutaneus brachii, .. 41.
musculo-cutaneus cruris, .. 163.
musculo-spiralis, 33, 39.
mylo-hyoideus, 398.
nasalis, 349, 387, 477.
maxillaris, 478.
naso-palatinus, 478.
obturatorius, 112, 259.
accessorius, 25..
occipitalis major, 323, 518.
minor, 322, 337.
œsophageales, 442.
olfactorius, 325, 477, 536.
ophthalmicus, 331, 385.
opticus, 325, 351, 536.
orbitalis, 321, 387, 472.
palatinus magnus, 478.
medius, 478.
minor, 478.
palmaris cutaneus, 52.
ulnaris profundus, ..
superficialis, 68, 70.
patellaris, 102.
perforans Casserii, 29, 41.
perinæales superficiales, 1.. 197.
peroneal, 125, 162.
petrosus superficialis magn.. 333, 480.
petrosus superficialis parv.. 481.
pharyngei, 461.
pharyngeus (8th) 459.
phrenicus, 371, 411.
plantaris externus, 149.
profundus, 153.
internus, 148.
pneumo-gastricus, 368, 444, 457, 539.
popliteus externus, 125, 162.
internus, 124.
portio dura, 325, 385, 529.
mollis, 325, 529.
pterygoidei, 398.
pterygo-palatinus, 480.
pudendus inferior, 122, 188, 19..
pudicus internus, 122, 195, 20.. 294.
pulmonares posteriores, 444.
radialis, 52, 53.
recurrens, 368, 444, 493.
articularis, 162.
respiratorius externus, 21, 343.
sacrales, rami anteriores, 297.
rami posteriores, 519.
saphenus externus, 132, 154.

Nerv: saphenus internus, 102, 154.
sciaticus magnus, 122, 128, 299.
parvus, 121, 299.
spheno-palatini, 472.
spinalis accessorius, 338, 370, 457, 539.
splanchnicus major, 261, 447.
minor, 262, 447.
minimus, 262, 448.
sternales cutanei, 337.
stylo-hyoideus, 385.
subcutaneus malæ, 387.
suboccipitalis, ramus anterior, 371, 457.
suboccipitalis, ramus posterior, 518.
subscapulares, 20, 37.
supra-maxillares nervi facialis, 385.
orbitalis, 320, 345, 386.
scapularis, 36, 343.
supra-trochlearis, 321, 345, 385.
sympatheticus abdominis, 261.
cervicis, 370, 456.
pelvis, 300.
thoracis, 447.
temporales nervi facialis, 321, 385.
profundi, 398.
temporo-facialis, 385.
temporo-orbitalis, 321, 387.
thoracici anteriores, 20.
laterales, 11.
thoracicus posterior, 21, 343.
tibialis anterior, 154, 161.
posterior, 141.
trigeminus, 324, 330, 538.
trochlearis, 330, 345, 537.
tympanicus, 460.
ulnaris, 30, 52, 68, 73.
uterini, 312.
Vidianus, 479.
Ninth nerve, 368, 403, 457, 540.
Nodule, 558.
Nose, cartilages, 389.
cavity of, 473.
meatus of, 474.
nerves of, 477.
septum of, 473.
Nuck, canal of, 313.
Nuclei of medulla oblongata, 559.
Nutritious artery of fibula, 140.
of femur, 111.
of humerus, 27.
of tibia, 140.
Nymphæ, 182.

Obliquus abdominis externus, 206, 208.
internus, 209.
capitis inferior muscle, 517.
superior muscle, 517.
oculi inferior, 352.
superior, 347.
Obturator artery, 97, 280, 293.
Obturator fascia, 281.
ligament, 280.
muscle, external, 113, 119.
internal, 118, 287.
nerve, 112, 259.
Occipital artery, 323, 365, 515.
lobe of brain, 544, 545.
sinus, 328.
nerve, 323, 518.
Occipito-atloid ligaments, 501.
axoid ligament, 503.
frontalis muscle, 319.
Odontoid ligaments, 504.
Œsophagus, connections of, 441.
structure, 442.
Œsophageal arteries, 445.
nerves, 442.
opening of diaphragm, 247.
Olfactory bulb, 535.
nerve, 325, 477, 536.
peduncle, 536.
region, 477.
Olivary body, 538, 561.
fasciculus, 555.
Omentum, great, 228.
small, 228.
splenic, 232.
Omo-hyoid muscle, 36, 339, 358.
Openings in diaphragm, 247.
Ophthalmic artery, 350.
ganglion, 349.
nerve, 331, 385.
vein, 351.
Opponens pollicis muscle, 72.
Optic commissure, 535.
nerve, 351, 536.
thalamus, 553.
tract, 536.
Ora serrata, 569.
Orbicular ligament of the radius, 82.
Orbicularis oris, 378.
palpebrarum, 377.
Orbit, 344.
muscles of, 347, 352.
nerves, 345, 349.
periosteum of, 345.
vessels, 346, 350.
Orbital nerve, 321, 387, 472.
Orifice of the urethra, 180, 306,
of the uterus, 312.
of the vagina, 182.
Ossicula-auditûs, 483.
Os tincæ, 312.
uteri externum, 312.
internum, 312.
Otic ganglion, 481.
Outlet of the pelvis, 296.
Ovary, 279, 313.
arteries of, 252.
bulb of, 313.

Palate, hard, 470.
soft, 465, 466.

Palatine arteries, posterior, 480.
artery, inferior, 407.
nerves, 478.
Palato-glossus, 469.
pharyngeus, 469.
Palm of the hand, 65.
Palmar arch, deep, 73.
superficial, 67.
nerve of the median, 52.
nerve of the ulnar, deep, 73.
superficial, 52.
fascia, 66.
Palmaris brevis muscle, 67.
longus muscle, 42.
Palpebræ, 388.
Palpebral arteries, 351.
ligament, 388.
Pancreas, connections, 244.
structure of, 269.
Pancreatic arteries, 241.
duct, 244, 269.
Pancreatico-duodenal arteries, 234, 240.
Papillæ of the tongue, 485.
Parallel fissure, 542.
Parietal lobe, 542.
lobule, 542.
Parotid gland, 381.
Parts beneath gluteus maximus, 116.
behind inner malleolus, 142.
triangular ligament, 191.
passing through superior aperture of thorax, 449.
Patellar nerve, 102.
plexus, 102.
Pectineus muscle, 108.
Pectoralis major muscle, 10.
minor muscle, 14.
Peduncle of the cerebellum, inferior, 557.
middle, 557.
superior, 556.
of the cerebrum, 537.
of the pineal body, 554.
Pelvis, female, dissection of, 279, 289
male, dissection of, 279.
Pelvic cavity, 279.
fascia, 281.
floor, 296.
plexus, 300.
side-view, 285.
viscera, male, 300.
female, 309.
Penis, 220.
structure of, 307.
vessels of, 195.
Perforating arteries of the femoral, 111, 128.
of internal mammary, 433.
of the palm, 74.
of the sole, 153.
Perforans Casserii nerve, 29, 41.
Pericardium, 441.
vessels of, 445.
Perinæum, both sexes, 181.
female, 195.
male, 186.
Perinæal artery, superficial, 187, [illegible]
fascia, deep, 190, 200.
superficial, 186, [illegible]
nerves. superficial, 188, [illegible]
Periosteum of the orbit, 345.
Peritoneum, 228.
of female pelvis, 279.
of male pelvis, 279.
Peroneal artery, 140.
anterior, 150.
nerve, 125, 162.
Peroneus brevis muscle, 162.
longus muscle, 162.
tertius muscle, 156.
Peroneo-tibial articulations, 171.
Pes accessorius, 551.
anserinus, 385.
hippocampi, 551.
Petrosal ganglion, 460.
sinus, inferior, 330.
superior, 330.
nerve, large, 333, 480.
small, 480.
Peyer's glands, 267.
Pharynx, fibrous bag of, 464.
interior, 465.
muscles of, 462.
openings of, 465.
Pharyngeal ascending artery, 462.
nerve, 459, 461.
plexus, 461.
Phrenic artery, 251.
nerve, 371, 411.
Pia mater of the brain, 528.
of the cord, 521.
Pillars of the abdominal ring, 208.
of the fauces, 465.
of the fornix, 555.
Pineal body, 553.
Pinna, or auricle of the ear, 375.
Pituitary body, 535.
Plantar aponeurosis, 143.
arch of artery, 153.
arteries, 147, 153.
ligament, long, 175.
nerve, external, 149, 153.
internal, 124.
Plantaris muscle, 134.
Platysma myoides muscle, 335, 364, 380.
Pleuræ, 409.
Plexus, aortic, 261.
brachial, 19, 342.
cardiac, superficial, 418.
deep, 437.
carotid, 333.
cavernous, 333.
cervical, 370.

Plexus, cervical, superficial, 335.
choroides cerebri, 548, 551.
cerebelli, 559.
gulæ, 442, 444.
hypogastric, 300.
lumbar, 259, 261.
mesenteric, inferior, 238.
superior, 236.
œsophageal, 442, 444.
patellar, 102.
pelvic, 300.
pharyngeal, 461.
pulmonary, posterior, 444.
sacral, 297.
solar, 261.
Plica semilunaris, 374.
Pneumo-gastric nerve, 368, 444, 457, 539.
Pomum Adami, 354, 494.
Pons Varolii, 537.
structure of, 540.
Popliteal artery, 125.
fascia, 123.
glands, 126.
nerve, external, 125, 162.
internal, 124.
space, 124.
vein, 126.
Popliteus muscle, 135.
Portio dura, 321, 385, 539.
mollis, 325, 539.
Porus opticus, 563.
Posterior chamber of eye, 565.
commissure, 552.
elastic layer of cornea, 564.
ligament of knee, 166.
medullary velum, 559.
pyramid, 525.
triangle of the neck, 334.
Pouch, laryngeal, 491.
Poupart's ligament, 95, 208.
Prepuce, 220.
Princeps cervicis artery, 216.
pollicis artery, 74.
of the foot, 153.
Processus cochleariformis, 482.
a cerebello ad testes, 556.
ad pontem, 557.
ad medullam, 557.
vermiformes, 556, 558.
Profunda artery, inferior, 27.
of the neck, 435, 518.
of the thigh, 101, 110.
superior, 27, 40.
Promontory, 482.
Pronator quadratus muscle, 48.
radii teres muscle, 41.
Prostate gland, connections, 288.
structure, 303.
Prostatic part of the urethra, 305.
sinuses, 305.
Psoas magnus muscle, 109, 257.
parvus muscle, 258.
Pterygoid arteries, 396.
nerves, 398.
Pterygoideus, externus muscle, 392.
internus muscle, 394.
Pterygo-maxillary region, 390.
ligament, 394.
palatine artery, 480.
nerve, 480.
Pubic symphysis, 316.
Pudendum, 181.
Pudendal inferior nerve, 122, 188, 193.
Pudic arteries, external, 91, 101, 208.
artery, internal, 120, 193, 201, 294.
nerve, internal, 122, 195, 201, 294.
Pulmonary artery, 421, 427.
nerves, 444.
veins, 412, 429.
Puncta lachrymalia, 374.
Pylorus, 264.
Pyloric arteries, 239.
Pyramid, anterior, 538.
of the cerebellum, 559.
of the tympanum, 482.
posterior, 525.
Pyramidal masses of kidney, 277.
Pyramidalis abdominis muscle, 221.
nasi muscle, 377.
Pyriformis muscle, 117, 300.

Quadratus femoris muscle, 119.
lumborum muscle, 258.

Radial artery, 43, 61.
nerve, 52, 53.
vein, 24, 40.
Radialis indicis artery, 74.
Radio-carpal articulation, 83.
ulnar articulations, 82.
Ranine artery, 406.
Raphé of the corpus callosum, 547.
of the perinæum, 181.
Receptaculum chyli, 262, 442.
Recto-vesical fascia, 282, 283.
pouch, 228, 287.
Rectus abdominis muscle, 221.
capitis anticus major, 453.
minor, 454.
posticus major, 517.
minor, 517.
lateralis, 454.
femoris, 104.
oculi externus, 351.
inferior, 351.
internus, 351.
superior, 347.
Rectum, connections of, in the female, 289.
in the male, 286.
structure, 301.
Recurrent interosseous artery, 60.
radial, 45.

Recurrent tibial, 159.
ulnar, anterior, 49.
posterior, 49.
nerve of pneumo-gastric, 368, 444, 493.
nerve of the tibial, 162.
Removal of the brain, 325.
Renal artery, 251.
vein, 252, 278.
Respiratory nerve, external, 21, 343.
Restiform body, 539.
Rete testis, 309.
Retina, 569.
Retrahens aurem, 319.
Rhomboideus major muscle, 511.
minor, 510.
Rima of the glottis, 487.
Ring, abdominal, external, 208.
internal, 213.
femoral, 94.
Risorius Santorini muscle, 380.
Root of the lung, 412.
Roots of the nerves, 525.
Round ligament of the hip joint, 131.
of the liver, 232, 271.
of the uterus, 291, 313.
Rugæ of the bladder, 304.
of the stomach, 264.
of the vagina, 311.

Sacculus laryngis, 491.
Sacral artery, lateral, 294.
middle, 252.
ganglia, 300.
nerves, anterior branches, 227.
posterior branches, 519.
plexus, 297.
Sacro-coccygeal articulation, 315.
iliac articulation, 314.
vertebral articulation, 314.
lumbalis muscle, 513.
sciatic ligament, large, 314.
small, 315.
Saphenous vein, external, 133.
internal, 92, 154.
opening, 92.
nerve, external, 132, 154.
internal, 102, 154.
Sartorius muscle, 103.
Scalenus anticus muscle, 452.
medius, 452.
posticus, 453.
Scalp, dissection of, 318.
Scapular artery, posterior, 37, 340.
ligaments, 76.
muscles, 32.
Scapulo-clavicular articulation, 76.
humeral articulation, 79.
Scarpa's triangle, 98.
Schneiderian membrane, 476.
Sciatic artery, 120, 294.
nerve, large, 122, 128.
small, 121.
Sclerotic coat of the eye, 562.
Scrotum, 181.
Second nerve, 351, 636.
Section of upper arm, 39.
fore-arm, 64.
wrist joint, 65.
thigh, 129.
ankle, 143.
leg, 163.
heart, 425.
the neck, 520.
spinal cord, 522, 526.
Semilunar cartilages, 169.
ganglia, 261.
valves of aorta, 424.
of pulmonary art
421.
Semi-bulbs of vagina, 201.
Semi-membranosus muscle, 127.
Seminal ducts, 303.
Seminiferal tubes, 309.
Semi-spinalis colli muscle, 519.
dorsi muscle, 519.
Semi-tendinosus muscle, 127.
Septum auricularum, 419.
crurale, 95.
intermuscular, of the ar
of the t
102.
lucidum, 549.
narium, 473.
pectiniforme, 307.
of the tongue, 486.
ventriculorum, 416.
Serratus magnus muscle, 21.
posticus inferior 512.
superior, 512.
Seventh nerve, 325, 539.
Sheath of the fingers, 46.
of the femoral vessels, 93.
of the rectus, 221.
of the toes, 146.
Shoulder joint, 78.
Sigmoid artery, 236.
flexure of the colon, 227.
valves, 421, 423.
Sinus, basilar, 333.
cavernous, 330.
circular, of Ridley, 333.
lateral, 329.
longitudinal, inferior, 328.
superior, 324.
occipital, 328.
petrosal, inferior, 330.
superior, 330.
pocularis, 305.
prostaticus, 305.
straight of the skull, 328.
torcular, 328.
transverse, 333.
of Valsalva, 424.
Sixth nerve, 330, 352, 538.
Small intestine, 226, 265.
omentum, 228.
Socia parotidis, 381.

ommissure, 553.
alate, 465, 466.
muscles of, 467.
plexus, 261.
f the foot, dissection of, 143.
s muscle, 134.
iatic artery, 252.
cord, 219.
fascia, 209.
veins, 252.
io-palatine artery, 480.
ganglion, 478.
nerves, 472.
cter ani externus, 183.
internus, 183.
vaginæ, 198, 201.
vesicæ, 304.
lian lobe, 273.
l accessory nerve, 370, 457, 539.
arteries, 524, 529.
cord, 524.
membranes of, 521.
structure, 526.
nerves, 525.
roots of, 525.
veins, 521.
alis dorsi muscle, 513.
ichnic nerve, large, 261, 447.
small, 262, 447.
smallest, 262, 448.
n, connections of, 245.
structure, 269.
ic artery, 239, 241.
omentum, 232.
vein, 241.
iius capitis muscle, 512.
colli muscle, 512.
gy bones, 474.
part of the urethra, 306.
edius muscle, 484.
es bone, 483.
ate ligament, 504.
son's duct, 381.
no-clavicular articulation, 356.
cleido-mastoid muscle, 355.
hyoid muscle, 358.
thyroid, 358.
nach, form and divisions, 263.
connections of, 226.
structure of, 264.
ight sinus, 328.
ate body, 547.
o-hyoid ligament, 407.
muscle, 358.
nerve, 385.
glossus muscle, 403.
mastoid artery, 365.
maxillary ligament, 407.
pharyngeus muscle, 404, 463.
anconeus muscle, 80.
arachnoid space, 528.
of the cord, 523.
clavian artery, 340, 434.
left, 431.
Subclavian artery, right, 432.
vein, 435.
Subclavius muscle, 14.
Subcrureus, 168.
Subcutaneous malar nerve, 387.
Sublingual artery, 406.
gland, 401.
Submaxillary ganglion, 403.
gland, 401.
region, 400.
Submental artery, 365.
Suboccipital nerve, anterior branch, 371, 457.
posterior branch, 518.
region, 517.
Subpubic ligament, 316.
Subscapular artery, 17, 37.
nerves, 20, 37.
Subscapularis muscle, 36.
Substantia perforata antica, 536.
postica, 536.
Sulci of brain, 540.
Sulcus, longitudinal, of the liver, 271.
transverse, 271.
of the spinal cord, anterior, 524.
lateral, 525.
posterior, 525.
Superficialis cervicis artery, 340.
nerve, 335, 355.
volæ artery, 45.
Superficial fascia of the groin, 206.
of the perinæum, 186, 195.
of the thigh, 91.
Supinator radii brevis, 57.
longus, 54.
Supra-marginal lobule, 542.
orbital artery, 321, 346, 350.
nerve, 320, 345, 386.
renal artery, 251.
capsule, 256, 278.
scapular artery, 36, 340.
nerve, 36, 343.
spinatus muscle, 33.
trochlear nerve, 321, 345, 385.
Sural arteries, 126.
Surgery of axillary artery, 13.
of brachial artery, 27.
of radial artery, 45.
of ulnar artery, 49.
strangulated femoral hernia, 96.
of femoral artery, 100.
fracture of the patella, 106.
of femoral artery in Hunter's canal, 107.
of popliteal artery, 126.
talipes varus, 138.
of posterior tibial artery, 139.
of anterior tibial artery, 159.
of dorsal artery of foot, 60.

Surgery, amputations of the foot, 177.
passing a male catheter, 179.
a female catheter, 180.
of the ischio-rectal fossa, 185.
lithotomy, 189.
strangulated inguinal hernia, 218.
radical cure of hernia, 219.
of abdominal aorta, 249.
of iliac arteries, 255.
of temporal artery, 322.
of subclavian artery, 340.
of common carotid artery, 362.
puncture of lachrymal sac, 373.
of facial artery, 384.
cleft palate, 468.
Suspensory ligament of the lens, 568.
of the liver, 232.
of the penis, 220.
Sympathetic nerve in the abdomen, 261.
in the head, 333.
in the neck, 376, 456.
in the pelvis, 300.
in the thorax, 447.
Symphysis pubis, 315.
Synovial membranes of wrist, 85.
of foot, 176.

Tænia hippocampi, 551.
semicircularis, 548.
Tapetum lucidum, 562.
Tarsal artery, 160.
cartilages, 388.
Tarso-metatarsal articulations, 176.
Teeth, 470.
Temporal artery, anterior, 322.
deep, 396.
middle, 322.
superficial, 322, 364.
fascia, 323.
lobe, 543.
muscle, 323, 391.
nerves, deep, 398.
superficial, 321, 385.
vein, 367.
Temporo-facial nerve, 385.
maxillary articulation, 392.
Tendo Achillis, 134.
Tendons of flexor sublimis, 70.
of flexor profundus, 71.
Tensor palati muscle, 468.
tarsi, 388.
tympani, 484.
vaginæ femoris, 103.
Tentorium cerebelli, 327.
Teres major muscle, 21, 34.
minor, 34.
Testis, descent of, 210.
structure of, 307.
Thalamus opticus, 548, 553.
Thebesian foramina, 419.
valve, 419.

Thigh, dissection of, back, 126.
front, 103.
inner side, 106
Third nerve, 330, 352, 537.
ventricle, 552.
Thoracic aorta, 444.
duct, 368, 442.
ganglia, 447.
nerves, anterior, 20.
nerve, posterior, 21.
Thoracica-acromialis artery, 16.
alaris, 16.
longa, 16.
superior, 16.
Thorax, 407.
superior aperture of, 449.
Thyro-arytænoid, ligaments, 491.
arytænoideus muscle, 490.
epiglottidean ligament, 49
hyoid membrane, 493.
muscle, 358.
nerve, 370.
Thyroid artery, inferior, 433.
lowest, 430.
superior, 364.
axis of artery, 433.
body, 439.
cartilage, 494.
vein, inferior, 434.
middle, 368.
superior, 368.
Tibial artery, anterior, 157.
posterior, 138.
nerve, anterior, 154, 161.
posterior, 141.
veins, anterior, 158.
posterior, 138.
Tibialis anticus muscle, 155.
posticus, 136.
Tibio-fibular articulations, 171.
tarsal articulation, 172.
Tongue, 484.
muscles of, 486.
nerves of, 486.
papillæ of, 485.
Tonsil, 470.
Tonsillitic artery, 407.
Torcular Herophili, 328.
Trachea, connections of, 438.
structure of, 450.
Trachelo-mastoid muscle, 514.
Tragus, 376.
Transverse arteries of pons, 529.
fissure of the cerebr 552.
of the liver, 271.
ligament of the acetab lum, 132.
of the atlas, 503.
of the knee, 169
of the metacarpu 75.
of the metatars 176.

Transverse ligament of the scapula, 77.
perinæal artery, 187, 197.
sinus, 333.
Transversalis abdominis muscle, 210.
colli artery, 339, 434.
cervicis muscle, 514.
faciei artery, 384.
fascia, 213.
humeri artery, 340, 434.
Transversus linguæ, 486.
pedis, 152.
perinæi, 189, 199.
deep, 193, 201.
Trapezius muscle, 508.
Trapezoid ligament, 77.
Triangle of the neck, anterior, 371.
posterior, 334.
Scarpa's, 98.
Triangular cartilage of the nose, 390.
fascia of groin, 209, 215.
fibro-cartilage of wrist, 86.
ligament of the urethra, 190, 200.
space of the thigh, 98.
bladder, 304.
Triangularis sterni muscle, 408.
Triceps extensor cubiti, 38.
Tricuspid valve, 420.
Trigeminal nerve, 330, 538.
Trigonum vesicæ, 304.
Trochlea, 347.
Trochlear nerve, infra, 349.
supra, 321, 345, 385.
Tuber cinereum, 535.
Tubercula quadrigemina, 555.
Tuberculum Loweri, 419.
Tubules of the stomach, 265.
of small intestine, 266.
of large intestine, 268.
Tubuli seminiferi, 309.
uriniferi, 277.
Tunica albuginea oculi, 563.
testis, 308.
of the ovary, 313.
conjunctiva, 388, 563.
Ruyschiana, 566.
vaginalis, 220, 308.
vasculosa testis, 308.
Turbinate bones, 474.
Tympanic nerve, 460.
Tympanum, 481.
muscles of, 484.
ossicles of, 483.

Ulnar artery, 49.
nerve, 30, 52, 68.
deep branch of, 73.
veins, anterior, 23.
posterior, 23, 53.
Umbilical hernia, 202.
region of the abdomen, 226.
Umbilicus, 202.
Ureter, 256, 287.
Urethra, female, connections, 289.
Urethra, female, orifice of, 182.
structure, 310.
male, interior, 304.
connections, 285.
Uriniferous tubes, 277.
Uterine arteries, 293.
plexus of nerves, 312.
Uterus, 311.
interior of, 312.
ligaments of, 279, 291.
connections of, 289.
structure of, 312.
Uvula cerebelli, 559.
palati, 465.
vesicæ, 304.

Vagina, connections, 291.
structure and form, 310.
Vaginal arteries, 293.
plexus, 300.
Vagus nerve, 368, 444, 457, 539.
Vallecula, 558.
Valve, coronary, 419.
Eustachian, 419.
of cæcum, 268.
mitral, 423.
pyloric, 264.
semilunar, 421, 424.
tricuspid, 420.
of Vieussens, 557.
Valvulæ conniventes, 266.
Vas deferens, 288, 308.
aberrans, 309.
Vasa brevia arteries, 241.
efferentia testis, 309.
recta testis, 339.
vorticosa, 566.
Vastus externus muscle, 105.
internus muscle, 105.
Vein, angular, 384.
ascending cervical, 434.
pharyngeal, 462.
auricular posterior, 367.
axillary, 19.
azygos, large, 262, 443.
small, 262, 446.
superior, 446.
basilic, 23.
brachial, 29.
brachio-cephalic, left, 435.
right, 435.
bronchial, 445.
cardiac, 418.
cava, inferior, 253, 419.
superior, 419, 430.
cephalic, 24.
cerebral, 552.
circumflex iliac, 254.
coronary, of the heart, 418.
of the stomach, 212.
deep cervical, 435.
diaphragmatic, inferior, 251.
dorsal, of the penis, 195, 220.
of the clitoris, 202.

Vein, dorsi-spinal, 521.
emulgent, 252, 278.
epigastric, deep, 254.
superficial, 91, 101.
facial, 384.
femoral, 101.
of Galen, 552.
gluteal, 295.
hæmorrhoidal, 186, 238.
hepatic, 274.
iliac, common, 253.
external, 254.
internal, 294.
infra-orbital, 473.
innominate, 435.
intercostal, 446.
superior, 435.
intraspinal, 521, 524.
interlobular, 274.
intralobular, 274.
jugular, anterior, 357.
external, 335.
internal, 367, 457.
lingual, 406.
longitudinal, of the spine, 521.
lumbar, 253.
mammary, internal, 432.
maxillary, internal, 396.
median, of the arm, 22, 40.
basilic, 24.
cephalic, 24.
meningo-rachidian, 521.
mesenteric, inferior, 238.
superior, 236.
occipital, 367.
ophthalmic, 351.
ovarian, 252.
pharyngeal, 462.
phrenic, inferior, 251.
popliteal, 126.
portal, 244.
posterior spinal, 521.
profunda, of the thigh, 111.
pudic, external, 91, 101.
internal, 120, 193, 295.
pulmonary, 412, 429.
radial, cutaneous, 24, 40.
renal, 252, 278.
sacral, lateral, 295.
middle, 252.
saphenous, external, 133.
internal, 92, 154.
sciatic, 295.
spermatic, 252.
spinal, 470.
splenic, 241.
subclavian, 435.
sublobular, 274.
supra-renal, 251.
scapular, 340.
Vein, temporal, superficial, 367.
thyroid, inferior, 434.
middle, 368.
superior, 368.
tibial, anterior, 158.
posterior, 138.
transverse cervical, 340.
ulnar anterior, 23, 40.
posterior, 23.
umbilical, 426.
vertebral, 432, 518.
of the vertebræ, 521.
Velum interpositum, 551.
pendulum *palati*, 465.
Vena cava, *inferior*, 253, *419*.
opening for, 247.
superior, 419, *430*.
portæ, 244.
Venæ cavæ hepaticæ, 274.
Ventricles of the brain, fifth, 549.
fourth, 549.
lateral, 547.
third, 552.
Ventricles of the heart, left, 423.
right, 420.
structure of, 420.
of the larynx, 491.
Vermiform appendix, 267.
processes, 556, 558.
Vertebral artery, 432, 455, 517, 529.
vein, 432, 518.
Veins of the brain, 552.
of the dura mater, 324, 328.
of the heart, 418.
Veru montanum, 305.
Vesica urinaria, 303.
Vesical artery, inferior, 293.
superior, 293.
Vesiculæ seminales, connections, 288.
structure, 303.
Vessels of root of neck, 430.
Vestibule of the vulva, 182.
Vidian artery, 480.
nerve, 479.
Villi, intestinal, 266.
Vitreous body, 568.
Vocal cords, 491.
Vulva, 181.

Wharton's duct, 401.
Willis, circle of, 532.
Winslow's foramen, 230.
Wrisberg's nerve, 20, 22.
Wrist joint, 83.

Yellow spot of eyeball, 569.

Zygomaticus major muscle, 380.
minor, 380.

THE END.